Student Solutions Manual for

CHEMICAL PRINCIPLES

Second Edition

Kenton H. Whitmire and **Charles Trapp**

W. H. FREEMAN AND COMPANY
NEW YORK

ISBN 0-7167-4435-X

Printed in the United States of America

Second printing 2002

PREFACE

This manual contains solutions and answers to the odd-numbered exercises in Atkins and Jones's *Chemical Principles, Second Edition.* The rules of significant figures have been adhered to in reporting the numerical answers for all exercises. In exercises with multiple parts, the properly rounded values were used for subsequent calculations.

Procedures and results that have been fully illustrated in the solutions to the exercises of an earlier chapter or in the introductory section, "Fundamentals," may not be repeated in the solutions to the exercises of later chapters. For example, the conversion between °C and K, the determination of molar mass from the formula of a compound, and other simple conversions such as mL to L or cm to m, are not usually worked out in the solutions.

Some abbreviations have been used to save space. These are usually defined at the start of a solution. Some that occur frequently are

n = neutron
p = proton
e = electron

I wish to thank Peter Atkins and Loretta Jones for their support and help during the production of this manual as well as for many stimulating exchanges about chemical principles and educational pedagogy. They, along with Max Bishop, Dave Price, and Jamie Nossal, have reviewed the solutions and made many valuable suggestions. I would also like to acknowledge the staff at W. H. Freeman, especially Charlie Van Wagner, Jodi Isman, and Jessica Fiorillo, who have been instrumental in getting this project to print.

K.H.W.

FUNDAMENTALS

A.1 (a) chemical; (b) physical; (c) physical

A.3 The temperature, humidity, and the evaporation of water are physical properties. The ripening of oranges is a chemical change.

A.5 (a) intensive; (b) intensive; (c) extensive; (d) extensive

A.7 $d = \dfrac{m}{V}$

$$= \left(\frac{112.32\ \text{g}}{29.27\ \text{mL} - 23.45\ \text{mL}}\right)\left(\frac{1\ \text{mL}}{1\ \text{cm}^3}\right)$$
$$= 19.30\ \text{g}\cdot\text{cm}^{-3}$$

A.9 $d = \dfrac{m}{V}$, rearranging gives $V = \dfrac{m}{d}$

$$= \left(\frac{0.750\ \text{carat}}{3.51\ \text{g}\cdot\text{cm}^{-3}}\right)\left(\frac{200\ \text{mg}}{1\ \text{carat}}\right)\left(\frac{1\ \text{g}}{1000\ \text{mg}}\right)$$
$$= 0.0427\ \text{cm}^3$$

A.11 $d = \dfrac{m}{V}$

$$= \left(\frac{3.95 \times 10^{-22}\ \text{g}}{\frac{4}{3}\pi(138\ \text{pm})^3}\right)\left(\frac{1\ \text{pm}}{1 \times 10^{-10}\ \text{cm}}\right)^3$$
$$= 35.9\ \text{g}\cdot\text{cm}^{-3}$$

Because the density of metallic uranium is much less than the density of a uranium atom, the metallic form of uranium must contain considerable empty space.

A.13 $E_K = \dfrac{1}{2}mv^2$

$$= \frac{1}{2}(4.2\ \text{kg})(14\ \text{km}\cdot\text{h}^{-1})^2\left(\frac{1\ \text{h}}{3600\ \text{s}}\right)^2\left(\frac{1000\ \text{m}}{1\ \text{km}}\right)^2$$
$$= 32\ \text{kg}\cdot\text{m}^2\cdot\text{s}^{-2}$$
$$= 32\ \text{J}$$

A.15 $E_K = \frac{1}{2}mv^2$

$$= \frac{1}{2}(2.5 \times 10^5\ \text{kg})(25\ \text{km}\cdot\text{s}^{-1})^2\left(\frac{1000\ \text{m}}{1\ \text{km}}\right)^2$$
$$= 7.8 \times 10^{13}\ \text{kg}\cdot\text{m}^2\cdot\text{s}^{-2}$$
$$= 7.8 \times 10^{13}\ \text{J}$$

A.17 $\mathcal{V} = mgh$

$$= (0.51\ \text{kg})(9.81\ \text{m}\cdot\text{s}^{-2})(3.0\ \text{m})$$
$$= 15\ \text{kg}\cdot\text{m}^2\cdot\text{s}^{-2}$$
$$= 15\ \text{J}$$

A.19 When the skier is at rest at the top of the slope, the kinetic energy is 0. The total energy is equal to the potential energy. At the bottom of the slope, the potential energy is 0; it has been converted into kinetic energy.

$$E = E_K + \mathcal{V}$$
$$= \mathcal{V}$$
$$= mgh$$
$$= (99.5\ \text{kg})(9.81\ \text{m}\cdot\text{s}^{-2})(244\ \text{m})$$
$$= 2.38 \times 10^5\ \text{kg}\cdot\text{m}^2\cdot\text{s}^{-2}$$
$$= 2.38 \times 10^5\ \text{J}$$
$$= 238\ \text{kJ}$$

A.21 $F = ma$

$$= (9.1 \times 10^{-31}\ \text{kg})(1.2 \times 10^6\ \text{m}\cdot\text{s}^{-2})$$
$$= 1.1 \times 10^{-24}\ \text{kg}\cdot\text{m}\cdot\text{s}^{-2}$$
$$= 1.1 \times 10^{-24}\ \text{N}$$

B.1 $$\text{number of beryllium atoms} = \frac{\text{mass of sample}}{\text{mass of one atom}}$$
$$= \left(\frac{0.210\ \text{g}}{1.50 \times 10^{-26}\ \text{kg}\cdot\text{atom}^{-1}}\right)\left(\frac{1\ \text{kg}}{1000\ \text{g}}\right)$$
$$= 1.40 \times 10^{22}\ \text{atoms}$$

B.3 (a) Radiation may pass through a metal foil. (b) All light (electromagnetic radiation) travels at the same speed; the slower speed supports the particle model.

(c) This observation supports the radiation model. (d) This observation supports the particle model; radiation has no mass and no charge.

B.5 (a) A deuterium atom has 1 proton, 1 neutron, and 1 electron. (b) A ^{127}I atom has 53 protons, 74 neutrons, and 53 electrons. (c) A ^{15}N atom has 7 protons, 8 neutrons, and 7 electrons. (d) A ^{204}Bi atom has 83 protons, 126 neutrons, and 83 electrons.

B.7 (a) ^{175}Lu; (b) ^{118}Sn; (c) ^{6}Li

B.9 (a) ^{12}C, ^{13}C, and ^{14}C all have 6 protons and 6 electrons. (b) They have different numbers of neutrons: an atom of ^{12}C has 6 neutrons, an atom of ^{13}C has 7 neutrons, and an atom of ^{14}C has 8 neutrons.

B.11 (a) Atoms of ^{55}Mn, ^{56}Fe, and ^{58}Ni all have 30 neutrons. (b) They have different masses, numbers of protons, and numbers of electrons.

B.13 (a) In NH_3, the nitrogen atom contributes 7 protons, 7 neutrons, and 7 electrons. Each hydrogen atom contributes 1 proton and 1 electron. There are 10 protons, 7 neutrons, and 10 electrons total.

(b) mass of protons $= (10)(1.673 \times 10^{-24}\text{ g}) = 1.673 \times 10^{-23}\text{ g}$

mass of neutrons $= (7)(1.675 \times 10^{-24}\text{ g}) = 1.172 \times 10^{-23}\text{ g}$

mass of electrons $= (10)(9.109 \times 10^{-28}\text{ g}) = 9.109 \times 10^{-27}\text{ g}$

B.15 (a) Silicon is a Group 14 metalloid. (b) Strontium is a Group 2 metal. (c) Tin is a Group 14 metal. (d) Antimony is a Group 15 metalloid.

B.17 (a) Ag, metal; (b) Ge, metalloid; (c) Kr, nonmetal

B.19 lithium, Li, 3; sodium, Na, 11; potassium, K, 19; rubidium, Rb, 37; cesium, Cs, 55. All the alkali metals react with water to produce hydrogen. Lithium reacts gently; the reaction becomes more vigorous down the periodic table. The melting points decrease down the periodic table.

B.21 (a) Metals are malleable, conduct electricity, and have luster. (b) Nonmetals are not malleable or ductile and do not conduct electricity.

C.1 (a) An ionic compound is made up of ions. Sodium chloride is an example of an ionic compound. (b) A molecular compound is made up of molecules. Sucrose (sugar) is an example of a molecular compound. Ionic compounds have higher melting points than solids made of molecules. Ionic compounds dissolve only in polar solvents if at all. Molecules are more likely to dissolve in nonpolar solvents. See Chapter 5.

C.3 (a)

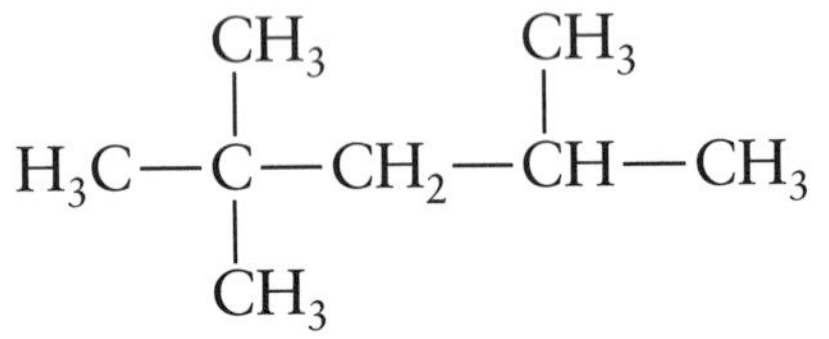

```
      CH3      CH3
      |        |
H3C—C—CH2—CH—CH3
      |
      CH3
```

(b)

```
      CH3
      |
H3C—CH—CH2—OH
```

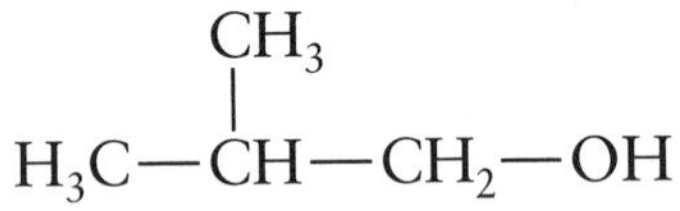

C.5 (a) Cesium is a metal in Group 1; it will form Cs^+ ions. (b) Iodine is a nonmetal in Group 17 and will form I^- ions. (c) Selenium is a Group 16 nonmetal and will form Se^{2-} ions. (d) Calcium is a Group 2 metal and will form Ca^{2+} ions.

C.7 (a) $^{4}He^{2+}$ has 2 protons, 2 neutrons, and no electrons. (b) $^{15}N^{3-}$ has 7 protons, 8 neutrons, and 10 electrons. (c) $^{127}I^-$ has 53 protons, 74 neutrons, and 54 electrons. (d) $^{80}Se^{2-}$ has 34 protons, 46 neutrons, and 36 electrons.

C.9 (a) $^{19}F^-$; (b) $^{24}Mg^{2+}$; (c) $^{128}Te^{2-}$; (d) $^{86}Rb^+$

C.11 (a) Aluminum forms Al^{3+} ions; tellurium forms Te^{2-} ions. Two aluminum atoms produce a charge of $2 \times +3 = +6$. Three tellurium atoms produce a charge of $3 \times -2 = -6$. The formula for aluminum telluride is Al_2Te_3. (b) Magnesium forms Mg^{2+}ions and oxygen forms O^{2-} ions. A magnesium ion produces a charge of +2, which is required to balance the charge on one O^{2-} ion. The formula for magnesium oxide is MgO. (c) Sodium forms +1 ions; sulfur forms −2 ions. The formula for sodium sulfide is Na_2S. (d) Rubidium forms +1 ions and iodine forms −1 ions. One iodide ions are required to balance the charge of one rubidium ion, so the formula is RbI.

D.1 (a) Al_2O_3. Aluminum forms 3+ ions and oxygen forms 2− ions. (b) Strontium forms +2 ions and the phosphate ion is PO_4^{3-}, so the formula is $Sr_3(PO_4)_2$.

(c) Aluminum forms +3 ions and the carbonate ion is CO_3^{2-}, giving a formula of $Al_2(CO_3)_3$. (d) Lithium forms Li^+ ions and the nitride ion is N^{3-}. The formula of lithium nitride is Li_3N.

D.3 (a) calcium phosphate; (b) tin(II) fluoride, stannous fluoride; (c) vanadium(V) oxide; (d) copper(I) oxide, cuprous oxide

D.5 (a) TiO_2; (b) $SiCl_4$; (c) CS_2; (d) SF_4; (e) Li_2S; (f) SbF_5; (g) N_2O_5; (h) IF_7

D.7 (a) sulfur hexafluoride; (b) dinitrogen pentoxide; (c) nitrogen triiodide; (d) xenon tetrafluoride; (e) arsenic tribromide; (f) chlorine dioxide

D.9 (a) hydrochloric acid; (b) sulfuric acid; (c) nitric acid; (d) acetic acid; (e) sulfurous acid; (f) phosphoric acid

D.11 (a) Na_2O; (b) K_2SO_4; (c) AgF; (d) $Zn(NO_3)_2$; (e) Al_2S_3

D.13 (a) sodium sulfite; (b) iron(III) oxide or ferric oxide; (c) iron(II) oxide or ferrous oxide; (d) magnesium hydroxide; (e) nickel(II) sulfate hexahydrate; (f) phosphorus(V) chloride; (g) chromium(III) dihydrogen phosphate; (h) diarsenic trioxide; (i) ruthenium(II) chloride

D.15 (a) heptane; (b) propane; (c) butanol; (d) chloropropane

D.17 (a) telluric acid; (b) sodium arsenate; (c) calcium selenite; (d) barium antimonate; (e) arsenic acid; (f) cobalt(III) tellurate

E.1 (a) $$\text{moles of people} = \frac{6.0 \times 10^9 \text{ people}}{6.022 \times 10^{23} \text{ people}\cdot\text{mol}^{-1}} = 1.0 \times 10^{-14} \text{ mol}$$

(b) $$\text{time} = \frac{1 \text{ mol peas}}{1.0 \times 10^{-14} \text{ mol}\cdot\text{s}^{-1}} = 1.0 \times 10^{14} \text{ s}$$

$$(1.0 \times 10^{14} \text{ s})\left(\frac{1 \text{ h}}{3600 \text{ s}}\right)\left(\frac{1 \text{ day}}{24 \text{ h}}\right)\left(\frac{1 \text{ yr}}{365 \text{ days}}\right) = 3.2 \times 10^6 \text{ years}$$

E.3 (a) mass of average Li atom

$$= \left(\frac{7.42}{100}\right)(9.988 \times 10^{-24}\ \text{g}) + \left(\frac{92.58}{100}\right)(1.165 \times 10^{-23}\ \text{g})$$
$$= 1.153 \times 10^{-23}\ \text{g}\cdot\text{atom}^{-1}$$

$$\text{molar mass} = (1.153 \times 10^{-23}\ \text{g}\cdot\text{atom}^{-1})(6.022 \times 10^{23}\ \text{atoms}\cdot\text{mol}^{-1})$$
$$= 6.94\ \text{g}\cdot\text{mol}^{-1}$$

(b) mass of average Li atom

$$= \left(\frac{5.67}{100}\right)(9.988 \times 10^{-24}\ \text{g}) + \left(\frac{100 - 5.67}{100}\right)(1.165 \times 10^{-23}\ \text{g})$$
$$= 1.1556 \times 10^{-23}\ \text{g}\cdot\text{atom}^{-1}$$

$$\text{molar mass} = (1.1556 \times 10^{-23}\ \text{g}\cdot\text{atom}^{-1})(6.022 \times 10^{23}\ \text{atoms}\cdot\text{mol}^{-1})$$
$$= 6.96\ \text{g}\cdot\text{mol}^{-1}$$

E.5 The percentage ^{10}B = 100 − percentage ^{11}B

$$\text{molar mass} = \left(\frac{\%\ ^{10}\text{B}}{100\%}\right)(\text{mass}\ ^{10}\text{B}) + \left(\frac{\%\ ^{11}\text{B}}{100\%}\right)(\text{mass}\ ^{11}\text{B})$$
$$= \left(\frac{100\% - \%\ ^{11}\text{B}}{100\%}\right)(\text{mass}\ ^{10}\text{B}) + \left(\frac{\%\ ^{11}\text{B}}{100\%}\right)(\text{mass}\ ^{11}\text{B})$$

Rearranging gives

$$\%\ ^{11}\text{B} = \frac{100\cdot\text{molar mass} - 100\cdot\text{mass}\ ^{10}\text{B}}{\text{mass}\ ^{11}\text{B} - \text{mass}\ ^{10}\text{B}}$$
$$= \frac{100(10.81\ \text{g}\cdot\text{mol}^{-1}) - 100(10.013\ \text{g}\cdot\text{mol}^{-1})}{11.093\ \text{g}\cdot\text{mol}^{-1} - 10.013\ \text{g}\cdot\text{mol}^{-1}}$$
$$= 73.8\%$$

$$\%\ ^{10}\text{B} = 26.2\%$$

E.7 (a) $\dfrac{75\ \text{g}}{114.82\ \text{g}\cdot\text{mol}^{-1}} = 0.65\ \text{mol In}$

$\dfrac{80\ \text{g}}{127.60\ \text{g}\cdot\text{mol}^{-1}} = 0.63\ \text{mol Te}$

80 g of tellurium contains more moles of atoms than does 75 g of indium.

(b) $\dfrac{15.0\ \text{g}}{30.97\ \text{g}\cdot\text{mol}^{-1}} = 0.484\ \text{mol P}$

$\dfrac{15.0\ \text{g}}{32.07\ \text{g}\cdot\text{mol}^{-1}} = 0.468\ \text{mol S}$

15.0 g of P has slightly more atoms than 15.0 g of S.

(c) Because the two samples have the same number of atoms, they will have the same number of moles, which is given by $\frac{2.49 \times 10^{22} \text{ atoms}}{6.022 \times 10^{23} \text{ atoms}\cdot\text{mol}^{-1}} =$ 0.0413 mol

E.9 (a) $m_{\text{Rh}} = \left(\frac{57 \text{ g N}}{14.01 \text{ g}\cdot\text{mol}^{-1} \text{ N}}\right)(102.91 \text{ g}\cdot\text{mol}^{-1} \text{ Rh})$
$= 4.1 \times 10^{2} \text{ g Rh}$

(b) $m_{\text{Rh}} = \left(\frac{57 \text{ g Zr}}{91.22 \text{ g}\cdot\text{mol}^{-1} \text{ Zr}}\right)(102.91 \text{ g}\cdot\text{mol}^{-1} \text{ Rh})$
$= 63 \text{ g Rh}$

E.11 (a) molar mass of $Al_2O_3 = 101.96 \text{ g}\cdot\text{mol}^{-1}$

$n_{Al_2O_3} = \frac{10.0 \text{ g}}{101.96 \text{ g}\cdot\text{mol}^{-1}} = 0.0981 \text{ mol}$

$N_{Al_2O_3} = (0.0981 \text{ mol})(6.022 \times 10^{23} \text{ atoms}\cdot\text{mol}^{-1}) = 5.91 \times 10^{22} \text{ molecules}$

(b) molar mass of HF $= 20.01 \text{ g}\cdot\text{mol}^{-1}$

$n_{\text{HF}} = \frac{25.92 \times 10^{-3} \text{ g}}{20.01 \text{ g}\cdot\text{mol}^{-1}} = 1.30 \times 10^{-3} \text{ mol}$

$N_{\text{HF}} = (1.30 \times 10^{-3} \text{ mol})(6.022 \times 10^{23} \text{ atoms}\cdot\text{mol}^{-1}) = 7.83 \times 10^{20} \text{ molecules}$

(c) molar mass of hydrogen peroxide $= 34.02 \text{ g}\cdot\text{mol}^{-1}$

$n_{H_2O_2} = \frac{1.55 \times 10^{-3} \text{ g}}{34.02 \text{ g}\cdot\text{mol}^{-1}} = 4.56 \times 10^{-5} \text{ mol}$

$N_{H_2O_2} = (4.56 \times 10^{-5} \text{ mol})(6.022 \times 10^{23} \text{ atoms}\cdot\text{mol}^{-1}) = 2.75 \times 10^{19} \text{ molecules}$

(d) molar mass of glucose $= 180.15 \text{ g}\cdot\text{mol}^{-1}$

$n_{\text{glucose}} = \frac{1250 \text{ g}}{180.15 \text{ g}\cdot\text{mol}^{-1}} = 6.94 \text{ mol}$

$N_{\text{glucose}} = (6.94 \text{ mol})(6.022 \times 10^{23} \text{ atoms}\cdot\text{mol}^{-1}) = 4.18 \times 10^{24} \text{ molecules}$

(e) molar mass of N atoms $= 14.01 \text{ g}\cdot\text{mol}^{-1}$

$n_{\text{N}} = \frac{4.37 \text{ g}}{14.01 \text{ g}\cdot\text{mol}^{-1}} = 0.312 \text{ mol}$

$N_{\text{N}} = (0.312 \text{ mol})(6.022 \times 10^{23} \text{ atoms}\cdot\text{mol}^{-1}) = 1.88 \times 10^{23} \text{ atoms}$

molar mass of N_2 molecules $= 28.02 \text{ g}\cdot\text{mol}^{-1}$

$n_{N_2} = \frac{4.37 \text{ g}}{28.02 \text{ g}\cdot\text{mol}^{-1}} = 0.156 \text{ mol}$

$N_{N_2} = (0.156 \text{ mol})(6.022 \times 10^{23} \text{ atoms}\cdot\text{mol}^{-1}) = 9.39 \times 10^{22} \text{ molecules}$

E.13 (a) molar mass of AgCl = 143.32 g·mol^{-1}

$$n_{AgCl} = \frac{2.00\text{ g}}{143.32\text{ g}\cdot\text{mol}^{-1}} = 0.0140\text{ mol}$$

The number of moles of Ag^+ ions equals the number of moles of AgCl.

(b) molar mass of UO_3 = 286.03 g·mol^{-1}

$$n_{UO_3} = \frac{600\text{ g}}{286.03\text{ g}\cdot\text{mol}^{-1}} = 2.10\text{ mol}$$

(c) molar mass of $FeCl_3$ = 162.20

$$n_{FeCl_3} = \left(\frac{4.19\text{ mg}}{162.20\text{ g}\cdot\text{mol}^{-1}}\right)\left(\frac{1\text{ g}}{1000\text{ mg}}\right) = 2.58 \times 10^{-5}\text{ mol}$$

The number of moles of Cl^- ions equals 3 times the number of moles of $FeCl_3$.

$n_{Cl^-} = 7.75 \times 10^{-5}$ mol

(d) molar mass of $AuCl_3 \cdot 2\,H_2O$ = 339.35 g·mol^{-1}

$$\left(\frac{2\text{ mol }H_2O}{1\text{ mol }AuCl_3\cdot 2\,H_2O}\right)(n_{AuCl_3\cdot 2\,H_2O}) = 2\left(\frac{1.00\text{ g}}{339.35\text{ g}\cdot\text{mol}^{-1}}\right)$$

$$= 5.89 \times 10^{-3}\text{ mol }H_2O$$

E.15 (a) number of formula units = (0.750 mol)(6.022×10^{23} formula units·mol^{-1})

= 4.52×10^{23} formula units

(b) molar mass of Ag_2SO_4 = 311.80 g·mol^{-1}

$$\left(\frac{2.39 \times 10^{20}\text{ formula units}}{6.022 \times 10^{23}\text{ formula units}\cdot\text{mol}^{-1}}\right)(311.80\text{ g}\cdot\text{mol}^{-1})\left(\frac{1000\text{ mg}}{1\text{ g}}\right) = 124\text{ mg}$$

(c) molar mass of $NaHCO_2$ = 68.01 g·mol^{-1}

$$\left(\frac{3.429\text{ g}}{68.01\text{ g}\cdot\text{mol}^{-1}}\right)\left(\frac{1000\text{ g}}{1\text{ kg}}\right)(6.022 \times 10^{23}\text{ formula units}\cdot\text{mol}^{-1})$$

$$= 3.036 \times 10^{25}\text{ formula units}$$

E.17 (a) molar mass of H_2O = 18.02 g·mol^{-1}

$$\left(\frac{18.02\text{ g}\cdot\text{mol}^{-1}}{6.022 \times 10^{23}\text{ molecules}\cdot\text{mol}^{-1}}\right) = 2.992 \times 10^{-23}\text{ g}\cdot\text{molecule}^{-1}$$

(b) $$N_{H_2O} = \left(\frac{1000\text{ g}}{18.02\text{ g}\cdot\text{mol}^{-1}}\right)(6.022 \times 10^{23}\text{ molecules}\cdot\text{mol}^{-1})$$

$$= 3.34 \times 10^{25}\text{ molecules}$$

E.19 (a) molar mass of $CuBr_2 \cdot 4\,H_2O$ = 295.42 g·mol^{-1}

$$\left(\frac{7.35\text{ g}}{295.42\text{ g}\cdot\text{mol}^{-1}}\right) = 2.49 \times 10^{-2}\text{ mol}$$

(b) Because there are 2 mol Br^- per mole of compound, the number of moles will be twice the amount in part (a), 4.98×10^{-2} mol.

(c) $\left(\dfrac{4 \text{ mol H}_2\text{O}}{1 \text{ mol CuBr}_2\cdot 4\text{ H}_2\text{O}}\right)(2.49 \times 10^{-2} \text{ mol})(6.022 \times 10^{23} \text{ molecules}\cdot\text{mol}^{-1})$

$= 6.00 \times 10^{22}$ molecules H_2O

(d) fraction of mass due to O $= \dfrac{4(16.00 \text{ g}\cdot\text{mol}^{-1})}{295.42 \text{ g}\cdot\text{mol}^{-1}} = 0.2166$

E.21 (a) moles of Cu $= \dfrac{43.4 \text{ g}}{63.54 \text{ g}\cdot\text{mol}^{-1}} = 0.683$ mol

S atoms required $= (0.683 \text{ mol})(6.022 \times 10^{23} \text{ atoms}\cdot\text{mol}^{-1}) = 4.11 \times 10^{23}$

(b) S_8 molecules required $= \dfrac{4.11 \times 10^{23}}{8} = 5.14 \times 10^{22}$

(c) mass of sulfur needed $= (0.683 \text{ mol})(32.06 \text{ g}\cdot\text{mol}^{-1}) = 21.9$ g

E.23 (a) $MgSO_4\cdot 7\,H_2O$ formula mass $= 246.48 \text{ g}\cdot\text{mol}^{-1}$

atoms of O $= \left(\dfrac{5.15 \text{ g}}{246.48 \text{ g}\cdot\text{mol}^{-1}}\right)\left(\dfrac{11 \text{ mol O atoms}}{\text{mol MgSO}_4\cdot 7\text{ H}_2\text{O}}\right)$

$(6.022 \times 10^{23} \text{ atoms}\cdot\text{mol}^{-1}) = 1.38 \times 10^{23}$

(b) formula units $= \left(\dfrac{5.15 \text{ g}}{246.48 \text{ g}\cdot\text{mol}^{-1}}\right)(6.022 \times 10^{23} \text{ atoms}\cdot\text{mol}^{-1})$

$= 1.26 \times 10^{22}$

(c) moles of H_2O $= 7\left(\dfrac{5.15 \text{ g}}{246.48 \text{ g}\cdot\text{mol}^{-1}}\right) = 0.146$ mol

F.1 molar mass of testosterone $= 288.41 \text{ g}\cdot\text{mol}^{-1}$

$\%\text{C} = \dfrac{19 \times 12.01 \text{ g}\cdot\text{mol}^{-1}}{288.41 \text{ g}\cdot\text{mol}^{-1}} \times 100 = 79.12\%$

$\%\text{H} = \dfrac{28 \times 1.0079 \text{ g}\cdot\text{mol}^{-1}}{288.41 \text{ g}\cdot\text{mol}^{-1}} \times 100 = 9.78\%$

$\%\text{O} = \dfrac{2 \times 16.00 \text{ g}\cdot\text{mol}^{-1}}{288.41 \text{ g}\cdot\text{mol}^{-1}} \times 100 = 11.10\%$

F.3 (a) For 100 g of compound,

moles of Na $= \dfrac{32.79 \text{ g}}{22.99 \text{ g}\cdot\text{mol}^{-1}} = 1.426$ mol

moles of Al $= \dfrac{13.02 \text{ g}}{26.98 \text{ g}\cdot\text{mol}^{-1}} = 0.4826$ mol

moles of F $= \dfrac{54.19 \text{ g}}{19.00 \text{ g}\cdot\text{mol}^{-1}} = 2.852$ mol

Dividing each number by 0.4826 gives a ratio of 1 Al : 2.95 Na : 5.91 F. The formula is Na_3AlF_6.

(b) For 100 g of compound,

$$\text{moles of K} = \frac{31.91\ \text{g}}{39.10\ \text{g}\cdot\text{mol}^{-1}} = 0.8161\ \text{mol}$$

$$\text{moles of Cl} = \frac{28.93\ \text{g}}{35.45\ \text{g}\cdot\text{mol}^{-1}} = 0.8161\ \text{mol}$$

mass of O is obtained by difference:

$$\text{moles of O} = \frac{100\ \text{g} - 31.91\ \text{g} - 28.93\ \text{g}}{16.00\ \text{g}\cdot\text{mol}^{-1}} = 2.448\ \text{mol}$$

Dividing each number by 0.8161 gives a ratio of 1.00 K : 1 Cl : 3.00 O. The formula is $KClO_3$.

(c) For 100 g of compound,

$$\text{moles of N} = \frac{12.2\ \text{g}}{14.01\ \text{g}\cdot\text{mol}^{-1}} = 0.871\ \text{mol}$$

$$\text{moles of H} = \frac{5.26\ \text{g}}{1.0079\ \text{g}\cdot\text{mol}^{-1}} = 5.22\ \text{mol}$$

$$\text{moles of P} = \frac{26.9\ \text{g}}{30.97\ \text{g}\cdot\text{mol}^{-1}} = 0.869\ \text{mol}$$

$$\text{moles of O} = \frac{55.6\ \text{g}}{16.00\ \text{g}\cdot\text{mol}^{-1}} = 3.475\ \text{mol}$$

Dividing each number by 0.869 gives a ratio of 1.00 N : 6.01 H : 1.00 P : 4.00 O. The formula is NH_6PO_4 or $[NH_4][H_2PO_4]$, ammonium dihydrogen phosphate.

F.5 $$\text{moles of P} = \frac{4.14\ \text{g}}{30.97\ \text{g}\cdot\text{mol}^{-1}} = 0.134\ \text{mol}$$

$$\text{moles of Cl} = \frac{27.8\ \text{g} - 4.14\ \text{g}}{35.45\ \text{g}\cdot\text{mol}^{-1}} = 0.667\ \text{mol}$$

Dividing each number by 0.134 mol gives a ratio of 4.98 Cl : 1 P. The formula is PCl_5.

F.7 For 100 g of compound,

$$\text{moles of C} = \frac{54.82\ \text{g}}{12.01\ \text{g}\cdot\text{mol}^{-1}} = 4.565\ \text{mol}$$

$$\text{moles of H} = \frac{5.62\ \text{g}}{1.0079\ \text{g}\cdot\text{mol}^{-1}} = 5.58\ \text{mol}$$

$$\text{moles of N} = \frac{7.10\ \text{g}}{14.01\ \text{g}\cdot\text{mol}^{-1}} = 0.507\ \text{mol}$$

$$\text{moles of O} = \frac{32.46\text{ g}}{16.00\text{ g}\cdot\text{mol}^{-1}} = 2.029\text{ mol}$$

Dividing each number by 0.507 mol gives a ratio of 9.00 C:11.01 H:1.00 N:4.00 O. The formula is $C_9H_{11}NO_4$.

F.9 For 100 g of the osmium carbonyl compound,

$$\text{moles of C} = \frac{15.89\text{ g}}{12.01\text{ g}\cdot\text{mol}^{-1}} = 1.323\text{ mol}$$

$$\text{moles of O} = \frac{21.18\text{ g}}{16.00\text{ g}\cdot\text{mol}^{-1}} = 1.324\text{ mol}$$

$$\text{moles of Os} = \frac{62.93\text{ g}}{190.2\text{ g}\cdot\text{mol}^{-1}} = 0.3309\text{ mol}$$

Dividing each number by 0.3309 mol gives a ratio of 4.00 C:4.00 O:1.00 Os. (a) The empirical formula is OsC_4O_4. (b) The formula mass of OsC_4O_4 is 302.24 $g\cdot mol^{-1}$. The molar mass is 907 $g\cdot mol^{-1}$ which is 3 times the formula mass, so the molecular formula is $Os_3C_{12}O_{12}$.

F.11 For 100 g of caffeine,

$$\text{moles of C} = \frac{49.48\text{ g}}{12.01\text{ g}\cdot\text{mol}^{-1}} = 4.12\text{ mol}$$

$$\text{moles of H} = \frac{5.19\text{ g}}{1.0079\text{ g}\cdot\text{mol}^{-1}} = 5.15\text{ mol}$$

$$\text{moles of N} = \frac{28.85\text{ g}}{14.01\text{ g}\cdot\text{mol}^{-1}} = 2.059\text{ mol}$$

$$\text{moles of O} = \frac{16.48\text{ g}}{16.00\text{ g}\cdot\text{mol}^{-1}} = 1.03\text{ mol}$$

Dividing each number by 1.03 mol gives a ratio of 4.00 C:5.00 H:2.00 N:1.00 O. The formula is $C_4H_5N_2O$ with a molar formula mass of 97.10 $g\cdot mol^{-1}$. Because the molecular molar mass is twice this value, the actual formula will be $C_8H_{10}N_4O_2$.

F.13 Glucose ($C_6H_{12}O_6$) has a molar mass of 180.15 $g\cdot mol^{-1}$ and will have the following composition:

$$\%\text{C} = \frac{6(12.01\text{ g}\cdot\text{mol}^{-1})}{180.15\text{ g}\cdot\text{mol}^{-1}} = 40.00\%$$

$$\%\text{H} = \frac{12(1.0079\text{ g}\cdot\text{mol}^{-1})}{180.15\text{ g}\cdot\text{mol}^{-1}} = 6.71\%$$

$$\%\text{O} = \frac{6(16.00\text{ g}\cdot\text{mol}^{-1})}{180.15\text{ g}\cdot\text{mol}^{-1}} = 53.29\%$$

Sucrose ($C_{12}H_{22}O_{11}$) has a molar mass of 342.29 $g\cdot mol^{-1}$ and will have the following composition:

$$\%C = \frac{12(12.01\ g\cdot mol^{-1})}{342.29\ g\cdot mol^{-1}} = 42.10\%$$

$$\%H = \frac{22(1.0079\ g\cdot mol^{-1})}{342.29\ g\cdot mol^{-1}} = 6.48\%$$

$$\%O = \frac{11(16.00\ g\cdot mol^{-1})}{342.29\ g\cdot mol^{-1}} = 51.42\%$$

While the %H values for glucose and sucrose are too close to allow us to distinguish between them by this value alone, %C (40.00 versus 42.10%) and %O (53.29 versus 51.42%) values are sufficient that, when taken together, can give us a reasonable amount of confidence in distinguishing between them.

G.1 (a) solubility; (b) the abilities of the components to adsorb; (c) boiling points

G.3 (a) homogeneous, distillation; (b) heterogeneous, dissolving followed by filtration and distillation; (c) homogeneous, distillation

G.5 mass of $AgNO_3 = (0.179\ mol\cdot L^{-1})(0.5000\ L)(169.88\ g\cdot mol^{-1}) = 15.2\ g$

G.7 (a) molarity of $Na_2CO_3 = \dfrac{2.111\ g}{(105.99\ g\cdot mol^{-1})(0.2500\ L)} = 0.079\ 67\ \text{M}\ Na_2CO_3$

$$V = \frac{(2.15 \times 10^{-3}\ mol\ Na^+(1\ mol\ Na_2CO_2)}{(0.079\ 67\ mol\cdot L^{-1}\ Na_2CO_3)(2\ mol\ Na^+)} = 1.35 \times 10^{-2}\ L \text{ or } 13.5\ mL$$

(b) $$V = \frac{(4.98 \times 10^{-3}\ mol\ CO_3^{2-})(1\ mol\ Na_2CO_2)}{(0.079\ 67\ mol\cdot L^{-1}\ Na_2CO_3)(1\ mol\ CO_3^{2-})} = 6.25 \times 10^{-2}\ L \text{ or } 62.5\ mL$$

(c) $$V = \frac{(50.0 \times 10^{-3}\ g\ Na_2CO_3)}{(105.99\ g\cdot mol^{-1})(0.079\ 67\ mol\cdot L^{-1}\ Na_2CO_3)}$$

$$= 5.92 \times 10^{-3}\ L \text{ or } 5.92\ mL$$

G.9 (a) Weigh 1.6 g (0.010 mol, molar mass of $KMnO_4$ = 158.04 $g\cdot mol^{-1}$) into a 1.0-L volumetric flask and add water to give a total volume of 1.0 L. Smaller (or larger) volumes could also be prepared by using a proportionally smaller (or larger) mass of $KMnO_4$.

(b) Starting with 0.050 $mol\cdot L^{-1}$ $KMnO_4$, you would need to add 4 volumes of water to 1 volume of starting solution, because the concentration desired is one-fifth of the starting solution. This relation can be derived from the expression

$$V_i \times \text{molarity}_i = V_f \times \text{molarity}_f$$

where i represents the initial solution and f the final solution. But $V_f = V_i + V_d$ where V_d represents the volume of solvent that must be added to dilute the initial solution. Rearranging the first equation gives

$$\frac{V_i}{V_f} = \frac{\text{molarity}_f}{\text{molarity}_i}$$

$$\frac{V_i}{V_i + V_d} = \frac{\text{molarity}_f}{\text{molarity}_i}$$

So if the ratio of final molarity to initial molarity is 1:5, we can write

$$\frac{V_i}{V_i + V_d} = \frac{1}{5}$$

$$5V_i = V_i + V_d$$

$$4V_i = V_d$$

For example, to prepare 50 mL of solution, you would add 40 mL of water to 10 mL of 0.050 $\text{mol}\cdot\text{L}^{-1}$ $KMnO_4$.

G.11 (a) $V(0.778\ \text{mol}\cdot\text{L}^{-1}) = (0.1500\ \text{L})(0.0234\ \text{mol}\cdot\text{L}^{-1})$

$V = 4.51 \times 10^{-3}$ L or 4.51 mL

(b) The concentration desired is one-fifth of the starting NaOH solution, so the stockroom attendant will need to add 4 volumes of water to 1 volume of the 2.5 $\text{mol}\cdot\text{L}^{-1}$ solution. To prepare 60.0 mL of solution, divide 60.0 by 5; so 12.0 mL of 2.5 $\text{mol}\cdot\text{L}^{-1}$ NaOH solution are added to 48.0 mL of water. See the solution to G.9.

G.13 (a) mass of $CuSO_4 = (0.20\ \text{mol}\cdot\text{L}^{-1})(0.250\ \text{L})(159.60\ \text{g}\cdot\text{mol}^{-1})$

$= 8.0$ g

(b) mass of $CuSO_4 \cdot 5\ H_2O = (0.20\ \text{mol}\cdot\text{L}^{-1})(0.250\ \text{L})(249.68\ \text{g}\cdot\text{mol}^{-1})$

$= 12$ g

G.15 (a) Chloride ions are supplied only by the $NiCl_2 \cdot 6\ H_2O$ complex:

$$[Cl^-] = \frac{(2\ \text{mol}\ Cl^-)(0.129\ \text{g}\ NiCl_2 \cdot 6\ H_2O)}{(1\ \text{mol}\ NiCl_2 \cdot 6\ H_2O)(237.70\ \text{g}\cdot\text{mol}^{-1}\ NiCl_2 \cdot 6\ H_2O)(0.250\ \text{L})}$$

$= 0.004\ 34$ M

(b) Ni^{2+} ions are present in both the $NiSO_4 \cdot 6\ H_2O$ and the $NiCl_2 \cdot 6\ H_2O$, so the final concentration will be the sum of the ions provided from the two sources:

Ni^{2+} from $NiCl_2 \cdot 6\ H_2O$

$$m_{Ni^{2+}\text{ from nickel chloride}} = \frac{(1\text{ mol Ni}^{2+})(0.129\text{ g NiCl}_2\cdot 6\text{ H}_2\text{O})}{(1\text{ mol NiCl}_2\cdot 6\text{ H}_2\text{O})(237.70\text{ g}\cdot\text{mol}^{-1}\text{ NiCl}_2\cdot 6\text{ H}_2\text{O})}$$

$$= 0.005\ 43\text{ mol}$$

Ni^{2+} from $NiSO_4 \cdot 6\ H_2O$

$$m_{Ni^{2+}\text{ from nickel chloride}} = \frac{(1\text{ mol Ni}^{2+})(0.376\text{ g NiSO}_4\cdot 6\text{ H}_2\text{O})}{(1\text{ mol NiSO}_4\cdot 6\text{ H}_2\text{O})(262.86\text{ g}\cdot\text{mol}^{-1}\text{ NiSO}_4\cdot 6\text{ H}_2\text{O})}$$

$$= 0.001\ 43\text{ mol}$$

total moles of Ni^{2+} = 0.005 43 mol + 0.001 43 mol = 0.006 86 mol

$$[\text{Ni}^{2+}] = \frac{0.006\ 86\text{ mol}}{0.2500\text{ L}} = 0.274\text{ M}$$

G.17 (a) mass of K_2SO_4 = $(0.125\text{ mol}\cdot\text{L}^{-1})(1.00\text{ L})(174.26\text{ g}\cdot\text{mol}^{-1}) = 21.8\text{ g}$
(b) mass of NaF = $(0.015\text{ mol}\cdot\text{L}^{-1})(0.375\text{ L})(41.99\text{ g}\cdot\text{mol}^{-1}) = 0.24\text{ g}$
(c) mass of $C_{12}H_{22}O_{11}$ = $(0.35\text{ mol}\cdot\text{L}^{-1})(0.500\text{ L})(342.29\text{ g}\cdot\text{mol}^{-1}) = 60\text{ g}$

H.1 (a) $BCl_3(g) + 3\ H_2O(l) \longrightarrow B(OH)_3(aq) + 3\ HCl(aq)$
(b) $2\ NaNO_3(s) \longrightarrow 2\ NaNO_2(s) + O_2(g)$
(c) $2\ Ca_3(PO_4)_2(s) + 6\ SiO_2(s) + 10\ C(s) \longrightarrow 6\ CaSiO_3(s) + 10\ CO(g) + P_4(s)$
(d) $4\ Fe_2P(s) + 18\ S(s) \longrightarrow P_4S_{10}(s) + 8\ FeS(s)$

H.3 (a) $2\ Na(s) + 2\ H_2O(l) \longrightarrow H_2(g) + 2\ NaOH(aq)$
(b) $Na_2O(s) + H_2O(l) \longrightarrow 2\ NaOH(aq)$
(c) $6\ Li(s) + N_2(g) \longrightarrow 2\ Li_3N(s)$
(d) $Ca(s) + 2\ H_2O(l) \longrightarrow H_2(g) + Ca(OH)_2(aq)$

H.5 (I) $3\ Fe_2O_3(s) + CO(g) \longrightarrow 2\ Fe_3O_4(s) + CO_2(g)$
(II) $Fe_3O_4(s) + 4\ CO(g) \longrightarrow 3\ Fe(s) + 4\ CO_2(g)$

H.7 (I) $N_2(g) + O_2(g) \longrightarrow 2\ NO(g)$
(II) $2\ NO(g) + O_2(g) \longrightarrow 2\ NO_2(g)$

H.9 $4\ HF(aq) + SiO_2(s) \longrightarrow SiF_4(aq) + 2\ H_2O(l)$

H.11 $2\ C_8H_{18}(l) + 25\ O_2(g) \longrightarrow 16\ CO_2(g) + 18\ H_2O(g)$

H.13 $4\ C_{10}H_{15}N(s) + 55\ O_2(g) \longrightarrow 40\ CO_2(g) + 30\ H_2O(l) + 2\ N_2(g)$

H.15 (I) $H_2S(g) + 2\ NaOH(aq) \longrightarrow Na_2S(aq) + 2\ H_2O(l)$

(II) $4\ H_2S(g) + Na_2S(alc) \xrightarrow{\text{alcohol}} Na_2S_5(alc) + 4\ H_2(g)$

(III) $2\ Na_2S_5(al) + 9\ O_2(g) + 5\ H_2O(l) \longrightarrow 2\ Na_2S_2O_3 \cdot 5\ H_2O(s) + 6\ SO_2(g)$

I.1 (a) CH_3OH, nonelectrolyte; (b) $CaBr_2$, strong electrolyte; (c) KI, strong electrolyte

I.3 (a) soluble; (b) slightly soluble; (c) insoluble; (d) insoluble

I.5 (a) $Na^+(aq)$ and $I^-(aq)$; (b) $Ag^+(aq)$ and $CO_3^{2-}(aq)$, Ag_2CO_3 is insoluble. The very small amount that does go into solution will be present as Ag^+ and CO_3^{2-} ions. (c) $NH_4^+(aq)$ and $PO_4^{3-}(aq)$ (d) $Fe^{2+}(aq)$ and $SO_4^{2-}(aq)$

I.7 (a) $Fe(OH)_3$, precipitate

$Fe_2(SO_4)_3(aq) + 6\ NaOH(aq) \longrightarrow 2\ Fe(OH)_3(s) + 3\ Na_2SO_4(aq)$

(b) Ag_2CO_3, precipitate forms

$2\ AgNO_3(aq) + K_2CO_3(aq) \longrightarrow Ag_2CO_3(s) + 2\ KNO_3(aq)$

(c) No precipitate will form because all possible products are soluble in water.

$Pb(NO_3)_2(aq) + 2\ CH_3COONa(aq) \longrightarrow 2\ NaNO_3(aq) + Pb(CH_3COO)_2(aq)$

I.9 (a) $FeCl_2(aq) + Na_2S(aq) \longrightarrow 2\ NaCl(aq) + FeS(s)$

net ionic equation: $Fe^{2+}(aq) + S^{2-}(aq) \longrightarrow FeS(s)$

spectator ions: Na^+, Cl^-

(b) $Pb(NO_3)_2(aq) + 2\ KI(aq) \longrightarrow PbI_2(s) + 2\ K^+(aq) + 2\ NO_3^-(aq)$

net ionic equation: $Pb^{2+}(aq) + 2\ I^-(aq) \longrightarrow PbI_2(s)$

spectator ions: K^+, NO_3^-

(c) $Ca(NO_3)_2(aq) + K_2SO_4 \longrightarrow CaSO_4(s) + 2\ K^+(aq) + 2\ NO_3^-(aq)$

net ionic equation: $Ca^{2+}(aq) + SO_4^{2-}(aq) \longrightarrow CaSO_4(s)$

spectator ions: NO_3^-, K^+

(d) $2\ Na_2CrO_4(aq) + Pb(NO_3)_2(aq) \longrightarrow$

$2\ Na^+(aq) + 2\ NO_3^-(aq) + PbCrO_4(s)$

net ionic equation: $Pb^{2+}(aq) + CrO_4^{2-}(aq) \longrightarrow PbCrO_4(s)$

spectator ions: Na^+, NO_3^-

(e) $Hg_2(NO_3)_2(aq) + K_2SO_4(aq) \longrightarrow Hg_2SO_4(s) + 2\ K^+(aq) + 2\ NO_3^-(aq)$

net ionic equation: $Hg_2^{2+}(aq) + SO_4^{2-}(aq) \longrightarrow Hg_2SO_4(s)$

spectator ions: K^+, NO_3^-

I.11 (a) overall equation: $(NH_4)_2CrO_4(aq) + BaCl_2(aq) \longrightarrow$
$BaCrO_4(s) + 2\ NH_4Cl(aq)$
complete ionic equation:
$2\ NH_4^+(aq) + CrO_4^{2-}(aq) + Ba^{2+}(aq) + 2\ Cl^-(aq) \longrightarrow$
$BaCrO_4(s) + 2\ NH_4^+(aq) + 2\ Cl^-(aq)$
net ionic equation: $Ba^{2+}(aq) + CrO_4^{2-}(aq) \longrightarrow BaCrO_4(s)$
spectator ions: NH_4^+, Cl^-
(b) $CuSO_4(aq) + Na_2S(aq) \longrightarrow CuS(s) + Na_2SO_4(aq)$
complete ionic equation:
$Cu^{2+}(aq) + SO_4^{2-}(aq) + 2\ Na^+(aq) + S^{2-}(aq) \longrightarrow$
$CuS(s) + 2\ Na^+(aq) + SO_4^{2-}(aq)$
net ionic equation: $Cu^{2+}(aq) + S^{2-}(aq) \longrightarrow CuS(s)$
spectator ions: Na^+, SO_4^{2-}
(c) $3\ FeCl_2(aq) + 2\ (NH_4)_3PO_4(aq) \longrightarrow Fe_3(PO_4)_2(s) + 6\ NH_4^+(aq) + 6\ Cl^-(aq)$
complete ionic equation:
$3\ Fe^{2+}(aq) + 6\ Cl^-(aq) + 6\ NH_4^+(aq) + 2\ PO_4^{3-}(aq) \longrightarrow$
$Fe_3(PO_4)_2(s) + 6\ NH_4^+(aq) + 6\ Cl^-(aq)$
net ionic equation: $3\ Fe^{2+}(aq) + 2\ PO_4^{3-}(aq) \longrightarrow Fe_3(PO_4)_2(s)$
spectator ions: Cl^-, NH_4^+
(d) $K_2C_2O_4(aq) + Ca(NO_3)_2(aq) \longrightarrow CaC_2O_4(s) + 2\ KNO_3(aq)$
complete ionic equation:
$2\ K^+(aq) + C_2O_4^{2-}(aq) + Ca^{2+}(aq) + 2\ NO_3^-(aq) \longrightarrow$
$CaC_2O_4(s) + 2\ K^+(aq) + 2\ NO_3^-(aq)$
net ionic equation: $Ca^{2+}(aq) + C_2O_4^{2-}(aq) \longrightarrow CaC_2O_4(s)$
spectator ions: K^+, NO_3^-
(e) $NiSO_4(aq) + Ba(NO_3)_2(aq) \longrightarrow Ni(NO_3)_2(aq) + BaSO_4(s)$
complete ionic equation:
$Ni^{2+}(aq) + SO_4^{2-}(aq) + Ba^{2+}(aq) + 2\ NO_3^-(aq) \longrightarrow$
$Ni^{2+}(aq) + 2\ NO_3^-(aq) + BaSO_4(s)$
net ionic equation: $Ba^{2+}(aq) + SO_4^{2-}(aq) \longrightarrow BaSO_4(s)$
spectator ions: Ni^{2+}, NO_3^-

I.13 (a) $Ba^{2+}(aq) + 2\ CH_3CO_2^-(aq) + 2\ Li^+(aq) + CO_3^{2-}(aq) \longrightarrow$
$BaCO_3(s) + 2\ Li^+(aq) + 2CH_3CO_2^-(aq)$
net ionic equation: $Ba^{2+}(aq) + CO_3^{2-}(aq) \longrightarrow BaCO_3(s)$
(b) $2\ NH_4^+(aq) + 2\ Cl^-(aq) + Hg_2^{2+} + 2\ NO_3^-(aq) \longrightarrow$
$Hg_2Cl_2(s) + 2\ NH_4^+(aq) + 2\ NO_3^-(aq)$
net ionic equation: $Hg_2^{2+}(aq) + 2\ Cl^-(aq) \longrightarrow Hg_2Cl_2(s)$

(c) $Cu^{2+}(aq) + 2\ NO_3^-(aq) + Ba^{2+}(aq) + 2\ OH^-(aq) \longrightarrow$
$Cu(OH)_2(s) + Ba^{2+}(aq) + 2\ NO_3^-(aq)$
net ionic equation: $Cu^{2+}(aq) + 2\ OH^-(aq) \longrightarrow Cu(OH)_2(s)$

I.15 (a) $AgNO_3$ and Na_2CrO_4
(b) $CaCl_2$ and Na_2CO_3
(c) $Cd(ClO_4)_2$ and $(NH_4)_2S$

I.17 (a) $2\ Ag^+(aq) + SO_4^{2-}(aq) \longrightarrow Ag_2SO_4(s)$
(b) $Hg^{2+}(aq) + S^{2-}(aq) \longrightarrow HgS(s)$
(c) $3\ Ca^{2+}(aq) + 2\ PO_4^{3-}(aq) \longrightarrow Ca_3(PO_4)_2(s)$
(d) $AgNO_3$ and Na_2SO_4; Na^+, NO_3^-
$Hg(CH_3CO_2)_2$ and Li_2S; Li^+, $CH_3CO_2^-$
$CaCl_2$ and K_3PO_4; K^+, Cl^-

J.1 (a) base; (b) acid; (c) base; (d) acid; (e) base

J.3 Complete and write the overall equation, the complete ionic equation, and the net ionic equation for the following acid-base reactions. If the substance is a weak acid or base, leave it in its molecular form in writing the equations.
(a) overall equation: $HF(aq) + NaOH(aq) \longrightarrow NaF(aq) + H_2O(l)$
total ionic equation:
$HF(aq) + Na^+(aq) + OH^-(aq) \longrightarrow Na^+(aq) + F^-(aq) + H_2O(l)$
net ionic equation:
$HF(aq) + OH^-(aq) \longrightarrow F^-(aq) + H_2O(l)$
(b) overall equation: $(CH_3)_3N(aq) + HNO_3(aq) \longrightarrow [(CH_3)_3NH]NO_3(aq)$
total ionic equation:
$(CH_3)_3N(aq) + H_3O^+(aq) + NO_3^-(aq) \longrightarrow$
$[(CH_3)_3NH]^+(aq) + NO_3^-(aq) + H_2O(l)$
net ionic equation: $(CH_3)_3N(aq) + H_3O^+(aq) \longrightarrow [(CH_3)_3NH]^+(aq) + H_2O(l)$
(c) overall equation: $LiOH(aq) + HI(aq) \longrightarrow LiI(aq) + H_2O(l)$
total ionic equation:
$Li^+(aq) + OH^-(aq) + H_3O^+(aq) + I^-(aq) \longrightarrow Li^+(aq) + I^-(aq) + 2\ H_2O(l)$

J.5 The overall equations are:
(a) $HBr(aq) + KOH(aq) \longrightarrow KBr(aq) + H_2O(l)$
(b) $Zn(OH)_2(aq) + 2\ HNO_2(aq) \longrightarrow Zn(NO_2)_2(aq) + 2\ H_2O(l)$

(c) $Ca(OH)_2(aq) + 2\,HCN(aq) \longrightarrow Ca(CN)_2(aq) + 2\,H_2O(l)$
(d) $3\,KOH(aq) + H_3PO_4(aq) \longrightarrow K_3PO_4(aq) + 3\,H_2O(l)$

J.7 (a) acid: $H_3O^+(aq)$; base: $CH_3NH_2(aq)$
(b) acid: $HCl(aq)$; base: $C_2H_5NH_2(aq)$
(c) acid: $HI(aq)$; base: $CaO(s)$

K.1 (a) $2\,NO_2(g) + O_3(g) \longrightarrow N_2O_5(g) + O_2(g)$
(b) $S_8(s) + 16\,Na(s) \longrightarrow 8\,NaS(s)$
(c) $2\,Cr^{2+}(aq) + Sn^{4+}(aq) \longrightarrow 2\,Cr^{3+}(aq) + Sn^{2+}(aq)$
(d) $2\,As(s) + 3\,Cl_2(g) \longrightarrow 2\,AsCl_3(l)$

K.3 (a) $Mg^0(s) + Cu^{2+}(aq) \longrightarrow Mg^{2+}(aq) + Cu^0(s)$
(b) $Fe^{2+}(aq) + Ce^{4+}(aq) \longrightarrow Fe^{3+}(aq) + Ce^{3+}(aq)$
(c) $H_2(g) + Cl_2(g) \longrightarrow 2\,HCl(g)$
(d) $4\,Fe(s) + 3\,O_2(g) \longrightarrow 2\,Fe_2O_3(s)$

K.5 (a) +4; (b) +4; (c) +2; (d) +5; (e) +1; (f) 0

K.7 (a) +2; (b) +2; (c) +6; (d) +4; (e) +1

K.9 (a) Methanol $CH_3OH(aq)$ is oxidized to formic acid (the carbon atom goes from an oxidation number of +2 to +4). The $O_2(g)$ is reduced to O^{2-} present in water.
(b) Mo is reduced from +5 to +4, while *some* sulfur (that which ends up as S(s)) is oxidized from −2 to 0. The sulfur present in $MoS_2(s)$ remains in the −2 oxidation state.
(c) Tl^+ is both oxidized and reduced. The product Tl(s) is a reduction of Tl^+ (from +1 to 0) while the Tl^{3+} is produced via an oxidation of Tl^+. A reaction in which a single substance is both oxidized and reduced is known as a *disproportionation reaction.*

K.11 (a) Cl_2 will more easily be reduced and is therefore a stronger oxidizing agent than Cl^-. (b) N_2O_5 will be a stronger oxidizing agent because it will be readily reduced. N^{5+} will accept e^- more readily than N^+.

K.13 (a) oxidizing agent: H^+ in HCl(aq); reducing agent: Zn(s)
(b) oxidizing agent: $SO_2(g)$; reducing agent: $H_2S(g)$
(c) oxidizing agent: $B_2O_3(s)$; reducing agent: Mg(s)

K.15 (a) $ClO_3^- \longrightarrow ClO_2$, Cl goes from $+5 \longrightarrow +4$; reducing agent

(b) $SO_4^{2-} \longrightarrow S^{2-}$, S goes from $+6 \longrightarrow -2$; reducing agent

(c) $Mn^{2+} \longrightarrow MnO_2$, Mn goes from $+2 \longrightarrow +4$; oxidizing agent

(d) $HCHO \longrightarrow HCOOH$, C goes from 0 to +2; oxidizing agent

K.17 (a) oxidizing agent: $WO_3(s)$; reducing agent: $H_2(g)$

(b) oxidizing agent: $HCl(aq)$; reducing agent: $Zn(s)$

(c) oxidizing agent: $SnO_2(s)$; reducing agent: $C(s)$

(d) oxidizing agent: $N_2O_4(g)$; reducing agent: $N_2H_4(g)$

L.1 $2\ Na_2S_2O_3(aq) + AgBr(s) \longrightarrow NaBr(aq) + Na_3[Ag(S_2O_3)_2](aq)$

(a) moles of $Na_2S_2O_3$ needed to dissolve 1.0 mg AgBr

$$= 1.0 \text{ mg AgBr}\left(\frac{1 \text{ g AgBr}}{1000 \text{ mg AgBr}}\right)\left(\frac{1 \text{ mol AgBr}}{187.78 \text{ g AgBr}}\right)\left(\frac{2 \text{ mol } Na_2S_2O_3}{1 \text{ mol AgBr}}\right)$$

$$= 1.1 \times 10^{-5} \text{ mol } Na_2S_2O_3$$

(b) mass of AgBr to produce 0.033 mol $Na_3[Ag(S_2O_3)_2]$

$$= 0.033 \text{ mol } Na_3[Ag(S_2O_3)_2]\left(\frac{1 \text{ mol AgBr}}{1 \text{ mol } Na_3[Ag(S_2O_3)_2]}\right)\left(\frac{187.78 \text{ g AgBr}}{1 \text{ mol AgBr}}\right)$$

$$= 6.2 \text{ g AgBr}$$

L.3 $6\ NH_4ClO_4(s) + 10\ Al(s) \longrightarrow 5\ Al_2O_3(s) + 3\ N_2(g) + 6\ HCl(g) + 9\ H_2O(g)$

(a) $$(1.325 \text{ kg } NH_4ClO_4)\left(\frac{1 \text{ mol } NH_4ClO_4}{117.49 \text{ g } NH_4ClO_4}\right)\left(\frac{1000 \text{ g}}{1 \text{ kg}}\right)$$

$$\left(\frac{10 \text{ mol Al}}{6 \text{ mol } NH_4ClO_4}\right)\left(\frac{26.98 \text{ g Al}}{1 \text{ mol Al}}\right) = 507 \text{ g Al}$$

(b) $$3500 \text{ kg Al}\left(\frac{1000 \text{ g Al}}{1 \text{ kg Al}}\right)\left(\frac{1 \text{ mol Al}}{26.98 \text{ g Al}}\right)\left(\frac{5 \text{ mol } Al_2O_3}{10 \text{ mol Al}}\right)\left(\frac{101.96 \text{ g } Al_2O_3}{1 \text{ mol } Al_2O_3}\right)$$

$$= 6.613 \times 10^6 \text{ g } Al_2O_3 \text{ or } 6.613 \times 10^3 \text{ kg } Al_2O_3$$

L.5 $2\ C_{57}H_{110}O_6(s) + 163\ O_2(g) \longrightarrow 114\ CO_2(g) + 110\ H_2O(l)$

(a) $$(454 \text{ g fat})\left(\frac{1 \text{ mol fat}}{891.44 \text{ g fat}}\right)\left(\frac{110 \text{ mol } H_2O}{2 \text{ mol fat}}\right)\left(\frac{18.02 \text{ g } H_2O}{1 \text{ mol } H_2O}\right) = 505 \text{ g } H_2O$$

(b) $$(454 \text{ g fat})\left(\frac{1 \text{ mol fat}}{891.44 \text{ g}}\right)\left(\frac{163 \text{ mol } O_2}{2 \text{ mol fat}}\right)\left(\frac{32.00 \text{ g } O_2}{1 \text{ mol } O_2}\right) = 1.33 \times 10^3 \text{ g } O_2$$

L.7 $2\ C_8H_{18}(l) + 25\ O_2(g) \longrightarrow 16\ CO_2(g) + 18\ H_2O(l)$

$d = 0.79 \text{ g·mL}^{-1}$, density of gasoline

$$(3.785\ \text{L gas})\left(\frac{1000\ \text{mL}}{1\ \text{L}}\right)\left(\frac{0.79\ \text{g gas}}{1\ \text{mL}}\right)\left(\frac{1\ \text{mol gas}}{114.22\ \text{g gas}}\right)\left(\frac{18\ \text{mol H}_2\text{O}}{2\ \text{mol C}_8\text{H}_{18}}\right)$$

$$\left(\frac{18.02\ \text{g H}_2\text{O}}{1\ \text{mol H}_2\text{O}}\right) = 4.246 \times 10^3\ \text{g H}_2\text{O or } 4.246\ \text{kg H}_2\text{O}$$

L.9 (a) $HCl + NaOH \longrightarrow NaCl + H_2O$

$$17.40\ \text{mL}\left(\frac{0.234\ \text{mol HCl}}{1000\ \text{mL}}\right)\left(\frac{1\ \text{mol NaOH}}{1\ \text{mol HCl}}\right) = 0.004\ 07\ \text{mol}$$

$$\text{concentration of NaOH} = \frac{0.004\ 07\ \text{mol}}{15.00 \times 10^{-3}\ \text{L}} = 0.271\ \text{M}$$

(b) $(0.271\ \text{mol}\cdot\text{L}^{-1})(0.01500\ \text{L})(40.00\ \text{g}\cdot\text{mol}^{-1}) = 0.163\ \text{g NaOH}$

L.11 (a) $Ba(OH)_2(aq) + 2\ HNO_3(aq) \longrightarrow Ba(NO_3)_2(aq) + 2\ H_2O(l)$

$$\text{molarity of HNO}_3 = \frac{\left(\frac{9.670\ \text{g Ba(OH)}_2}{171.36\ \text{g}\cdot\text{mol}^{-1}}\right)}{0.250\ \text{L}}\left(\frac{11.56\ \text{mL}}{1000\ \text{mL}\cdot\text{L}^{-1}}\right)\left(\frac{2\ \text{mol HNO}_3}{1\ \text{mol Ba(OH)}_2}\right)$$

$$\div\ 0.0250\ \text{L} = 0.209\ \text{mol}\cdot\text{L}^{-1}$$

(b) mass of HNO_3 in solution:

$$\left(\frac{0.209\ \text{mol}}{1000\ \text{mL}}\right)(25.0\ \text{mL})\left(\frac{63.02\ \text{g HNO}_3}{1\ \text{mol HNO}_3}\right) = 0.329\ \text{g}$$

L.13 $HX(aq) + NaOH(aq) \longrightarrow NaX(aq) + H_2O(l)$

$$(68.8\ \text{mL})\left(\frac{0.750\ \text{mol NaOH}}{1000\ \text{mL NaOH}}\right) = 0.0516\ \text{mol NaOH}$$

3.25 g HX corresponds to 0.0516 mol NaOH used

$$\frac{3.25\ \text{g}}{0.0516\ \text{mol}} = 63.0\ \text{g}\cdot\text{mol}^{-1} = \text{molar mass of acid}$$

L.15 (a) $Na_2CO_3(aq) + 2\ HCl(aq) \longrightarrow 2\ NaCl(aq) + H_2CO_3(aq)$

(b) $$(0.832\ \text{g Na}_2\text{CO}_3)\left(\frac{1\ \text{mol Na}_2\text{CO}_3}{105.99\ \text{g Na}_2\text{CO}_3}\right)\left(\frac{0.025\ \text{L}}{0.100\ \text{L}}\right)\left(\frac{2\ \text{mol HCl}}{1\ \text{mol Na}_2\text{CO}_3}\right)$$

$$\left(\frac{1}{0.031\ 25\ \text{L}}\right) = 0.126\ \text{mol}\cdot\text{L}^{-1}\ \text{HCl (diluted)}$$

The original HCl solution is 100 times more concentrated than the solution used for titration (diluted 10.00 mL to 1000 mL), so the original concentration of the HCl solution is $9.43\ \text{mol}\cdot\text{L}^{-1}$.

M.1 $CaCO_3(s) \longrightarrow CaO(s) + CO_2(g)$

theoretical yield:

$$(42.73 \text{ g CaCO}_3)\left(\frac{1 \text{ mol CaCO}_3}{100.09 \text{ g CaCO}_3}\right)\left(\frac{1 \text{ mol CO}_2}{1 \text{ mol CaCO}_3}\right)\left(\frac{44.01 \text{ g CO}_2}{1 \text{ mol CO}_2}\right)$$

$$= 18.79 \text{ g CO}_2$$

actual yield:

$$\frac{17.5 \text{ g}}{18.79 \text{ g}} \times 100\% = 93.1\% \text{ yield}$$

M.3 (a) $C_2H_5OH(aq) + O_2(g) \longrightarrow CH_3COOH(aq) + H_2O(l)$

The amount of acetic acid produced is

$$n_{\text{acetic acid}} = \frac{(0.0275 \text{ g}\cdot\text{mL}^{-1})(1000 \text{ mL})}{60.05 \text{ g}\cdot\text{mol}^{-1}} = 0.458 \text{ mol acetic acid}$$

According to the reaction stoichiometry, one mole of oxygen is required to produce one mole of acetic acid from one mole of ethanol.

$$n_{\text{oxygen}} = (0.458 \text{ mol acetic acid})\left(\frac{1 \text{ mol O}_2}{1 \text{ mol acetic acid}}\right) = 0.458 \text{ mol O}_2$$

$$m_{\text{oxygen}} = (0.458 \text{ mol O}_2)(32.00 \text{ g}\cdot\text{mol}^{-1} \text{ O}_2) = 14.6 \text{ g O}_2$$

(b) The reaction stoichiometry indicates that 1 mole of ethanol produces 1 mole of acetic acid. The theoretical yield is calculated as follows:

The mass of ethanol present in the bottle of wine is given by

$$n_{\text{ethanol}} = \frac{(1000 \text{ mL wine})(0.085 \text{ mL ethanol}\cdot\text{mL}^{-1} \text{ wine})(0.816 \text{ g}\cdot\text{mol}^{-1} \text{ ethanol})}{(46.07 \text{ g}\cdot\text{mol}^{-1} \text{ ethanol})}$$

$$= 1.5 \text{ mol ethanol}$$

$$\% \text{ yield} = \frac{0.458 \text{ mol}}{1.5 \text{ mol}} \times 100\% = 30\%$$

M.5 (a) $P_4(s) + 3\, O_2(g) \longrightarrow P_4O_6(s)$

$P_4O_6(s) + 2\, O_2(g) \longrightarrow P_4O_{10}(s)$

In the first reaction, 5.77 g P_4 uses

$$(5.77 \text{ g P}_4)\left(\frac{1 \text{ mol P}_4}{123.88 \text{ g P}_4}\right)\left(\frac{3 \text{ mol O}_2}{1 \text{ mol P}_4}\right)\left(\frac{32.00 \text{ g O}_2}{1 \text{ mol O}_2}\right) = 4.47 \text{ g O}_2 \text{ (g)}$$

excess O_2 = 5.77 g − 4.47 g O_2 = 1.30 g O_2

In the second reaction, 5.77 g P_4 uses

$$\left(\frac{5.77 \text{ g P}_4}{123.88 \text{ g}\cdot\text{mol}^{-1} \text{ P}_4}\right)\left(\frac{1 \text{ mol P}_4\text{O}_6}{1 \text{ mol P}_4}\right)\left(\frac{2 \text{ mol O}_2}{1 \text{ mol P}_4\text{O}_6}\right)\left(\frac{32.00 \text{ g O}_2}{1 \text{ mol O}_2}\right) = 2.98 \text{ g O}_2$$

limiting reagent: O_2

(b) $$\left(\frac{1.30 \text{ g O}_2}{32.00 \text{ g}\cdot\text{mol}^{-1} \text{ O}_2}\right)\left(\frac{1 \text{ mol P}_4\text{O}_{10}}{2 \text{ mol O}_2}\right)\left(\frac{283.88 \text{ g P}_4\text{O}_{10}}{1 \text{ mol P}_4\text{O}_{10}}\right) = 5.77 \text{ g P}_4\text{O}_{10}$$

(c) $\left(\frac{1.30 \text{ g O}_2}{32.00 \text{ g}\cdot\text{mol}^{-1} \text{ O}_2}\right)\left(\frac{1 \text{ mol P}_4\text{O}_6}{2 \text{ mol O}_2}\right)\left(\frac{219.88 \text{ g P}_4\text{O}_6}{1 \text{ mol P}_4\text{O}_6}\right) = 4.47 \text{ g P}_4\text{O}_6$ used

In the first reaction, 5.77 g P_4 produces

$$\left(\frac{5.77 \text{ g P}_4}{123.88 \text{ g}\cdot\text{mol}^{-1}}\right)\left(\frac{219.88 \text{ g P}_4\text{O}_6}{1 \text{ mol P}_4\text{O}_6}\right)\left(\frac{1 \text{ mol P}_4\text{O}_6}{1 \text{ mol P}_4}\right) = 10.2 \text{ g P}_4\text{O}_6$$

excess reagent: 10.2 g − 4.47 g = 5.7 g P_4O_6

M.7 $C_{63}H_{88}CoN_{14}O_{14}P$. The molar mass of cobalamin is 1355.37 $\text{g}\cdot\text{mol}^{-1}$.

$$n_{\text{cobalamin}} = \frac{0.1674 \text{ g}}{1355.37 \text{ g}\cdot\text{mol}^{-1}} = 1.235 \times 10^{-4} \text{ mol}$$

1 mole of cobalamin will produce 63 moles of CO_2 and 44 moles of H_2O.

$$m_{\text{CO}_2} = (1.235 \times 10^{-4} \text{ mol cobalamin})\left(\frac{63 \text{ mol CO}_2}{1 \text{ mol cobalamin}}\right)(44.01 \text{ g}\cdot\text{mol}^{-1} \text{ CO}_2)$$

$$= 0.3424 \text{ g CO}_2$$

$$m_{\text{H}_2\text{O}} = (1.235 \times 10^{-4} \text{ mol cobalamin})\left(\frac{44 \text{ mol H}_2\text{O}}{1 \text{ mol cobalamin}}\right)(18.02 \text{ g}\cdot\text{mol}^{-1} \text{ CO}_2)$$

$$= 0.097\ 92 \text{ g H}_2\text{O}$$

M.9 $(0.682 \text{ g CO}_2)\left(\frac{1 \text{ mol CO}_2}{44.01 \text{ g CO}_2}\right)\left(\frac{1 \text{ mol C}}{1 \text{ mol CO}_2}\right) = 0.0155 \text{ mol C}$

$(0.0155 \text{ mol C})(12.01 \text{ g}\cdot\text{mol}^{-1} \text{ C}) = 0.186 \text{ g C}$

$(0.174 \text{ g H}_2\text{O})\left(\frac{1 \text{ mol H}_2\text{O}}{18.02 \text{ g H}_2\text{O}}\right)\left(\frac{2 \text{ mol H}}{1 \text{ mol H}_2\text{O}}\right) = 0.0193 \text{ mol H}$

$(0.0193 \text{ mol H})(1.0079 \text{ g}\cdot\text{mol}^{-1} \text{ H}) = 0.0195 \text{ g H}$

$(0.110 \text{ g N}_2)\left(\frac{1 \text{ mol N}_2}{28.02 \text{ g N}_2}\right)\left(\frac{2 \text{ mol N}}{1 \text{ mol N}_2}\right) = 0.007\ 85 \text{ mol N}$

$(0.007\ 85 \text{ mol N})(14.01 \text{ g}\cdot\text{mol}^{-1} \text{ N}) = 0.110 \text{ g N}$

mass of O = 0.376 g − (0.186 g + 0.0193 g + 0.110 g) = 0.061 g O

$\frac{0.061 \text{ g O}}{16.00 \text{ g O}} = 0.0038 \text{ mol O}$

Dividing each amount by 0.0038 gives C:H:N:O ratios = 4.1:5.1:2.1:1. The empirical formula is $C_4H_5N_2O$.

The molecular mass of caffeine is 194 $\text{g}\cdot\text{mol}^{-1}$. Its empirical mass is 97.10 $\text{g}\cdot\text{mol}^{-1}$.

molecular formula = 2 × empirical formula = $C_8H_{10}N_4O_2$

$2\ C_8H_{10}N_4O_2(s) + 19\ O_2(g) \longrightarrow 16\ CO_2(g) + 10\ H_2O(l) + 4\ N_2(g)$

M.11 $3\ Ca(NO_3)_2(aq) + 2\ H_3PO_4(aq) \longrightarrow Ca_3(PO_4)_2(s) + 6\ HNO_3(aq)$

(a) The solid is calcium phosphate, $Ca_3(PO_4)_2$.

(b) $(206\ g\ Ca(NO_3)_2)\left(\frac{1\ mol\ Ca(NO_3)_2}{164.10\ g\ Ca(NO_3)_2}\right)\left(\frac{2\ mol\ H_3PO_4}{3\ mol\ Ca(NO_3)_2}\right)$

$\left(\frac{97.99\ g\ H_3PO_4}{1\ mol\ H_3PO_4}\right) = 82.01\ g\ H_3PO_4$

Therefore $Ca(NO_3)_2$ is the limiting reagent.

$(206\ g\ Ca(NO_3)_2)\left(\frac{1\ mol\ Ca(NO_3)_2}{164.10\ g\ Ca(NO_3)_2}\right)\left(\frac{1\ mol\ Ca_3(PO_4)_2}{3\ mol\ Ca(NO_3)_2}\right)$

$\left(\frac{310.18\ g\ Ca_3(PO_4)_2}{1\ mol\ Ca_3(PO_4)_2}\right) = 130\ g\ Ca_3(PO_4)_2$

M.13 If the 2-naphthol ($144.16\ g\cdot mol^{-1}$) were pure, it would give the following combustion analysis:

$$\%C = \frac{10(12.01\ g\cdot mol^{-1})}{(144.16\ g\cdot mol^{-1}\ 2 - naphthol)} \times 100\% = 83.31\%\ C$$

$$\%H = \frac{8(1.0079\ g\cdot mol^{-1})}{(144.16\ g\cdot mol^{-1}\ 2 - naphthol)} \times 100\% = 5.59\%\ H$$

The observed percentages are low as is expected for a sample contaminated with a substance that contains no C or H. Because the sample does not contain C or H, the percent purity can be easily obtained by

$$\%\text{purity (based on C)} = \frac{\%\ \text{found}}{\%\ \text{theoretical}} = \frac{77.48\%}{83.31\%} \times 100\% = 93.00\%$$

$$\%\text{purity (based on H)} = \frac{\%\ \text{found}}{\%\ \text{theoretical}} = \frac{5.20\%}{5.59\%} \times 100\% = 93.0\%$$

CHAPTER 1
ATOMS: THE QUANTUM WORLD

1.1 ultraviolet radiation < visible light < infrared radiation < radio waves

1.3 microwaves < visible light < ultraviolet radiation < gamma rays ≈ x-rays

1.5 (a) If wavelength is known, the frequency can be obtained from the relation $c = \nu\lambda$:

$$2.997\,92 \times 10^{8}\ \text{m}\cdot\text{s}^{-1} = (\nu)(925 \times 10^{-9}\ \text{m})$$

$$\nu = \frac{2.997\,92 \times 10^{8}\ \text{m}\cdot\text{s}^{-1}}{925 \times 10^{-9}\ \text{m}}$$

$$= 3.24 \times 10^{14}\ \text{s}^{-1}$$

(b) $2.997\,92 \times 10^{8}\ \text{m}\cdot\text{s}^{-1} = (\nu)(4.15 \times 10^{-3}\ \text{m})$

$$\nu = \frac{2.997\,92 \times 10^{8}\ \text{m}\cdot\text{s}^{-1}}{4.15 \times 10^{-3}\ \text{m}}$$

$$= 7.22 \times 10^{10}\ \text{s}^{-1}$$

1.7 Wien's law states that $T\lambda_{max}$ = constant = 2.88×10^{-3} K·m.

If T/K = 1540°C + 273°C = 1813 K, then $\lambda_{max} = \dfrac{2.88 \times 10^{-3}\ \text{K}\cdot\text{m}}{1813\ \text{K}}$

$\lambda_{max} = 1.59 \times 10^{-6}$ m, or 1590 nm

1.9 (a) From $c = \nu\lambda$ and $E = h\nu$, we can write

$$E = hc\lambda^{-1}$$

$$= (6.626\,08 \times 10^{-34}\ \text{J}\cdot\text{s})(2.997\,92 \times 10^{8})(589 \times 10^{-9}\ \text{m})^{-1}$$

$$= 3.37 \times 10^{-19}\ \text{J}$$

(b) $$E = \left(\frac{5.00 \times 10^{-3}\ \text{g Na}}{22.99\ \text{g}\cdot\text{mol}^{-1}\ \text{Na}}\right)(6.022 \times 10^{23}\ \text{atoms}\cdot\text{mol}^{-1})(3.37 \times 10^{-19}\ \text{J}\cdot\text{atom}^{-1})$$

$$= 44.1\ \text{J}$$

(c) $$E = (6.022 \times 10^{23}\ \text{atoms}\cdot\text{mol}^{-1})(3.37 \times 10^{-19}\ \text{J}\cdot\text{atom}^{-1})$$

$$= 2.03 \times 10^{5}\ \text{J or 203 kJ}$$

1.11 The energy is first converted from eV to joules:

$$E = (140.511 \times 10^3\ \text{eV})(1.6022 \times 10^{-19}\ \text{J}\cdot\text{eV}^{-1}) = 2.2513 \times 10^{-14}\ \text{J}$$

From $E = h\nu$ and $c = \nu\lambda$ we can write

$$\lambda = \frac{hc}{E}$$

$$= \frac{(6.626\,09 \times 10^{-34}\ \text{J}\cdot\text{s})(2.997\,92 \times 10^8\ \text{m}\cdot\text{s}^{-1})}{2.2513 \times 10^{-14}\ \text{J}}$$

$$= 8.8236 \times 10^{-12}\ \text{m or } 8.8236\ \text{pm}$$

1.13 (a) false. The total intensity is proportional to T^4. (b) true; (c) false. Photons of radio-frequency radiation are lower in energy than photons of ultraviolet radiation.

1.15 (a) Use the de Broglie relationship, $\lambda = hp^{-1} = h(mv)^{-1}$.

$$m_e = (9.109\,39 \times 10^{-28}\ \text{g})(1\ \text{kg}/1000\ \text{g}) = 9.109\,39 \times 10^{-31}\ \text{kg}$$

$$(2.2 \times 10^3\ \text{km}\cdot\text{s}^{-1})(1000\ \text{m}\cdot\text{km}^{-1}) = 2.2 \times 10^6\ \text{m}\cdot\text{s}^{-1}$$

$$\lambda = h(mv)^{-1}$$

$$= \frac{6.626\,08 \times 10^{-34}\ \text{J}\cdot\text{s}}{(9.109\,39 \times 10^{-31}\ \text{kg})(2.2 \times 10^6\ \text{m}\cdot\text{s}^{-1})}$$

$$= 3.3 \times 10^{-10}\ \text{m}$$

(b) $E = h\nu$

$$= (6.626\,08 \times 10^{-34}\ \text{J}\cdot\text{s})(1.00 \times 10^{15}\ \text{s}^{-1})$$

$$= 6.63 \times 10^{-19}\ \text{J}$$

(c) The photon needs to contain enough energy to eject the electron from the surface as well as to cause it to move at $2.2 \times 10^3\ \text{km}\cdot\text{s}^{-1}$. The energy involved is the kinetic energy of the electron, which equals $\frac{1}{2}mv^2$.

$$E_{\text{photon}} = 6.63 \times 10^{-19}\ \text{J} + \frac{1}{2}mv^2$$

$$= 6.63 \times 10^{-19}\ \text{J} + \frac{1}{2}(9.109\,39 \times 10^{-31}\ \text{kg})(2.2 \times 10^6\ \text{m}\cdot\text{s}^{-1})^2$$

$$= 6.63 \times 10^{-19}\ \text{J} + 2.2 \times 10^{-18}\ \text{J}$$

$$= 2.9 \times 10^{-18}\ \text{J}$$

But we are asked for the wavelength of the photon, which we can get from $E = h\nu$ and $c = \nu\lambda$ or $E = hc\lambda^{-1}$.

$$2.9 \times 10^{-18}\ \text{kg}\cdot\text{m}^2\cdot\text{s}^{-2} = (6.626\,08 \times 10^{-34}\ \text{kg}\cdot\text{m}^2\cdot\text{s}^{-1})(2.997\,92\cdot10^8\ \text{m}\cdot\text{s}^{-1})\lambda^{-1}$$

$$\lambda = 6.8 \times 10^{-8}\ \text{m}$$

$$= 68\ \text{nm}$$

(d) 68 nma is in the far uv, almost to the x-ray region.

1.17 To answer this question, we need to convert the quantities to a consistent set of units, in this case, SI units.

$(5.15 \text{ ounce})(28.3 \text{ g}\cdot\text{ounce}^{-1})(1 \text{ kg}/1000 \text{ g}) = 0.146 \text{ kg}$

$(92 \text{ mi}\cdot\text{h}^{-1})[(3600 \text{ s}\cdot\text{h}^{-1})(0.6214 \text{ mi}\cdot\text{km}^{-1})(1 \text{ km}/1000 \text{ m})]^{-1} = 41 \text{ m}\cdot\text{s}^{-1}$

Use the de Broglie relationship.

$$\lambda = hp^{-1} = h(mv)^{-1}$$

$$= h(mv)^{-1}$$

$$= \frac{6.626\,08 \times 10^{-34} \text{ J}\cdot\text{s}}{(0.146 \text{ kg})(0.041 \text{ km}\cdot\text{s}^{-1})}$$

$$= \frac{6.626\,08 \times 10^{-34} \text{ kg}\cdot\text{m}^2\cdot\text{s}^{-1}}{(0.146 \text{ kg})(41 \text{ m}\cdot\text{s}^{-1})}$$

$$= 1.1 \times 10^{-34} \text{ m}$$

1.19 From the de Broglie relationship, $p = h\lambda^{-1}$ or $h = mv\lambda$, we can calculate the velocity of the neutron:

$$v = \frac{h}{m\lambda}$$

$$= \frac{(6.626\,08 \times 10^{-34} \text{ kg}\cdot\text{m}^2\cdot\text{s}^{-1})}{(1.674\,93 \times 10^{-27} \text{ kg})(100 \times 10^{-12} \text{ m})} \quad \text{(remember that } 1 \text{ J} = 1 \text{ kg}\cdot\text{m}^2\cdot\text{s}^{-2}\text{)}$$

$$= 3.96 \times 10^{3} \text{ m}\cdot\text{s}^{-1}$$

1.21 $E = h\nu = hc\lambda^{-1} = (6.626\,08 \times 10^{-34} \text{ J}\cdot\text{s})(2.997\,92 \times 10^{8} \text{ m}\cdot\text{s}^{-1})(589 \times 10^{-9} \text{ m})^{-1}$

$= 3.37 \times 10^{-19} \text{ J}$

1.23 (a) The Rydberg equation gives ν when $\mathcal{R} = 3.29 \times 10^{15} \text{ s}^{-1}$, from which one can calculate λ from the relationship $c = \nu\lambda$.

$$\nu = \mathcal{R}\left(\frac{1}{n_2^{\,2} - n_1^{\,2}}\right)$$

and $c = \nu\lambda = 2.997\,92 \times 10^{8} \text{ m}\cdot\text{s}^{-1}$

$$c = \mathcal{R}\left(\frac{1}{n_2^{\,2} - n_1^{\,2}}\right)\lambda$$

$$2.997\,92 \times 10^{8} \text{ m}\cdot\text{s}^{-1} = (3.29 \times 10^{15} \text{ s}^{-1})\left(\frac{1}{4} - \frac{1}{36}\right)\lambda$$

$\lambda = 4.10 \times 10^{-7} \text{ m} = 410 \text{ nm}$

(b) Balmer series

(c) violet

1.25 (a) This problem is the same as that solved in Example 1.5, but the electron is moving between different energy levels. For movement between energy levels separated by a difference of 1 in principal quantum number, the expression is

$$\Delta E = E_{n+1} - E_n = \frac{(n+1)^2h^2}{8mL^2} - \frac{n^2h^2}{8mL^2} = \frac{(2n+1)h^2}{8mL^2}$$

For $n = 2$ and $n + 1 = 3$, $\Delta E = \dfrac{5h^2}{8mL^2}$

Then $\lambda_{3,2} = \dfrac{hc}{E} = \dfrac{8mhcL^2}{5h^2} = \dfrac{8mcL^2}{5h}$

For an electron in a 150-pm box, the expression becomes

$$\lambda_{3,2} = \frac{8(9.109\,39 \times 10^{-31}\ \text{kg})(2.997\,92 \times 10^{8}\text{m}\cdot\text{s}^{-1})(150 \times 10^{-12}\ \text{m})^2}{5(6.626\,08 \times 10^{-34}\ \text{J}\cdot\text{s})}$$

$$= 1.48 \times 10^{-8}\ \text{m}$$

(b) We need to remember that the equation for ΔE was originally determined for energy separations between successive energy levels, so the expression needs to be altered to make it general for energy levels two units apart:

$$\Delta E = E_{n+2} - E_n = \frac{(n+2)^2h^2}{8mL^2} - \frac{n^2h^2}{8mL^2} = \frac{(n^2 + 4n + 4 - n^2)h^2}{8mL^2} = \frac{(4n+4)h^2}{8mL^2}$$

$$\lambda = \frac{hc}{\Delta E} = \frac{hc(8mL^2)}{h^2[4n+4]} = \frac{8mcL^2}{h^2[4n+4]}$$

For $n = 2$, the expression becomes

$$\lambda = \frac{8mcL^2}{h[(4 \times 2) + 4]} = \frac{8mcL^2}{9h}$$

$$= \frac{8(9.109\,39 \times 10^{-31}\ \text{kg})(2.997\,92 \times 10^{8}\ \text{m}\cdot\text{s}^{-1})(150 \times 10^{-12}\ \text{m})^2}{12(6.626\,08 \times 10^{-34}\ \text{kg}\cdot\text{m}^2\cdot\text{s}^{-1})}$$

$$= 6.18 \times 10^{-9}\ \text{m}$$

1.27 In each of these series, the principal quantum number for the lower energy level involved is the same for each absorption line. Thus, for the Lyman series, the lower energy level is $n = 1$; for the Balmer series, $n = 2$; for the Paschen series, $n = 3$; and for the Brackett series, $n = 4$.

1.29 (a) See Figures 1.27, 1.30, and 1.31. (b) A node is a region in space where the wavefunction ψ passes through 0. (c) The simplest *s*-orbital has 0 nodes, the simplest *p*-orbital has 1 nodal plane, and the simplest *d*-orbital has 2 nodal planes. (d) Given the increase in number of nodes, an *f*-orbital would be expected to have 3 nodal planes.

1.31 The p_x orbital will have its lobes oriented along the x axis, the p_y orbital will have its lobes oriented along the y axis, and the p_z orbital will have its lobes oriented along the z axis.

1.33 The equation derived in Illustration 1.4 can be used:

$$\frac{\psi^2(r = 0.35a_0,\theta,\phi)}{\psi^2(0,\theta,\phi)} = \frac{\dfrac{e^{-2(0.35a_0)/a_0}}{\pi a_0^3}}{\left(\dfrac{1}{\pi a_0^3}\right)} = 0.50$$

1.35 (a) For $n = 2$, there are two subshells because l can be 0 or 1 (*s*- or *p*-orbitals).
(b) For $n = 3$, there are three possibilities because $l = 0, 1, 2$ (*s*-, *p*-, or *d*-orbitals).
(c) When $n = 3$, l can be 0, 1, 2.

1.37 (a) 1 orbital; (b) 5 orbitals; (c) 3 orbitals; (d) 7 orbitals

1.39 (a) 7 values: 0, 1, 2, 3, 4, 5, 6; (b) 5 values; $-2, -1, 0, 1, 2$; (c) 3 values: $-1, 0, 1$; (d) 4 subshells: 4*s*, 4*p*, 4*d*, and 4*f*

1.41 (a) $n = 7; l = 0$; (b) $n = 5; l = 3$; (c) $n = 3; l = 2$; (d) $n = 2; l = 1$

1.43 (a) 0; (b) $-3, -2, -1, 0, 1, 2, 3$; (c) $-2, -1, 0, 1, 2$; (d) $-1, 0, 1$

1.45 (a) 6 electrons; (b) 10 electrons; (c) 2 electrons; (d) 14 electrons

1.47 (a) 5*d*, five; (b) 1*s*, one; (c) 6*f*, seven; (d) 2*p*, three

1.49 (a) $n = 2, l = 1$; (b) $n = 5, l = 3$; (c) $n = 3, l = 0$; (d) $n = 4, l = 2$

1.51 (a) six; (b) two; (c) eight; (d) two

1.53 (a) cannot exist; (b) exists; (c) cannot exist; (d) exists

1.55 (a) The total Coulomb potential energy $V(r)$ is the sum of the individual coulombic attractions and repulsions. There will be one attraction between the nucleus and each electron plus a repulsive term to represent the interaction between each pair of electrons. For lithium, there are three protons in the nucleus and three electrons. Each attractive Coulomb potential will be equal to

$$\frac{(-e)(+3e)}{4\pi\epsilon_0 r} = \frac{-3e^2}{4\pi\epsilon_0 r}$$

where $-e$ is the charge on the electron and $+3e$ is the charge on the nucleus, ϵ_0 is the vacuum permittivity, and r is the distance from the electron to the nucleus. The total attractive potential will thus be

$$\left(\frac{-3e^2}{4\pi\epsilon_0 r_1}\right) + \left(\frac{-3e^2}{4\pi\epsilon_0 r_2}\right) + \left(\frac{-3e^2}{4\pi\epsilon_0 r_3}\right) = \left(\frac{-3e^2}{4\pi\epsilon_0}\right)\left(\frac{1}{r_1} + \frac{1}{r_2} + \frac{1}{r_3}\right)$$

The repulsive terms will have the form

$$\frac{(-e)(-e)}{4\pi\epsilon_0 r_{ab}} = \frac{e^2}{4\pi\epsilon_0 r_{ab}}$$

where r_{ab} represents the distance between two electrons a and b. The total repulsive term will thus be

$$\frac{e^2}{4\pi\epsilon_0 r_{12}} + \frac{e^2}{4\pi\epsilon_0 r_{13}} + \frac{e^2}{4\pi\epsilon_0 r_{23}} = \frac{e^2}{4\pi\epsilon_0}\left(\frac{1}{r_{12}} + \frac{1}{r_{13}} + \frac{1}{r_{23}}\right)$$

This gives

$$V(r) = \left(\frac{-3e^2}{4\pi\epsilon_0}\right)\left(\frac{1}{r_1} + \frac{1}{r_2} + \frac{1}{r_3}\right) + \frac{e^2}{4\pi\epsilon_0}\left(\frac{1}{r_{12}} + \frac{1}{r_{13}} + \frac{1}{r_{23}}\right)$$

(b) The first term represents the coulombic attractions between the nucleus and each electron, and the second term represents the coulombic repulsions between each pair of electrons.

1.57 (a) false. Z_{eff} is considerably affected by the total number of electrons present in the atom because the electrons in the lower energy orbitals will "shield" the electrons in the higher energy orbitals from the nucleus. This effect arises because the *e-e* repulsions tend to offset the attraction of the electron to the nucleus. (b) true; (c) false. The electrons are increasingly less able to penetrate to the nucleus as l increases. (d) true.

1.59 Only (d) is the configuration expected for a ground-state atom; the others all represent excited-state configurations.

1.61 (a) This configuration is possible. (b) This configuration is not possible because $l = 0$ here, so m_l must also equal 0. (c) This configuration is not possible because the maximum value l can have is $n - 1$; $n = 4$, so $l_{max} = 3$.

1.63

Ga	$[Ar]3d^{10}4s^2 4p^1$	one unpaired electron
Ge	$[Ar]3d^{10}4s^2 4p^2$	two unpaired electrons
As	$[Ar]3d^{10}4s^2 4p^3$	three unpaired electrons
Se	$[Ar]3d^{10}4s^2 4p^4$	two unpaired electrons
Br	$[Ar]3d^{10}4s^2 4p^5$	one unpaired electron

1.65

(a) silver	$[Kr]4d^{10}5s^1$
(b) beryllium	$[He]2s^2$
(c) antimony	$[Kr]4d^{10}5s^2 5p^3$
(d) gallium	$[Ar]3d^{10}4s^2 4p^1$
(e) tungsten	$[Xe]4f^{14}5d^4 6s^2$
(f) iodine	$[Kr]4d^{10}5s^2 5p^5$

1.67 (a) Fe; (b) P; (c) At; (d) Pu

1.69 (a) $4p$; (b) $4s$; (c) $6s$; (d) $6s$

1.71 (a) 5; (b) 3; (c) 16 (including the d electrons), 6 (if the d electrons are not included). For Te the d electrons are not available for bonding and are often not included in the valence electron count. (d) 8

1.73 (a) 3; (b) 2; (c) 3; (d) 2

1.75 (a) ns^1; (b) ns^2np^1; (c) $(n - 1)d^5ns^2$; (d) $(n - 1)d^{10}ns^1$

1.77 (a) oxygen ($1310\ kJ\cdot mol^{-1}$) > selenium ($941\ kJ\cdot mol^{-1}$) > tellurium ($870\ kJ\cdot mol^{-1}$); ionization energies generally decrease as one goes down a group. (b) gold ($890\ kJ\cdot mol^{-1}$) > osmium ($840\ kJ\cdot mol^{-1}$) > tantalum ($761\ kJ\cdot mol^{-1}$); ionization energies generally decrease as one goes from right to left in the periodic table. (c) lead ($716\ kJ\cdot mol^{-1}$) > barium ($502\ kJ\cdot mol^{-1}$) > cesium ($376\ kJ\cdot mol^{-1}$); ionization energies generally decrease as one goes from right to left in the periodic table.

1.79 The atomic radii (in pm) are

Sc	164	Fe	124
Ti	147	Co	125
V	135	Ni	125
Cr	129	Cu	128
Mn	137	Zn	137

The major trend is for decreasing radius as the nuclear charge increases, with the exception that Cu and Zn begin to show the effects of electron-electron repulsions and become larger as the *d*-subshell becomes filled. Mn is also an exception as found for other properties; this may be attributed to having the *d*-shell half-filled.

1.81 $P^{3-} > S^{2-} > Cl^{-}$

1.83 (a) fluorine; (b) carbon (c) chlorine; (d) lithium

1.85 A diagonal relationship is a similarity in chemical properties between an element in the periodic table and one lying one period lower and one group to the right. It is caused by the similarity in size of the ions. The lower-right element in the pair would generally be larger because it lies in a higher period, but it also will have a higher oxidation state, which will cause the ion to be smaller. For example, Al^{3+} and Ge^{4+} compounds show the diagonal relationship, as do Li^{+} and Mg^{2+}.

1.87 (a) N and S; (b) Li and Mg

1.89 The ionization energies of the *s*-block metals are considerably lower, thus making it easier for them to lose electrons in chemical reactions.

1.91 (a) metal; (b) nonmetal; (c) metal; (d) metalloid; (e) metalloid; (f) metal

1.93 (a) $\dfrac{\nu}{c} = 2143\ \text{cm}^{-1}$

$\nu = c(2143\ \text{cm}^{-1})$

$\nu = (2.997\,92 \times 10^{8}\ \text{m}\cdot\text{s}^{-1})(2143\ \text{cm}^{-1})$

$\nu = (2.997\,92 \times 10^{10}\ \text{cm}\cdot\text{s}^{-1})(2143\ \text{cm}^{-1})$

$\nu = 6.424 \times 10^{13}\ \text{s}^{-1}$

(b) From $E = h\nu$: $E = (6.626\,08 \times 10^{-34}\ \text{J}\cdot\text{s})(6.424 \times 10^{13}\ \text{s}^{-1})$

$= 4.257 \times 10^{-20}\ \text{J}$.

(c) 1.00 mol of molecules $= 6.022 \times 10^{23}$ molecules, so the energy absorbed by 1.00 mol will be $(4.25 \times 10^{-20}\ \text{J}\cdot\text{molecule}^{-1})(6.022 \times 10^{23}\ \text{molecules}\cdot\text{mol}^{-1})$ $= 2.564 \times 10^{4}\ \text{J}\cdot\text{mol}^{-1}$ or $25.64\ \text{kJ}\cdot\text{mol}^{-1}$.

1.95 The ground-state configuration for Cr is $[Ar]3d^{5}4s^{1}$ and that for Cu is $[Ar]3d^{10}4s^{1}$. This exception apparently arises because the formation of either a half-filled or a filled subshell. This results in a more stable electron configuration than normally encountered. In Cr, the *d*-orbitals are half-filled; and for Cu, they are completely filled. Thus, the energy of both the half-filled and the filled shells gives rise to the observed configurations rather than to $[Ar]3d^{4}4s^{2}$ and $[Ar]3d^{9}4s^{2}$ as expected.

1.97 This trend is attributed to the inert-pair effect, which states that the *s*-electrons are less available for bonding in the heavier elements. Thus, there is an increasing trend as we descend the periodic table for the preferred oxidation number to be 2 units lower than the maximum one.

As one descends the periodic table, ionization energies tend to decrease. For Tl, however, the values are slightly higher than those of its lighter analogues.

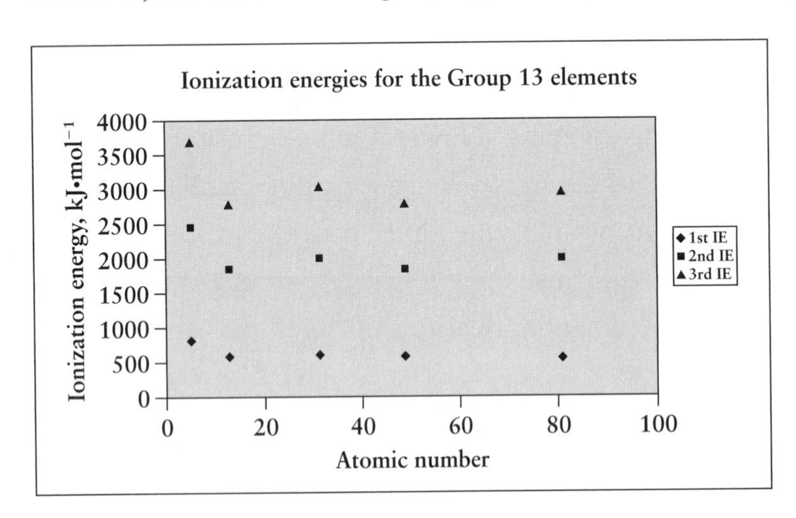

1.99 From $E = hc\lambda^{-1}$, we can calculate the wavelength of light that matches the $348\ \text{kJ}\cdot\text{mol}^{-1}$ necessary to break the C—C bond. First we calculate the energy needed to break one bond and use that value in the equation above to calculate λ:

$$E = \frac{3.48 \times 10^{5}\ \text{J}\cdot\text{mol}^{-1}}{6.022 \times 10^{23}\ \text{molecules}\cdot\text{mol}^{-1}} = 5.78 \times 10^{-19}\ \text{J}$$

$$= hc\lambda^{-1} = \frac{(6.626 \times 10^{-34}\ \text{J}\cdot\text{s}^{-1})(2.997\ 92 \times 10^{8}\ \text{m}\cdot\text{s}^{-1})}{\lambda}$$

$\lambda = 3.44 \times 10^{-7}$ m or 344 nm

334-nm light is in the ultraviolet region, so visible light will not have enough energy to break the C—C bond. UV radiation of 344 nm would be suitable, however.

1.101 (a) The relation is derived as follows: the energy of the photon entering, E_{total}, must be equal to the energy to eject the electron, E_{ejection}, plus the energy that ends up as kinetic energy, E_{kinetic}, in the movement of the electron:

$E_{\text{total}} = E_{\text{ejection}} + E_{\text{kinetic}}$

But E_{total} for the photon $= h\nu$ and $E_{\text{kinetic}} = (\frac{1}{2})mv^2$, where m is the mass of the object and v is its velocity. E_{ejection} corresponds to the ionization energy, I, so we arrive at the final relationship desired.

(b) $E_{\text{total}} = h\nu = hc\lambda^{-1}$

$$= (6.626\ 08 \times 10^{-34}\ \text{J}\cdot\text{s})(2.997\ 92 \times 10^{8}\ \text{m}\cdot\text{s}^{-1})(58.4 \times 10^{-9}\ \text{m})^{-1}$$
$$= 3.40 \times 10^{-18}\ \text{J}$$
$$= E_{\text{ejection}} + E_{\text{kinetic}}$$

$$E_{\text{kinetic}} = (\tfrac{1}{2})mv^2 = (\tfrac{1}{2})(9.109\ 39 \times 10^{-28}\ \text{g})(2450\ \text{km}\cdot\text{s}^{-1})^2$$
$$= (\tfrac{1}{2})(9.109\ 39 \times 10^{-31}\ \text{kg})(2.450 \times 10^{6}\ \text{m}\cdot\text{s}^{-1})^2$$
$$= 2.73 \times 10^{-18}\ \text{kg}\cdot\text{m}^2\cdot\text{s}^{-2} = 2.73 \times 10^{-18}\ \text{J}$$

$$3.40 \times 10^{-18}\ \text{J} = E_{\text{ejection}} + 2.73 \times 10^{-18}\ \text{J}$$
$$E_{\text{ejection}} = 6.7 \times 10^{-19}\ \text{J}$$

1.103 By the time we get to the lanthanides and actinides—the two series of *f*-orbital filling elements—the energy levels become very close together and minor changes in environment cause the different types of orbitals to switch in energy-level ordering. For the elements mentioned, the electronic configurations are

La	$[\text{Xe}]6s^2 5d^1$	Lu	$[\text{Xe}]6s^2 4f^{14} 5d^1$
Ac	$[\text{Rn}]7s^2 6d^1$	Lr	$[\text{Rn}]7s^2 5f^{14} 6d^1$

As can be seen, all these elements have one electron in a *d*-orbital, so placement in the third column of the periodic table could be considered appropriate for either, depending on what aspects of the chemistry of these elements we are comparing. The choice is not without argument, and it is discussed by W. B. Jensen (1982), *J. Chem. Ed.* **59**, 634.

1.105 (a)—(c) We can use the hydrogen 2*s* wavefunction found in Table 1.2. Remember that the probability of locating an electron at a small region in space is proportional to ψ^2, not ψ.

$$\psi_{2s} = \frac{1}{4}\left(\frac{1}{2\pi a_0^{\,3}}\right)^{\frac{1}{2}}\left(2 - \frac{r}{a_0}\right) e^{-\frac{r}{2a_0}}$$

$$\psi_{2s}^{\,2} = \left(\frac{1}{32\pi a_0^{\,3}}\right)\left(2 - \frac{r}{a_0}\right)^2 e^{-\frac{r}{a_0}}$$

$$\frac{\psi_{2s}^{\,2}(r,\theta,\phi)}{\psi_{2s}^{\,2}(0,\theta,\phi)} = \frac{\left(\frac{1}{32\pi a_0^{\,3}}\right)\left(2 - \frac{r}{a_0}\right)^2 e^{-\frac{r}{a_0}}}{\left(\frac{1}{32\pi a_0^{\,3}}\right)2^2\, e^{-\frac{0}{a_0}}}$$

$$= \frac{\left(2 - \frac{r}{a_0}\right)^2 e^{-\frac{r}{a_0}}}{4}$$

Because r will be equal to some fraction x times a_0, the expression will simplify further:

$$\frac{\psi_{2s}^{2}(r,\theta,\phi)}{\psi_{2s}^{2}(0,\theta,\phi)} = \frac{\left(2 - \frac{xa_0}{a_0}\right)^2 e^{-\frac{xa_0}{a_0}}}{4} = \frac{(2 - x)^2 e^{-x}}{4}$$

Carrying out this calculation for the other points, we obtain:

x	relative probability
0.1	0.82
0.2	0.66
0.3	0.54
0.4	0.43
0.5	0.34
0.6	0.27
0.7	0.21
0.8	0.16
0.9	0.12
1	0.092
1.1	0.067
1.2	0.048
1.3	0.033
1.4	0.022
1.5	0.014
1.6	0.0081
1.7	0.0041
1.8	0.0017
1.9	0.0004
2	0.0000
2.1	0.00031
2.2	0.0011
2.3	0.0023
2.4	0.0036
2.5	0.0051
2.6	0.0067
2.7	0.0082
2.8	0.0097
2.9	0.011
3	0.012

This can be most easily carried out graphically by simply plotting the function

$$f(x) = \frac{(2 - x)^2 e^{-x}}{4}$$

from 0 to 3.

The node occurs when $x = 2$, or when $r = 2a_0$. This is exactly what is obtained by setting the radial part of the equation equal to 0.

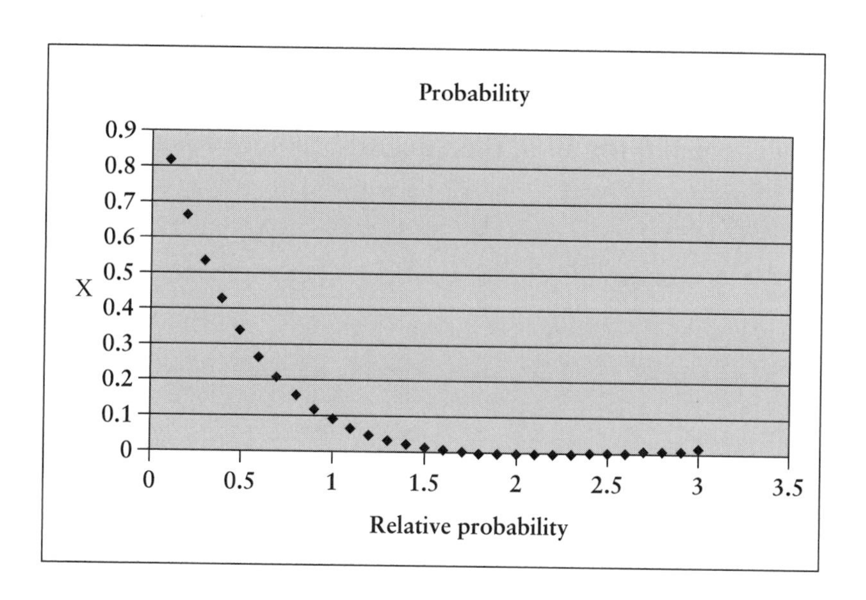

1.107 The approach to showing that this is true involves summing the probability function over all space. The probability function is given by the square of the wave function, so that for the particle in the box we have

$$\psi = \left(\frac{2}{L}\right)^{\frac{1}{2}} \sin\left(\frac{n\pi x}{L}\right)$$

and the probability function will be given by

$$\psi^2 = \left(\frac{2}{L}\right) \sin^2\left(\frac{n\pi x}{L}\right)$$

Because x can range from 0 to L (the length of the box), we can write the integration as

$$\int_0^x \psi^2 dx = \int_0^x \left(\frac{2}{L}\right) \sin^2\left(\frac{n\pi x}{L}\right) dx$$

for the entire box, we write

probability of finding the particle somewhere in the box =

$$\int_0^L \left(\frac{2}{L}\right) \sin^2\left(\frac{n\pi x}{L}\right) dx$$

$$
\begin{aligned}
probability &= \left(\frac{2}{L}\right)\int_0^x \sin^2\left(\frac{n\pi x}{L}\right)dx \\
&= \left(\frac{2}{L}\right)\left(\frac{1}{2}x - \frac{L}{4n\pi}\sin 2n\pi x\right)_0^L \\
&= \left(\frac{2}{L}\right)\left(\frac{1}{2}L - \frac{L}{4n\pi}\sin 2n\pi L\right) - \left(\frac{2}{L}\right)\left(\left(\frac{1}{2}\right)0 - \frac{L}{4n\pi}\sin 2n\pi(0)\right) \\
&= \left(\frac{2}{L}\right)\left(\frac{1}{2}L - \frac{L}{4n\pi}\sin 2n\pi\, L\right) \\
&= \frac{L}{L} - \left(\frac{2}{L}\right)\left(\frac{L}{4n\pi}\sin 2n\pi L\right)
\end{aligned}
$$

Because $\sin 2n\pi x$ will equal 0 for all n, the integral becomes equal to 1.

CHAPTER 2
CHEMICAL BONDS

2.1 The coulombic attraction is directly proportional to the charge on each ion (Equation 1) so the ions with the higher charges will give the greater coulombic attraction. The answer is therefore (b) Ga^{3+}, O^{2-}.

2.3 The Li^+ ion is smaller than the Rb^+ ion (58 vs 149 pm). Because the lattice energy is related to the coulombic attraction between the ions, it will be inversely proportional to the distance between the ions (see Equation 2). Hence the larger rubidium ion will have the lower lattice energy for a given anion.

2.5 (a) 5; (b) 4; (c) 7; (d) 3

2.7 (a) [Ar]; (b) [Ar]$3d^{10}4s^2$; (c) [Kr]$4d^5$; (d) [Ar]$3d^{10}4s^2$

2.9 (a) [Kr]; (b) [Xe]$4f^{14}5d^{10}6s^2$; (c) [Ar]$3d^{10}$; (d) [Xe]$4f^{14}5d^{10}$

2.11 (a) [Xe]$5d^8$; (b) [Ar]$3d^3$; (c) [Ar]$3d^9$; (d) [Ar]$3d^2$

2.13 (a) Co^{2+}; (b) Fe^{2+}; (c) Mo^{2+}; (d) Nb^{2+}

2.15 (a) Co^{3+}; (b) Fe^{3+}; (c) Ru^{3+}; (d) Mo^{3+}

2.17 (a) 4*s*; (b) 3*p*; (c) 3*p*; (d) 4*s*

2.19 (a) −1; (b) −2; (c) +1; (d) +3, although +1 is sometimes observed; (e) +1, because of the inert effect. +3 is also observed.

2.21 (a) 3; (b) 6; (c) 6; (d) 2

2.23 (a) [Xe]$4f^4 5d^{10}6s^2$; no unpaired electrons; (b) [Kr]$4d^{10}$; no unpaired electrons; (c) [Kr]$4d^4$; four unpaired electrons; (d) [Kr]; no unpaired electrons; (e) [Ar]$3d^8$; 2 unpaired electrons

2.25 (a) 3*d*; (b) 5*s*; (c) 5*p*; (d) 4*d*

2.27 (a) +7; (b) −1; (c) [Kr] for +7, [Xe] for −1; (d) Electrons are lost or added to give noble gas configuration.

2.29 (a) Mg_3As_2; (b) In_2S_3. In_2S is also possible but not as common as In_2S_3.
(c) AlH_3; (d) H_2Te; (e) BiF_3

(BiF_5 is also known, but BiF_3 is the more stable compound due to the inert pair effect.)

2.31 (a) GaAs; (b) MgO; (c) Al_2Te_3; (d) RuO_2; (e) V_2O_5

2.33 (a) Bi_2O_3; (b) PbO_2; (c) Tl_2O_3

2.35 (a) :C̈l: above and :C̈l: below C, :C̈l—C—C̈l: (b) :O: double-bonded (‖) to C, :C̈l—C—C̈l: (c) Ö=N̈—F̈: (d) :F̈—N̈—F̈: with :F̈: below N

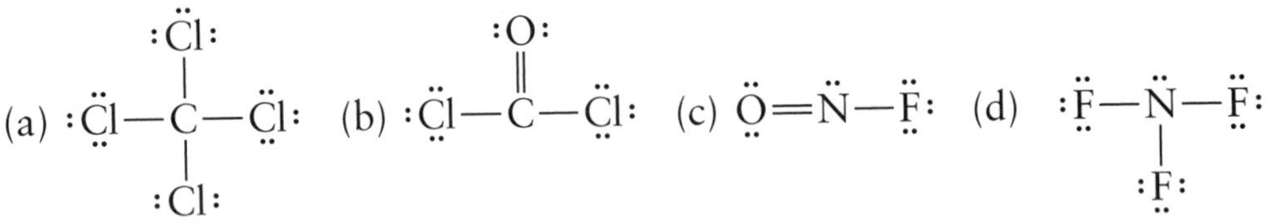

2.37 (a) H above and H below B, H—B—H, charge − (b) :B̈r—Ö:$^-$ (c) :N̈—H with H below N, charge −

2.39 (a) [H—N—H with H above and H below N]$^+$:C̈l:$^-$ (b) 3 K$^+$ [:P̈:]$^{3-}$ (c) Na$^+$ [:C̈l—Ö:]$^-$

2.41 (a) :O: double-bonded (‖) to C, H—C—H (b) :Ö—H above C and H below C, H—C—H (c) H—N̈—C—C—Ö: with H below N; H above and H below the first C; :O: double-bonded (‖) to the second C; H below Ö

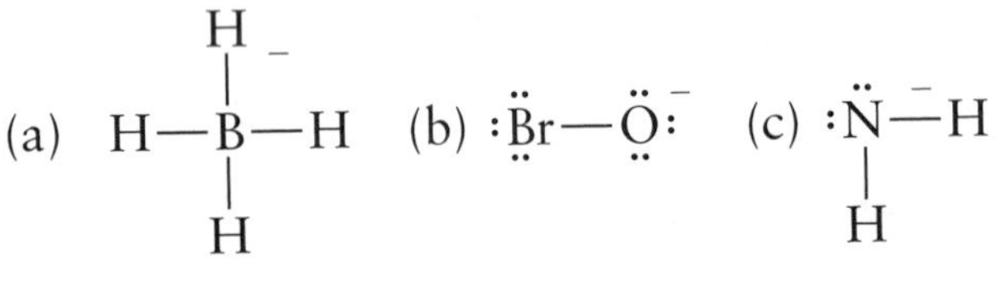

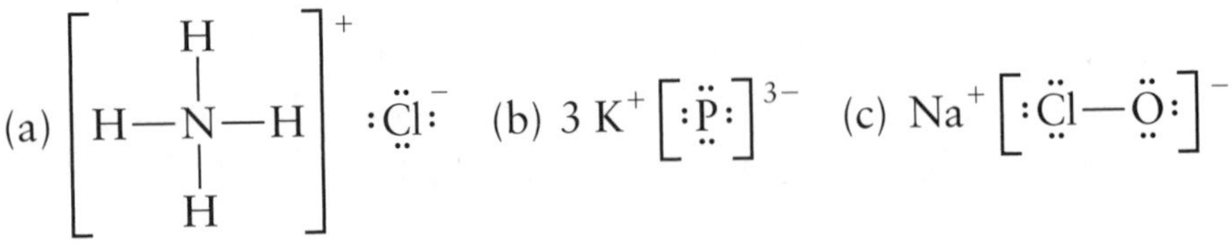

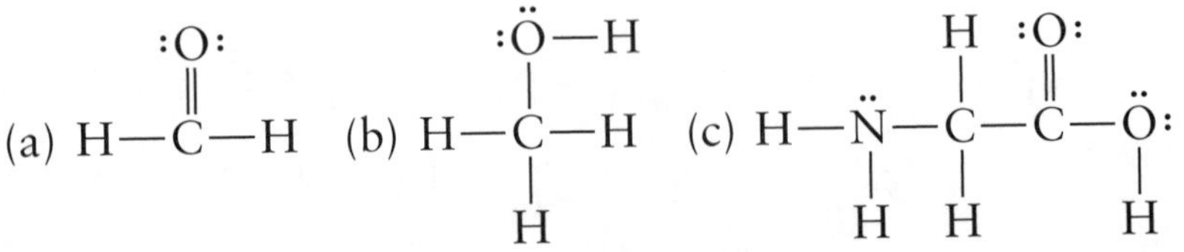

2.43 Anthracene has four resonance structures:

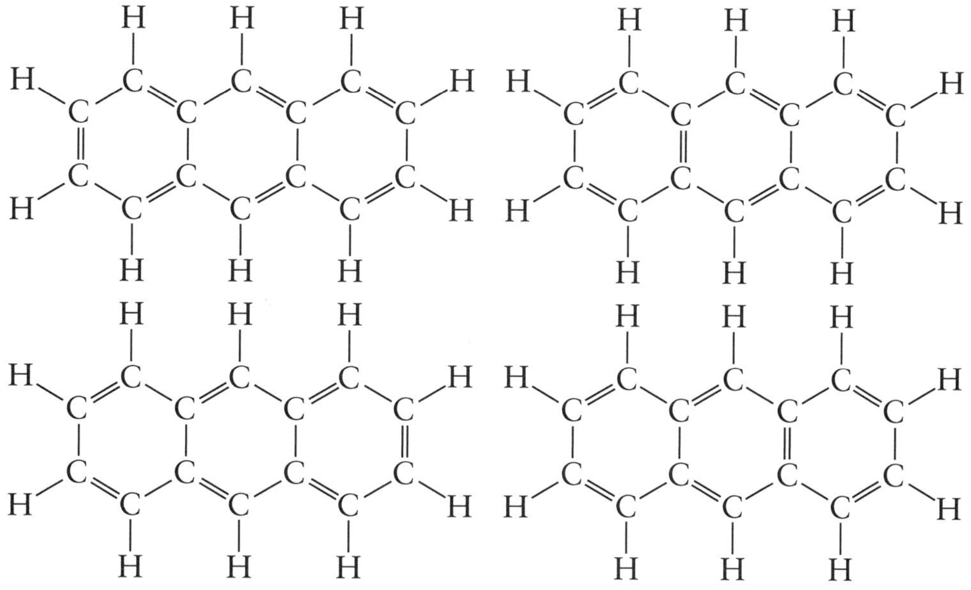

```
     H     H     H                 H     H     H
     |     |     |                 |     |     |
H    C     C     C    H      H     C     C     C     H
  C     C     C     C          C     C     C     C
  ||    |     |     |          |     ||    |     |
  C     C     C     C          C     C     C     C
H    C     C     C    H      H     C     C     C     H
     |     |     |                 |     |     |
     H     H     H                 H     H     H

     H     H     H                 H     H     H
     |     |     |                 |     |     |
H    C     C     C    H      H     C     C     C     H
  C     C     C     C          C     C     C     C
  |     |     |     ||         |     |     ||    |
  C     C     C     C          C     C     C     C
H    C     C     C    H      H     C     C     C     H
     |     |     |                 |     |     |
     H     H     H                 H     H     H
```

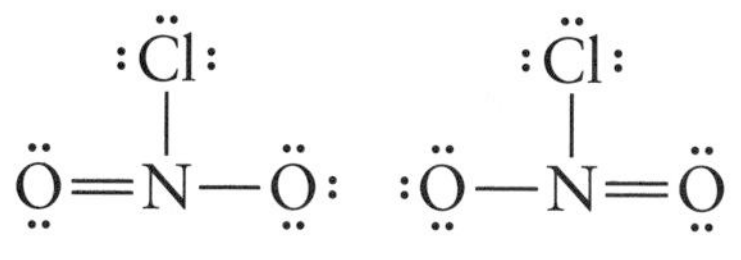

2.45 $\ddot{\underset{..}{O}}=N(-\ddot{Cl}:)-\ddot{\underset{..}{O}}:$ $:\ddot{\underset{..}{O}}-N(-\ddot{Cl}:)=\ddot{\underset{..}{O}}$

```
       :Cl:              :Cl:
        |                 |
2.45  O=N—O:          :O—N=O
```

2.47 (a) $:N\equiv O:^{+}$ (b) $:N\equiv N:$ (c) $:C\equiv O:$ (d) $:C\equiv C:^{2-}$ (e) $:C\equiv N:^{-}$

	(a) $:N\equiv O:^{+}$	(b) $:N\equiv N:$	(c) $:C\equiv O:$	(d) $:C\equiv C:^{2-}$	(e) $:C\equiv N:^{-}$
↑ ↑	0, +1	0, 0	−1, +1	−1, −1	−1, 0

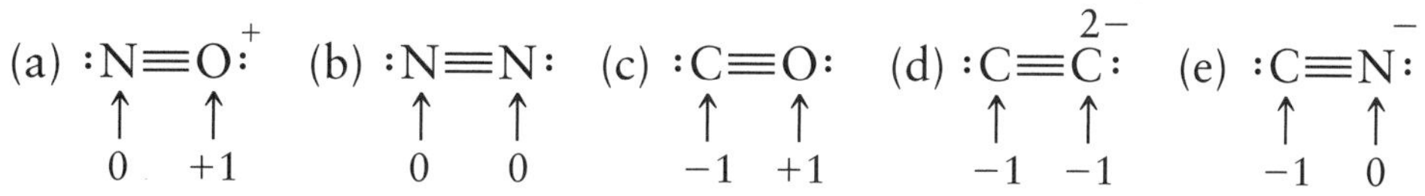

2.49 (a) The first structure has a formal charge of 0 at S, 0 at two of the oxygen atoms and −1 at the third. The S atom in the second structure has a formal charge of +1, 0 at one of the oxygen atoms and −1 at the other two. The first structure would be expected to be lower in energy. (b) The formal charge at S in the first structure is +1, at two O atoms it is 0, and at the other two it is −1. In the second structure, the formal charge at S is 0, at three O atoms it is also 0, and at the fourth O atom it is −1. The second structure would be expected to be lower in energy.

2.51 (a)

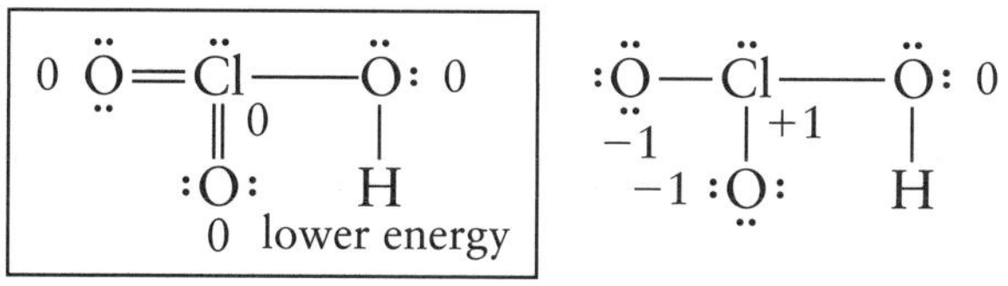

```
 ┌──────────────────────┐
 │ 0 O=Cl——O: 0         │     :O—Cl——O: 0
 │      ||0   |         │     −1   |+1   |
 │     :O:    H         │     −1 :O:     H
 │     0 lower energy   │
 └──────────────────────┘
```

(b) 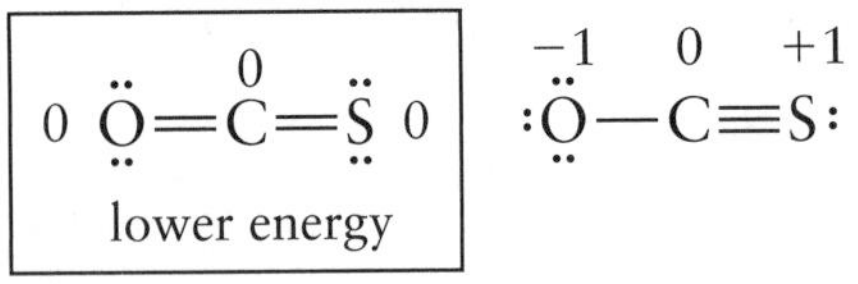 | O=C=S (formal charges: O 0, C 0, S 0) — lower energy | :O—C≡S: (formal charges: O −1, C 0, S +1) |

(c) | H—C≡N: (formal charges: H 0, C 0, N 0) — lower energy | H—C=N: (formal charges: H 0, C −1, N +1) |

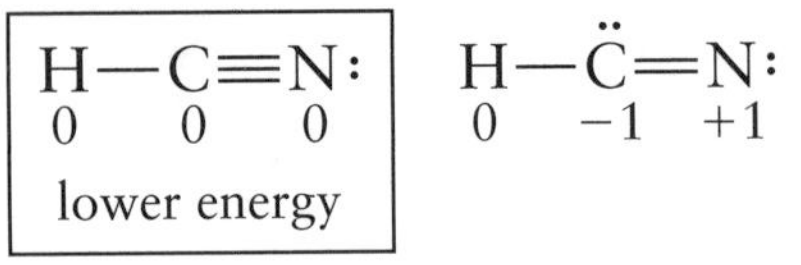

2.53 (a) The formal charges at Xe and F are 0 in the first structure, whereas in the second structure Xe is −1, one F is 0, and one F is +1. The first structure is favored, based on formal charges. (b) In the first structure, all of the atoms have formal charges of 0, whereas in the second, one oxygen atom has a formal charge of +1 and one has a formal charge of −1. The first structure is thus preferred.

2.55 (a) The sulfite ion has one Lewis structure that obeys the octet rule. The formal charge of +1 at S can be reduced to 0 by including 1 double bond to oxygen, a form with three resonance structures.

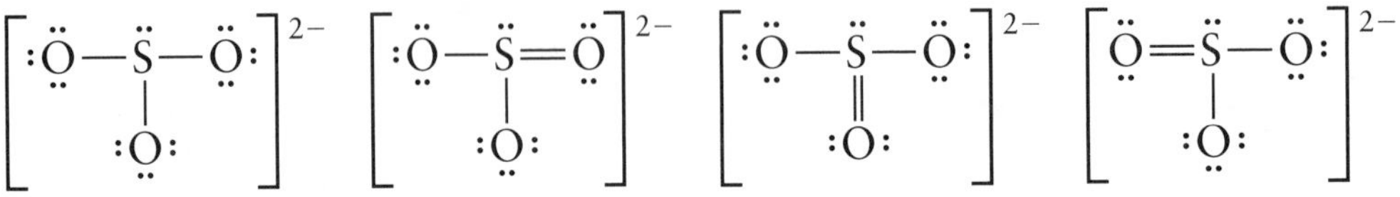

$[:O—S(—O:)—O:]^{2-}$ $[:O—S(—O:)=O]^{2-}$ $[:O—S(=O:)—O:]^{2-}$ $[O=S(—O:)—O:]^{2-}$

(b) There is one Lewis structure that obeys the octet rule shown below. The formal charge at sulfur can be reduced to 0 by including one double bond contribution. This gives rise to two expanded octet structures. Notice that, unlike the case of the sulfite ion where there are three resonance forms, the presence of the hydrogen ion restricts the electrons on the oxygen atom to which it is attached. Because H is electropositive, its placement near an oxygen atom makes it less likely for that oxygen to donate a lone pair to an adjacent atom.

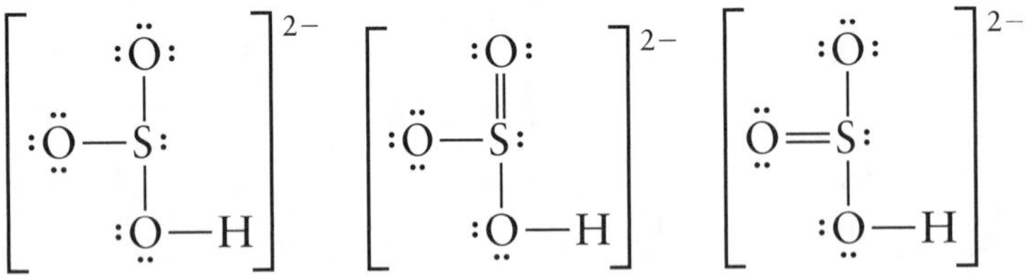

$[:O—S(—O:)—O—H]^{2-}$ $[:O—S(=O:)—O—H]^{2-}$ $[O=S(—O:)—O—H]^{2-}$

(c) There is one Lewis structure for the perchlorate ion that obeys the octet rule. The formal charge at Cl can be reduced to 0 by including three double bond contributions, giving rise to four resonance forms.

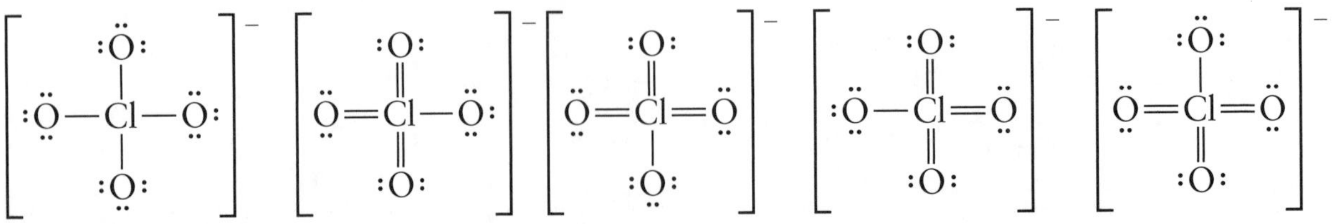

$[:O—Cl(—O:)(—O:)—O:]^{-}$ $[O=Cl(=O:)(=O:)—O:]^{-}$ $[O=Cl(=O:)(—O:)=O]^{-}$ $[:O—Cl(=O:)(=O:)=O]^{-}$ $[O=Cl(—O:)(=O:)=O]^{-}$

(d) For the nitrite ion, there are two resonance forms, which both obey the octet rule.

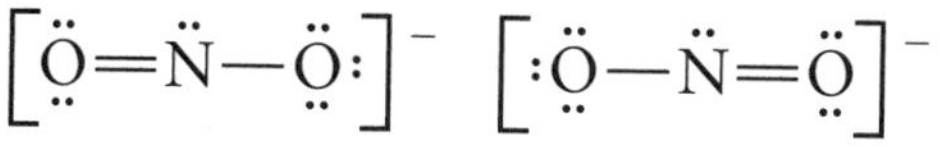

$$\left[\ddot{\underset{\cdot\cdot}{O}}{=}\ddot{N}{-}\ddot{\underset{\cdot\cdot}{O}}{:}\right]^- \quad \left[{:}\ddot{\underset{\cdot\cdot}{O}}{-}\ddot{N}{=}\ddot{\underset{\cdot\cdot}{O}}\right]^-$$

2.57 The Lewis structures are

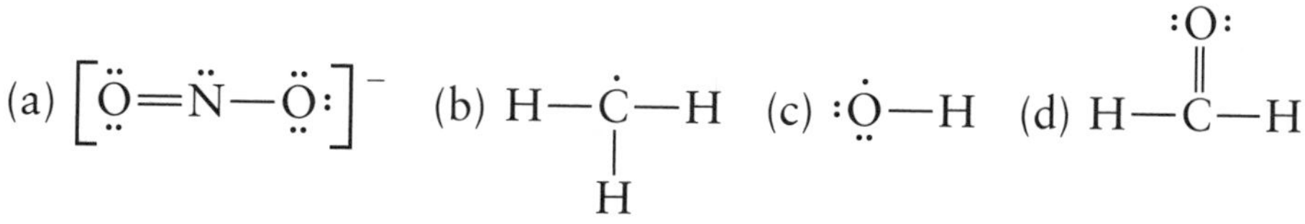

(a) $\left[\ddot{\underset{\cdot\cdot}{O}}{=}\ddot{N}{-}\ddot{\underset{\cdot\cdot}{O}}{:}\right]^-$ (b) H—Ċ(H)—H (c) ${:}\dot{\underset{\cdot\cdot}{O}}{-}H$ (d) H—C(=O:)—H

Radicals are species with an unpaired electron, therefore (b) and (c) are radicals.

2.59 (a) ${:}\dot{\underset{\cdot\cdot}{Cl}}{-}\ddot{\underset{\cdot\cdot}{O}}{:}$ radical (b) ${:}\ddot{\underset{\cdot\cdot}{Cl}}{-}\ddot{\underset{\cdot\cdot}{O}}{-}\ddot{\underset{\cdot\cdot}{O}}{-}\ddot{\underset{\cdot\cdot}{Cl}}{:}$ not a radical

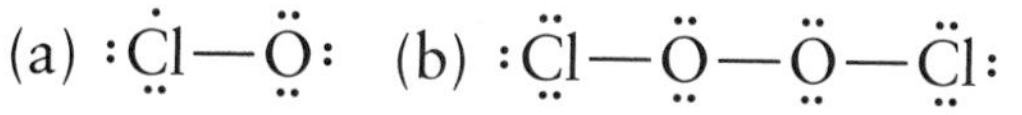

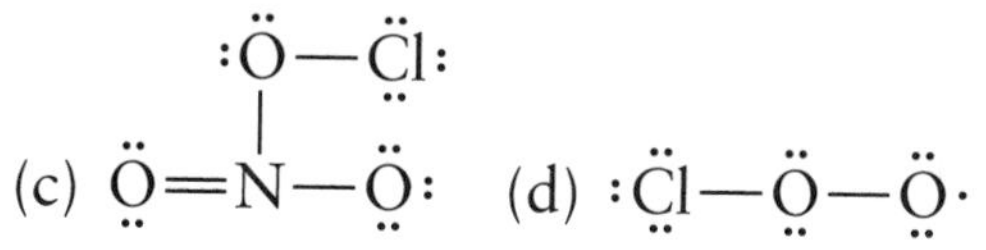

(c) $\ddot{\underset{\cdot\cdot}{O}}{=}N(-\ddot{O}-\ddot{\underset{\cdot\cdot}{Cl}}{:}){-}\ddot{\underset{\cdot\cdot}{O}}{:}$ not a radical (d) ${:}\ddot{\underset{\cdot\cdot}{Cl}}{-}\ddot{\underset{\cdot\cdot}{O}}{-}\ddot{\underset{\cdot\cdot}{O}}{\cdot}$ radical

2.61 (a) SF_6 12 electrons (b) XeF_2 10 electrons

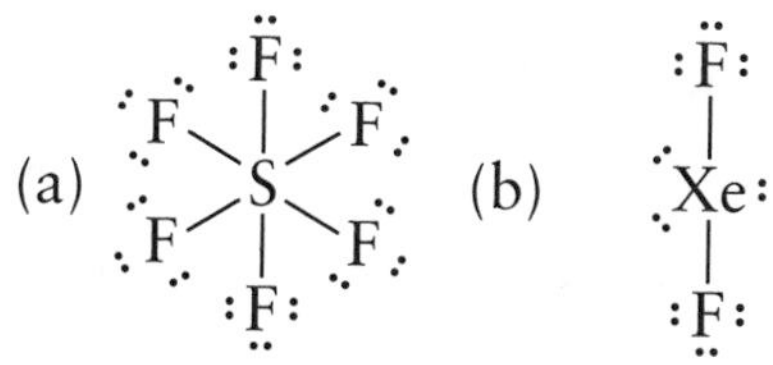

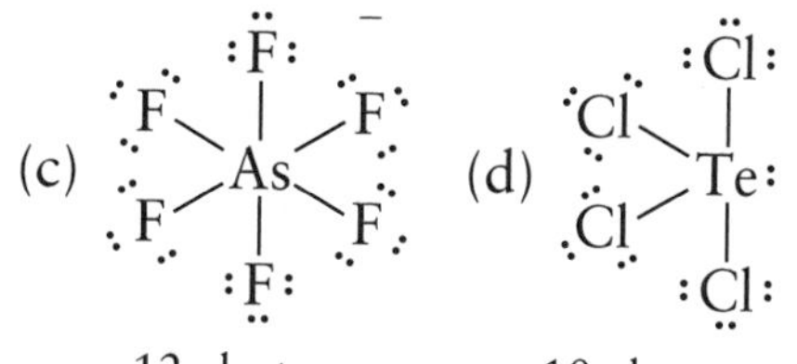

(c) $[AsF_6]^-$ 12 electrons (d) $TeCl_4$ 10 electrons

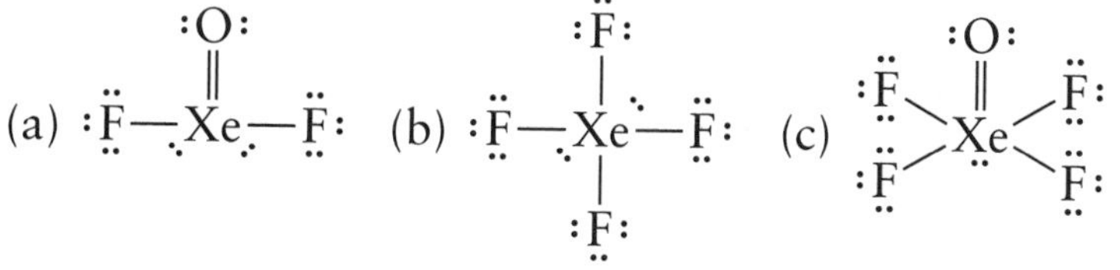

2.63 (a) ${:}\ddot{\underset{\cdot\cdot}{F}}{-}Xe(=O{:}){-}\ddot{\underset{\cdot\cdot}{F}}{:}$ 12 electrons, 2 lone pairs (b) XeF_4 12 electrons, 2 lone pairs (c) $XeOF_4$ 14 electrons, 1 lone pair

2.65 (a) Lewis base; (b) Lewis acid; (c) Lewis base; (d) Lewis acid

2.67 (a) PF_5 is the Lewis acid; F^- is the Lewis base; the product is PF_6^-.
(b) SO_2 is the Lewis acid; Cl^- is the Lewis base; the product is SO_2Cl^-.
(c) SO_3 is the Lewis acid; OH^- is the Lewis base; the product is HSO_4^-.

2.69 (a) CH_3^- is more basic than CH_4 because it has a lone pair of electrons, whereas CH_4 does not. (b) H_2O is more basic than H_2S because O is more electronegative than S and has a higher partial negative charge. (c) NH_2^- is more basic than NH_3 due to the higher negative charge.

2.71 I (2.7) < Br (3.0) < Cl (3.2) < F (4.0); electronegativity generally increases as one goes higher in the periodic table.

2.73 In (1.8) < Sn (2.0) < Sb (2.1) < Se (2.6). Generally electronegativity increases as one goes from left to right across the periodic table and as one goes from heavier to lighter elements within a group.

2.75 (a) Iodide is more polarizable than Cl^-, so HI would be more covalent than HCl. (b) The bonds in CF_4 would be more ionic. The electronegativity difference is greater between C and F than between C and H, making the C—F bonds more ionic. (c) C and S have nearly identical electronegativities, so the C—S bonds would be expected to be almost completely covalent, whereas the C—O bonds would be more ionic.

2.77 $Rb^+ < Sr^{2+} < Be^{2+}$; the smaller, more highly charged cations will have the greater polarizing power. The ionic radii are 149 pm, 116 pm, 27 pm, respectively.

2.79 $O^{2-} < N^{3-} < Cl^- < Br^-$; the polarizability increases as the ion gets larger and less electronegative. The ionic radii for these species are 140 pm, 171 pm, 181 pm, 196 pm, respectively.

2.81 (a) $CO_3^{2-} > CO_2 > CO$
CO_3^{2-} will have the longest C—O bond length. In CO there is a triple bond and in CO_2 the C—O bonds are double bonds. In carbonate, the bond is an average of three Lewis structures in which the bond is double in one form and single in two

of the forms. We would thus expect the bond order to be approximately 1.3. Because the bond length is inversely related to the number of bonds between the atoms, we expect the bond length to be longest in carbonate.

(b) $SO_3^{2-} > SO_2 \sim SO_3$

Similar arguments can be used for these molecules as in part (a). In SO_2 and SO_3, the Lewis structures with the lowest formal charge at S have double bonds between S and each O. In the sulfite ion, however, there are three Lewis structures that have a 0 formal charge at S. Each has one S—O double bond and two S—O single bonds. Because these S—O bonds would have a substantial amount of single bond character, they would be expected to be longer than those in SO_2 or SO_3. This is consistent with the experimental data that show the S—O bond lengths in SO_2 and SO_3 to be 143 pm, whereas those in SO_3^{2-} range from about 145 pm to 152 pm depending on the compound.

(c) $CH_3NH_2 > CH_2NH > HCN$

The C—N bond in HCN is a triple bond, in CH_2NH it is a double bond, and in CH_3NH_2 it is a single bond. The C—N bond in the last molecule would, therefore, be expected to be the longest.

2.83 (a) The covalent radius of N is 75 pm, so the N—N single bond in hydrazine would be expected to be ca. 150 pm. The experimental value is 145 pm. (b) The C—O bonds in carbon dioxide are double bonds. The covalent radius for doubly bonded carbon is 67 pm and that of O is 60 pm. Thus we predict the C=O in CO_2 to be ca. 127 pm. The experimental bond length is 116.3 pm. (c) The C—O bond is a double bond so it would be expected to be the same as in (b). This is the experimentally found value. The C—N bonds are single bonds and so one might expect the bond distance to be the sum of the single bond C radius and the single bond N radius (77 plus 75 pm) which is 152 pm. However, because the C atom is involved in a multiple bond, its radius is actually smaller. The sum of that radius (67 pm) and the N single bond radius gives 132 pm, which is close to the experimental value of 133 pm. (d) The N—N bond is a double bond so we expect the bond distance to be two times the double bond covalent radius of N, which is $2 \times (60 \text{ pm})$ or 120 pm. The experimental value is 123.0 pm.

2.85 (a) 77 pm + 72 pm = 149 pm (b) 111 pm + 72 pm = 183 pm

(c) 141 pm + 72 pm = 213 pm; the bond distance increases as you go down the periodic table because the atoms are getting larger.

2.87 (a)

$$\left[\begin{array}{c} 0 \quad -1 \\ :\!O\!: \;\; :\!\ddot{O}\!: \\ \| \qquad | \\ {}^{0}C - C^{0} \\ | \qquad \| \\ :\!\ddot{O}\!: \;\; :\!O\!: \\ -1 \quad 0 \end{array}\right]^{2-} \quad \left[\begin{array}{c} :\!\ddot{O}\!: \;\; :\!O\!: \\ | \qquad \| \\ C - C \\ \| \qquad | \\ :\!O\!: \;\; :\!\ddot{O}\!: \end{array}\right]^{2-} \quad \left[\begin{array}{c} :\!O\!: \;\; :\!O\!: \\ \| \qquad \| \\ C - C \\ | \qquad | \\ :\!\ddot{O}\!: \;\; :\!\ddot{O}\!: \end{array}\right]^{2-} \quad \left[\begin{array}{c} :\!\ddot{O}\!: \;\; :\!\ddot{O}\!: \\ | \qquad | \\ C - C \\ \| \qquad \| \\ :\!O\!: \;\; :\!O\!: \end{array}\right]^{2-}$$

(b)

$$\left[\begin{array}{c} +1 \quad 0 \\ \ddot{Br} = \ddot{O} \end{array}\right]^{+}$$

(c)

$$\left[\begin{array}{c} -1 \quad -1 \\ :\!C \equiv C\!: \end{array}\right]^{2-}$$

2.89 (a)

$$\left[\begin{array}{c} :\!O\!: \\ \| \\ CH_3 - C - \ddot{O}\!: \end{array}\right]^{-} \quad \left[\begin{array}{c} :\!\ddot{O}\!: \\ | \\ CH_3 - C = \ddot{O} \end{array}\right]^{-}$$

(b)

$$\left[\begin{array}{c} H \\ | \\ H - C = C - \ddot{O}\!: \\ | \\ CH_3 \end{array}\right]^{-} \quad \left[\begin{array}{c} H \\ | \\ H - \ddot{C} - C = \ddot{O} \\ | \\ CH_3 \end{array}\right]^{-}$$

(c)

$$\left[\begin{array}{c} H \;\; H \;\; H \\ | \quad | \quad | \\ H - C = C - C - H \end{array}\right]^{+} \quad \left[\begin{array}{c} H \;\; H \;\; H \\ | \quad | \quad | \\ H - C - C = C - H \end{array}\right]^{+}$$

(d)

$$\left[\begin{array}{c} :\!O\!: \;\; H \\ \| \quad | \\ CH_3 - C - \ddot{N}\!: \end{array}\right]^{-} \quad \left[\begin{array}{c} :\!\ddot{O}\!: \;\; H \\ | \quad | \\ CH_3 - C = N\!: \end{array}\right]^{-}$$

2.91 The chemical formula for phosgene is $COCl_2$ as determined in section F-2 and F-3. The Lewis structure is

$$\begin{array}{c} :\!O\!: \\ \| \\ :\!\ddot{Cl} - C - \ddot{Cl}\!: \end{array} \quad \text{molar mass} = 98.91\ \text{g·mol}^{-1}$$

2.93 P and S are larger atoms that are less able to form multiple bonds to themselves. Unlike the small N and O atoms. All bonds in P_4 and S_8 are single bonds, whereas N_2 has a triple bond and O_2 a double bond.

2.95 (a) H—C≡C—H H—C≡Si—H H—Si≡Si—H

H—C≡N: :N≡N:

(b)

Note: The ability to draw the structures suggests that the compounds are feasible but it does not indicate stability. For example, the N_6 ring gives a reasonable Lewis structure but would be very unstable.

2.97 The Lewis structures for NO and NO_2 are

$$\dot{N}=O \quad O=\dot{N}-O: \longleftrightarrow :O-\dot{N}=O$$

Both compounds are radicals.

(a) NO has a double bond, but NO_2 has an N—O bond that is the average of a single bond and a double bond. Thus NO would be expected to have a shorter, stronger bond. This is indicated by the bond energies.

(b) The fact that the two N—O bonds in NO_2 are equal is a result of the two available resonance forms.

2.99 (a)

Metal Iodide	d(M − I), pm	Lattice Energy, kJ/mol
LiI	278	761
NaI	322	705
KI	358	649
RbI	369	632
CsI	390	601
AgI	353	886

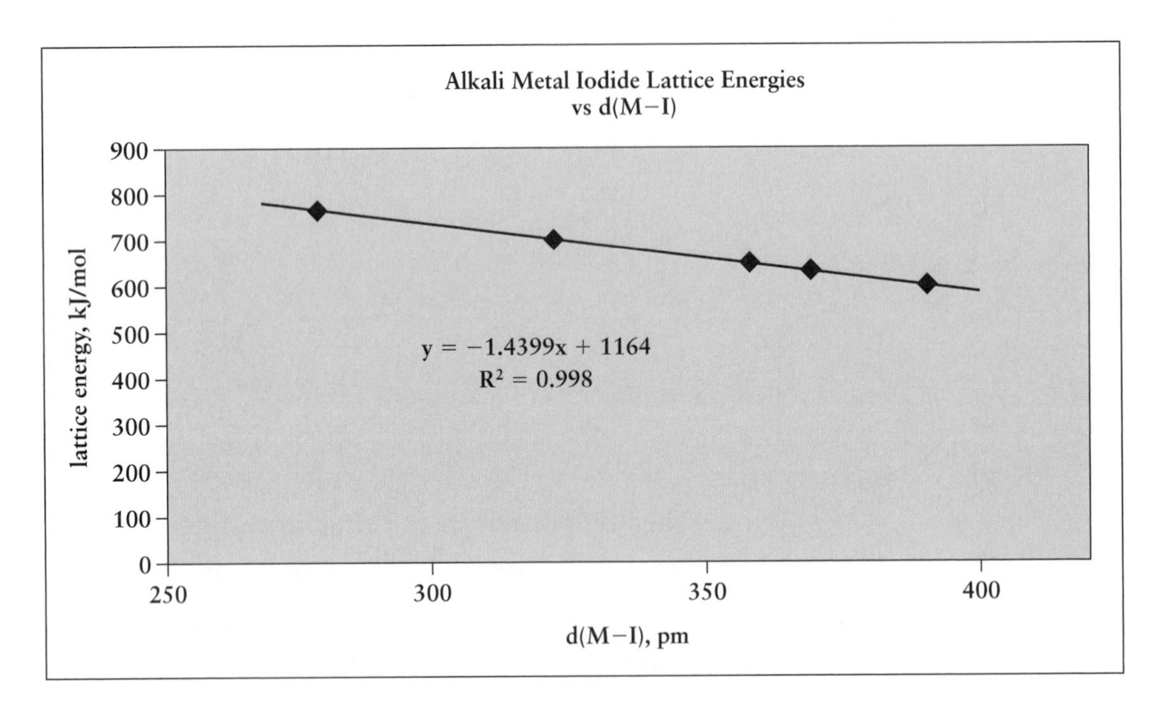

The correlation is excellent with an agreement coefficient of greater than 99%.

(b) From the equation of this line

Lattice Energy $= -1.984\ d_{M-X} + 1356$

and the Ag-I distance of 353 pm, we can estimate the AgI lattice energy to be 656 $kJ \cdot mol^{-1}$. This is not very good agreement with the experimental value of 886 $kJ \cdot mol^{-1}$. One possible explanation is that the structure of silver iodide is different from that of the alkali metal iodides and therefore would have a different Madelung constant. This is the case as AgI crystallizes in a type of lattice known as the Wurtzite structure. However, the Madelung constant for the Wurtzite structure is 1.641 versus 1.748 for the rock salt structure of all the alkali metal iodides (except CsI, which adopts the CsCl structure with a very similar Madelung constant of 1.763). One would therefore predict that the lattice energy of AgI would be less than the calculated value based upon the alkali metal series, rather than greater. A better explanation is that that Ag^+ ion is much more polarizable than the alkali metal cations of similar size and therefore the bonding in AgI is much more covalent.

2.101 (a)

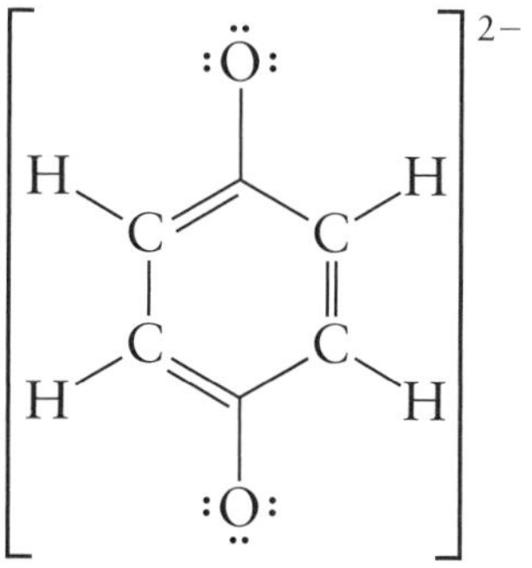

```
         ..        2-
        :O:
         |
  H      C      H
   \   //  \   /
    C         C
    |         ||
    C         C
   /  \\   /   \
  H      C      H
         |
        :O:
         ..
```

(b) All the atoms have 0 formal charge except the two oxygen atoms which are −1. The negative charge is mostly likely concentrated at the oxygen atoms.

(c) Because the oxygen atoms are the most negative sites and the lone pairs on the oxygen atoms are good Lewis bases, the protons will bond to the oxygen atoms. This molecule is known as hydroquinone.

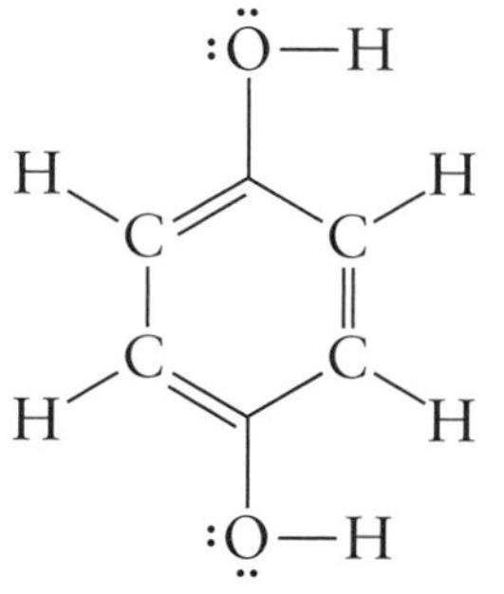

```
         ..
        :O—H
         |
  H      C      H
   \   //  \   /
    C         C
    |         ||
    C         C
   /  \\   /   \
  H      C      H
         |
        :O—H
         ..
```

hydroquinone

2.103 The resonance forms are achieved by moving the location of the radical from one carbon atom to the next as shown.

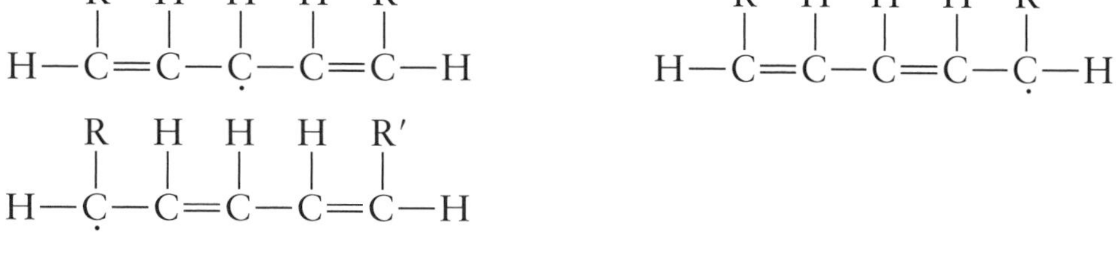

```
   R  H  H  H  R'
   |  |  |  |  |
H—C=C—Ċ—C=C—H
```

```
   R  H  H  H  R'
   |  |  |  |  |
H—C=C—C=C—Ċ—H
```

```
   R  H  H  H  R'
   |  |  |  |  |
H—Ċ—C=C—C=C—H
```

(the dot below a C marks the radical carbon)

2.105 (a) C_2H_2

H—C≡C—H

The hydrogen atoms are equivalent.

(b) C_2H_4

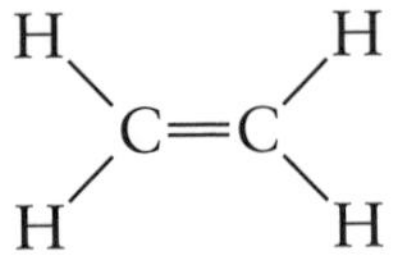

```
H         H
 \       /
  C == C
 /       \
H         H
```

All of the hydrogen atoms are equivalent.

(c) C_2H_3Cl

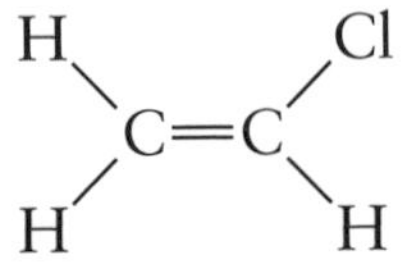

```
H         Cl
 \       /
  C == C
 /       \
H         H
```

All of the hydrogen atoms are in different environments.

(d) *cis*-$C_2H_2Cl_2$

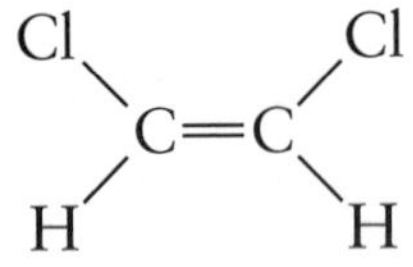

```
Cl        Cl
  \      /
   C == C
  /      \
H         H
```

The two hydrogen atoms are equivalent.

(e) *trans*-$C_2H_2Cl_2$

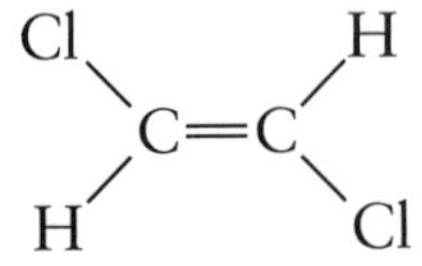

```
Cl        H
  \      /
   C == C
  /      \
H         Cl
```

The two hydrogen atoms are equivalent.

(f) C_2H_5Cl

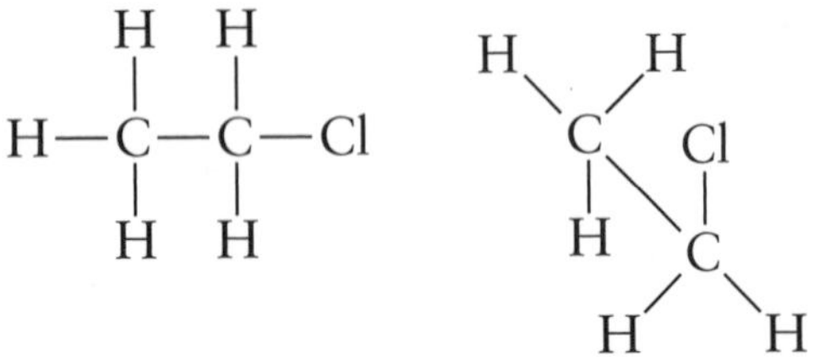

```
    H   H              H     H
    |   |               \   /
H — C — C — Cl            C      Cl
    |   |                 | \    |
    H   H                 H   \  |
                               C
                              / \
                             H   H
```

Although it may appear from a drawing of the molecule that the hydrogen atoms on the CH_3 group are inequivalent because they have different orientations relative to the chlorine atom in the CH_2Cl group, they always appear to have the same chemical environment because there is rapid rotation about the C—C bond.

2.107 The alkyne group will have the stiffer C—H bond because a large force constant k results in a higher frequency absorption.

CHAPTER 3
MOLECULAR SHAPE AND STRUCTURE

3.1 $:\ddot{O}:$ double-bonded to S in $:\ddot{Cl}-\ddot{S}-\ddot{Cl}:$

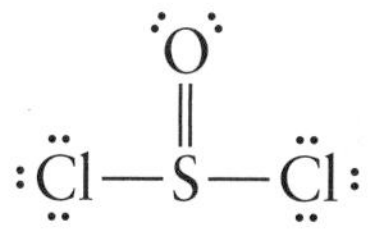

(a) The shape should be trigonal pyramidal based upon the presence of four electron groups (one lone pair and three bonds) attached to the central atom. The arrangement of the electrons is tetrahedral. (b) The O—S—Cl angles will be the same so there is only one bond angle. (c) The bond angles should be slightly less than 109.5°.

3.3 $\ddot{O}=\ddot{Cl}^{+}=\ddot{O}$

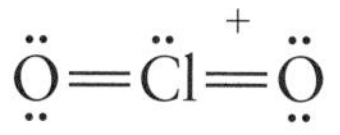

(a) The ClO_2^+ ion should be bent or angular. (b) The bond angle should be close to 120° as predicted by the presence of three groups (one lone pair and two bonds) attached to the central atom. The electron arrangement should be trigonal planar.

3.5 $\left[\ddot{O}=C(-\ddot{O}:)-\ddot{O}:\right]^{2-}$ + 2 resonance structures

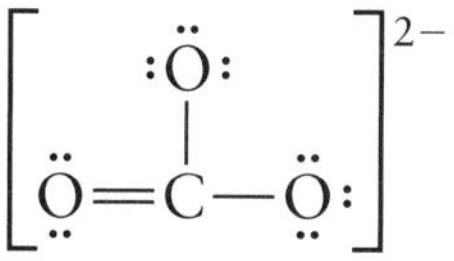

(a) trigonal planar; (b) all O—C—O bond angles are equivalent (c) 120°

3.7 $:N\equiv\ddot{S}-\ddot{F}:$

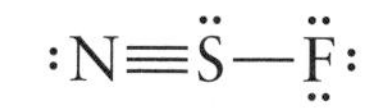

(a) angular or bent; (b) slightly less than 120°

3.9 The Lewis structures are

(a) 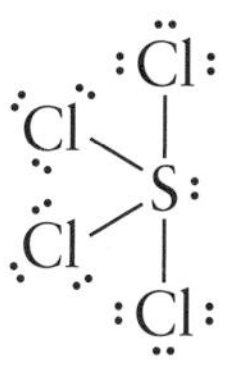 SCl_4 (b) 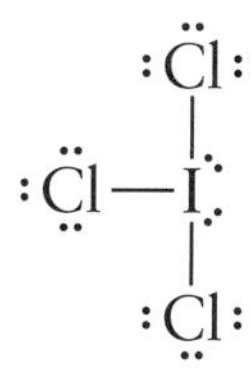 ICl_3 (c) 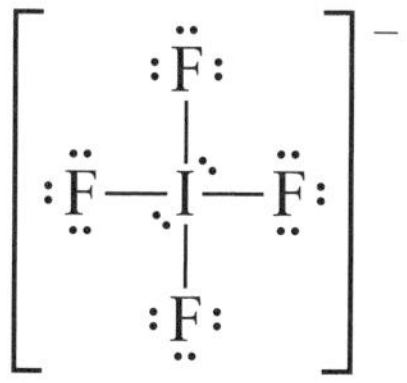 $[IF_4]^-$ (d) 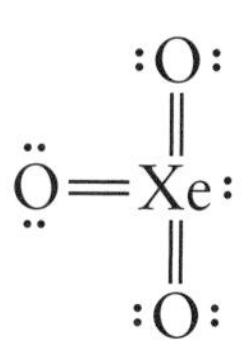 $\ddot{O}=XeO_2$ ($:\ddot{Xe}$ with three O)

(a) The sulfur atom will have five pairs of electrons about it: one nonbonding pair and four bonding pairs to chlorine atoms. The arrangement of electron pairs will be trigonal bipyramidal; the nonbonding pair of electrons will prefer to lie in an equatorial position, because in that location the e-e repulsions will be lowest. The actual structure is described as a seesaw. AX_4E

(b) Like the sulfur atom in (a), the iodine in iodine trichloride has five pairs of electrons about it, but here there are two lone pairs and three bonding pairs. The arrangement of electron pairs will be the same as in (a), and again the lone pairs will occupy the equatorial positions. Because the name of the molecule ignores the lone pairs, it will be classified as T-shaped. AX_3E_2

(c) There are six pairs of electrons about the central iodine atom in IF_4^-. Of these, two are lone pairs and four are bonding pairs. The pairs will be placed about the central atom in an octahedral arrangement with the lone pairs opposite each other. This will minimize repulsions between them. The name given to the structure is square planar. AX_4E_2

(d) In determining the shape of a molecule, double bonds count the same as single bonds. The XeO_3 structure has four "objects" about the central Xe atom: three bonds and one lone pair. These will be placed in a tetrahedral arrangement. Because the lone pair is ignored in naming the molecule, it will be classified as trigonal pyramidal. AX_3E

3.11 The Lewis structures for the ions and molecules are

(a) $[:\ddot{I}-\ddot{I}-\ddot{I}:]^-$ (b) IF_3: :F̈—Ï(—F̈:)(—F̈:) (c) $[IO_4]^-$: Ö=I(=Ö)(=O)—Ö: (d) TeF_6

(a) The I_3^- molecule is predicted to be linear, so the < I—I—I should equal 180°. AX_2E_3

(b) The IF_3 molecule is T-shaped. There should be two F—I—F angles of 90° and one of 180°. AX_3E_2

(c) The structure of IO_4^- will be tetrahedral, so the O—I—O bond angles should be 109.5°. AX_4

(d) The structure of TeF_6 is octahedral, so the F—Te—F angles should be either 90° or 180°. AX_6

3.13 The Lewis structures are

(a) $CClF_3$: :F̈—C(—Cl:)(—F:)—F̈: (b) GaI_3 (c) $XeOF_4$ (d) $[H-\ddot{C}(-H)-H]^-$

(a) The shape of CF_3Cl is tetrahedral; all halogen—C—halogen angles should be approximately 109.5°. AX_4; (b) GaI_3 molecules will be trigonal planar with I—Ga—I bond angles of approximately 120°. AX_3; (c) $XeOF_4$ molecules will be square pyramidal with O—Xe—F angles of 90°, and F—Xe—F angles of 90° and 180°. AX_5E; (d) CH_3^- ions will be trigonal pyramidal with H—C—H angles of slightly less than 109.5°. AX_3E

3.15 (a) a and b are expected to be about 120°, c is expected to be about 109.5° in 2,4-pentanedione. All of the angles are expected to be about 120° in the acetylacetonate ion. (b) The major difference arises at the C of the original sp^3-hybridized CH_2 group, which upon deprotonation goes to sp^2 hybrization with only three groups attached.

3.17 (a) slightly less than 120°; (b) 180°; (c) 180°; (d) slightly less than 109.5°

3.19 The Lewis structures are

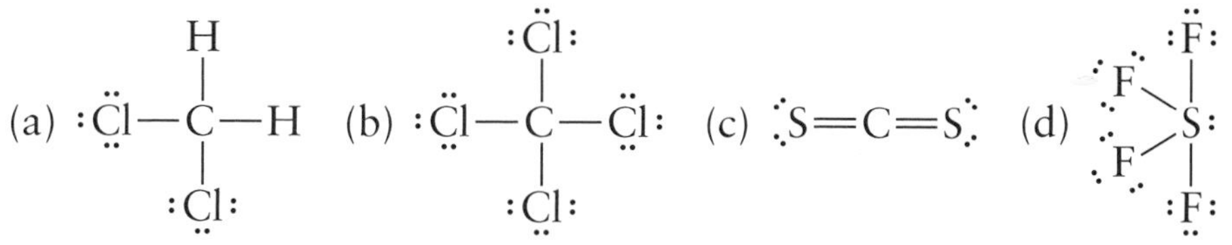

(a) :Cl—C(H)(Cl:)—H (b) :Cl—C(:Cl:)(:Cl:)—Cl: (c) :S=C=S: (d) SF_4 with four F atoms and a lone pair on S

Molecules (a) and (d) will be polar; (b) and (c) will be nonpolar.

3.21 (a) pyridine

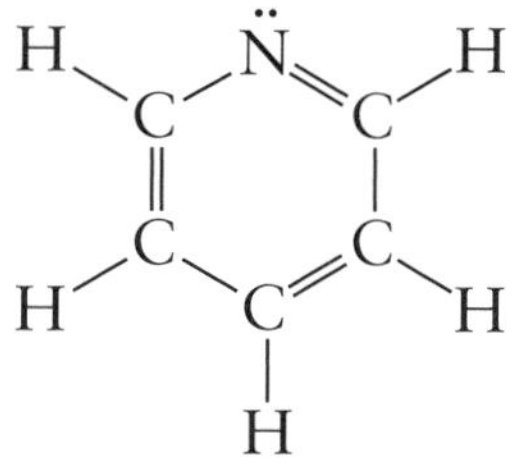

C_5H_5N ring structure (N with lone pair, each C bearing one H)

polar

(b) ethane

H—C(H)(H)—C(H)(H)—H

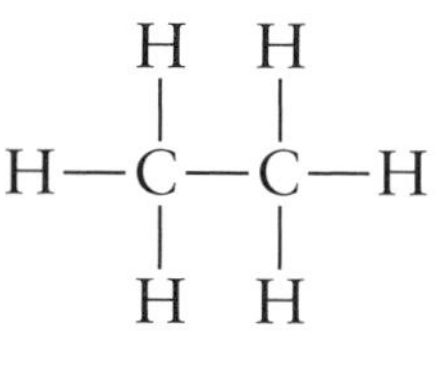

nonpolar

(c) trichloromethane

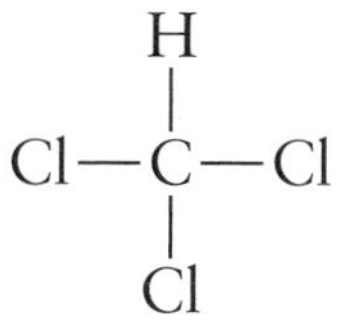

Cl—C(H)(Cl)—Cl

polar

3.23 Of the three forms, only **3** is nonpolar. This is because the C—Cl bond dipoles are pointing in exactly opposite directions in **3**. The dipole moment for **1** would be the largest because the C—Cl bond vectors are pointing most nearly in the same direction in **1** (60° apart) whereas in **2** the C—Cl vectors point more away from each other (120°), giving a larger cancellation of dipole.

3.25

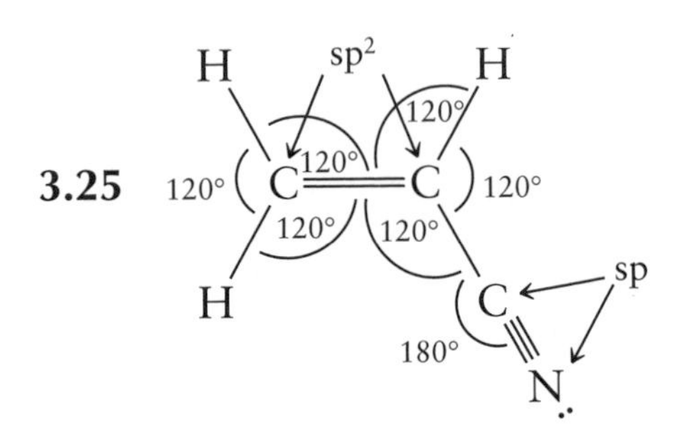

3.27 (a) sp^3, orbitals oriented toward corners of a tetrahedron (109.5° apart); (b) sp, orbitals oriented directly opposite to each other (180° apart); (c) sp^3d^2, orbitals oriented toward the corners of an octahedron (interorbital angles of 90° and 180°); (d) sp^2, orbitals oriented toward the corners of an equilateral triangle trigonal planar array (angles = 120°); trigonal planar.

3.29 (a) sp^3d; (b) sp^2; (c) sp^3; (d) sp

3.31 (a) sp^2; (b) sp^3; (c) sp^3d; (d) sp^3

3.33 (a) sp^3; (b) sp^3d^2; (c) sp^3d; (d) sp^3

3.35 (a) :S̤—S(=O:)(—:O̤:)=Ö $^{2-}$ (b) H_3C—Be—CH_3 (c) H—B^-—H (d) :Cl̤—S̈n—Cl̤:

(a) Thiosulfate is tetrahedral with bond angles approximately equal to 109.5°.
(b) Dimethyl beryllium is linear with a C—Be—C bond angle of 180°; the H—C—H bond angles are approximately 109.5°.
(c) BH_2^- will be angular with an H—B—H bond angle of slightly less than 120°.
(d) $SnCl_2$ will be angular with a Cl—Sn—Cl bond angle of slightly less than 120°.

3.37 (a) $H_2C{=}CH_2$ (b) :Cl—C≡N: (c) $POCl_3$ (O=P with three P—Cl) (d) H—N(H)—N(H)—H

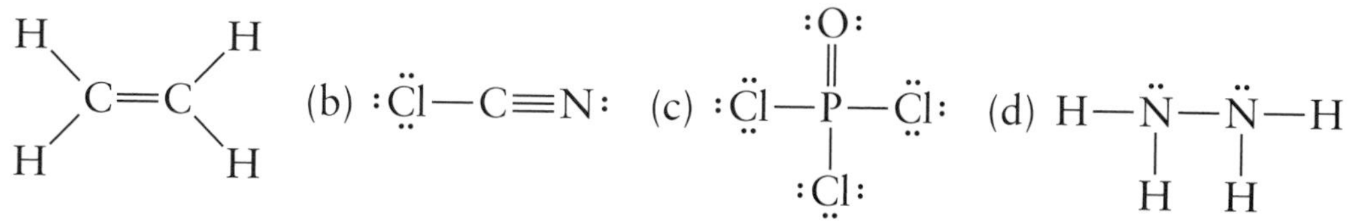

(a) bond angles are all ca. 120°; (b) bond angle = 180°; (c) bond angles = 109.5°; (d) bond angles about both nitrogen atoms are slightly less than 109.5°

3.39 $SbOCl_3$ (O=Sb, Sb=Cl, two Sb—Cl) $SOCl_2$ (O=S=O with two S—Cl) IO_2F_2 (two I=O, two I—F)

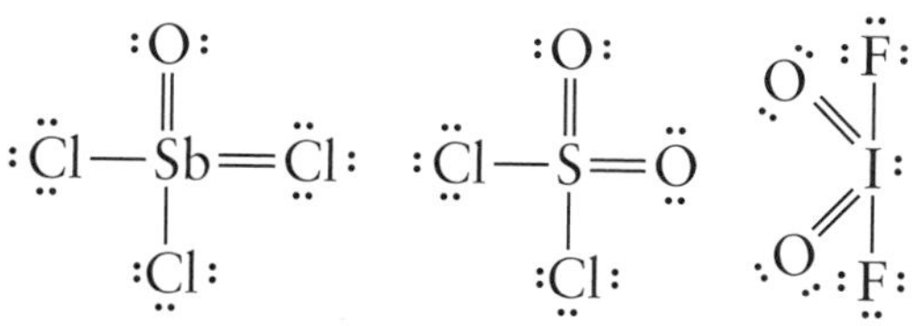

(a) tetrahedral, bond angles are all ca. 109.5°; (b) tetrahedral, bond angles are all ca. 109.5°; (c) Seesaw-shaped, angles are 90°, 180° and 120°.

3.41 (a) Li_2 BO = $\frac{1}{2}(2 + 2 - 2) = 1$

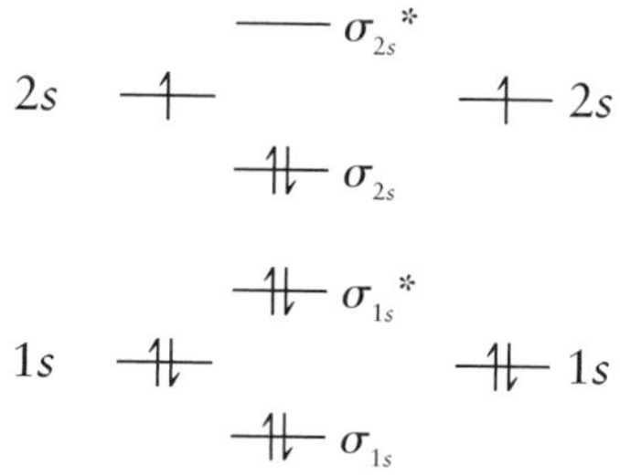

— σ_{2s}^*

2s ↑ ↑ 2s

↑↓ σ_{2s}

↑↓ σ_{1s}^*

1s ↑↓ ↑↓ 1s

↑↓ σ_{1s}

diamagnetic, no unpaired electrons

Li_2^+

(b) Li_2^+ BO = $\frac{1}{2}(2 + 2 - 2 - 1) = \frac{1}{2}$

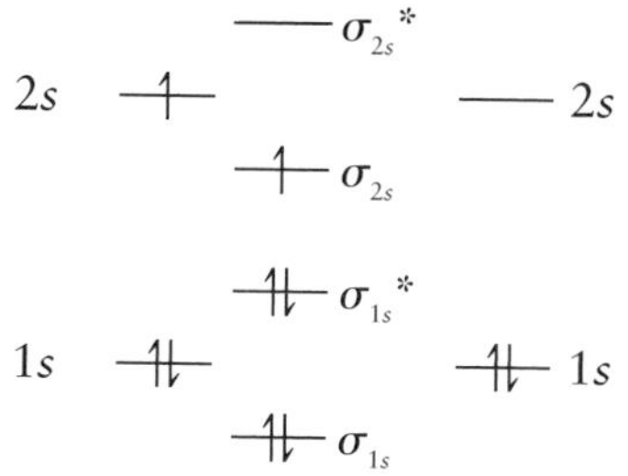

— σ_{2s}^*

2s ↑ — 2s

↑ σ_{2s}

↑↓ σ_{1s}^*

1s ↑↓ ↑↓ 1s

↑↓ σ_{1s}

paramagnetic, one unpaired electron

(c) Li_2^- BO = $\frac{1}{2}(2 + 2 - 2 - 1) = \frac{1}{2}$

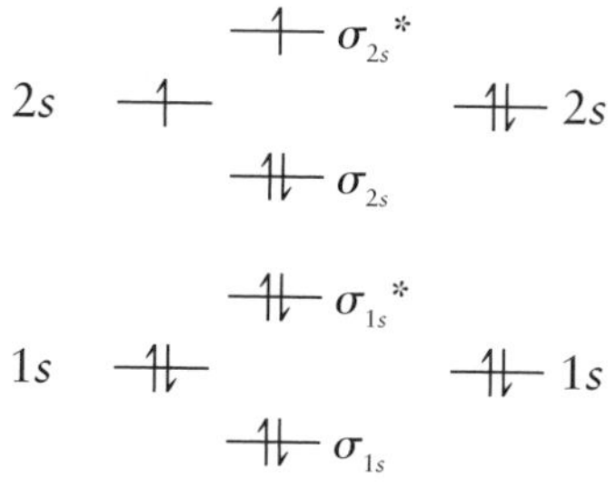

↑ σ_{2s}^*

2s ↑ ↑↓ 2s

↑↓ σ_{2s}

↑↓ σ_{1s}^*

1s ↑↓ ↑↓ 1s

↑↓ σ_{1s}

paramagnetic, one unpaired electron

3.43 (a) (1) $(\sigma_{2s})^2(\sigma_{2s}{}^*)^2(\sigma_{2p})^2(\pi_{2p_x})^2(\pi_{2p_y})^2(\pi_{2p_x}{}^*)^2(\pi_{2p_y}{}^*)^1$
(2) $(\sigma_{2s})^2(\sigma_{2s}{}^*)^2(\sigma_{2p})^2(\pi_{2p_x})^2(\pi_{2p_y})^2(\pi_{2p}{}^*)^1$
(3) $(\sigma_{2s})^2(\sigma_{2s}{}^*)^2(\sigma_{2p})^2(\pi_{2p_x})^2(\pi_{2p_y})^2(\pi_{2p_x}{}^*)^2(\pi_{2p_y}{}^*)^2$

(b) (1) 1.5; (2) 2.5; (3) 1

(c) (1) and (2) are paramagnetic with one unpaired electron each

(d) π in all three cases

3.45 An N—N double bond is composed of a σ and a π bond, whereas the N—N single bond is a σ-type bond. Because the types of orbitals that compose the σ and π bonds are different, it is reasonable to expect that the resulting molecular orbitals would have different energies.

3.47 (a) See Figure 3.34 for the energy level diagram for N_2. (b) The nitrogen atom is more electronegative, which will make its orbitals lower in energy than those of C. The revised energy-level diagram is shown below. This will make all of the bonding orbitals closer to N than to C in energy and will make all the antibonding orbitals closer to C than to N in energy.

Energy level diagram for CN^-

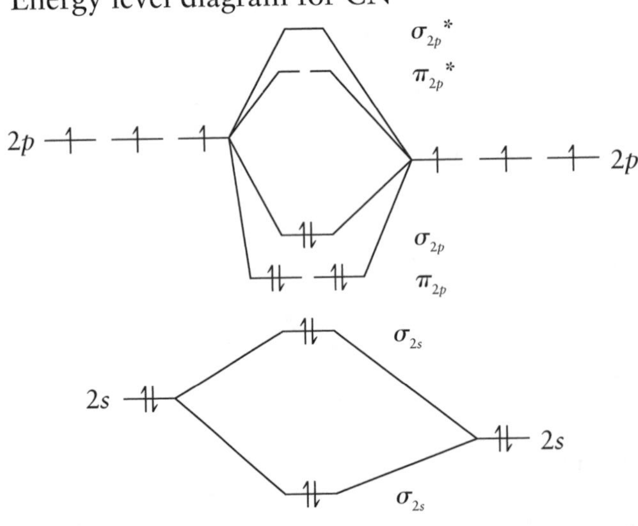

energy levels on C energy levels on N

(c) The electrons in the bonding orbitals will have a higher probability of being at N because it is the more electronegative atom and its orbitals are lower in energy.

3.49 (a) C_2 (8 valence electrons): $(\sigma_{2s})^2(\sigma_{2s}{}^*)^2(\sigma_{2p})^2(\pi_{2p_x})^1(\pi_{2p_y})^1$; bond order = 1. (b) Be_2 (4 valence electrons): $(\sigma_{2s})^2(\sigma_{2s}{}^*)^2$, bond order = 0. (c) Ne_2 (16 valence electrons): $(\sigma_{2s})^2(\sigma_{2s}{}^*)^2(\sigma_{2p})^2(\pi_{2p_x})^2(\pi_{2p_y})^2(\pi_{2p_x}{}^*)^2(\pi_{2p_y}{}^*)^2(\sigma_{2p}{}^*)^2$; bond order = 0.

3.51 All these molecules are paramagnetic. O_2^- and O_2^+ have an odd number of electrons and must, therefore, have at least one unpaired electron. O_2 has an even number of electrons, but in its molecular orbital energy level diagram, the HOMO is a degenerate set of orbitals that are each singly occupied, giving this molecule two unpaired electrons. For O_2^-, one more electron will be placed in this degenerate set of orbitals, causing one of the original unpaired electrons to now be paired. O_2^- will therefore have one unpaired electron. Likewise, O_2^+ will have one less electron than O_2; thus one of the originally unpaired electrons will be removed, leaving one unpaired electron in this molecule.

3.53 (a) F_2 with 14 valence electrons has a valence electron configuration of $(2s\sigma)^2$ $(2s\sigma^*)^2$ $(2p\sigma)^2(2p\pi_x)^2(2p\pi_y)^2(2p\pi_x{}^*)^2(2p\pi_y{}^*)^2$ with a bond order of 1. After forming F_2^- from F_2, an electron is added into a $2p\sigma^*$ orbital. The addition of an electron to this antibonding orbital will result in a reduction of the bond order to 1/2 (See 51). F_2 will have the stronger bond. (b) B_2 will have an electron configuration of $(2s\sigma)^2(2s\sigma^*)^2(2p\pi_x)^1(2p\pi_y)^1$ with a bond order of 1. Removing one electron to form B_2^+ will eliminate one electron in the bonding orbitals, creating a bond order of 1/2. B_2 will have the stronger bond.

3.55 The conductivity of a semiconductor increases with temperature as increasing numbers of electrons are promoted into the conduction band, whereas the conductivity of a metal will decrease as the motion of the atoms will slow down the migration of electrons.
Note: The electron due to the charge has arbitrarily been placed in a $2p$ orbital on C on the left-hand side of the diagram.

3.57 (a) In and Ga; (b) P and Sb

3.59 (a)

```
         ..
        :Cl:
         |   -
  ..     |     ..
 :Cl — Ga — Cl:
  ..     |     ..
         |
        :Cl:
         ..
```

tetrahedral, sp^3, all Cl—Ga—Cl bond angles = 109.5°, nonpolar

(b) :F:
:F—Te:—F:
:F—
:F:

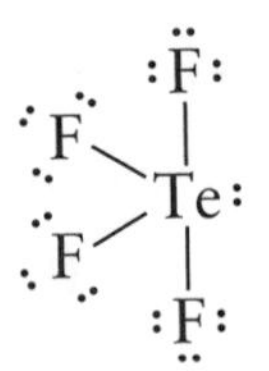

Seesaw, sp^3d, F—S—F bond angles = 90°, 120° and 180°, polar

(c) :Cl: $^-$
:Cl—Sb:
:Cl—
:Cl:

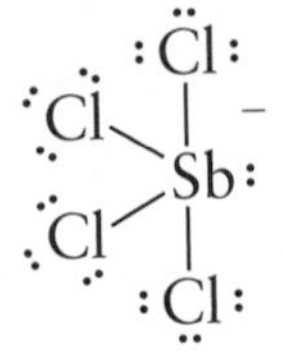

Seesaw, sp^3d, Cl—Sb—Cl bond angles = 90°, 120° and 180°, polar

(d) :Cl:
:Cl—Si—Cl:
:Cl:

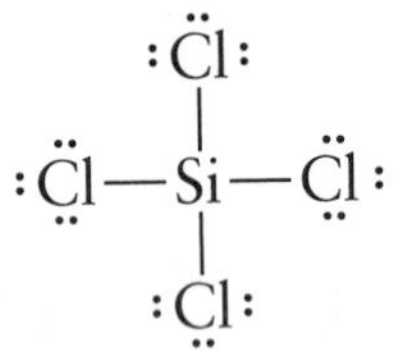

tetrahedral, sp^3, all Cl—Si—Cl bond angles = 109.5°, nonpolar

3.61 (a) :Cl: $^-$
:Cl—In—Cl:
:Cl:

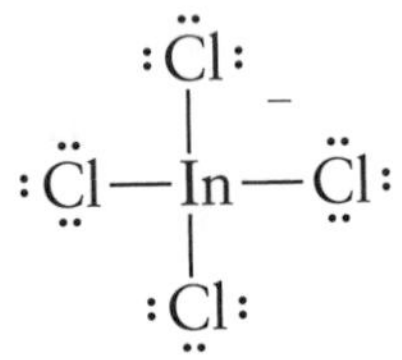

tetrahedral, sp^3, all Cl—In—Cl bond angles = 109.5°, nonpolar

(b) :Cl: $^+$
:Cl—P—Cl:
:Cl:

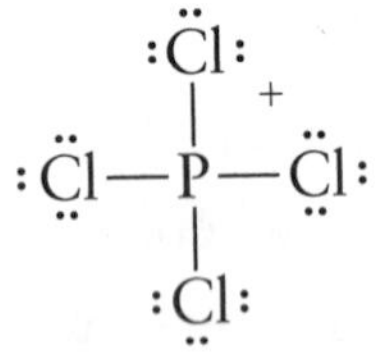

tetrahedral, sp^3, all Cl—P—Cl bond angles = 109.5°, nonpolar

(c) :Cl: $^+$
:Cl—I:
:Cl—
:Cl:

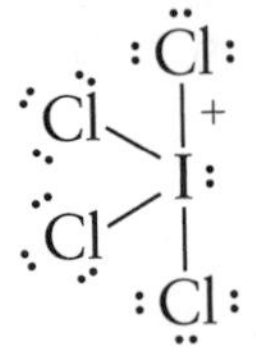

Seesaw, sp^3d, Cl—I—Cl bond angles = 90°, 120° and 180°, polar

(d) O
‖
:F—Se—F:

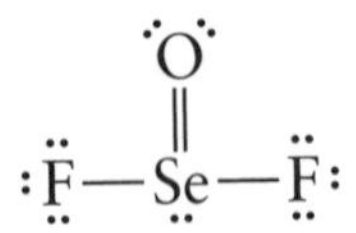

Trigonal pyramidal, F—Se—F and F—S—O bond angles should be slightly less than 109.5°.

3.63 (a) SiF_4: SiF_4 is nonpolar (tetrahedral AX_4 structure) but PF_3 is polar (trigonal pyramidal AX_3E structure); (b) SF_6: SF_6 is nonpolar (octahedral AX_6 structure) whereas SF_4 is polar (seesaw, AX_4E structure); (c) AsF_5: IF_5 is polar (square pyramidal, AX_5E structure) whereas AsF_5 is nonpolar (AX_5, trigonal bipyramidal structure).

3.65 The elemental composition gives an empirical formula of CH_4O, which agrees with the molar mass. There is only one reasonable Lewis structure; this corresponds to the compound methanol. Both carbon and oxygen are sp^3 hybridized. All of the bond angles at carbon should be 109.5°. The bond angles about oxygen should be close to 109.5° but will be somewhat less, due to the repulsions by the lone pairs. The molecule is polar.

```
      H
      |     ..
  H — C — O:
      |    |
      H    H
```

3.67 (a)

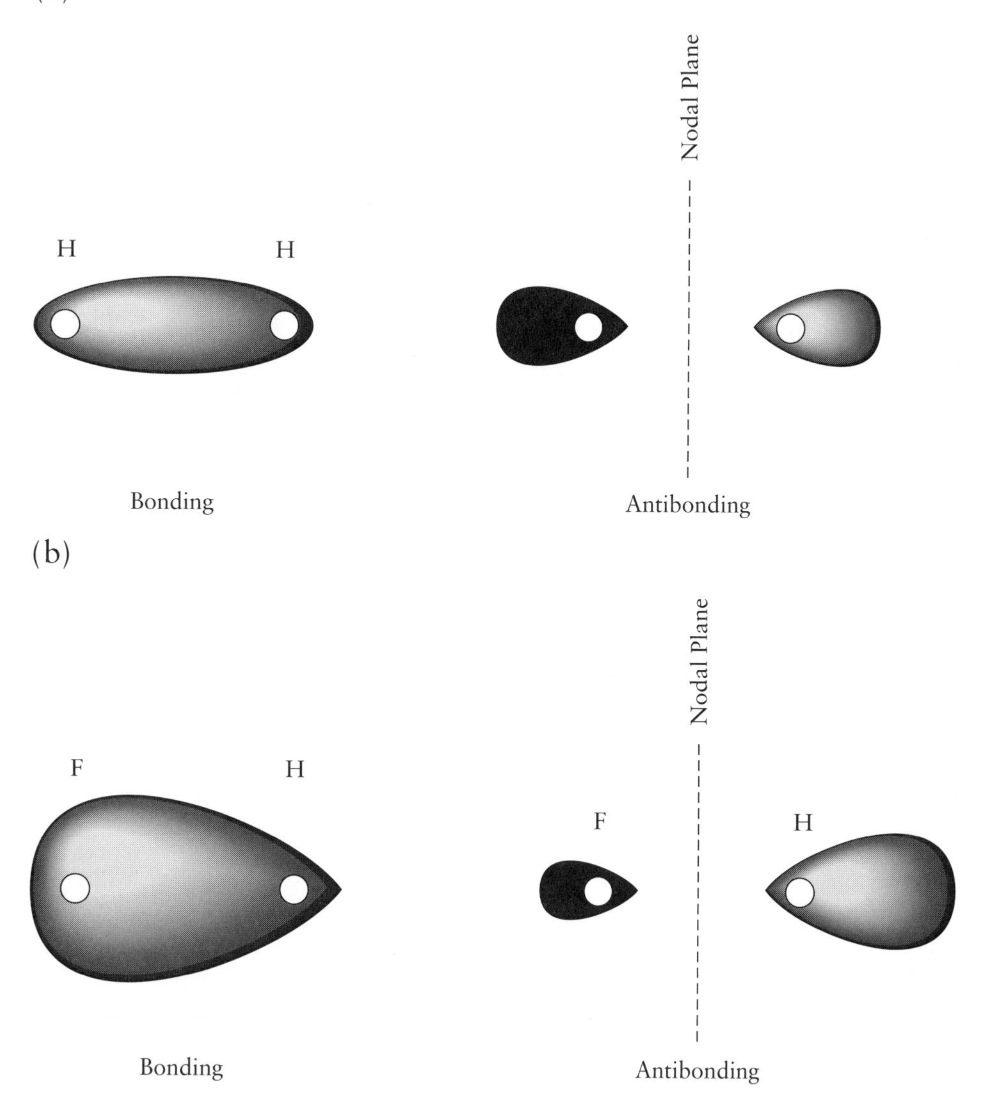

3.69 The expected molecular orbital diagram for CF is

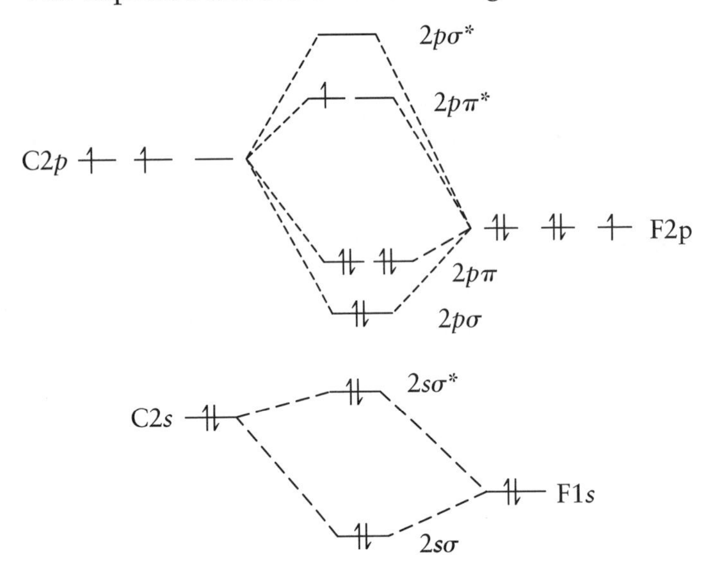

The bond order for the neutral species is 2.5 because one electron occupies a $2p\pi^*$ orbital. Adding an electron to form CF^- will reduce the bond order by 1/2 to 2, while removing an electron from form CF^+ will increase the bond order to 3. The bond lengths will increase as the bond order decreases: $CF^- < CF < CF^+$. The CF^+ ion will be diamagnetic but both CF and CF^- will have unpaired electrons (one in the case of CF and two in the case of CF^-).

3.71 The Lewis structure of borazine is nearly identical to that of benzene. It is obtained by replacing alternating C atoms in the benzene structure with B and N, as shown. The orbitals at each B and N atom will be sp^2 hybridized.

```
            H
            |
    H     N       H
     \  /   \\   /
      B       B
      ||      |
      N       N
     /  \   //  \
    H     B       H
          |
          H
```

3.73 The Lewis structures are:

```
     H  +            H               -
     |               |          ..
     C           H—C—H      H—C—H
   /   \             |           |
  H     H            H           H

   ..              2+         ..
 H—C          H—C—H       H—C:2
   |                          |
   H                          H
```

The predicted bond angles in each species based upon the Lewis structure and VSEPR theory will be

CH_3^+	AX_3	trigonal planar	120°
CH_4	AX_4	tetrahedral	109.5°
CH_3^-	AX_3E	pyramidal	slightly less than 109.5°
CH_2	AX_2E	angular	slightly less than 120°
CH_2^{2+}	AX_2	linear	180°
CH_2^{2-}	AX_2E_2	angular	less than 109.5°, more so than CH_3^- due to the presence of two lone pairs

The order of increasing H—C—H bond angle will be

$$CH_2^{2-} < CH_3^- < CH_4 < CH_2 < CH_3^+ < CH_2^{2+}$$

All of these species are expected to be diamagnetic. None are radicals.

3.75 $H-C\equiv C-H \longrightarrow$ (—CH=CH—CH=CH—CH=CH—CH=CH—)

Polyacetylene retains multiple bonds along the chain. It is through the series of orbitals that electrons can be conducted. A resonance form of the Lewis structure can be drawn showing that the electrons may be delocalized along the polyacetylene chain. No such resonance form is possible for polyethylene.

Note: The dark color of the material results from the formation of a large number of molecular orbitals that are not very different in energy. These molecular orbitals are made up of combinations of the *p*-orbitals on the carbons that make up the double bonds. Because the orbitals are closely spaced in energy, electrons in them can readily absorb visible light to be promoted to a higher energy orbital. See section 3.13.

3.77 (a)

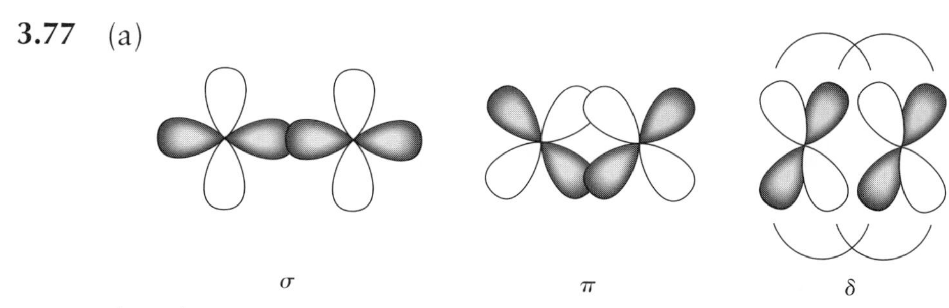

σ π δ

(b) The overlap between the orbitals will decrease as one goes from σ to π to δ, so we expect the bond strengths to decline in the same order.

3.79 The effect of changes (a) and (b) will be similar. The overall bond order will change. In the first case, electrons will be removed from π orbitals so that the net π-bond order will drop from three to two. The same thing will happen in (b), but because two electrons are added to antibonding orbitals, a net total of one π-bond will be broken. Based on this simple model, the ions formed should be paramagnetic because the electrons are added to or taken from doubly degenerate orbitals.

3.81 (a) The Lewis structure of benzyne is

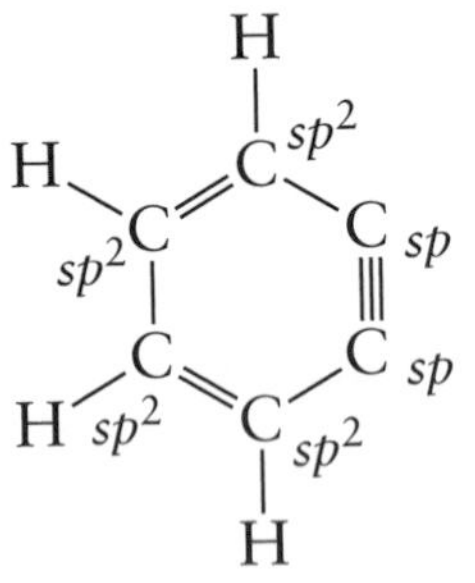

(b) Benzyne would be highly reactive because the two carbons that are *sp* hybridized are constrained to have a very strained structure compared to what their hybridization would like to adopt—namely a linear arrangement. Instead of 100° angles at these carbon atoms, the angles by necessity of being in a six-membered ring are constrained to be close to 120°. A possibility that allows the carbons to adopt more reasonable angles is the formation of a diradical:

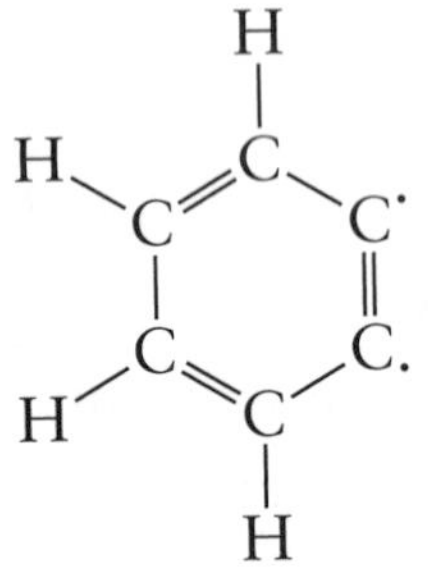

3.83 (a) ClO_2F and AsF_5 react in a Lewis acid/base reaction to give $[ClO_2]^+[AsF_6]^-$.
(b) The complete Lewis structure is

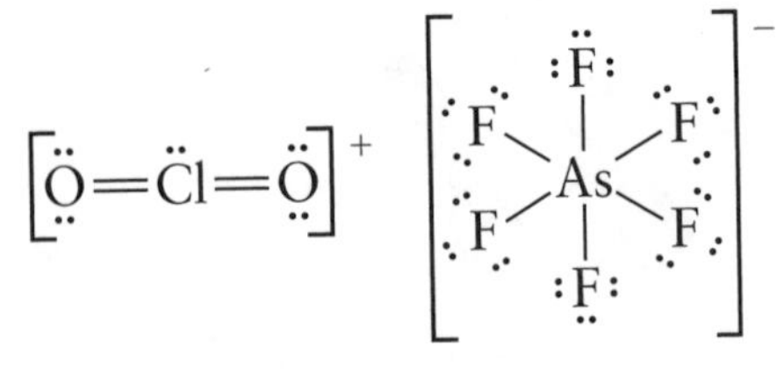

$$[O=Cl=O]^+ \quad [AsF_6]^-$$

The formal charges are: oxygen atoms, 0; chlorine, +1; arsenic, −1; fluorine, 0. The ClO_2^+ ion is bent or angular (based upon a trigonal planar arrangement of electrons) with an expected O—Cl—O of slightly less than 120°. The AsF_6^- ion will be a regular octahedron with F—As—F angles of 90° and 180°.

3.85 (a) The carbon atoms are all sp^3 hybridized. (b) The C—C—C, H—C—H and H—C—C bond angles should be 109.5° based upon the answer to (a). (c) Because of the ring structure, however, the C—C—C bond angles must be 60°. (d) The σ-bond will have the electron density of the bond located on a line between the two atoms that it joins. (e) If the C atoms are truly sp^3 hybridized, then the bonding orbitals will not necessarily point directly between the C atoms. (f) The sp^3 hybridized orbitals can still overlap even if they do not point directly between the atoms as shown. Such bonds are sometimes called "bent" bonds, or "banana" bonds. As a result of the situation in the C—C—C bond angles, the H—C—H bond angles are also distorted from 109.5°.

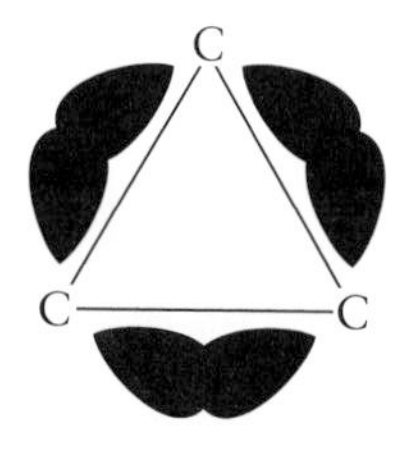

(g) The values found on the CD are <C—C—C = 60°; <H—C—H = 115.3°.

3.87 (a) Methane is a tetrahedral molecule.

H
C
H H
H

If we use the covalent radii given in Figure 2.19 (37 pm for H and 77 pm for C), we arrive at a C—H bond distance of 114 pm. The size of methane will be somewhat larger, however, as the bond distance is from the center of the C atom to the center of the H atom. Thus, from the center of the methane molecule which is taken as the location of the carbon atom, to the far edge of the H atom would be 114 pm + 37 pm = 151 pm. We can consider this distance to be the dimension for the radius of a sphere that would be necessary to encapsulate the CH_4 molecule. The volume of this sphere is given by

$$V = \frac{4}{3}\pi r^3$$

$$= \frac{4}{3}\pi(151\ \text{pm})^3$$

$$= 1.44 \times 10^7\ \text{pm}^3$$

(b) Ethane can similarly be inscribed in a cylinder.

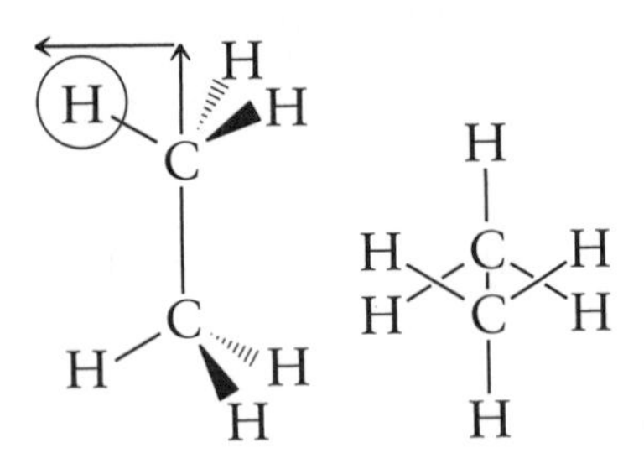

The axis of the cylinder lies along the C—C bond and the circular faces will be described by the H atoms on each carbon. The difficulty then becomes to determine the height of the cylinder and the radius of the circular faces. We can set up the following diagram to help us with the calculations.

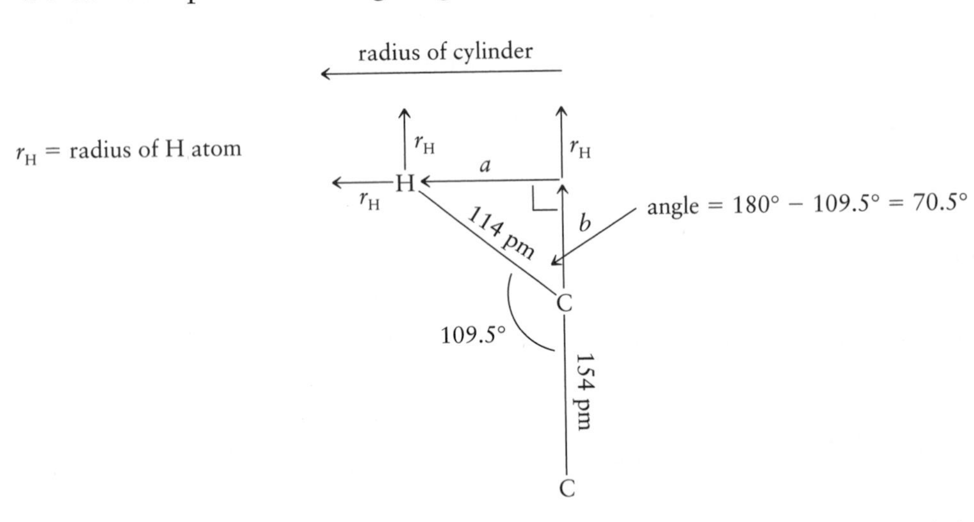

As can be seen, the radius of the cylinder will be equal to $a + r_H$; while the height of the cylinder will be given by the sum of the C—C bond length plus $2b$ plus $2\ r_H$.

$r_{cylinder} = a + r_H$

$h = 154\text{ pm} + 2\ a + 2\ r_H$

$$\sin 70.5° = \frac{a}{114\text{ pm}}$$

$a = 108\text{ pm}$

$$\cos 70.5° = \frac{b}{114\text{ pm}}$$

$b = 38.0\text{ pm}$

$r_{cylinder} = 108\text{ pm} + 37\text{ pm} = 145\text{ pm}$

$h = 154\text{ pm} + 216\text{ pm} + 74\text{ pm} = 444\text{ pm}$

The volume of a cylinder is given by

$V = \pi r^2 h$

$V = (3.141)(145\text{ pm})^2(444\text{ pm}) = 2.93 \times 10^7\text{ pm}^3$

(c) Benzene can also be represented by a cylinder whose height is equal to the diameter of a carbon atom. We could use the value from Table 2.3 for the aromatic C—C bond distance (139 pm) because that value will also be equal to twice the radius of the aromatic C atom, or we could approximate this value by using the average of the C—C single bond and C=C double bond radii given in Figure 2.19 ({67 pm + 77 pm}/2 = 72 pm) for a diameter of 144 pm. The remainder of the problem is then to find the diameter of a circle that would encompass the benzene molecule.

The radius of the cylinder will be given by the distance from the center of the benzene ring to the center of the H atom plus an additional radius of the H atom. Because the benzene ring is a regular hexagon, the distance from the center of the hexagon to the center of a carbon atom will be equal to the C—C bond distance (139 pm). As seen in the diagram below, the radius of a circle that would encompass the benzene ring will be given by r = 139 pm + 109 pm + 37 pm = 285 pm.

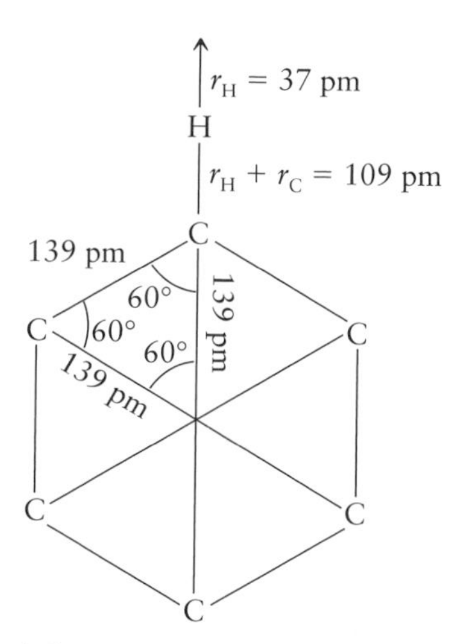

The volume of the cylinder is then given by

$V = (3.141)(285\ \text{pm})^2(139\ \text{pm}) = 1.06 \times 10^7\ \text{pm}^3$

Please note that these values are only rough approximations because, as we have seen from quantum mechanics, an atom is not a hard sphere with a sharp boundary located at the value of the atomic radius. We will return to this later (see Exercise 4.99).

3.89 (a)

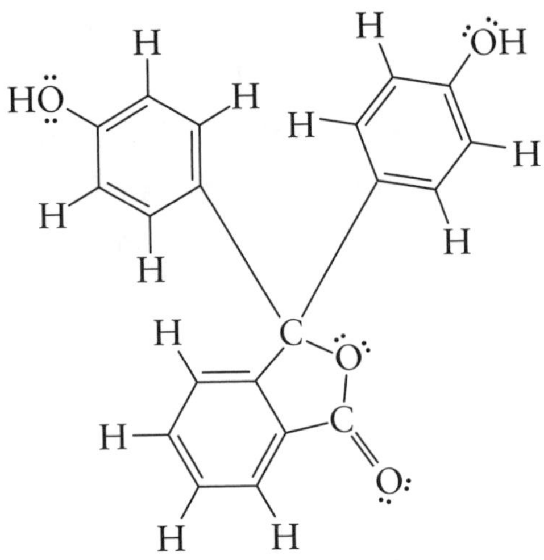

phenolphthalein (acid form)

There are two possible arrangements of the double bonds in each benzene ring giving a total of 2^3 possible resonance forms for the acid form.

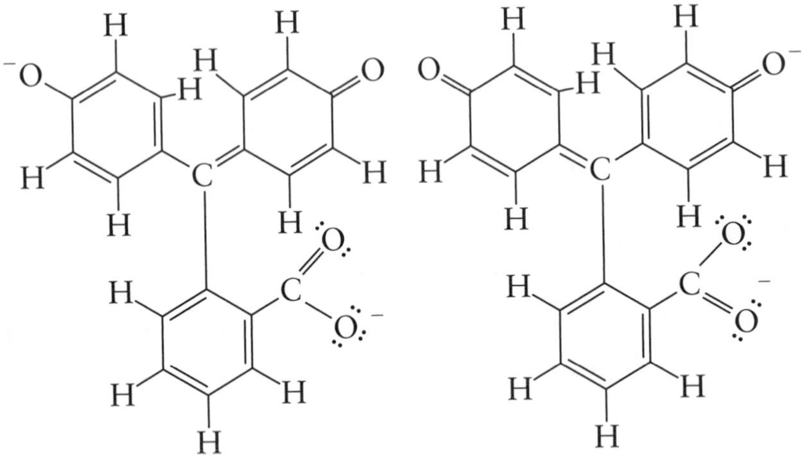

(b) The basic form is red; this suggests a more extensive conjugation between the atoms of the phenolphthalein. More delocalization leads to molecular orbitals that are more closely spaced and which can, therefore, possibly absorb light in the visible region of the electromagnetic spectrum. The more extensive π-network is allowed in the base form because the central carbon atom becomes sp^2 hybridized. This means that all the carbon atoms are sp^2 hybridized in the base form and all will have p orbitals that can overlap to form a set of molecular orbitals that extend over the entire molecule. This is indicated by the resonance forms shown above.

3.91 (a) and (d) have the possibility of n-to-π^* transitions because these molecules possess both an atom with a lone pair of electrons (on O in HCOOH and on N in HCN) and a π-bond to that atom. The other molecules have either a lone pair or a π-bond, but not both.

CHAPTER 4
THE PROPERTIES OF GASES

4.1 (a) 8×10^9 Pa; (b) 80 kilobars; (c) 6×10^7 Torr; (d) 1×10^6 $\text{lb}\cdot\text{in}^{-2}$

4.3 (a) The difference in column height will be equal to the difference in pressure between atmospheric pressure and pressure in the gas bulb. If the pressures were equal, the height of the mercury column on the air side and on the apparatus side would be the same. The pressure in the gas bulb is 1.200 atm or 1.200×760 $\text{Torr}\cdot\text{atm}^{-1}$ = 912 Torr. The difference in pressure will, therefore, be 912 Torr − 754 Torr = 158 Torr. A Torr corresponds almost exactly to 1 mmHg, so the height difference will be approximately 158 mm. (b) The atmosphere side will be higher because the pressure in the apparatus is greater than the pressure of the atmosphere. (c) If the sides were switched and the difference in heights were the same, then the calculation should give 754 Torr − 158 Torr = 596 Torr for the pressure of neon. In this case, the mercury level on the side of the manometer attached to the glass bulb would be higher.

4.5 $d_1h_1 = d_2h_2$

$$73.5\text{ cm} \times \frac{13.6\text{ g}\cdot\text{cm}^{-3}}{1.1\text{ g}\cdot\text{cm}^{-3}} = 909\text{ cm or }9.09\text{ m}$$

4.7 $(20\text{ in})(10\text{ in})(14.7\text{ lb}\cdot\text{in}^2) = 2.9 \times 10^3\text{ lb}$

4.9 (a)

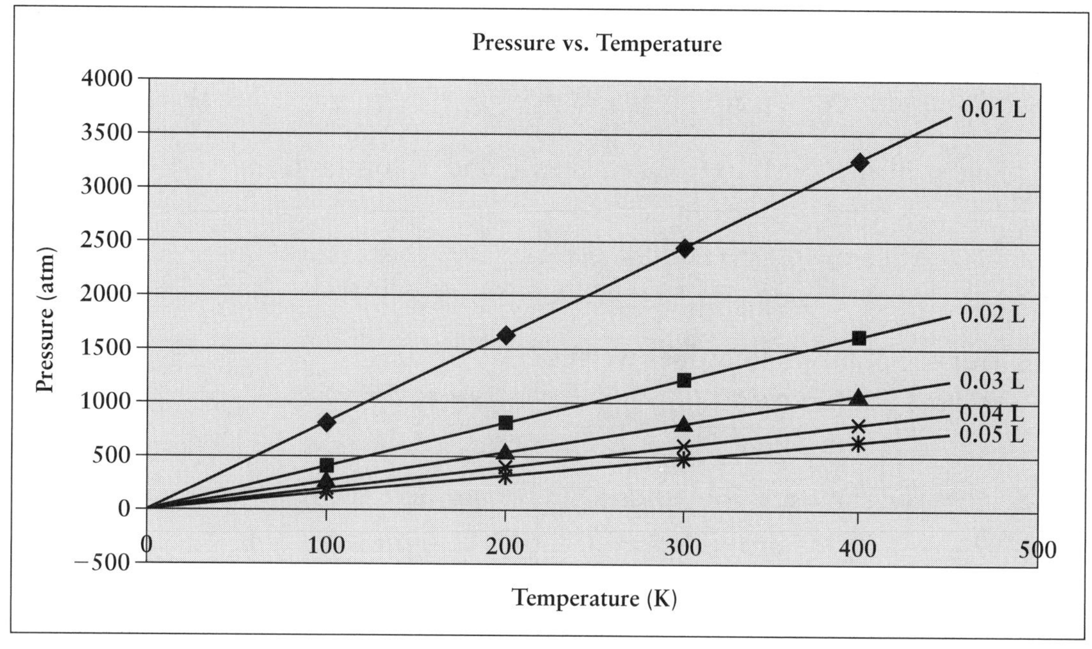

(b) The slope is equal to $\frac{nR}{V}$

Volume, L	$\frac{nR}{V}$, atm·K^{-1}
0.01	8.21
0.02	4.10
0.03	2.74
0.04	2.05
0.05	1.64

(c) The intercept is equal to 0 for all the plots.

4.11 (a) from $P_1V_1 = P_2V_2$, we have $(4.9 \times 10^5 \text{ kPa})(7.36 \text{ mL}) = (P_2)(1000 \text{ mL})$; solving for P_2 we get 3.6×10^3 kPa; (b) similar to (a), $P_1V_1 = P_2V_2$ or $(537 \text{ Torr})(67.3 \text{ cm}^3) = (P_2)(9.5 \text{ cm}^3)$, $P_2 = 3.8 \times 10^3$ Torr

4.13 Using $\frac{P_1}{T_1} = \frac{P_2}{T_2}$ and expressing T in Kelvins, $\frac{1.10 \text{ atm}}{298 \text{ K}} = \frac{P_2}{898 \text{ K}}$; $P_2 = 3.31$ atm

4.15 Using $\frac{P_1}{T_1} = \frac{P_2}{T_2}$ and expressing T in Kelvins, $\frac{1.5 \text{ atm}}{283 \text{ K}} = \frac{P_2}{303 \text{ K}}$; $P_2 = 1.6$ atm

4.17 Because we want P and T to remain constant, we can use the relation $\frac{V_1}{n_1} = \frac{V_2}{n_2}$. Substituting the numbers that we have, we get $\frac{V_1}{0.100 \text{ mol}} = \frac{V_2}{0.110 \text{ mol}}$. Solving for V_2 in terms of V_1, we obtain $V_2 = \frac{n_2V_1}{n_1} = \frac{0.110V_1}{0.10} = 1.10V_1$. Thus, the volume must be increased by 10% to keep P and T constant.

4.19 (a) Because P, V, and T all change, we use the relation $\frac{P_1V_1}{T_1} = \frac{P_2V_2}{T_2}$. Substituting for the appropriate values we get

$$\frac{(0.255 \text{ atm})(35.5 \text{ mL})}{228 \text{ K}} = \frac{(1.00 \text{ atm})(V_2)}{298 \text{ K}}; V_2 = 11.8 \text{ mL}$$

(b) The same relation holds as in (a) but here the final temperature and volume are known: $\frac{(0.255 \text{ atm})(35.5 \text{ mL})}{228 \text{ K}} = \frac{(P_2)(12.0 \text{ mL})}{293 \text{ K}}$. $P_2 = 0.969$ atm (c) Similarly, we can use the same expression, with P and V known and T wanted.

$$\frac{(0.255 \text{ atm})(35.5 \text{ mL})}{228 \text{ K}} = \frac{\left(\dfrac{(500 \text{ Torr})}{760 \text{ Torr}\cdot\text{atm}^{-1}}\right)(12.0 \text{ mL})}{T_2}; T_2 = 199 \text{ K}$$

4.21 (a) Using the ideal gas law with the gas constant R expressed in kPa:

$P(0.3500 \text{ L}) = (0.1500 \text{ mol})(8.314\ 51 \text{ L}\cdot\text{kPa}\cdot\text{K}^{-1}\cdot\text{mol}^{-1})(297 \text{ K});$

$P = 1.06 \times 10^3 \text{ kPa}$

(b) BrF_3 has a molar mass of 136.91 $\text{g}\cdot\text{mol}^{-1}$. We then substitute into the ideal gas equation:

$$\left(\frac{10.0 \text{ Torr}}{760 \text{ Torr}\cdot\text{atm}^{-1}}\right)V = \left(\frac{23.9 \times 10^{-3}\text{g}}{136.91 \text{ g}\cdot\text{mol}^{-1}}\right)(0.082\ 06 \text{ L}\cdot\text{atm}\cdot\text{K}^{-1}\cdot\text{mol}^{-1})(373 \text{ K})$$

$V = 4.06 \times 10^2 \text{ mL}$

(c) $$(0.77 \text{ atm})(0.1000 \text{ L}) = \left(\frac{m}{64.06 \text{ g}\cdot\text{mol}^{-1}}\right)(0.082\ 06 \text{ L}\cdot\text{atm}\cdot\text{K}^{-1}\cdot\text{mol}^{-1})$$
(303 K)

$m = 0.20 \text{ g}$

(d) $$(129 \text{ kPa})(6.00 \times 10^3 \text{ m}^3)\left(\frac{1 \times 10^6 \text{ cm}^3}{\text{m}^3}\right)\left(\frac{1 \text{ L}}{1000 \text{ cm}^3}\right) =$$
$n(8.314\ 51 \text{ L}\cdot\text{kPa}\cdot\text{K}^{-1}\cdot\text{mol}^{-1})(287 \text{ K})$

$n = 3.24 \times 10^5 \text{ mol } CH_4$

(e) The number of He atoms is the Avogadro constant N_A multiplied by the number of moles. The number of moles is obtained from the ideal gas equation:

$PV = nRT$

$$n = \frac{PV}{RT}; n = \frac{N}{N_A}$$

so the number of atoms N will be given by

$$N = N_A\left(\frac{PV}{RT}\right)$$

$$= (6.022 \times 10^{23} \text{ atoms}\cdot\text{mol}^{-1})\left(\frac{(2.00 \text{ kPa})(1.0 \times 10^{-6} \text{ L})}{(8.314\ 51 \text{ L}\cdot\text{kPa}\cdot\text{K}^{-1}\cdot\text{mol}^{-1})(158 \text{ K})}\right)$$

$$= 9.2 \times 10^{14} \text{ atoms}$$

4.23 For all problems, the number of molecules will be given by the number of moles n multiplied by the Avogadro constant N_A.

$$N_A\cdot n = N_A\left(\frac{PV}{RT}\right)$$

$$= (6.022 \times 10^{23} \text{ molecules}\cdot\text{mol}^{-1})\left(\frac{(1.00 \text{ atm})V}{(0.082\ 06 \text{ L}\cdot\text{atm}\cdot\text{K}^{-1}\cdot\text{mol}^{-1})(298 \text{ K})}\right)$$

(a) $N = (6.022 \times 10^{23}\ \text{molecules}\cdot\text{mol}^{-1})\left(\dfrac{(1.00\ \text{atm})(74.57 \times 10^{-9}\ \text{L})}{(0.082\ 06\ \text{L}\cdot\text{atm}\cdot\text{K}^{-1}\cdot\text{mol}^{-1})(298\ \text{K})}\right)$

$= 1.83 \times 10^{14}$ molecules

(b) $N = (6.022 \times 10^{23}\ \text{molecules}\cdot\text{mol}^{-1})\left(\dfrac{(1.00\ \text{atm})\left(\dfrac{\frac{4}{3}\pi(7.75\ \text{cm})^3}{1000\ \text{cm}^3\cdot\text{L}^{-1}}\right)}{(0.082\ 06\ \text{L}\cdot\text{atm}\cdot\text{K}^{-1}\cdot\text{mol}^{-1})(298\ \text{K})}\right)$

$= 4.80 \times 10^{22}$ molecules

(c) $N = (6.022 \times 10^{23}\ \text{molecules}\cdot\text{mol}^{-1})\left(\dfrac{(1.00\ \text{atm})\left(\dfrac{\pi(0.125\ \text{cm})^2(1.00\ \text{cm})}{1000\ \text{cm}^3\cdot\text{L}^{-1}}\right)}{(0.082\ 06\ \text{L}\cdot\text{atm}\cdot\text{K}^{-1}\cdot\text{mol}^{-1})(298\ \text{K})}\right)$

$= 1.21 \times 10^{18}$ molecules

4.25 (a) $V = \dfrac{nRT}{P} = \dfrac{(1\ \text{mol})(0.082\ 06\ \text{L}\cdot\text{atm}\cdot\text{K}^{-1}\cdot\text{mol}^{-1})(773\ \text{K})}{1\ \text{atm}} = 63.4\ \text{L}$

(b) $V = \dfrac{nRT}{P} = \dfrac{(1\ \text{mol})(0.082\ 06\ \text{L}\cdot\text{atm}\cdot\text{K}^{-1}\cdot\text{mol}^{-1})(77\ \text{K})}{1\ \text{atm}} = 6.32\ \text{L}$

4.27 Because P, V, and T are state functions, the intermediate conditions are irrelevant to the final states. We can simply use the ideal gas law in the form

$$\frac{P_1V_1}{T_1} = \frac{P_2V_2}{T_2}$$

$$\frac{\left(\dfrac{759\ \text{Torr}}{760\ \text{Torr}\cdot\text{atm}^{-1}}\right)(1.00\ \text{L})}{253\ \text{K}} = \frac{\left(\dfrac{252\ \text{Torr}}{760\ \text{Torr}\cdot\text{atm}^{-1}}\right)(V_2)}{1523\ \text{K}}$$

$V_2 = 18.1\ \text{L}$

4.29 Because T is constant, we can use

$P_1V_1 = P_2V_2$

$(1.00\ \text{atm})(1.00\ \text{L}) = P_2(0.239\ \text{L})$

$P_2 = 4.18\ \text{atm}$

4.31 $PV = nRT$

$$\left(\frac{24.5\ \text{kPa}}{101.325\ \text{kPa}\cdot\text{atm}^{-1}}\right)(0.2500\ \text{L}) = n(0.082\ 06\ \text{L}\cdot\text{atm}\cdot\text{K}^{-1}\cdot\text{mol}^{-1})(292.7\ \text{K})$$

$n = 2.52 \times 10^{-3}\ \text{mol}$

4.33 (a) $\dfrac{P_1V_1}{T_1} = \dfrac{P_2V_2}{T_2}$

$$\frac{(104 \text{ kPa})(2.0 \text{ m}^3)}{294.3 \text{ K}} = \frac{(52 \text{ kPa})V_2}{268.2 \text{ K}}$$

$V_2 = 3.6 \text{ m}^3$

(b) $\frac{P_1V_1}{T_1} = \frac{P_2V_2}{T_2}$

$$\frac{(104 \text{ kPa})(2.0 \text{ m}^3)}{294.3 \text{ K}} = \frac{(0.880 \text{ kPa})V_2}{221.2 \text{ K}}$$

$V_2 = 1.8 \times 10^2 \text{ m}^3$

4.35 From $P_1V_1 = P_2V_2$ we can calculate the *total pressure* of the final sample:

$(765 \text{ Torr})(555 \text{ mL}) = (P_2)(125 \text{ mL})$

$P_2 = 3.40 \times 10^3 \text{ Torr}$

3.40×10^3 Torr should be the final pressure. The additional pressure needed will be $3.40 \times 10^3 \text{ Torr} - 765 \text{ Torr} = 2.63 \times 10^3 \text{ Torr}$.

4.37 The pressure of the Ar sample will be given by

$$P_{Ar} = \frac{nRT}{V} = \frac{\left(\frac{2.00 \times 10^{-3} \text{ g}}{39.95 \text{ g·mol}^{-1}}\right)(0.082\ 06 \text{ L·atm·K}^{-1}\text{·mol}^{-1})(293 \text{ K})}{0.050\ 0 \text{ L}}$$

$$P_{Kr} = \frac{\left(\frac{2.00 \times 10^{-3} \text{ g}}{83.80 \text{ g·mol}^{-1}}\right)(0.082\ 06 \text{ L·atm·K}^{-1}\text{·mol}^{-1})(\text{T}_2)}{0.050\ 0 \text{ L}}$$

Because we want the pressure to be the same, we can set these two equal to each other. Because volume, mass of the gases, and the gas constant R are the same on both sides of the equation, they will cancel.

$$\left(\frac{1}{83.80 \text{ g·mol}^{-1}}\right)(T_2) = \left(\frac{1}{39.95 \text{ g·mol}^{-1}}\right)(293 \text{ K})$$

Solving for T_2, we obtain temperature = 615 K or 342°C.

4.39 Density is proportional to the molar mass of the gas as seen from the ideal gas law:

$PV = nRT$

$PV = \frac{m}{M}RT$

$$\text{density} = \text{mass per unit volume} = \frac{m}{V} = \frac{MP}{RT}$$

The molar masses of the gases in question are 28.01 g·mol^{-1} for CO(g), 44.01 g·mol^{-1} for CO_2(g), and 34.01 g·mol^{-1} for H_2S(g). The most dense will be the one with the highest molar mass, which in this case is CO_2.

The order of increasing density will be $CO < H_2S < CO_2$.

4.41 (a) Density is proportional to the molar mass of the gas as seen from the ideal gas law. See Section 4.9.

$$d = \frac{(119.37\ \text{g}\cdot\text{mol}^{-1})\left(\dfrac{200\ \text{Torr}}{760\ \text{Torr}\cdot\text{atm}^{-1}}\right)}{(0.080\ 26\ \text{L}\cdot\text{atm}\cdot\text{K}^{-1}\cdot\text{mol}^{-1})(298\ \text{K})} = 1.28\ \text{g}\cdot\text{L}^{-1}$$

(b) $$d = \frac{(119.37\ \text{g}\cdot\text{mol}^{-1})(1.00\ \text{atm})}{(0.080\ 26\ \text{L}\cdot\text{atm}\cdot\text{K}^{-1}\cdot\text{mol}^{-1})(373\ \text{K})} = 3.90\ \text{g}\cdot\text{L}^{-1}$$

4.43 (a) $$M = \frac{dRT}{P} = \frac{(8.0\ \text{g}\cdot\text{L}^{-1})(0.082\ 06\ \text{L}\cdot\text{atm}\cdot\text{K}^{-1}\cdot\text{mol}^{-1})(300\ \text{K})}{2.81\ \text{atm}} = 70\ \text{g}\cdot\text{mol}^{-1}$$

(b) The compound is most likely CHF_3, for which $M = 70\ \text{g}\cdot\text{mol}^{-1}$. It might also be $C_2H_4F_2$, for which $M = 66\ \text{g}\cdot\text{mol}^{-1}$.

(c) You can use the relationship in (a) to calculate the new density, or you can apply the proportionality changes expected from the change in pressure and temperature to the original density:

$$d_2 = (8.0\ \text{g}\cdot\text{L}^{-1})\left(\frac{1.00\ \text{atm}}{2.81\ \text{atm}}\right)\left(\frac{300\ \text{K}}{298\ \text{K}}\right) = 2.9\ \text{g}\cdot\text{L}^{-1}$$

4.45 From the analytical data, an empirical formula of CHCl is calculated. The empirical formula mass is $48.47\ \text{g}\cdot\text{mol}^{-1}$. The problem may be solved using the ideal gas law:

$$PV = nRT$$

$$PV = \frac{m}{M}RT$$

$$M = \frac{mRT}{PV}$$

$$M = \frac{(3.557\ \text{g})(0.082\ 06\ \text{L}\cdot\text{atm}\cdot\text{K}^{-1}\cdot\text{mol}^{-1})(273\ \text{K})}{(1.10\ \text{atm})(0.755\ \text{L})} = 95.9\ \text{g}\cdot\text{mol}^{-1}$$

The value of n in the formula $(CHCl)_n$ is, therefore, equal to $95.9\ \text{g}\cdot\text{mol}^{-1} \div 1.98$. The formula is $C_2H_2Cl_2$.

4.47 Density is proportional to the molar mass of the gas as seen from the ideal gas law:

$$PV = nRT$$

$$PV = \frac{m}{M}RT$$

$$\text{density} = \text{mass per unit volume} = \frac{m}{V} = \frac{MP}{RT}$$

$$1.23\ \text{g}\cdot\text{L}^{-1} = \frac{M\left(\dfrac{25.5\ \text{kPa}}{101.325\ \text{kPa}\cdot\text{atm}^{-1}}\right)}{(0.082\ 06\ \text{L}\cdot\text{atm}\cdot\text{K}^{-1}\cdot\text{mol}^{-1})(330\ \text{K})}$$

$$M = 132\ \text{g}\cdot\text{mol}^{-1}$$

4.49 (a) The number of moles of H_2 needed will be 1.5 times the amount of NH_3 produced, as seen from the balanced equation:

$\frac{1}{2}\ N_2(g) + \frac{3}{2}\ H_2(g) \longrightarrow NH_3(g)$

or

$N_2(g) + 3\ H_2(g) \longrightarrow 2\ NH_3(g)$

Once the number of moles is known, the volume can be obtained from the ideal gas law.

$$V = \frac{n_{H_2}RT}{P} = \frac{(\frac{3}{2}n_{NH_3})RT}{P} = \frac{\left(\left(\dfrac{3\ \text{mol}\ H_2}{2\ \text{mol}\ NH_3}\right)\dfrac{(10^3\ \text{kg})(10^3\ \text{g}\cdot\text{kg}^{-1})}{17.03\ \text{g}\cdot\text{mol}^{-1}}\right)RT}{P}$$

$$= \frac{\left(\dfrac{3}{2}\right)\left(\dfrac{10^6\ \text{g}}{17.03\ \text{g}\cdot\text{mol}^{-1}}\right)(0.082\ 06\ \text{L}\cdot\text{atm}\cdot\text{K}^{-1}\cdot\text{mol}^{-1})(623\ \text{K})}{15.00\ \text{atm}}$$

$$= 3.00 \times 10^5\ \text{L}$$

(b) The ideal gas equation

$$\frac{P_1V_1}{n_1RT_1} = \frac{P_2V_2}{n_2RT_2} \quad \text{simplifies to} \quad \frac{P_1V_1}{T_1} = \frac{P_2V_2}{T_2}$$

because R and n are constant for this problem.

$$\frac{(15.00\ \text{atm})(3.00 \times 10^5\ \text{L})}{623\ \text{K}} = \frac{(376\ \text{atm})V_2}{(523\ \text{K})}$$

$$V_2 = 1.00 \times 10^4\ \text{L}$$

4.51 To answer this, we need to know the number of moles of $CH_4(g)$ present in each case. Because the combustion reaction is the same in both cases, as are the temperature and pressure, the larger number of moles of $CH_4(g)$ should produce the larger volume of $CO_2(g)$. We will use the ideal gas equation to solve for n in the first case:

$$n = \frac{PV}{RT} = \frac{(1.00\ \text{atm})(2.00\ \text{L})}{(0.082\ 06\ \text{L}\cdot\text{atm}\cdot\text{K}^{-1}\cdot\text{mol}^{-1})(348\ \text{K})} = 0.0700\ \text{mol}\ CH_4$$

2.00 g of CH_4 will be $\dfrac{2.00\ \text{g}}{16.04\ \text{g}\cdot\text{mol}^{-1}} = 0.124$ mol

The latter case will have the greater number of moles of CH_4 and should produce the larger amount of $CO_2(g)$.

4.53 The molar mass of glucose is $180.15\ \text{g}\cdot\text{mol}^{-1}$. From this, we can calculate the number of moles of glucose formed and, using the reaction stoichiometry, determine the number of moles of CO_2 needed. With that information and the other information provided in the problem, we can use the ideal gas law to calculate the volume of air that is needed:

$PV = nRT$

$V =$

$$\frac{\left[\left(\dfrac{10.0\ \text{g glucose}}{180.15\ \text{g glucose}\cdot\text{mol}^{-1}\ \text{glucose}}\right)\left(\dfrac{6\ \text{mol CO}_2}{1\ \text{mol glucose}}\right)\right](0.082\ 06\ \text{L}\cdot\text{atm}\cdot\text{K}^{-1}\cdot\text{mol}^{-1})(298\ \text{K})}{\left(\dfrac{0.26\ \text{Torr}}{760\ \text{Torr}\cdot\text{atm}^{-1}}\right)}$$

$= 2.4 \times 10^4\ \text{L}$

4.55 (a) First, we need to determine the number of moles of TiO_2 formed, which can be obtained from the number of moles of $TiCl_4$ that reacted. The moles of $TiCl_4$ can be calculated from the ideal gas law:

$$n = \frac{PV}{RT} = \frac{\left(\dfrac{11\ \text{Torr}}{760\ \text{Torr}\cdot\text{atm}^{-1}}\right)(1.00\ \text{L})}{(0.082\ 06\ \text{L}\cdot\text{atm}\cdot\text{K}^{-1}\cdot\text{mol}^{-1})(298\ \text{K})} = 5.9 \times 10^{-4}\ \text{mol}$$

5.9×10^{-4} mol of $TiCl_4$ will produce 5.9×10^{-4} mol $TiO_2(s)$. The molar mass of $TiO_2(s)$ is $79.88\ \text{g}\cdot\text{mol}^{-1}$.

mass of TiO_2 produced $= 5.9 \times 10^{-4}\ \text{mol} \times 79.88\ \text{g}\cdot\text{mol}^{-1} = 0.047\ \text{g}$

(b) The number of moles of HCl produced will be 4 times the number of moles of $TiCl_4$ reacted, or $4 \times 5.9 \times 10^{-4}\ \text{mol} = 2.36 \times 10^{-3}$ mol. The volume of HCl produced can then be determined from the ideal gas law:

$PV = nRT$

$(1.00\ \text{atm})(V) = (2.36 \times 10^{-3}\ \text{mol})(0.082\ 06\ \text{L}\cdot\text{atm}\cdot\text{K}^{-1}\cdot\text{mol}^{-1})(298\ \text{K})$

$V = 0.058$ L or 58 mL

4.57 (a) This is a limiting reactant problem. Our first task is to determine the number of moles of NH_3 and HCl that are present to start with. This can be done from the ideal gas equation:

$PV = nRT$

$$n_{NH_3} = \frac{PV}{RT} = \frac{\left(\dfrac{100\ \text{Torr}}{760\ \text{Torr}\cdot\text{atm}^{-1}}\right)(0.0150\ \text{L})}{(0.082\ 06\ \text{L}\cdot\text{atm}\cdot\text{K}^{-1}\cdot\text{mol}^{-1})(303\ \text{K})} = 7.94 \times 10^{-5}\ \text{mol}$$

$$n_{HCl} = \frac{PV}{RT} = \frac{\left(\frac{150 \text{ Torr}}{760 \text{ Torr}\cdot\text{atm}^{-1}}\right)(0.0250 \text{ L})}{(0.082\ 06 \text{ L}\cdot\text{atm}\cdot\text{K}^{-1}\cdot\text{mol}^{-1})(298 \text{ K})} = 2.02 \times 10^{-4} \text{ mol}$$

The ammonia is the limiting reactant. The number of moles of $NH_4Cl(s)$ that form will be equal to the number of moles of NH_3 that react. From the molar mass of NH_4Cl (53.49 $g\cdot mol^{-1}$) and the number of moles, we can calculate the mass of NH_4Cl that forms:

$(7.94 \times 10^{-5} \text{ mol } NH_4Cl(s))(53.49 \text{ g}\cdot\text{mol}^{-1}) = 4.25 \times 10^{-3} \text{ g}$

(b) There will be $(2.02 \times 10^{-4} \text{ mol} - 7.94 \times 10^{-5} \text{ mol}) = 1.23 \times 10^{-4}$ mol HCl left after the reaction. This quantity will exist in a total volume after mixing of 40.0 mL or 0.0400 L. Again, we use the ideal gas law to determine the final pressure:

$PV = nRT$

$$P = \frac{(1.23 \times 10^{-4} \text{ mol})(0.082\ 06 \text{ L}\cdot\text{atm}\cdot\text{K}^{-1}\cdot\text{mol}^{-1})(300 \text{ K})}{0.0400 \text{ L}} = 0.0757 \text{ atm}$$

4.59 (a) The molar volume of an ideal gas is 22.4 L at 273.15 K. 1.0 mol of ideal gas will exert a pressure of 1.0 atm under those conditions. The partial pressure of $N_2(g)$ will be 1.0 atm. Because there are 2.0 mol of $H_2(g)$, the partial pressure of $H_2(g)$ will be 2.0 atm. (b) The total pressure will be 1.0 atm + 2.0 atm = 3.0 atm.

4.61 (a) We find the pressure of $SO_2(g)$ originally present by difference. The initial data gives us the total number of moles present, whereas the data for the gas sample after being passed over $CaSO_3(s)$ represents the number of moles of $N_2(g)$.

$PV = nRT$

$$n_{total} = \frac{PV}{RT} = \frac{(1.09 \text{ atm})(0.500 \text{ L})}{(0.082\ 06 \text{ L}\cdot\text{atm}\cdot\text{K}^{-1}\cdot\text{mol}^{-1})(298 \text{ K})} = 0.0223 \text{ mol}$$

$$n_{N_2} = \frac{PV}{RT} = \frac{(1.09 \text{ atm})(0.150 \text{ L})}{(0.082\ 06 \text{ L}\cdot\text{atm}\cdot\text{K}^{-1}\cdot\text{mol}^{-1})(323 \text{ K})} = 0.006\ 17 \text{ mol}$$

The number of moles of SO_2 gas is 0.0223 − 0.006 17 mol = 0.0161 mol. The partial pressure will be given by the mole fraction multiplied by the total pressure. The mole fraction of $SO_2(g)$ will be 0.0161 mol ÷ 0.0223 mol = 0.722. The pressure due to SO_3 in the original mixture is (0.722)(1.09 atm) = 0.787 atm.

(b) The mass of SO_2 will be obtained by multiplying the number of moles of SO_2 by the molar mass of SO_2:

$m_{SO_2} = (0.0161 \text{ mol})(64.06 \text{ g}\cdot\text{mol}^{-1}) = 1.03 \text{ g}$

4.63 (a) Of the 756.7 Torr measured, 17.54 Torr will be due to water vapor. The pressure due to $H_2(g)$ will, therefore, be 756.7 Torr − 17.54 Torr = 739.2 Torr. (b) $H_2O(l) \longrightarrow H_2(g) + \frac{1}{2}O_2(g)$; (c) To answer this question, we must determine the number of moles of H_2 produced in the reaction. Using the partial pressure of H_2 calculated in part (a) and the ideal gas equation, we can set up the following:

$$\left(\frac{739.2\ \text{Torr}}{760\ \text{Torr}\cdot\text{atm}^{-1}}\right)(0.220\ \text{L}) = n\ (0.082\ 06\ \text{L}\cdot\text{atm}\cdot\text{K}^{-1}\cdot\text{mol}^{-1})(293\ \text{K})$$

Solving for n, we obtain n = 0.008 90 mol. According to the stoichiometry of the reaction, half as much oxygen as hydrogen should be produced, so the number of moles of O_2 = 0.004 45 mol. The mass of O_2 will be given by
$(0.004\ 45\ \text{mol})(32.00\ \text{g}\cdot\text{mol}^{-1}) = 0.142\ \text{g}$.

4.65 Graham's law of effusion states that the rate of effusion of a gas is inversely proportional to the square root of its molar mass:

$$\text{rate of effusion} = \frac{1}{\sqrt{M}}$$

If we have two different gases whose rates of effusion are measured under identical conditions, we can take the ratio

$$\frac{\text{rate}_1}{\text{rate}_2} = \frac{\dfrac{1}{\sqrt{M_1}}}{\dfrac{1}{\sqrt{M_2}}} = \sqrt{\frac{M_2}{M_1}}$$

If a compound takes 1.24 times as long to effuse as Kr gas, the rate of effusion of Kr is 1.24 times that of the unknown. We can now use the expression to calculate the molar mass of the unknown, given the mass of Kr:

$$\frac{1.24}{1} = \sqrt{\frac{M_2}{83.80\ \text{g}\cdot\text{mol}^{-1}}}$$

$M_2 = 129\ \text{g}\cdot\text{mol}^{-1}$

A mass of $129\ \text{g}\cdot\text{mol}^{-1}$ corresponds to a molecular formula of $C_{10}H_{10}$.

4.67 The rate of effusion is inversely proportional to the square root of the molar mass. Using a ratio as follows allows us to calculate the time of effusion without knowing the exact conditions of pressure and temperature:

$$\frac{\text{rate}_1}{\text{rate}_2} = \frac{\dfrac{1}{\sqrt{M_1}}}{\dfrac{1}{\sqrt{M_2}}} = \sqrt{\frac{M_2}{M_1}}$$

The rate will be equal to the number of molecules N that effuse in a given time interval. For the conditions given, N will be the same for argon and for the second gas chosen.

$$\frac{\dfrac{N}{\text{time}}}{\dfrac{N}{147\text{ s}}} = \frac{\dfrac{1}{\text{time}}}{\dfrac{1}{147\text{ s}}} = \sqrt{\frac{39.95\text{ g}\cdot\text{mol}^{-1}}{M_1}}$$

In order to calculate the time of effusion, we need to know only the molar mass of the gases.

(a) For CO_2 with a molar mass of $44.01\text{ g}\cdot\text{mol}^{-1}$: $\dfrac{\dfrac{1}{\text{time}_{CO_2}}}{\dfrac{1}{147\text{ s}}} = \sqrt{\dfrac{39.95\text{ g}\cdot\text{mol}^{-1}}{44.01\text{ g}\cdot\text{mol}^{-1}}}$

time = 154 s

(b) For C_2H_4 with a molar mass of $28.05\text{ g}\cdot\text{mol}^{-1}$: $\dfrac{\dfrac{1}{\text{time}_{C_2H_4}}}{\dfrac{1}{147\text{ s}}} = \sqrt{\dfrac{39.95\text{ g}\cdot\text{mol}^{-1}}{28.05\text{ g}\cdot\text{mol}^{-1}}}$

time = 123 s

(c) For H_2 with a molar mass of $2.01\text{ g}\cdot\text{mol}^{-1}$: $\dfrac{\dfrac{1}{\text{time}_{CO_2}}}{\dfrac{1}{147\text{ s}}} = \sqrt{\dfrac{39.95\text{ g}\cdot\text{mol}^{-1}}{2.01\text{ g}\cdot\text{mol}^{-1}}}$

time = 33.0 s

(d) For SO_2 with a molar mass of $64.06\text{ g}\cdot\text{mol}^{-1}$: $\dfrac{\dfrac{1}{\text{time}_{CO_2}}}{\dfrac{1}{147\text{ s}}} = \sqrt{\dfrac{39.95\text{ g}\cdot\text{mol}^{-1}}{64.06\text{ g}\cdot\text{mol}^{-1}}}$

time = 186 s

4.69 The formula mass of C_2H_3 is $27.04\text{ g}\cdot\text{mol}^{-1}$. From the effusion data, we can calculate the molar mass of the sample.

$$\frac{\text{rate}_1}{\text{rate}_2} = \frac{\dfrac{1}{\sqrt{M_1}}}{\dfrac{1}{\sqrt{M_2}}} = \sqrt{\frac{M_2}{M_1}}$$

Because time is inversely proportional to rate, we can write alternatively

$$\frac{\frac{1}{349\text{ s}}}{\frac{1}{210\text{ s}}} = \sqrt{\frac{39.95\text{ g}\cdot\text{mol}^{-1}}{M_1}}$$

$$\frac{210}{349} = \sqrt{\frac{39.95\text{ g}\cdot\text{mol}^{-1}}{M_1}}$$

$$M_1 = 110\text{ g}\cdot\text{mol}^{-1}$$

The molar mass is 4.8 times that of the empirical formula mass, so the molecular formula is C_8H_{12}.

4.71 (a) The average kinetic energy is obtained from the expression: average kinetic energy $= \frac{3}{2}RT$. The value is independent of the nature of the monatomic ideal gas. The numerical values are:

(a) $4103.2\text{ J}\cdot\text{mol}^{-1}$; (b) $4090.7\text{ J}\cdot\text{mol}^{-1}$;

(c) $4103.2\text{ J}\cdot\text{mol}^{-1} - 4090.7\text{ J}\cdot\text{mol}^{-1} = 12.5\text{ J}\cdot\text{mol}^{-1}$

4.73 The root mean square speed is calculated from the following equation:

$$c = \sqrt{\frac{3\,RT}{M}}$$

(a) methane, CH_4, $M = 16.04\text{ g}\cdot\text{mol}^{-1}$

$$c = \sqrt{\frac{3(8.314\text{ kg}\cdot\text{m}^2\cdot\text{s}^{-2}\cdot\text{K}^{-1}\cdot\text{mol}^{-1})(253\text{ K})}{1.604\times10^{-2}\text{ kg}\cdot\text{mol}^{-1}}}$$

$$= 627\text{ m}\cdot\text{s}^{-1}$$

(b) ethane, C_2H_6, $M = 30.07\text{ g}\cdot\text{mol}^{-1}$

$$c = \sqrt{\frac{3(8.314\text{ kg}\cdot\text{m}^2\cdot\text{s}^{-2}\cdot\text{K}^{-1}\cdot\text{mol}^{-1})(253\text{ K})}{3.007\times10^{-2}\text{ kg}\cdot\text{mol}^{-1}}}$$

$$= 458\text{ m}\cdot\text{s}^{-1}$$

(c) propane, C_3H_8, $M = 44.09\text{ g}\cdot\text{mol}^{-1}$

$$c = \sqrt{\frac{3(8.314\text{ kg}\cdot\text{m}^2\cdot\text{s}^{-2}\cdot\text{K}^{-1}\cdot\text{mol}^{-1})(253\text{ K})}{4.409\times10^{-2}\text{ kg}\cdot\text{mol}^{-1}}}$$

$$= 378\text{ m}\cdot\text{s}^{-1}$$

4.75 (a) The most probable speed is the one that corresponds to the maximum on the distribution curve. (b) The percentage of molecules having the most probable speed decreases as the temperature is raised (the distribution spreads out).

4.77 Hydrogen bonding is important in HF. At low temperatures, this hydrogen bonding causes the molecules of HF to be attracted to each other more strongly, thus lowering the pressure. As the temperature is increased, the hydrogen bonds are broken and the pressure rises more quickly than for an ideal gas. Dimers (2 HF molecules bonded to each other) and chains of HF molecules are known to form.

4.79 The pressures are calculated very simply from the ideal gas law:

$$P = \frac{nRT}{V} = \frac{(1.00\ \text{mol})(0.082\ 06\ \text{L}\cdot\text{atm}\cdot\text{K}^{-1}\cdot\text{mol}^{-1})(298\ \text{K})}{V}$$

Calculating for the volumes requested, we obtain $P =$ (a) 1.63 atm; (b) 48.9 atm; (c) 489 atm. The calculations can now be repeated using the van der Waals equation:

$$\left(P + \frac{an^2}{V^2}\right)(V - nb) = nRT$$

We can rearrange this to solve for P:

$$P = \left(\frac{nRT}{V - nb}\right) - \left(\frac{an^2}{V^2}\right)$$

$$= \left(\frac{(1.00\ \text{mol})(0.082\ 06\ \text{L}\cdot\text{atm}\cdot\text{K}^{-1}\cdot\text{mol}^{-1})(298\ \text{K})}{V - (1.00\ \text{mol})(0.04267\ \text{L}\cdot\text{mol}^{-1})}\right) - \left(\frac{(3.640\ \text{L}^2\cdot\text{atm}\cdot\text{mol}^{-2})(1.00^2)}{V^2}\right)$$

Using the three values for V, we calculate for $P =$ (a) 1.62; (b) 38.9; (c) 1.88×10^3 atm. Note that at low pressures, the ideal gas law gives essentially the same values as the van der Waals equation, but at high pressures there is a very significant difference.

4.81 The values for the pressure of gas with varying numbers of moles of CO_2 present are calculated as follows:

The ideal gas law values are calculated from

$$P = \frac{nRT}{V}$$

values for the van der Waals equation can be obtained by rearranging the equation:

$$\left(P + \frac{an^2}{V^2}\right)(V - nb) = nRT$$

$$P = \left(\frac{nRT}{V - nb}\right) - \left(\frac{an^2}{V^2}\right)$$

$$= \left(\frac{(n)(0.082\ 06\ \text{L}\cdot\text{atm}\cdot\text{K}^{-1}\cdot\text{mol}^{-1})(300\ \text{K})}{1.00\ \text{L} - (n)(0.042\ 67\ \text{L}\cdot\text{mol}^{-1})}\right) - \left(\frac{(3.640\ \text{L}^2\cdot\text{atm}\cdot\text{mol}^{-2})(n^2)}{(1.00\ \text{L})^2}\right)$$

The resulting values are

n	P_{ideal}	$P_{\text{van der Waals}}$	% deviation*
0.100	2.46	2.44	0.8
0.200	4.92	4.82	2.1
0.300	7.38	7.15	3.2
0.400	9.85	9.44	4.3
0.500	12.31	11.67	5.5

$$^{*}\%\ \text{deviation} = \frac{P\ \text{ideal} - P\ \text{van der Waals}}{P\ \text{van der Waals}} \times 100$$

(b) Consider one point, for example, the case for $n = 0.400$ mol. The term $V - nb$ will increase the ideal value $\frac{nRT}{V}$ by 1.7% of the ideal value (10.02 atm versus 9.85 atm) whereas the correction from $\frac{an^2}{V^2}$ will decrease the value by 0.58 atm, a change of 5.9% over the ideal gas value. The second effect, which is due to the intermolecular attractions, dominates in this case.
(c) The gas starts to deviate from ideality by more than 5% at pressures above about 10 atm.

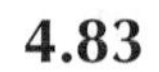

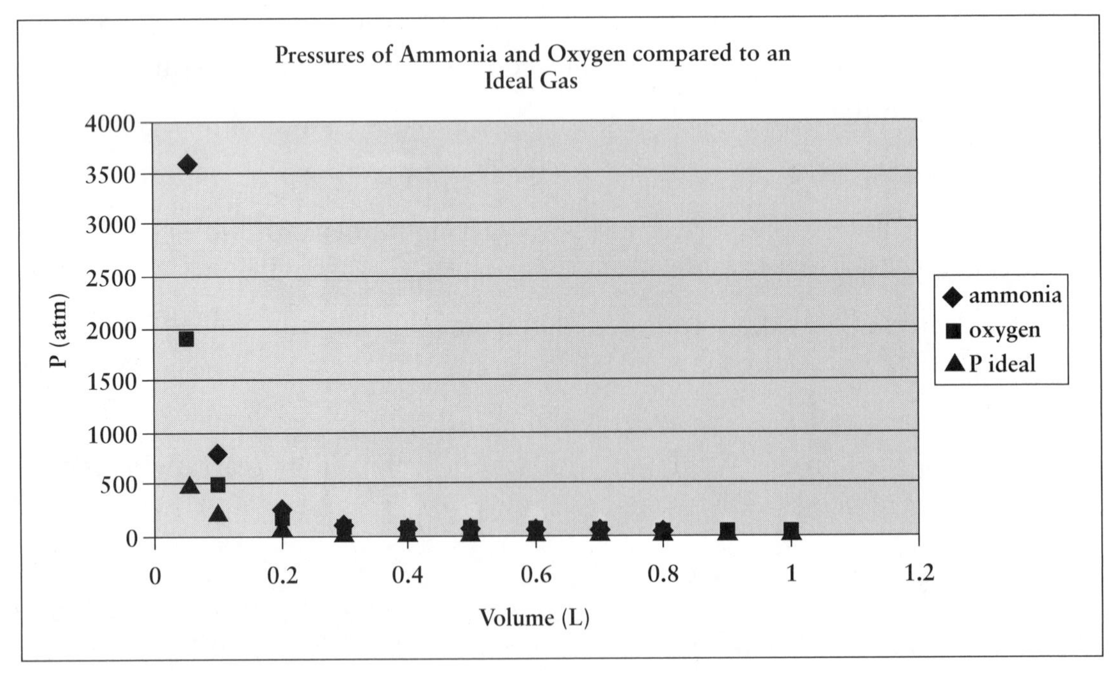

Ammonia: $a = 4.225\ \text{L}^2\cdot\text{atm}\cdot\text{mol}^{-2}$; $b = 0.037\ 07\ \text{L}\cdot\text{atm}^{-1}$
Oxygen: $a = 1.378\ \text{L}^2\cdot\text{atm}\cdot\text{mol}^{-2}$; $b = 0.0318\ 83\ \text{L}\cdot\text{atm}^{-1}$

Volume	P, ammonia	P, oxygen	P, ideal
0.05	3581	1897	489
0.1	811	497	245
0.2	256	180	122
0.3	140	106	82
0.4	94	75	61
0.5	70	58	49
0.6	55	47	41
0.7	46	39	35
0.8	39	34	31
0.9	34	30	27
1	30	27	24

Clearly, the most deviation from the ideal gas law values occurs at low volumes or higher pressures. Ammonia deviates more strongly and its van der Waals constants are larger than those for oxygen. This may likely arise because ammonia is more polar and will have stronger intermolecular interactions.

4.85 (a) From $P_1V_1 = P_2V_2$

$(764\ \text{Torr})(545\ \text{cm}^3) = (P_2)(58.8\ \text{cm}^3)$

$P_2 = 7.08 \times 10^3$ Torr, or 9.32 atm

(b) If the gas mixture is 97.5% by volume air, octane is present at 2.5 volume %. Because the number of moles of a gas is directly proportional to the volume of the gas, the mole fraction of octane will be 2.5/100 = 0.025.

(c) Oxygen makes up 20.95% of the volume of air, so its mole fraction will be given by 20.95/100 = 0.2095.

(d) First, we must write the balanced equation for the combustion of octane:

$C_8H_{18}(g) + 12.5\ O_2(g) \longrightarrow 8\ CO_2(g) + 9\ H_2O(g)$

We have a mole fraction of 0.2095 for O_2 and 0.025 for octane. To combust all of the octane, we would need a ratio of 12.5 mole O_2 per mole octane. The ratio is only 8.38, so there is not enough oxygen gas present to burn all of the octane. Oxygen is, therefore, the limiting reagent. For ease of calculation, we will now assume that we have a total of one mole of substances present. The 0.2095 mol of O_2 that is present will completely burn 0.017 mol of octane. This will leave $0.025 - 0.017 = 0.008$ mol. So after the combustion is over, we will have no oxygen and 0.008 mole octane. There will still be 0.7655 mole of N_2 (and other inert gases). The reaction will have produced $0.017 \times 8 = 0.14$ mol CO_2 and $0.017 \times 9 = 0.15$ mol H_2O. At the end of the reaction there will be

0.7655 mol N_2 (and other inert gases)

0 mol O_2

0.008 mol octane

0.14 mol CO_2

0.15 mol H_2O

Total: 1.06 mol

The mole fractions will be

0.72 N_2

0.008 octane

0.13 CO_2

0.14 H_2O

4.87 (a) $N_2O_4(g) \longrightarrow 2\ NO_2(g)$;

(b) If all the gas were $N_2O_4(g)$, then the moles can be calculated from the ideal gas equation:

$$P = \frac{nRT}{V}$$

$$= \frac{\left(\dfrac{43.78\ g}{92.02\ g\cdot mol^{-1}}\right)(0.082\ 06\ L\cdot atm\cdot K^{-1}\cdot mol^{-1})(298\ K)}{5.00\ L}$$

$$= 2.33\ atm$$

(c) The only difference in the calculation between part (b) and part (c) is that the molar mass of NO_2 is half that of N_2O_4.

$$P = \frac{nRT}{V}$$

$$= \frac{\left(\dfrac{43.78\ g}{46.01\ g\cdot mol^{-1}}\right)(0.082\ 06\ L\cdot atm\cdot K^{-1}\cdot mol^{-1})(298\ K)}{5.00\ L}$$

$$= 4.65\ atm$$

(d) Because both N_2O_4 and NO_2 are present, we need to determine some way of calculating the relative amounts of each present. This can be done by taking advantage of the gas law relationships. The total pressure at the end of the reaction will give us the total number of moles present:

$$P_{total} = 2.96\ atm$$

$$(2.96\ atm)(5.00\ L) = n_{total}(0.082\ 06\ L\cdot atm\cdot K^{-1}\cdot mol^{-1})(298\ K)$$

$$n_{total} = 0.605\ mol$$

$$\therefore n_{N_2O_4} + n_{NO_2} = 0.605\ mol$$

This gives us one equation, but we have two unknowns, so another relationship is needed. We can take advantage of knowing the stoichiometry of the reaction. If we assume that all of the gas begins as N_2O_4 and we allow some to react, we can write the following:

Initial amount of N_2O_4	0.476 mol
Amount of N_2O_4 that reacts	x (mol)
Amount of NO_2 formed	$2x$ (mol)

When the reaction is completed, there will be $0.476 - x$ mole of N_2O_4 and $2x$ mole NO_2. The total number of moles will be given by:

$$(0.476 - x) + 2x = n_{\text{total}}$$

$$0.605 \text{ mol} = 0.476 \text{ mol} + x$$

$$x = 0.129 \text{ mol}$$

$$\begin{aligned} n_{NO_2} &= 2x \\ &= 2(0.129 \text{ mol}) \\ &= 0.258 \text{ mol} \\ n_{N_2O_4} &= 0.476 \text{ mol} - x \\ &= 0.347 \text{ mol} \end{aligned}$$

$$X_{NO_2} = \frac{0.258 \text{ mol}}{0.605 \text{ mol}} = 0.426$$

$$X_{N_2O_4} = \frac{0.347 \text{ mol}}{0.605 \text{ mol}} = 0.574$$

4.89 (a) The elemental analyses yield an empirical formula of NH_2. The formula unit has a mass of 16.02 $\text{g}\cdot\text{mol}^{-1}$. The mass, volume, pressure, and temperature data will allow us to calculate the molar mass, using the ideal gas equation:

$$PV = nRT$$

$$PV = \frac{m}{M}RT$$

$$M = \frac{mRT}{PV} = \frac{(0.473 \text{ g})(0.082\,06 \text{ L}\cdot\text{atm}\cdot\text{K}^{-1}\cdot\text{mol}^{-1})(298 \text{ K})}{(1.81 \text{ atm})(0.200 \text{ L})} = 31.9 \text{ g}\cdot\text{mol}^{-1}$$

The molar mass divided by the mass of the empirical formula mass will give the value of n in the formula $(NH_2)_n$. $31.9 \text{ g}\cdot\text{mol}^{-1} \div 16.02 \text{ g}\cdot\text{mol}^{-1} = 1.99$, so the molecular formula is N_2H_4, which corresponds to the molecule known as hydrazine.

(b)
```
    ..  ..
H — N — N — H
    |   |
    H   H
```

(c) $$\frac{rate_A}{rate_B} = \sqrt{\frac{M_B}{M_A}}$$

$$\frac{\dfrac{3.5 \times 10^{-4}\text{ mol}}{15.0\text{ min}}}{\dfrac{X}{25.0\text{ min}}} = \sqrt{\frac{32.05\text{ g}\cdot\text{mol}^{-1}}{17.03\text{ g}\cdot\text{mol}^{-1}}}$$

$$X = \left(\frac{3.5 \times 10^{-4}\text{ mol}}{15.0\text{ min}}\right)(25.0\text{ min})\sqrt{\frac{17.03\text{ g}\cdot\text{mol}^{-1}}{32.05\text{ g}\cdot\text{mol}^{-1}}}$$

$$= 4.2 \times 10^{-4}\text{ mol}$$

4.91 (a)

```
      ..  ..
      S — S
     /..  ..\
  :S:        :S:
   |          |
  :S:        :S:
     \..  ../
      S — S
      ..  ..
```

(b) bent or angular (each S atom has 4 electron pairs around it)

(c) Yes. Each sulfur atom has eight electrons around it.

(d) The overly high pressure obtained when S_8 is vaporized is an indication that S_8 units have decomposed to a greater number of units, S_6, S_4, S_2, etc. This dissociation increases with temperature. In other words, the number of moles of gas increases upon vaporization.

4.93 The gases react according to the equation:

$CO(g) + Cl_2(g) \longrightarrow COCl_2(g)$

(a) We can write the following relationship based upon the stoichiometry of the reaction:

$P_{\text{final}} = P_{\text{final, CO}} + P_{\text{final, chlorine}} + P_{\text{final, phosgene}}$

By the stoichiometry, we can write

$P_{\text{final, phosgene}} = x$

$P_{\text{final, CO}} = P_{\text{initial, CO}} - x$

$P_{\text{final, chlorine}} = P_{\text{initial, chlorine}} - x$

$P_{\text{final}} = P_{\text{initial, CO}} - x + P_{\text{initial, CO}} - x + x = P_{\text{initial, CO}} + P_{\text{initial, CO}} - x$

The initial pressures, however, must be adjusted to the new temperature:

$$P_{\text{initial, CO, 223°C}} = (3.59\text{ atm})\left(\frac{500\text{ K}}{298\text{ K}}\right) = 6.02\text{ atm}$$

$$P_{\text{initial, Cl}_2\text{, 223°C}} = (2.75\text{ atm})\left(\frac{500\text{ K}}{298\text{ K}}\right) = 4.61\text{ atm}$$

$$P_{\text{final}} = 6.02\text{ atm} + 4.61\text{ atm} - x = 9.75\text{ atm}$$

$x = 0.88$ atm

$P_{\text{final, CO}} = 6.02 \text{ atm} - 0.88 \text{ atm} = 5.14 \text{ atm}$

$P_{\text{final, chlorine}} = 4.61 \text{ atm} - 0.88 \text{ atm} = 3.73 \text{ atm}$

The mole fractions are proportional to the pressure so we can write

$$X_{COCl_2} = \frac{0.88}{9.75} = 0.090$$

$$X_{CO} = \frac{5.14}{9.75} = 0.527$$

$$X_{Cl_2} = \frac{3.73}{9.75} = 0.383$$

(b) The gas density will not change over the course of the reaction because the steel cylinder is a fixed size. The density can be calculated from the relationship

$$d = \frac{m}{V}$$

$$= \frac{PM}{RT}$$

We do not know the mass of samples added, nor the volume of the container, but we can calculate the density from the individual densities of the gases put into the cylinder initially. Because no mass is added or subtracted from the cylinder and its volume does not change, the density will be the same at the end of the reaction as at the beginning.

$$d_{COCl_2} = \frac{PM}{RT}$$

$$= \frac{(3.51 \text{ atm})(28.01 \text{ g}\cdot\text{mol}^{-1})}{(0.082\,06 \text{ L}\cdot\text{atm}\cdot\text{K}^{-1}\cdot\text{mol}^{-1})(298 \text{ K})}$$

$$= 2.45 \text{ g}\cdot\text{L}^{-1}$$

$$d_{Cl_2} = \frac{(2.75 \text{ atm})(70.90 \text{ g}\cdot\text{mol}^{-1})}{(0.082\,06 \text{ L}\cdot\text{atm}\cdot\text{K}^{-1}\cdot\text{mol}^{-1})(298 \text{ K})}$$

$$= 4.75 \text{ g}\cdot\text{L}^{-1}$$

$d_{\text{total}} = 2.45 \text{ g}\cdot\text{L}^{-1} + 4.75 \text{ g}\cdot\text{L}^{-1} = 7.20 \text{ g}\cdot\text{L}^{-1}$

One could do a similar calculation for all three gases at 500 K to obtain the same answer.

4.95 The molar mass calculation follows from the ideal gas law:

$$PV = nRT$$

$$PV = \frac{m}{M}RT$$

$$M = \frac{mRT}{PV} = \frac{(1.509\ \text{g})(0.082\ 06\ \text{L}\cdot\text{atm}\cdot\text{K}^{-1}\cdot\text{mol}^{-1})(473\ \text{K})}{\left(\dfrac{745\ \text{Torr}}{760\ \text{Torr}\cdot\text{atm}^{-1}}\right)(0.235\ \text{L})} = 254\ \text{g}\cdot\text{mol}^{-1}$$

If the molecular formula is OsO_x, then the molar mass will be given by:

$190.2\ \text{g}\cdot\text{mol}^{-1} + x(16.00\ \text{g}\cdot\text{mol}^{-1}) = 254\ \text{g}\cdot\text{mol}^{-1}$

$x = 3.99$

The formula is OsO_4.

4.97 In this problem, the volume, pressure, and molar mass of the substance stay constant. In order to calculate the new mass with the same conditions, we can resort to using the ideal gas equation rearranged to group the constant terms on one side of the equation:

$PV = nRT$

$$PV = \frac{m}{M}RT$$

but M, P, and V are constants, so we can write

$$\frac{MPV}{R} = mT$$

Now we have two sets of conditions, 1 and 2, for which $\dfrac{MPV}{R}$ is constant so we can set them equal:

$m_1T_1 = m_2T_2$

$(32.5\ \text{g})(295\ \text{K}) = (m_2)(485\ \text{K})$; therefore $m_2 = 19.8\ \text{g}$

The mass of gas released must therefore be $32.5\ \text{g} - 19.8\ \text{g} = 12.7\ \text{g}$.

4.99 (a) volume of one atom = molar volume ÷ Avogadro's number

$2.370 \times 10^{-2}\ \text{L}\cdot\text{mol}^{-1} \div 6.022 \times 10^{23}\ \text{atoms}\cdot\text{mol}^{-1-} = 3.936 \times 10^{-26}\ \text{L}\cdot\text{atom}^{-1}$

$3.936 \times 10^{-26}\ \text{L}\cdot\text{atom}^{-1} \times 1000\ \text{cm}^3\cdot\text{L}^{-1} = 3.936 \times 10^{-23}\ \text{cm}^3\cdot\text{atm}^{-1}$

$3.936 \times 10^{-23}\ \text{cm}^3\cdot\text{atm}^{-1} \times (10^{10}\ \text{pm}\cdot\text{cm}^{-1})^3 = 3.936 \times 10^7\ \text{pm}^3$

$$3.936 \times 10^7\ \text{pm}^3 = \frac{4}{3}\pi r^3$$

$$r = 211\ \text{pm}$$

(b) The atomic radius of He is 128 pm (Appendix 2D).

The volume of the He atom, based upon this radius, is

$$V = \frac{4}{3}\pi r^3$$

$$= \frac{4}{3}\pi(128\ \text{pm})^3$$

$$= 8.78 \times 10^6\ \text{pm}^3$$

(c) The difference in these values illustrates that there is no easy definition for the boundaries of an atom. The van der Waals value obtained from the correction for molar volume is considerably larger than the atomic radius, owing perhaps to longer range and weak interactions between atoms. One should also bear in mind that the value for the van der Waals b is a parameter used to obtain a good fit to a curve, and its interpretation is more complicated than a simple molar volume.

CHAPTER 5
LIQUIDS AND SOLIDS

5.1 (a) London forces, dipole-dipole; (b) London forces, dipole-dipole; (c) London forces, dipole-dipole, hydrogen bonding; (d) London forces

5.3 Only (b) CH_3Cl, (c) CH_2Cl_2, and (d) $CHCl_3$ will have dipole-dipole interactions. The molecules CH_4 and CCl_4 do not have dipole moments.

5.5 The interaction energies can be ordered based on the relationship the energy has to the distance separating the interacting species. Thus ion-ion interactions are the strongest and are directly proportional to the distance separating the two interacting species. Ion-dipole energies are inversely proportional to d^2, whereas dipole-dipole for constrained molecules (i.e., solid state) is inversely proportional to d^3. Dipole-dipole interactions where the molecules are free to rotate become comparable to induced dipole-induced dipole interactions, which are both inversely related to d^6. The order thus derived is: (b) dipole-induced dipole $\cong$ (c) dipole-dipole in the gas phase $<$ (e) dipole-dipole in the solid phase $<$ (a) ion-dipole $<$ (d) ion-ion.

5.7 Only molecules with H attached to the electronegative atoms F, N, and O can hydrogen bond. Additionally, there must be lone pairs available for the H's to bond to. This is true only of (c) H_2SO_3.

5.9 (a) NaCl (801°C vs. −114.8°C) because it is an ionic compound as opposed to a molecular compound; (b) butanol (−90°C vs. −116°C) due to hydrogen bonding in butanol that is not possible in diethyl ether; (c) HF because it exhibits strong hydrogen bonding, whereas HCl does not; (d) H_2O because the number of possible hydrogen bonds is greater than the number in methanol.

5.11 (a) PF_3 and PCl_3 are both trigonal pyramidal and should have similar intermolecular forces, but PCl_3 has the greater number of electrons and should have the higher boiling point. The boiling point of PF_3 is −101.5°C and that of PCl_3 is 75.5°C. (b) SO_2 is bent and has a dipole moment whereas CO_2 is linear and will be nonpolar. SO_2 should have the higher boiling point. SO_2 boils at −10°C, whereas CO_2

sublimes at −78°C. (c) BF_3 and BCl_3 are both trigonal planar, so the choice of higher boiling point depends on the difference in total number of electrons. BCl_3 should have the higher boiling point (12.5°C vs. −99.9°C). (d) AsF_3 is pyramidal and has a dipole moment, whereas AsF_5 is a trigonal bipyramid and is nonpolar. AsF_5, on the other hand, has more electrons, so the two effects oppose each other. The actual boiling points of AsF_3 and AsF_5 are 63°C and −53°C, respectively, so the effect of the polarity is greater than the effect of the increased mass, in this case.

5.13 The ionic radius of Al^{3+} is 53 pm and that of Be^{2+} is 27 pm. The ratio of energies will be given by

$$V \propto \frac{-|z|\mu}{d^2}$$

$$V_{Al^{3+}} \propto \frac{-|z|\mu}{d^2} = \frac{-|3|\mu}{(53)^2}$$

$$V_{Be^{2+}} \propto \frac{-|z|\mu}{d^2} = \frac{-|2|\mu}{(27)^2}$$

The electric dipole moment of the water molecule (μ) will cancel:

$$\text{ratio}\left(\frac{V_{Al^{3+}}}{_{Be^{2+}}}\right) = \frac{-|3|\mu/(53)^2}{-|2|\mu/(27)^2} = \frac{3(27)^2}{2(53)^2} = 0.39$$

The attraction of the Be^{2+} ion will be greater than that of the Al^{3+} ion. Even though the Be^{2+} ion has a lower charge, its radius is much smaller than that of Al^{3+}, making the attraction greater.

5.15 The ionic radius of Al^{3+} is 53 pm and that of Ga^{3+} is 62 pm. The ratio of energies will be given by

$$V \propto \frac{-|z|\mu}{d^2}$$

$$V_{Al^{3+}} \propto \frac{-|3|\mu}{(53\text{ pm})^2}$$

$$V_{Ga^{3+}} \propto \frac{-|3|\mu}{(62\text{ pm})^2}$$

The electric dipole moment of water (μ) will cancel:

$$\text{ratio}\left(\frac{V_{Al^{3+}}}{V_{Ga^{3+}}}\right) = \frac{\dfrac{-|3|\mu}{(53\text{ pm})^2}}{\dfrac{-|3|\mu}{(62\text{ pm})^2}} = \frac{(62\text{ pm})^2}{(53\text{ pm})^2} = 1.4$$

The water molecule will be more strongly attracted to the Al^{3+} ion because of its smaller radius.

5.17 (a) Xenon is larger, with more electrons, giving rise to larger London forces that increase the melting point. (b) Hydrogen bonding in water causes the molecules to be held together more tightly than in diethyl ether. (c) Both molecules have the same molar mass, but pentane is a linear molecule compared to dimethylpropane, which is a compact, spherical molecule. The compactness of the dimethyl propane gives it a lower surface area. That means that the intermolecular attractive forces, which are of the same type (London forces) for both molecules, will have a larger effect for pentane.

5.19 (a) *cis*-Dichloroethene is polar, whereas *trans*-dichloroethene, whose individual bond dipole moments cancel, is nonpolar. Therefore, *cis*-dichloroethene has the greater intermolecular forces and the greater surface tension. (b) Surface tension of liquids decreases with increasing temperature as a result of thermal motion as temperature rises. Increased thermal motion allows the molecules to more easily break away from each other, which manifests itself as decreased surface tension.

5.21 At 50°C all three compounds are liquids. C_6H_6 (nonpolar) < C_6H_5SH (polar, but no hydrogen bonding) < C_6H_5OH (polar and with hydrogen bonding). The viscosity will show the same ordering as the boiling points, which are 80°C for C_6H_6, 169°C for C_6H_5SH, 182° for C_6H_5OH.

5.23 Using $h = \dfrac{2\gamma}{gdr}$ we can calculate the height. For water:

$$r = \frac{1}{2}\,\textit{diameter} = \frac{1}{2}(0.15\text{ mm})\left(\frac{1\text{ m}}{1000\text{ mm}}\right) = 7.5 \times 10^{-5}\text{ m}$$

$$d = 0.997\text{ g}\cdot\text{cm}^{-3}\left(\frac{1\text{ kg}}{1000\text{ g}}\right)\left(\frac{10^6\text{ cm}^3}{\text{m}^3}\right) = 9.97 \times 10^2\text{ kg}\cdot\text{m}^{-3}$$

$$h = \frac{2(72.75 \times 10^{-3}\text{ N}\cdot\text{m}^{-1})}{(9.81\text{ m}\cdot\text{s}^{-1})(9.97 \times 10^2\text{ kg}\cdot\text{m}^{-3})(7.5 \times 10^{-5}\text{ m})} = 0.20\text{ m or }200\text{ mm}$$

Remember that $1\text{ N} = 1\text{ kg}\cdot\text{m}^{-1}\cdot\text{s}^{-2}$

For ethanol:

$$d = 0.79\text{ g}\cdot\text{cm}^{-3}\left(\frac{1\text{ kg}}{1000\text{ g}}\right)\left(\frac{10^6\text{ cm}^3}{\text{m}^3}\right) = 7.9 \times 10^2\text{ kg}\cdot\text{m}^{-3}$$

$$h = \frac{2(22.8 \times 10^{-3}\text{ N}\cdot\text{m}^{-1})}{(9.81\text{ m}\cdot\text{s}^{-1})(7.9 \times 10^2\text{ kg}\cdot\text{m}^{-3})(7.5 \times 10^{-5}\text{ m})} = 0.078\text{ m or }78\text{ mm}$$

Water will rise to a higher level than ethanol. There are two opposing effects to consider. While the greater density of water, as compared to ethanol, acts against it rising as high, it has a much higher surface tension.

5.25 (a) At center: 1 center × 1 atom·center^{-1} = 1 atom; at 8 corners, 8 corners × $\frac{1}{8}$ atom·corner^{-1} = 1 atom; total = 2 atoms; (b) There are eight nearest neighbors, hence a coordination number of 8; (c) The direction along which atoms touch each other is the body diagonal of the unit cell. This body diagonal will be composed of four times the radius of the atom. In terms of the unit cell edge length a, the body diagonal will be $\sqrt{3}a$. The unit cell edge length will, therefore, be given by

$$4r = \sqrt{3}a \text{ or } a = \frac{4r}{\sqrt{3}} = \frac{4\cdot(235 \text{ pm})}{\sqrt{3}} = 543 \text{ pm}$$

5.27 (a) a = length of side for a unit cell; for an fcc unit cell, $a = \sqrt{8}\,r$ or $2\sqrt{2}\,r$ = 393 pm. $V = a^3 = (393 \text{ pm} \times 10^{-12} \text{ m}\cdot\text{pm}^{-1})^3 = 6.07 \times 10^{-29} \text{ m}^3 = 6.07 \times 10^{-23} \text{ cm}^3$. Because for a fcc unit cell there are 4 atoms per unit cell, we have

$$\text{mass(g)} = 4 \text{ Pt atoms} \times \frac{1 \text{ mol Pt atoms}}{6.022 \times 10^{23} \text{ atoms}\cdot\text{mol}^{-1}} \times \frac{195.09 \text{ g}}{\text{mol Pt atoms}}$$

$$= 1.30 \times 10^{-21} \text{ g}$$

$$d = \frac{1.30 \times 10^{-21} \text{ g}}{6.07 \times 10^{-23} \text{ cm}^3} = 21.4 \text{ g}\cdot\text{cm}^{-3}$$

(b) $$a = \frac{4r}{\sqrt{3}} = \frac{4 \times 272 \text{ pm}}{\sqrt{3}} = 628 \text{ pm}$$

$$V = (628 \times 10^{-12} \text{ m})^3 = 2.48 \times 10^{-28} \text{ m}^3 = 2.48 \times 10^{-22} \text{ cm}^3$$

There are 2 atoms per bcc unit cell:

$$\text{mass(g)} = 2 \text{ Cs atoms} \times \frac{1 \text{ mol Cs atoms}}{6.022 \times 10^{23} \text{ atoms}\cdot\text{mol}^{-1}} \times \frac{132.91 \text{ g}}{\text{mol Cs atoms}}$$

$$= 4.41 \times 10^{-22} \text{ g}$$

$$d = \frac{4.41 \times 10^{-22} \text{ g}}{2.48 \times 10^{-22} \text{ cm}^3} = 1.78 \text{ g}\cdot\text{cm}^{-3}$$

5.29 a = length of unit cell edge

$$V = \frac{\text{mass of unit cell}}{d}$$

(a) $$V = a^3 = \frac{(1 \text{ unit cell})\left(\frac{195.09 \text{ g Pt}}{\text{mol Pt}}\right)\left(\frac{1 \text{ mol Pt}}{6.022 \times 10^{23} \text{ atoms Pt}}\right)\left(\frac{4 \text{ atoms}}{1 \text{ unit cell}}\right)}{21.450 \text{ g}\cdot\text{cm}^3}$$

$a = 3.92 \times 10^{-8}$ cm

Because for an fcc cell, $a = \sqrt{8}\, r$, $r = \dfrac{\sqrt{2}\, a}{4} = \dfrac{\sqrt{2}\,(3.92 \times 10^{-8}\ \text{cm})}{4}$

$= 1.39 \times 10^{-8}\ \text{cm} = 139\ \text{pm}$

(b) $V = a^3$

$$= \frac{(1\ \text{unit cell})\left(\dfrac{180.95\ \text{g Ta}}{1\ \text{mol Ta}}\right)\left(\dfrac{1\ \text{mol Ta}}{6.022 \times 10^{23}\ \text{atoms Ta}}\right)\left(\dfrac{2\ \text{atoms}}{1\ \text{unit cell}}\right)}{16.654\ \text{g}\cdot\text{cm}^{3}}$$

$= 3.61 \times 10^{-23}\ \text{cm}^3$

$a = 3.30 \times 10^{-8}$ cm

$$r = \frac{\sqrt{3}\, a}{4} = \frac{\sqrt{3}(3.30 \times 10^{-8}\ \text{cm})}{4} = 1.43 \times 10^{-8}\ \text{cm} = 143\ \text{pm}$$

5.31 (a) There are three types of cubic unit cells that are possible choices. These include the simple cubic cell, the body-centered cubic cell, and the face-centered cubic cell. They differ in that the simple cubic cell has one atom per unit cell, the body-centered has a total of two atoms per unit cell, and the fcc cell has four atoms. The mass of an Al atom will be

$26.98\ \text{g}\cdot\text{mol}^{-1} \div (6.022 \times 10^{23})\ \text{atoms}\cdot\text{mol}^{-1} = 4.480 \times 10^{-23}\ \text{g}\cdot\text{atom}^{-1}$

The volume of the unit cell will equal

$$V_{\text{unit cell}} = \left(404\ \text{pm} \times \frac{10^{-12}\ \text{m}}{\text{pm}} \times \frac{100\ \text{cm}}{\text{m}}\right)^3 = 6.594 \times 10^{-23}\ \text{cm}^3\cdot\text{unit cell}^{-1}$$

The density will be given by

$$d = \frac{\text{mass in unit cell}}{\text{volume of unit cell}} = \frac{Z\cdot(4.480 \times 10^{-23}\ \text{g}\cdot\text{unit cell}^{-1})}{6.594 \times 10^{-23}\ \text{cm}^3\cdot\text{unit cell}^{-1}} = Z\cdot 0.6794\ \text{g}\cdot\text{cm}^{3}$$

where Z is the number of atoms in the unit cell.

Because the observed density is $2.70\ \text{g}\cdot\text{cm}^3$, we can calculate Z

$Z\cdot 0.6794\ \text{g}\cdot\text{cm}^3 = 2.70\ \text{g}\cdot\text{cm}^3$

$$Z = \frac{2.70\ \text{g}\cdot\text{cm}^3}{0.6794\ \text{g}\cdot\text{cm}^3} = 3.97$$

Because Z is 4, the lattice type must be fcc.

(b) For a metal packing in fcc unit cell (cubic close-packing), the coordination number will be 12.

5.33 (a) To answer this question, we will need to have the dimensions in comparable units. The edge length of the unit cell will be

$$a = 562.8\ \text{pm} \times \frac{1\ \text{m}}{10^{12}\ \text{pm}} = 5.628 \times 10^{-10}\ \text{m}$$

$V_{\text{unit cell}} = a^3 = (5.628 \times 10^{-10}\ \text{m})^3 = 1.783 = 10^{-28}\ \text{m}^3\cdot\text{unit cell}^{-1}$

The volume of the crystal will be given by

$$V_{\text{crystal}} = \left(1.00\ \text{mm} \times \frac{1\ \text{m}}{10^3\ \text{mm}}\right)^3 = 1.00 \times 10^{-9}\ \text{m}^3$$

The number of units cells in the crystal will then be given by $V_{\text{crystal}} \div V_{\text{unit cell}}$:

$$\#\ \text{of unit cells in crystal} = \frac{1.00 \times 10^{-9}\ \text{m}^3\cdot\text{crystal}^{-1}}{1.783 \times 10^{-28}\ \text{m}^3\cdot\text{unit cell}^{-1}}$$

$$= 5.61 \times 10^{18}\ \text{unit cells}\cdot\text{crystal}^{-1}$$

(b) Because the unit cell is fcc, there will be four formula units of NaCl per unit cell, so the number of NaCl units will be given by $4(5.61 \times 10^{18}\ \text{unit cells}) = 2.24 \times 10^{19}$ formula units. To get the number of moles of NaCl, we divide this number by Avogadro's number: $(2.24 \times 10^{19}) \div (6.022 \times 10^{23}) = 3.72 \times 10^{-5}$ mol.

5.35 (a) There are eight chloride ions at the eight corners, giving a total of

$8\ \text{corners} \times \frac{1}{8}\ \text{atom}\cdot\text{corner}^{-1} = 1\ \text{Cl}^-\ \text{ion}$

There is one Cs^+ that lies at the center of the unit cell. All of this ion belongs to the unit cell. The ratio is thus 1:1 for an empirical formula of CsCl, with one formula unit per unit cell.

(b) The titanium atoms lie at the corners of the unit cell and at the body center:

$8\ \text{corners} \times \frac{1}{8}\ \text{atom}\cdot\text{corner}^{-1} + 1\ \text{at body center} = 2\ \text{atoms per unit cell}$

Four oxygen atoms lie on the faces of the unit cell and two lie completely within the unit cell, giving:

$4\ \text{atoms in faces} \times \frac{1}{2}\ \text{atom}\cdot\text{face}^{-1} + 2\ \text{atoms wholly within cell} = 4\ \text{atoms}$

The ratio is thus two Ti per four O, or an empirical formula of TiO_2 with two formula units per unit cell (c). The Ti atoms are 6-coordinate and the O atoms are 3-coordinate.

5.37 Y: $8\ \text{atoms} \times \frac{1}{8}\ \text{atom}\cdot\text{corner}^{-1} = 1\ \text{Y atom}$

Ba: $8\ \text{atoms} \times \frac{1}{4}\ \text{atom}\cdot\text{edge}^{-1} = 2\ \text{Ba atoms}$

Cu: 3 Cu atoms completely inside unit cell = 3 Cu atoms

O: $10\ \text{atoms on faces} \times \frac{1}{2}\ \text{atom}\cdot\text{face}^{-1} + 2\ \text{atoms completely inside unit cell}$ $= 7$ O atoms

Formula = $YBa_2Cu_3O_7$

5.39 (a) ratio $= \dfrac{149\ \text{pm}}{133\ \text{pm}} = 1.12$, predict cesium-chloride structure with (8,8) coordination; however, rubidium fluoride actually adopts the rock-salt structure

(b) ratio $= \dfrac{72\text{ pm}}{140\text{ pm}} = 0.51$, predict rock-salt structure with (6,6) coordination

(c) ratio $= \dfrac{102\text{ pm}}{196\text{ pm}} = 0.520$, predict rock-salt structure with (6,6) coordination

5.41 (a) In the rock-salt structure, the unit cell edge length is equal to two times the radius of the cation plus two times the radius of the anion. Thus for CaO, a = 2(100 pm) + 2(140 pm) = 480 pm. The volume of the unit cell will be given by (converting to cm^3 because density is normally given in terms of $g{\cdot}cm^{-3}$)

$$V = \left(480\text{ pm} \times \frac{10^{-12}\text{ m}}{\text{pm}} \times \frac{100\text{ cm}}{\text{m}}\right)^3 = 1.11 \times 10^{-22}\text{ cm}^3$$

There are four formula units in the unit cell, so the mass in the unit cell will be given by

$$\text{mass in unit cell} = \frac{\left(\frac{4\text{ formula units}}{1\text{ unit cell}}\right) \times \left(\frac{56.08\text{ g CaO}}{1\text{ mol CaO}}\right)}{6.022 \times 10^{23}\text{ molecules}\cdot\text{mol}^{-1}} = 3.725 \times 10^{-22}\text{ g}$$

The density will be given by the mass in the unit cell divided by the volume of the unit cell:

$$d = \frac{3.725 \times 10^{-22}\text{ g}}{1.11 \times 10^{-22}\text{ cm}^3} = 3.37\text{ g}\cdot\text{cm}^{-3}$$

(b) For a cesium chloride-like structure, it is the body diagonal that represents two times the radius of the cation and plus two times the radius of the anion. Thus the body diagonal for CsBr is equal to 2(170 pm) + 2(196) = 732 pm.
For a cubic cell, the body diagonal $= \sqrt{3}\,a = 732$ pm

$$a = 423\text{ pm}$$

$$a^3 = V = \left(423\text{ pm} \times \frac{10^{-12}\text{ m}}{\text{pm}} \times \frac{100\text{ cm}}{\text{m}}\right)^3 = 7.57 \times 10^{-23}\text{ cm}^3$$

There is one formula unit of CsBr in the unit cell, so the mass in the unit cell will be given by

$$\text{mass in unit cell} = \frac{\left(\frac{1\text{ formula units}}{1\text{ unit cell}}\right)\left(\frac{212.82\text{ g CsBr}}{1\text{ mol CsBr}}\right)}{6.022 \times 10^{23}\text{ molecules}\cdot\text{mol}^{-1}} = 3.534 \times 10^{-22}\text{ g}$$

$$d = \frac{3.534 \times 10^{-22}\text{ g}}{7.57 \times 10^{-23}\text{ cm}^3} = 4.67\text{ g}\cdot\text{cm}^3$$

5.43 Glucose will be held in the solid by London forces, dipole-dipole interactions, and hydrogen bonds; benzophenone will be held in the solid by dipole-dipole interactions and London forces; methane will be held together by London forces only.

London forces are strongest in benzophenone, but glucose can experience hydrogen bonding, which is a strong interaction and dominates intermolecular forces. Methane has few electrons so experiences only weak London forces. We would expect the melting points to increase in the order CH_4 (m.p. = −182°C) < benzophenone (m.p. = 48°C) < glucose (m.p. = 148 − 155°C).

5.45 one form of boron nitride, silicon dioxide, plus many others

5.47 (a) The alloy is undoubtedly interstitial, because the atomic radius of nitrogen is much smaller (74 pm vs. 124 pm) than that of iron. The rule of thumb is that the solute atom be less than 60% the solvent atom in radius, in order for an interstitial alloy to form. That criterion is met here. (b) We expect that nitriding will make iron harder and stronger, with a lower electrical conductivity.

5.49 Graphite is a metallic conductor parallel to the planes; the electrons are quite free to move within them. Between planes, however, there is an energy barrier to conduction, though this barrier can be partially overcome by raising the temperature. Thus, graphite is a semiconductor perpendicular to the planes and a conductor parallel to the planes.

5.51 (a) These problems are most easily solved by assuming 100 g of substance. In 100 g of Ni-Cu alloy there will be 25 g Ni and 75 g Cu, corresponding to 0.43 mol Ni and 1.2 mol Cu. The atom ratio will be the same as the mole ratio:

$$\frac{1.2 \text{ mol Cu}}{0.43 \text{ mol Ni}} = 2.8 \text{ Cu per Ni}$$

(b) Pewter, which is 7% Sb, 3% Cu, and 90% Sn, will contain 7 g Sb, 3 g Cu, and 90 g Sn per 100 g of alloy. This will correspond to 0.06 mol Sb, 0.05 mol Cu, and 0.76 mol Sn per 100 g. The atom ratio will be 15 Sn : 1.2 Sb : 1 Cu.

5.53 There are too many ways that these molecules can rotate and twist so that they do not remain rod-like. The molecular backbone when the molecule is stretched out is rod-like, but the molecules tend to curl up on themselves, destroying any possibility of long range order with neighboring molecules. This is partly due to the fact that the molecules have only single bonds that allow rotation about the bonds, so that each molecule can adopt many configurations. If multiple bonds are present, the bonding is more rigid.

5.55 Use of a nonpolar solvent such as hexane or benzene (etc.) in place of water should give rise to the formation of inverse micelles.

5.57 (a) Pentane and 2,2-dimethylpropane are isomers; both have the chemical formula C_5H_{12}. We will assume that 2,2-dimethylpropane is roughly spherical and that all the hydrogen atoms lie on this sphere. The surface area of a sphere is given by $A = 4\pi r^2$. For this particular sphere, $A = 4\pi(254\text{ pm})^2 = 8.11 \times 10^5\text{ pm}^2$.

For pentane, the surface area of the rectangular prism is $2(295\text{ pm} \times 766\text{ pm}) + 2(295\text{ pm} \times 254\text{ pm}) + 2(254\text{ pm} \times 766\text{ pm}) = 9.91 \times 10^5\text{ pm}^2$.

(b) The pentane should have the higher boiling point. It has a significantly larger surface area and should have stronger intermolecular forces between the molecules.

5.59 (a) anthracene, $C_{14}H_{10}$

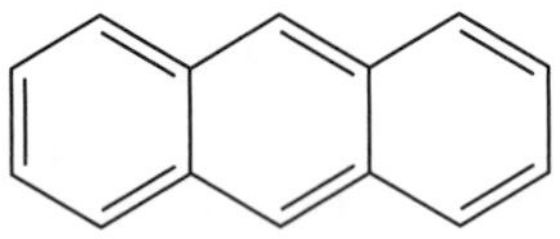

London forces

(b) phosgene, $COCl_2$

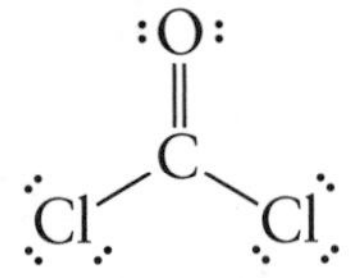

:O:
‖
C
Cl Cl

Dipole-dipole forces, London forces

(c) glutamic acid, $C_5H_9NO_4$

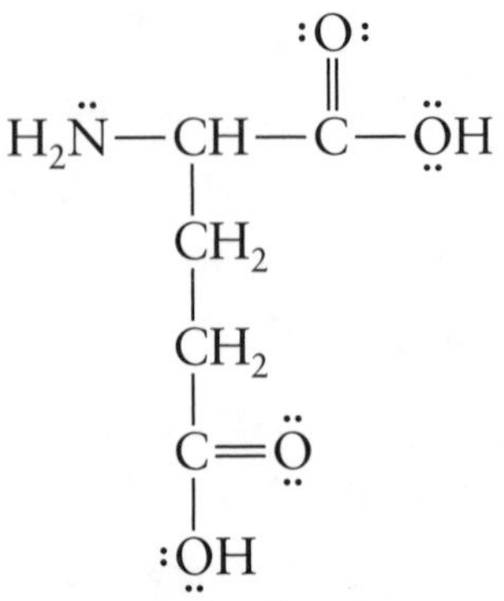

:O:
‖
H₂N—CH—C—OH
|
CH₂
|
CH₂
|
C=O
|
:OH

Hydrogen bonding, dipole-dipole forces, London forces

5.61 The unit cell for a cubic close-packed lattice is the fcc unit cell. For this cell, the relation between the radius of the atom r and the unit cell edge length a is

$$4r = \sqrt{2}\,a$$

$$a = \frac{4r}{\sqrt{2}}$$

The volume of the unit cell is given by

$$V = a^3 = \left(\frac{4r}{\sqrt{2}}\right)^3$$

If r is given in pm, then a conversion factor to cm is required:

$$V = a^3 = \left(\frac{4r}{\sqrt{2}} \times \frac{10^{-12}\ \text{m}}{\text{pm}} \times \frac{100\ \text{cm}}{\text{m}}\right)^3$$

Because there are four atoms per fcc unit cell, the mass in the unit cell will be given by

$$\text{mass} = \left(\frac{4\ \text{atoms}}{\text{unit cell}}\right)\left(\frac{M}{6.022 \times 10^{23}\ \text{atoms}\cdot\text{mol}^{-1}}\right)$$

The density will be given by

$$d = \frac{\text{mass of unit cell}}{\text{volume of unit cell}} = \frac{\left(\frac{4\ \text{atoms}}{\text{unit cell}}\right) \times \left(\frac{M}{6.022 \times 10^{23}\ \text{atoms}\cdot\text{mol}^{-1}}\right)}{\left(\frac{4r}{\sqrt{2}} \times \frac{10^{-12}\ \text{m}}{\text{pm}} \times \frac{100\ \text{cm}}{\text{m}}\right)^3}$$

$$= \frac{(2.936 \times 10^5)M}{r^3}$$

or

$$r = \sqrt[3]{\frac{(2.936 \times 10^5)M}{d}}$$

where M is the atomic mass in $\text{g}\cdot\text{mol}^{-1}$ and r is the radius in pm.

For the different gases we calculate the results given in the following table:

Gas	**Density ($\text{g}\cdot\text{cm}^3$)**	**Molar mass ($\text{g}\cdot\text{mol}^{-1}$)**	**Radius (pm)**
Neon	1.20	20.18	170
Argon	1.40	39.95	203
Krypton	2.16	83.80	225
Xenon	2.83	131.30	239
Radon	4.4	222	246

5.63 There are two approaches to this problem. The information given does not specify the radius of the tungsten atom. This value can be looked up in Appendix 2D. We can calculate an answer, however, based simply upon the fact that the density is $19.3\ \text{g}\cdot\text{cm}^{-3}$ for the bcc cell, by taking the ratio between the expected densities, based upon the assumption that the atomic radius of tungsten will be the same for both. The unit cell for a cubic close-packed lattice is the fcc unit cell. For this cell, the relation between the radius of the atom r and the unit cell edge length a is

$$4\,r = \sqrt{2}\,a$$

$$a = \frac{4\,r}{\sqrt{2}}$$

The volume of the unit cell is given by

$$V = a^3 = \left(\frac{4\,r}{\sqrt{2}}\right)^3$$

If r is given in pm, then a conversion factor to cm is required:

$$V = a^3 = \left(\frac{4\,r}{\sqrt{2}} \times \frac{10^{-12}\ \text{m}}{\text{pm}} \times \frac{100\ \text{cm}}{\text{m}}\right)^3$$

Because there are four atoms per fcc unit cell, the mass in the unit cell is given by

$$\text{mass} = \left(\frac{4\ \text{atoms}}{\text{unit cell}}\right)\left(\frac{M}{6.022 \times 10^{23}\ \text{atoms}\cdot\text{mol}^{-1}}\right)$$

The density is given by

$$d = \frac{\text{mass of unit cell}}{\text{volume of unit cell}} = \frac{\left(\frac{4\ \text{atoms}}{\text{unit cell}}\right)\left(\frac{M}{6.022 \times 10^{23}\ \text{atoms}\cdot\text{mol}^{-1}}\right)}{\left(\frac{4\,r}{\sqrt{2}} \times \frac{10^{-12}\ \text{m}}{\text{pm}} \times \frac{100\ \text{cm}}{\text{m}}\right)^3}$$

$$= \frac{(2.936 \times 10^5)M}{r^3}$$

or

$$r = \sqrt[3]{\frac{(2.936 \times 10^5)M}{d}}$$

where M is the atomic mass in $\text{g}\cdot\text{mol}^{-1}$ and r is the radius in pm.

Likewise, for a body-centered cubic lattice there will be two atoms per unit cell. For this cell, the relationship between the radius of the atom r and the unit cell edge length a is derived from the body diagonal of the cell, which is equal to 4 times the radius of the atom. The body diagonal is found from the Pythagorean theorem to be equal to the $\sqrt{3}\,a$.

$$4\,r = \sqrt{3}\,a$$

$$a = \frac{4\,r}{\sqrt{3}}$$

The volume of the unit cell is given by

$$V = a^3 = \left(\frac{4\,r}{\sqrt{3}}\right)^3$$

If r is given in pm, then a conversion factor to cm is required:

$$V = a^3 = \left(\frac{4\,r}{\sqrt{3}} \times \frac{10^{-12}\text{ m}}{\text{pm}} \times \frac{100\text{ cm}}{\text{m}}\right)^3$$

Because there are two atoms per bcc unit cell, the mass in the unit cell will be given by

$$\text{mass} = \left(\frac{2\text{ atoms}}{\text{unit cell}}\right) \times \left(\frac{M}{6.022 \times 10^{23}\text{ atoms}\cdot\text{mol}^{-1}}\right)$$

The density will be given by

$$d = \frac{\text{mass of unit cell}}{\text{volume of unit cell}} = \frac{\left(\frac{2\text{ atoms}}{\text{unit cell}}\right)\left(\frac{M}{6.022 \times 10^{23}\text{ atoms}\cdot\text{mol}^{-1}}\right)}{\left(\frac{4\,r}{\sqrt{3}} \times \frac{10^{-12}\text{ m}}{\text{pm}} \times \frac{100\text{ cm}}{\text{m}}\right)^3}$$

$$= \frac{\left(\frac{2\text{ atoms}}{\text{unit cell}}\right)\left(\frac{M}{6.022 \times 10^{23}\text{ atoms}\cdot\text{mol}^{-1}}\right)}{(2.309 \times 10^{-10}\,r)^3}$$

$$= \frac{(2.698 \times 10^5)M}{r^3}$$

or

$$r = \sqrt[3]{\frac{(2.698 \times 10^5)M}{d}}$$

Setting these two expressions equal and cubing both sides, we obtain

$$\frac{(2.936 \times 10^5)M}{d_{\text{fcc}}} = \frac{(2.698 \times 10^5)M}{d_{\text{bcc}}}$$

The molar mass of tungsten M is the same for both ratios and will cancel from the equation.

$$\frac{(2.936 \times 10^5)}{d_{\text{fcc}}} = \frac{(2.698 \times 10^5)}{d_{\text{bcc}}}$$

Rearranging, we get

$$d_{\text{fcc}} = \frac{(2.936 \times 10^5)}{(2.698 \times 10^5)}\,d_{\text{bcc}}$$

$$= 1.088\,d_{\text{bcc}}$$

For W, $d_{\text{fcc}} = 1.088 \times 19.6\text{ g}\cdot\text{cm}^3$

$$= 21.3\text{ g}\cdot\text{cm}^{-3}$$

5.65 (a) The oxidation state of the titanium atoms must balance the charge on the oxide ions, O^{2-}. The presence of 1.18 O^{2-} ions means that the Ti present must have a charge to compensate the -2.36 charge on the oxide ions. The average oxidation state of Ti is thus $+2.36$. (b) This is most easily solved by setting up a set of two equations in two unknowns. We know that the total charge on the titanium atoms present must equal 2.36, so if we multiply the charge on each type of titanium by the fraction of titanium present in that oxidation state and sum the values, we should get 2.36:

let x = fraction of Ti^{2+}, y = fraction of Ti^{3+}, then

$2x + 3y = 2.36$

Also, because we are assuming all the titanium is either +2 or +3, the fractions of each present must add up to 1:

$x + y = 1$

Solving these two equations simultaneously, we obtain y = 0.36, x = 0.64.

5.67 (a) true. If this is not the case, the unit cell will not match with other unit cells of the same type when stacked to form the entire lattice.

(b) false. Unit cells do not have to have atoms at the corners.

(c) true. In order for the unit cell to repeat properly, opposite faces must have the same composition.

(d) false. If one face is centered, the opposing face must be centered, but the other faces do not necessarily have to be centered.

5.69 The density is calculated as for cubic structures, but the volume = a^2c for a tetragonal unit cell.

$V = a^2c = (459\ \text{pm})^2(296\ \text{pm}) = 6.24 \times 10^7\ \text{pm}^3$ or $6.24 \times 10^{-23}\ \text{cm}^3$

Density is then calculated as before:

density =

$$\frac{(2\ \text{TiO}_2\ \text{formula units}\cdot\text{unit cell}^{-1})(79.88\ \text{g}\cdot\text{mol}^{-1}\ \text{TiO}_2)\left(\dfrac{1\ \text{mol}}{6.022 \times 10^{23}\ \text{formula units}}\right)}{6.24 \times 10^{-23}\ \text{cm}^3}$$

$= 4.25\ \text{g}\cdot\text{cm}^{-3}$

5.71 There are several ways to draw unit cells that will repeat to generate the entire lattice. Some examples are shown below. The choice of unit cell is determined by conventions that are beyond the scope of this text (the smallest unit cell that indicates all of the symmetry present in the lattice is typically the one of choice).

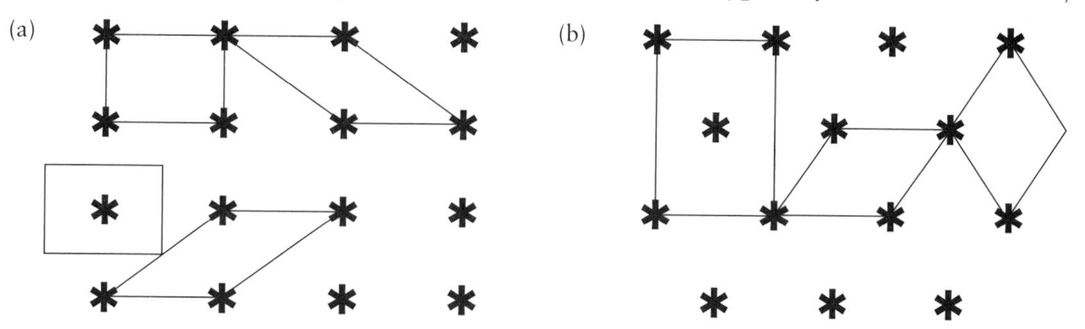

5.73 In a salt such as sodium chloride or sodium bromide, all the interactions will be ionic. In sodium acetate or sodium methoxide, however, the organic anions CH_3COO^- and CH_3O^- will have the negative charges localized largely on the oxygen atoms, which are the most electronegative atoms. The carbon part of the molecule will interact with other parts of the crystal lattice via London forces rather than ionic interactions. In general, solid organic salts will not be as hard as purely ionic substances and will also have lower melting points.

5.75 Fused silica, also known as fused quartz, is predominantly SiO_2 with very few impurities. This glass is the most refractory, which means it can be used at the highest temperatures. Quartz vessels are routinely used for reactions that must be carried out at temperatures up to 1000°C. Vycor, which is 96% SiO_2, can be used at up to ca. 900°C, and normal borosilicate glasses at up to approximately 550°C. This is due to the fact that the glasses melt at lower temperatures, as the amount of materials other than SiO_2 increases. The borosilicate glasses (Pyrex or Kimax) are commonly used because the lower softening points allow them to be more easily molded and shaped into different types of glass objects, such as reaction flasks, beakers, and other types of laboratory and technical glassware.

5.77 For M, there is a total of one atom in the unit cell from the corners;
cations: 8 corners $\times \frac{1}{8}$ atom$\cdot$corner^{-1} + 6 faces + $\frac{1}{2}$ atom$\cdot$face^{-1} = 4 atoms
anions: 8 tetrahedral holes $\times$ 1 atom$\cdot$tetrahedral hole^{-1} = 8 atoms
The cation to anion ratio is thus 4 : 8 or 1 : 2; the empirical formula is MA_2.

5.79 The cesium chloride lattice is a simple cubic lattice of Cl^- ions with a Cs^+ at the center of the unit cell (See 5.43). In the unit cell, there is a total of one Cl^- ion and one Cs^+ ion. If the density is 3.988 g·cm^{-3}, then we can determine the volume and unit cell edge length. The molar mass of CsCl is 168.36 g·mol^{-1}.

$$3.988\ \text{g}\cdot\text{cm}^{-3} = \frac{\left(\dfrac{168.36\ \text{g}\cdot\text{mol}^{-1}}{6.022 \times 10^{23}\ \text{formula units}\cdot\text{mol}^{-1}}\right)}{a^3}$$

$$a^3 = \frac{\left(\dfrac{168.36\ \text{g}\cdot\text{mol}^{-1}}{6.022 \times 10^{23}\ \text{formula units}\cdot\text{mol}^{-1}}\right)}{3.988\ \text{g}\cdot\text{cm}^{-3}}$$

$$a = 4.12 \times 10^{-8}\ \text{cm}$$
$$= 412\ \text{pm}$$

The volume of the unit cell is $(412\ \text{pm})^3 = 6.99 \times 10^7\ \text{pm}^3$.

We will determine the size of the Cs^+ and Cl^- ions from ionic radii given in Appendix 2D, but we can check these values against the unit cell dimensions. For this type of unit cell, the body diagonal will be equal to $2\,r(Cs^+) + 2\,r(Cl^-) = a\sqrt{3} =$ 714 pm. The sum of the ionic radii gives us 2 (170 pm) + 2 (181 pm) = 702 pm, which is in very good agreement. Note that we cannot calculate the size of these ions independently from the unit cell data without more information, because this lattice is not close-packed.

We will assume that the ions are spherical. The volume occupied in the unit cell will be

$$V_{Cs^+} = \frac{4}{3}\pi r^3 = \frac{4}{3}\pi(170\ \text{pm})^3 = 2.06 \times 10^7\ \text{pm}^3$$

$$V_{Cl^-} = \frac{4}{3}\pi r^3 = \frac{4}{3}\pi(181\ \text{pm})^3 = 2.48 \times 10^7\ \text{pm}^3$$

The total occupied volume in the cell is $2.06 \times 10^7\ \text{pm}^3 + 2.48 \times 10^7\ \text{pm}^3 = 4.54 \times 10^7\ \text{pm}^3$. The empty space is $6.99 \times 10^7\ \text{pm}^3 - 4.54 \times 10^7\ \text{pm}^3 = 2.45 \times 10^7\ \text{pm}^3$. The percent empty space is $2.45 \times 10^7\ \text{pm}^3 \div 6.99 \times 10^7\ \text{pm}^3 \times 100 =$ 35%.

5.81 (a) Because the face-centered cubic lattice is a close-packed lattice, we expect it to be the more dense, because its packing efficiency is greater than that of the body-centered cubic lattice. The more dense phase should be more stable under pressure. We can substantiate this by calculating the densities of the two phases. Face-center cubic phase:

The volume of the unit cell will be

$$a^3 = (529.5\ \text{pm})^3 = 1.485 \times 10^8\ \text{pm}^3 \text{ or } 1.485 \times 10^{-22}\ \text{cm}^3$$

The mass in the unit cell will be given by

$138.91\ \text{g}\cdot\text{mol}^{-1} \div 6.022 \times 10^{23}\ \text{atoms}\cdot\text{mol}^{-1} = 2.307 \times 10^{-22}\ \text{g}\cdot\text{atom}^{-1}$

$$d = \frac{(2.307 \times 10^{-22}\ \text{g}\cdot\text{atom}^{-1})(4\ \text{atoms}\cdot\text{unit cell}^{-1})}{1.485 \times 10^{-22}\ \text{cm}^3\cdot\text{unit cell}^{-1}} = 6.214\ \text{g}\cdot\text{cm}^3$$

Body-centered cell:

The volume of the unit cell will be

$a^3 = (426\ \text{pm})^3 = 7.73 \times 10^7\ \text{pm}^3$ or $7.73 \times 10^{-23}\ \text{cm}^3$

$$d = \frac{(2.307 \times 10^{-22}\ \text{g}\cdot\text{atom}^{-1})(2\ \text{atoms}\cdot\text{unit cell}^{-1})}{7.73 \times 10^{-23}\ \text{cm}^3\cdot\text{unit cell}^{-1}} = 5.97\ \text{g}\cdot\text{cm}^3$$

The face-centered cubic form is about 4% more dense than the body-centered cubic form.

(b) The density of the hexagonal phase is $6.145\ \text{g}\cdot\text{cm}^{-3}$, whereas that of the face-centered cubic phase is $6.214\ \text{g}\cdot\text{cm}^{-3}$. Because both are theoretically close-packed, the densities of the two phases should be the same, but experimentally there is a slightly higher density for the cubic close-packed form.

5.83 (a) The unit cell information shows that calcium fluoride exists in a fcc unit cell, with the Ca^{2+} ions occupying the corners and face centers of the cube, and the fluoride ions found in all of the tetrahedral holes produced in the lattice of Ca^{2+} ions. This arrangement will contain four formula units of CaF_2 in the unit cell. From this information and the density, we can calculate the unit cell edge length a:

a = length of unit cell edge

$$V = \frac{\text{mass of unit cell}}{\text{density}}$$

$$V = a^3 = \frac{(1\ \text{unit cell})\left(\frac{4\ \text{atoms}}{1\ \text{unit cell}}\right)\left(\frac{78.08\ \text{g CaF}_2}{\text{mol CaF}_2}\right)\left(\frac{1\ \text{mol CaF}_2}{6.022 \times 10^{23}\ \text{molecules CaF}_2}\right)}{3.180\ \text{g}\cdot\text{cm}^3}$$

$a = 5.46 \times 10^{-8}$ cm or 546 pm

(b) There are several approaches that one can take to analyze the geometry in order to calculate the Ca-F distance. Two will be presented here. By looking at the structure, one can see that the fluoride ion sits in the center of a regular tetrahedron of Ca^{2+} ions. The distance between the Ca^{2+} ions is given by half the length of the face diagonal of the unit cell. The face diagonal of the fcc unit cell is equal to $a\sqrt{2} = 772$ pm. The Ca^{2+}-Ca^{2+} separation is, therefore, 386 pm. Because the F^- sits at the center of a regular tetrahedron, the Ca-F-Ca angle will be 109.47°. From these two pieces of information we can construct the triangle shown below.

A right triangle can be constructed by dropping a line from the F^- ion to the midpoint of the line connecting the two Ca^{2+} ions.

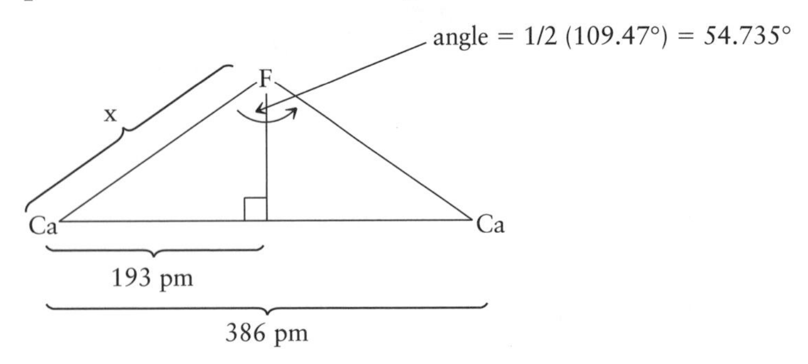

The Ca-F distance (labeled x) can be calculated using simple trigonometry:

$$\sin 54.735° = \frac{193 \text{ pm}}{x}$$

$$x = 236 \text{ pm}$$

An alternative approach is to realize that the Ca-F separation is one-fourth the body diagonal. Because the body diagonal is $\sqrt{3}\,a$, the Ca-F separation will be 236 pm.

(c) This is in good agreement with the sum of the ionic radii found in the Appendix, which gives $r(Ca^{2+}) + r(F^-) = 100 \text{ pm} + 133 \text{ pm} = 233 \text{ pm}$.

5.85 In order to work this problem, we must first realize that the unit cell dimension does not correspond to the direction of closest packing. The close-packed layers in a fcc cell will lie perpendicular to the body diagonal, which will be given by $\sqrt{3}\,a = \sqrt{3}\,(408 \text{ pm}) = 707 \text{ pm}$. In examining the unit cell along the body diagonal, it is apparent that there are four layers of close-packing that lie perpendicular to the body diagonal within one unit cell; the separation between these layers will be $(707 \text{ pm}) \div 3$. Each layer will add 236 pm. In a deposit that is 0.125 mm thick, we will then expect to find

$$\text{number of layers} = \frac{0.125 \text{ mm} \times \dfrac{1 \text{ m}}{10^3 \text{ mm}}}{236 \text{ pm} \times \dfrac{1 \text{ m}}{10^{12} \text{ pm}}} = 5.30 \times 10^5 \text{ layers}$$

5.87 The lattice layers from which constructive x-ray diffraction occurs are parallel. First draw perpendicular lines from the point of intersection of the top x-ray with the lattice plane to the lower x-ray for both the incident and diffracted rays. The x-rays are in phase and parallel at point A. If we want them to be still in phase and parallel when they exit the crystal, then they must still be in phase when they reach point C. In order for this to be true, the extra distance that the second beam

travels with respect to the first must be equal to some integral number of wavelengths. The total extra distance traveled, A $\rightarrow$ B $\rightarrow$ C, is equal to $2x$.
From the diagram, we can see that the angle A-D-B must also be equal to θ. The angles θ and α sum to 90°, as do the angles α and A-D-B. We can then write

$$\sin\theta = \frac{x}{d} \quad \text{and} \quad x = d\sin\theta$$

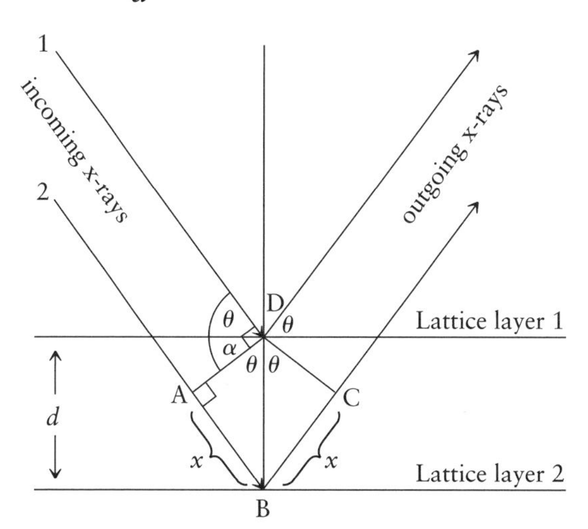

The total distance traveled is $2x = 2d\sin\theta$. So, for the two x-rays to be in phase as they exit the crystal, $2d\sin\theta$ must be equal to an integral number of wavelengths.

5.89 The answer to this problem is obtained from Bragg's law: $\lambda = 2d\sin\theta$ where λ is the wavelength of radiation, d is the interplanar spacing, and θ is the angle of incidence of the x-ray beam. Here $d = 401.8$ pm and $\lambda = 71.037$ pm.
$71.07 \text{ pm} = 2(401.8 \text{ pm})\sin\theta$
$\theta = 5.074°$

5.91 (a) Because the bisulfate anions occupy spaces between the planes of the graphite layers, they will separate the layers more than they are separated in pure graphite. The conduction between layers will be reduced. The conduction within the layers, however, is enhanced because the partial oxidation produces some empty spaces in the delocalized molecular orbitals that extend over the graphite sheet. The presence of these empty spaces makes it easier for the electrons to migrate through the solid along the planar carbon networks.
(b) The layers of carbon atoms in graphite will constitute a series of lattice planes. In graphite the distance between these layers is 335 pm and they give rise to a

characteristic diffraction of x-rays according to the Bragg relation $n\lambda = 2d \sin \theta$, or $d = \frac{n\lambda}{2 \sin \theta}$. From this we can see that the spacing d is inversely proportional to $\sin \theta$. So, when we intercalate atoms or molecules between the graphite layers, we force the layers apart. In the bisulfate, the maximum distance is about 800 pm. The spacing d for this set of lattice planes will become larger, and consequently the angle θ will become smaller. To illustrate this, we can calculate the values for θ for the normal graphite and the graphite bisulfate. We will use Cu Kα radiation (λ = 154 pm) and look at the first order diffraction pattern ($n = 1$).

For graphite:

$154 \text{ pm} = 2\,(335 \text{ pm}) \sin \theta$

$\theta = 13.3°$

For graphite bisulfate:

$154 \text{ pm} = 2\,(800 \text{ pm}) \sin \theta$

$\theta = 5.52°$

The result of this is that the diffracted x-rays get closer together as the lattice spacing increases.

CHAPTER 6
THERMODYNAMICS: THE FIRST LAW

6.1 (a) isolated; (b) closed; (c) isolated; (d) open; (e) closed; (f) open

6.3 The change in internal energy ΔU is given simply by summing the two energy terms involved in this process. We must be careful, however, that the signs on the energy changes are appropriate. In this case, internal energy will be added to the gas sample both by heating and by compression, so both are positive numbers.
$\Delta U = 437\ \text{kJ} + 295\ \text{kJ} = 732\ \text{kJ}$

6.5 (a) and (b) $w = \Delta U - q = 982\ \text{J} - 492\ \text{J} = +490\ \text{J}$. Work was done on the system, as indicated by the positive value.

6.7 To get the entire internal energy change, we must sum the changes due to heat and work. In this problem, $q = +5500\ \text{kJ}$. Work will be given by $w = -P_{ext}\Delta V$ because it is an expansion against a constant opposing pressure:

$$w = -\left(\frac{750\ \text{Torr}}{760\ \text{Torr}\cdot\text{atm}^{-1}}\right)\left(\frac{1846\ \text{mL} - 345\ \text{mL}}{1000\ \text{mL}\cdot\text{L}^{-1}}\right) = -1.48\ \text{L}\cdot\text{atm}$$

To convert to J we use the equivalency of the ideal gas constants:

$$w = -(1.48\ \text{L}\cdot\text{atm})\left(\frac{8.314\ \text{J}\cdot\text{K}^{-1}\cdot\text{mol}^{-1}}{0.082\,06\ \text{L}\cdot\text{atm}\cdot\text{K}^{-1}\cdot\text{mol}^{-1}}\right) = -150\ \text{J}$$

$\Delta U = q + w = 5500\ \text{kJ} - 0.150\ \text{kJ} = 5500\ \text{kJ}$

The energy change due to the work term turns out to be negligible in this problem.

6.9 Using $\Delta U = q + w$ where $\Delta U = -2573\ \text{kJ}$ and
$q = -947\ \text{kJ} - 2573\ \text{kJ} = -947\ \text{kJ} + w$
$w = -1626\ \text{kJ}$
1626 kJ of work can be done by the system on its surroundings.

6.11 From $\Delta H = \Delta U + P\Delta V$ at constant pressure, or $\Delta U = \Delta H - P\Delta V$. Because $w = -P\Delta V = +22\ \text{kJ}$, we get $-15\ \text{kJ} + 22\ \text{kJ} = \Delta U = +7\ \text{kJ}$.

6.13 (a) The irreversible work of expansion against a constant opposing pressure is given by

$$w = -P_{ex}\Delta V$$

$$w = -(1.00 \text{ atm})(6.52 \text{ L} - 4.29 \text{ L})$$
$$= -2.23 \text{ L·atm}$$
$$= -2.23 \text{ L·atm} \times 101.325 \text{ J·L}^{-1}\text{·atm}^{-1} = -226 \text{ J}$$

(b) An isothermal expansion will be given by

$$w = -nRT\frac{V_2}{V_1}$$

n is calculated from the ideal gas law:

$$n = \frac{PV}{RT} = \frac{(1.79 \text{ atm})(4.29 \text{ L})}{(0.082\,06 \text{ L·atm·K}^{-1}\text{·mol}^{-1})(305 \text{ K})} = 0.307 \text{ mol}$$

$$w = -(0.307 \text{ mol})(8.314 \text{ J·K}^{-1}\text{·mol}^{-1})(305 \text{ K}) \ln\frac{6.52}{4.29}$$
$$= -326 \text{ J}$$

Note that the work done is greater when the process is carried out reversibly.

6.15 (a) $q = (150.0 \text{ g})(16.6°\text{C} - 50.0°\text{C})(2.42 \text{ J·(°C)}^{-1}\text{·g}^{-1}) = -12.1 \text{ kJ}$
(12.1 kJ of heat removed)
(b) $1.00 \text{ L} \times 0.794 \text{ g·cm}^{-3} \times (78°\text{C} - 25°\text{C}) \times 2.42 \text{ J·(°C)}^{-1}\text{·g}^{-1} = 102 \text{ kJ}$

6.17 (a) $0.38 \text{ J·(°C)}^{-1}\text{·g}^{-1} \times 63.45 \text{ g·mol}^{-1} = 24 \text{ J·(°C)}^{-1}\text{·mol}^{-1}$
(b) $q = (\text{mass})(205°\text{C} - 15°\text{C})(0.38 \text{ J·(°C)}^{-1}\text{·g}^{-1}) = 425 \times 10^3 \text{ J}$
mass = 5.9 kg

6.19 (a) The heat change will be made up of two terms: one term to raise the temperature of the copper and the other to raise the temperature of the water:

$$q = (450.0 \text{ g})(4.18 \text{ J·(°C)}^{-1}\text{·g}^{-1})(100.0°\text{C} - 25.0°\text{C})$$
$$+ (500.0 \text{ g})(0.38 \text{ J·(°C)}^{-1}\text{·g}^{-1})(100°\text{C} - 25°\text{C})$$
$$= 141 \text{ kJ} + 14 \text{ kJ} = 155 \text{ kJ}$$

(b) The percentage of heat attributable to raising the temperature of water will be

$$\left(\frac{141 \text{ kJ}}{155 \text{ kJ}}\right)(100) = 91.1\%$$

6.21 heat lost by metal = −heat gained by water

$$(20.0 \text{ g})(T_{\text{final}} - 100.0°\text{C})(0.38 \text{ J·(°C)}^{-1}\text{·g}^{-1})$$
$$= -(50.7 \text{ g})(4.18 \text{ J·(°C)}^{-1}\text{·g}^{-1})(T_{\text{final}} - 22.0°\text{C})$$
$$(T_{\text{final}} - 100.0°\text{C})(7.6 \text{ J·(°C)}^{-1}) = -(212 \text{ J·(°C)}^{-1})(T_{\text{final}} - 22.0°\text{C})$$

$$T_{final} - 100.0°C = -28(T_{final} - 22.0°C)$$
$$T_{final} + 28\ T_{final} = 100.0°C + 616°C$$
$$29\ T_{final} = 716°C$$
$$T_{final} = 25°C$$

6.23 $C_{cal} = \dfrac{22.5\ kJ}{23.97°C - 22.45°C} = 14.8\ kJ\cdot(°C)^{-1}$

6.25 (a) The molar heat capacity of a monatomic ideal gas at constant volume is given by $C_{V,m} = \frac{3}{2}R = 12.5\ J\cdot mol^{-1}\cdot K^{-1}$, which is derived from the kinetic molecular theory of gases.

$\Delta U = (0.325\ mol)\ (12.5\ J\cdot mol^{-1}\cdot(°C)^{-1})\ (50°C - (-25°C)) = 305\ J$

(b) The molar heat capacity of a monatomic ideal gas at constant pressure is given by $C_{P,m} = \frac{5}{2}R = 20.8\ J\cdot mol^{-1}\cdot K^{-1}$

$\Delta H = (0.325\ mol)(20.8\ J\cdot mol^{-1}\cdot(°C)^{-1})\ (50°C - (-25°C)) = 507\ J$

Note that the heat capacities can be expressed in terms of °C as well as K because a Celsius degree is the same size as a kelvin degree.

(c) At constant pressure, the gas is expanding and does work on its surroundings. Some of the extra heat thus spills over into the surroundings in the form of expansion work.

6.27 NO_2. The heat capacity increases with molecular complexity—as more atoms are present in the molecule, there are more possible bond vibrations that can absorb added energy.

6.29 (a) The molar heat capacity of a monatomic ideal gas at constant pressure is $C_{P,m} = \frac{5}{2}R$. The heat released will be given by

$$q = \left(\frac{1.078\ g}{83.80\ g\cdot mol^{-1}}\right)(25.0°C - 97.6°C)(20.8\ J\cdot mol^{-1}\cdot(°C)^{-1}) = -19.4\ J$$

(b) Similarly, the molar heat capacity of a monatomic ideal gas at constant volume is $C_{V,m} = \frac{3}{2}R$. The heat released will be given by

$$q = \left(\frac{1.078\ g}{83.80\ g\cdot mol^{-1}}\right)(25.0°C - 97.6°C)(12.5\ J\cdot mol^{-1}\cdot(°C)^{-1}) = -11.7\ J$$

6.31 (a) HCN is a linear molecule. The contribution from molecular motions will be $5/2\ R$.

(b) C_2H_6 is a polyatomic, nonlinear molecule. The contribution from molecular motions will be $3R$.

(c) Ar is a monoatomic ideal gas. The contribution from molecular motions to the heat capacity will be 3/2 R.

(d) HBr is a polyatomic, linear molecule. The contribution from molecular motions will be 5/2 R.

6.33 (a) $\Delta H_{vap} = \dfrac{1.93\ \text{kJ}}{0.235\ \text{mol}} = 8.21\ \text{kJ}\cdot\text{mol}^{-1}$

(b) $\Delta H_{vap} = \dfrac{21.2\ \text{kJ}}{\left(\dfrac{22.45\ \text{g}}{46.07\ \text{g}\cdot\text{mol}^{-1}}\right)} = 43.5\ \text{kJ}\cdot\text{mol}^{-1}$

6.35 (a) $\Delta H = \left(\dfrac{100.0\ \text{g H}_2\text{O}}{18.02\ \text{g}\cdot\text{mol}^{-1}\ \text{H}_2\text{O}}\right)(40.7\ \text{kJ}\cdot\text{mol}^{-1}) = 226\ \text{kJ}$

(b) $\Delta H = \left(\dfrac{612\ \text{g NH}_3}{17.03\ \text{g}\cdot\text{mol}^{-1}\ \text{NH}_3}\right)(5.65\ \text{kJ}\cdot\text{mol}^{-1}) = 203\ \text{kJ}$

6.37 This process is composed of two steps: melting the ice at 0°C and then raising the temperature of the liquid water from 0°C to 25°C:

Step 1: $\Delta H = \left(\dfrac{50.0\ \text{g}}{18.02\ \text{g}\cdot\text{mol}^{-1}}\right)(6.01\ \text{kJ}\cdot\text{mol}^{-1}) = 16.7\ \text{kJ}$

Step 2: $\Delta H = (50.0\ \text{g})(4.18\ \text{J}\cdot(°\text{C})^{-1}\cdot\text{g}^{-1})(25°\text{C} - 0°\text{C}) = 5.2\ \text{kJ}$

Total heat required $= 16.7\ \text{kJ} + 5.2\ \text{kJ} = 21.9\ \text{kJ}$

6.39 The heat gained by the water in the ice cube will be equal to the heat lost by the initial sample of hot water. The enthalpy change for the water in the ice cube will be composed of two terms: the heat to melt the ice at 0°C and the heat required to raise the ice from 0°C to the final temperature.

$$\begin{aligned}\text{heat (ice cube)} &= \left(\frac{50.0\ \text{g}}{18.02\ \text{g}\cdot\text{mol}^{-1}}\right)(6.01 \times 10^3\ \text{J}\cdot\text{mol}^{-1}) \\ &\quad + (50.0\ \text{g})(4.184\ \text{J}\cdot(°\text{C})^{-1}\cdot\text{g}^{-1})(T_f - 0°) \\ &= 1.67 \times 10^4\ \text{J} + (209\ \text{J}\cdot(°\text{C})^{-1})(T_f - 0°) \\ \text{heat (water)} &= (400\ \text{g})(4.184\ \text{J}\cdot(°\text{C})^{-1}\cdot\text{g}^{-1})(T_f - 45°) \\ &= (1.67 \times 10^3\ \text{J}\cdot(°\text{C})^{-1})(T_f - 45°)\end{aligned}$$

Setting these equal:

$-(1.67 \times 10^3\ \text{J}\cdot(°\text{C})^{-1})T_f + 7.5 \times 10^4\ \text{J} = 1.67 \times 10^4\ \text{J} + (209\ \text{J}\cdot(°\text{C})^{-1})T_f$

Solving for T_f:

$$T_f = \frac{5.8 \times 10^4\ \text{J}}{1.88 \times 10^3\ \text{J}\cdot(°\text{C})^{-1}} = 31\ °\text{C}$$

6.41 (a) $\Delta H = (1.25\text{ mol})(+358.8\text{ kJ}\cdot\text{mol}^{-1}) = 448\text{ kJ}$

(b) $\Delta H = \left(\dfrac{197\text{ g C}}{12.01\text{ g}\cdot\text{mol}^{-1}\text{ C}}\right)\left(\dfrac{358.8\text{ kJ}}{4\text{ mol C}}\right) = 1.47 \times 10^2\text{ kJ}$

(c) $\Delta H = 415\text{ kJ} = (n_{CS_2})\left(\dfrac{358.8\text{ kJ}\cdot\text{mol}^{-1})}{4\text{ mol }CS_2}\right)$

$n_{CS_2} = 4.63\text{ mol }CS_2$ or $(4.63\text{ mol})(76.13\text{ g}\cdot\text{mol}^{-1}) = 352\text{ g }CS_2$

6.43 (a) $(12\text{ ft} \times 12\text{ ft} \times 8\text{ ft})\left(\dfrac{30.48\text{ cm}}{1\text{ ft}}\right)^3 = 3.26 \times 10^7\text{ cm}^3$

The heat capacity of air is $1.01\text{ J}\cdot(°\text{C})^{-1}\cdot\text{mol}^{-1}$ and the average molar mass of air is $28.97\text{ g}\cdot\text{mol}^{-1}$ (see Table 4.1). The density of air can be calculated from the ideal gas law:

$$d = \frac{P}{MRT} = \frac{1.00\text{ atm}}{(28.97\text{ g}\cdot\text{mol}^{-1})(0.082\ 06\text{ L}\cdot\text{atm}\cdot\text{K}^{-1}\cdot\text{mol}^{-1})(277.6\text{ K})}$$

$$d = 0.001\ 52\text{ g}\cdot\text{cm}^{-3}$$

$40°\text{F} = 4.4°\text{C}$, $78°\text{F} = 26°\text{C}$

$\Delta T = 26°\text{C} - 4.4°\text{C} = 22°$

The heat required is

$(3.26 \times 10^7\text{ cm}^3)(0.001\ 52\text{ g}\cdot\text{cm}^{-3})(1.01\text{ J}\cdot(°\text{C})^{-1}\cdot\text{mol}^{-1})(22°\text{C}) = 1.1 \times 10^3\text{ kJ}$

The mass of octane required to produce this much heat will be given by

$$\left(\frac{-1.1 \times 10^3\text{ kJ}}{-5471\text{ kJ}\cdot\text{mol}^{-1}}\right)(114.22\text{ g}\cdot\text{mol}^{-1}) = 23.0\text{ g}$$

(b) $\Delta H = \left(\dfrac{(1.0\text{ gal})(3.785 \times 10^3\text{ mL}\cdot\text{gal}^{-1})(0.70\text{ g}\cdot\text{mL}^{-1})}{114.22\text{ g}\cdot\text{mol}^{-1}}\right)\left(\dfrac{-10\ 942\text{ kJ}}{2\text{ mol octane}}\right)$

$= -1.3 \times 10^5\text{ kJ}$

6.45 (a) mass of fat burned in 1 hour $= \dfrac{2000\text{ kJ}}{38\text{ kJ}\cdot\text{g}^{-1}} = 53\text{ g}$

(b) he would need to run $\dfrac{1.0\text{ lb} \times 454\text{ g}\cdot\text{lb}^{-1}}{53\text{ g}\cdot\text{hr}^{-1}} = 8.6\text{ hr}$

6.47 (a) For $H_2O(l)$, we want to find the enthalpy of the reaction

$H_2(g) + \frac{1}{2}O_2(g) \longrightarrow H_2O(l)$

The enthalpy change can be estimated from bond enthalpies. We will need to put in $(1\text{ mol})(436\text{ kJ}\cdot\text{mol}^{-1})$ to break the H—H bonds in 1 mol $H_2(g)$, $(\frac{1}{2}\text{ mol})(496\text{ kJ}\cdot\text{mol}^{-1})$ to break the O—O bonds in $\frac{1}{2}$ mol $O_2(g)$; we will get back $(2\text{ mol})(463\text{ kJ}\cdot\text{mol}^{-1})$ for the formation of 2 mol O—H bonds. This will give $\Delta H = -242\text{ kJ}\cdot\text{mol}^{-1}$. This value, however, will be to produce water in the gas phase. In order to get the value for the liquid, we will need to take into account the amount of heat given off when the gaseous water condenses to the liquid phase. This is $44.0\ \text{kJ}\cdot\text{mol}^{-1}$ at 298 K:

$$\Delta H^\circ_{\text{f, water(l)}} = \Delta H^\circ_{\text{f, water(g)}} - \Delta H^\circ_{\text{vap}} = -242\ \text{kJ} - (1\ \text{mol})(44.0\ \text{kJ}\cdot\text{mol}^{-1})$$
$$= -286\ \text{kJ}\cdot\text{mol}^{-1}$$

(b) The calculation for methanol is done similarly:

$$\text{C(gr)} + 2\ \text{H}_2\text{(g)} + \tfrac{1}{2}\ \text{O}_2\text{(g)} \longrightarrow \text{CH}_3\text{OH(l)}$$

ΔH for individual bond contributions:

atomize 1 mol C(gr)	$(1\ \text{mol})(717\ \text{kJ}\cdot\text{mol}^{-1})$
break 2 mol H—H bonds	$(2\ \text{mol})(436\ \text{kJ}\cdot\text{mol}^{-1})$
break $\frac{1}{2}$ mol O_2 bonds	$(\frac{1}{2}\ \text{mol})(496\ \text{kJ}\cdot\text{mol}^{-1})$
form 3 mol C—H bonds	$-(3\ \text{mol})(412\ \text{kJ}\cdot\text{mol}^{-1})$
form 1 mol C—O bonds	$-(1\ \text{mol})(360\ \text{kJ}\cdot\text{mol}^{-1})$
form 1 mol O—H bonds	$-(1\ \text{mol})(463\ \text{kJ}\cdot\text{mol}^{-1})$
Total	-222 kJ

$$\Delta H^\circ_{\text{f, methanol(l)}} = \Delta H^\circ_{\text{f, methanol(g)}} - \Delta H^\circ_{\text{vap}} = -222\ \text{kJ} - (1\ \text{mol})(35.3\ \text{kJ}\cdot\text{mol}^{-1})$$
$$= -257\ \text{kJ}$$

(c) $6\ \text{C(gr)} + 3\ \text{H}_2\text{(g)} \longrightarrow \text{C}_6\text{H}_6\text{(l)}$

Without resonance, we do the calculation considering benzene to have three double and three single C—C bonds:

atomize 6 mol C(gr)	$(6\ \text{mol})(717\ \text{kJ}\cdot\text{mol}^{-1})$
break 3 mol H—H bonds	$(3\ \text{mol})(436\ \text{kJ}\cdot\text{mol}^{-1})$
form 3 mol C=C bonds	$-(3\ \text{mol})(612\ \text{kJ}\cdot\text{mol}^{-1})$
form 3 mol C—C bonds	$-(3\ \text{mol})(348\ \text{kJ}\cdot\text{mol}^{-1})$
form 6 mol C—H bonds	$-(6\ \text{mol})(412\ \text{kJ}\cdot\text{mol}^{-1})$
Total	$+258$ kJ

$$\Delta H^\circ_{\text{f, benzene(l)}} = \Delta H^\circ_{\text{f, benzene(g)}} - \Delta H^\circ_{\text{vap}} = +258\ \text{kJ} - (1\ \text{mol})(30.8\ \text{kJ}\cdot\text{mol}^{-1})$$
$$= +227\ \text{kJ}$$

(d) $6\ \text{C(gr)} + 3\ \text{H}_2\text{(g)} \longrightarrow \text{C}_6\text{H}_6\text{(l)}$

With resonance, we repeat the calculation considering benzene to have six resonance-stabilized C—C bonds:

atomize 6 mol C(gr)	$(6\ \text{mol})(717\ \text{kJ}\cdot\text{mol}^{-1})$
break 3 mol H—H bonds	$(3\ \text{mol})(436\ \text{kJ}\cdot\text{mol}^{-1})$
form 6 mol C—C bonds, resonance	$-(6\ \text{mol})(518\ \text{kJ}\cdot\text{mol}^{-1})$
form 6 mol C—H bonds	$-(6\ \text{mol})(412\ \text{kJ}\cdot\text{mol}^{-1})$
Total	$+30$ kJ

$$\Delta H^\circ_{f,\ benzene(l)} = \Delta H^\circ_{f,\ benzene(g)} - \Delta H^\circ_{vap} = +30\ kJ\cdot mols^{-1} - (1\ mol)(30.8\ kJ\cdot mol^{-1})$$
$$= -1\ kJ$$

6.49 The combustion reaction of diamond is reversed and added to the combustion reaction of graphite to give the desired reaction:

$C(gr) + O_2(g) \longrightarrow CO_2(g)$ $\quad \Delta H^\circ = -393.51\ kJ$

$CO_2(g) \longrightarrow C(dia) + O_2(g)$ $\quad \Delta H^\circ = +395.41\ kJ$

$C(gr) \longrightarrow C(dia)$ $\quad \Delta H^\circ = +1.90\ kJ$

6.51 The first reaction is doubled, reversed, and added to the second to give the desired total reaction:

$2[SO_2(g) \longrightarrow S(s) + O_2(g)]$ $\quad (2)[+296.83\ kJ]$

$2\ S(s) + 3\ O_2(g) \longrightarrow 2\ SO_3(g)$ $\quad -791.44\ kJ$

$2\ SO_2(g) + O_2(g) \longrightarrow 2\ SO_3(g)$

$\Delta H^\circ = (2)(+296.83\ kJ\cdot mol^{-1}) - (791.44\ kJ\cdot mol^{-1}) = -197.78\ kJ$

6.53 The equations are arranged to add to the desired reaction. The second equation is multiplied by 4 and added to the first:

$P_4(s) + 6\ Cl_2(g) \longrightarrow 4\ PCl_3(l)$ $\quad -1278.8\ kJ$

$4[PCl_3(l) + Cl_2(g) \longrightarrow PCl_5(s)]$ $\quad 4[-124\ kJ]$

$P_4(s) + 10\ Cl_2(g) \longrightarrow 4\ PCl_5(s)$

$\Delta H^\circ = -1278.8\ kJ + 4\ (-124\ kJ) = -1775\ kJ$ or $-1.775\ MJ$

6.55 First, write the balanced equations for the reaction given:

$C_2H_2(g) + \frac{5}{2}\ O_2(g) \longrightarrow 2\ CO_2(g) + H_2O(l)$ $\quad \Delta H^\circ = -1300\ kJ$

$C_2H_6(g) + \frac{7}{2}\ O_2(g) \longrightarrow 2\ CO_2(g) + 3\ H_2O(l)$ $\quad \Delta H^\circ = -1560\ kJ$

$H_2(g) + \frac{1}{2}\ O_2(g) \longrightarrow H_2O(l)$ $\quad \Delta H^\circ = -286\ kJ$

The second equation is reversed and added to the first, plus two times the third:

$C_2H_2(g) + \frac{5}{2}\ O_2(g) \longrightarrow 2\ CO_2(g) + H_2O(l)$ $\quad \Delta H^\circ = -1300\ kJ$

$2\ CO_2(g) + 3\ H_2O(l) \longrightarrow C_2H_6(g) + \frac{7}{2}\ O_2(g)$ $\quad \Delta H^\circ = +1560\ kJ$

$2[H_2(g) + \frac{1}{2}\ O_2(g) \longrightarrow H_2O(l)]$ $\quad 2[\Delta H^\circ = -286\ kJ]$

$C_2H_2(g) + 2\ H_2(g) \longrightarrow C_2H_6(g)$

$\Delta H^\circ = -1300\ kJ + 1560\ kJ + 2\ (-286\ kJ) = -312\ kJ$

6.57 $\Delta H° = 12\ \Delta H°_f(H_2O, l) - [4(\Delta H°_f[HNO_3, l]) + 5(\Delta H°_f[N_2H_4, l])]$
$= 12(-285.83\ kJ{\cdot}mol^{-1}) - [4(-174.10\ kJ{\cdot}mol^{-1}) + 5\ (+50.63\ kJ{\cdot}mol^{-1})]$
$= -2987\ kJ{\cdot}mol^{-1}$

6.59 The desired reaction may be obtained by reversing the first reaction and multiplying it by 2, reversing the second reaction, and adding these to the third:

$2[NH_4Cl(s) \longrightarrow NH_3(g) + HCl(g)]$ $\quad 2[\Delta H° = +176.0\ kJ]$
$2\ NH_3(g) \longrightarrow N_2(g) + 3\ H_2(g)$ $\quad \Delta H° = +92.22\ kJ$
$N_2(g) + 4\ H_2(g) + Cl_2(g) \longrightarrow 2\ NH_4Cl(s)$ $\quad \Delta H° = -628.86\ kJ$

$H_2(g) + Cl_2(g) \longrightarrow 2\ HCl(g)$
$\Delta H° = 2(+176.0\ kJ) + 92.22\ kJ - 628.86\ kJ = -184.6\ kJ$

6.61 (a) $2\ C(s) + 2\ H_2(g) + \frac{1}{2}\ O_2(g) \longrightarrow CH_3CHO(l)$
(b) $Cu(s) + S(s) + \frac{9}{2}\ O_2(g) + 5\ H_2(g) \longrightarrow CuSO_4{\cdot}5\ H_2O(s)$
(c) $\frac{1}{2}\ N_2(g) + 2\ H_2(g) + 2\ O_2(g) + \frac{1}{2}\ Cl_2(g) \longrightarrow NH_4ClO_4(s)$

6.63 From Appendix 2A, $\Delta H°_f\ (NO) = +90.25\ kJ$
The reaction we want is
$N_2(g) + \frac{5}{2}\ O_2(g) \longrightarrow N_2O_5(g)$
gAdding the first reaction to half of the second gives

$2\ NO(g) + O_2(g) \longrightarrow 2\ NO_2(g)$ $\quad \Delta H° = -114.1\ kJ$
$2\ NO_2(g) + \frac{1}{2}\ O_2(g) \longrightarrow N_2O_5(g)$ $\quad \Delta H° = -55.1\ kJ$

$2\ NO(g) + \frac{3}{2}\ O_2(g) \longrightarrow N_2O_5(g)$ $\quad -169.2\ kJ$
The enthalpy of this reaction equals the enthalpy of formation of N_2O_5 (g) minus twice the enthalpy of formation of NO, so we can write
$-169.2\ kJ = \Delta H°_f\ (N_2O_5) - 2(+90.25\ kJ)$
$\Delta H°_f\ (N_2O_5) = +11.3\ kJ$

6.65 The enthalpy of the reaction
$PCl_3(l) + Cl_2(g) \longrightarrow PCl_5(s)$ $\quad \Delta H° = -124\ kJ$
is $\Delta H°_r = \Sigma\ \Delta H°_f\ (products) - \Sigma\ \Delta H°_f\ (reactants)$
$-124\ kJ = \Delta H°_f\ (PCl_5, s) - \Delta H°_f\ (PCl_3, l)$
Remember that the standard enthalpy of formation of $Cl_2(g)$ will be 0 by definition because this is an element in its reference state. From the Appendix we find that
$\Delta H°_f\ (PCl_3, l) = -319.7\ kJ{\cdot}mol^{-1}$
$-124\ kJ = \Delta H°_f\ (PCl_5, s) - (-319.7\ kJ)$
$\Delta H°_f\ (PCl_5, s) = -444\ kJ{\cdot}mol^{-1}$

6.67 The lattice energy corresponds to the process

$CaS(s) \longrightarrow Ca^{2+}(g) + S^{2-}(g)$

To calculate this, we set up a Born-Haber cycle, which is essentially an application of Hess's law. Perhaps the best way to do this is to write out the reactions that correspond to each of the pieces of data that we have and then add them up to give the net reaction we want.

$Ca(s) \longrightarrow Ca(g)$	+178 kJ
$S(s) \longrightarrow S(g)$	+279 kJ
$CaS(s) \longrightarrow Ca(s) + S(s)$	+482 kJ
$Ca(g) \longrightarrow Ca^{+}(g) + e^{-}$	+590 kJ
$Ca^{+}(g) \longrightarrow Ca^{2+}(g) + e^{-}$	+1150 kJ
$S(g) + e^{-} \longrightarrow S^{-}(g)$	−200 kJ
$S^{-}(g) + e^{-} \longrightarrow S^{2-}(g)$	+532 kJ
$CaS(s) \longrightarrow Ca^{2+}(g) + S^{2-}(g)$	+3011 kJ

6.69 For the reaction $Na_2O(s) \longrightarrow 2\,Na^{+}(g) + O^{2-}(g)$

$$\Delta H_L = 2\,\Delta H^\circ_f(\text{Na, g}) + \Delta H^\circ_f(\text{O, g}) + 2\,I_1(\text{Na}) - E_{ea1}(\text{O}) - E_{ea2}(\text{O}) - \Delta H_f(\text{Na}_2\text{O(s)})$$

$$\Delta H_L = 2(107.32\ \text{kJ}\cdot\text{mol}^{-1}) + 249\ \text{kJ}\cdot\text{mol}^{-1} + 2(494\ \text{kJ}\cdot\text{mol}^{-1}) - 141\ \text{kJ}\cdot\text{mol}^{-1} + 844\ \text{kJ}\cdot\text{mol}^{-1} + 409\ \text{kJ}\cdot\text{mol}^{-1}$$

$$\Delta H_L = 2564\ \text{kJ}\cdot\text{mol}^{-1}$$

6.71 (a) $\Delta H_L = \Delta H^\circ_f(\text{Na, g}) + \Delta H^\circ_f(\text{Cl, g}) + I_1(\text{Na}) - E_{ea}\text{ of Cl} - \Delta H_f(\text{NaCl(s)})$

$$787\ \text{kJ}\cdot\text{mol}^{-1} = 108\ \text{kJ}\cdot\text{mol}^{-1} + 122\ \text{kJ}\cdot\text{mol}^{-1} + 494\ \text{kJ}\cdot\text{mol}^{-1} - 349\ \text{kJ}\cdot\text{mol}^{-1} - \Delta H_f(\text{NaCl(s)})$$

$$\Delta H_f(\text{NaCl(s)}) = -412\ \text{kJ}\cdot\text{mol}^{-1}$$

(b) $\Delta H_L = \Delta H^\circ_f(\text{K, g}) + \Delta H^\circ_f(\text{Br, g}) + I_1(\text{K}) - E_{ea}(\text{Br}) - \Delta H_f(\text{KBr(s)})$

$$\Delta H_L = 89\ \text{kJ}\cdot\text{mol}^{-1} + 97\ \text{kJ}\cdot\text{mol}^{-1} + 418\ \text{kJ}\cdot\text{mol}^{-1} - 325\ \text{kJ}\cdot\text{mol}^{-1} + 394\ \text{kJ}\cdot\text{mol}^{-1} = 673\ \text{kJ}\cdot\text{mol}^{-1}$$

(c) $\Delta H_L = \Delta H^\circ_f(\text{Rb, g}) + \Delta H^\circ_f(\text{F, g}) + I_1(\text{Rb}) - E_{ea}(\text{F}) - \Delta H_f(\text{RbF(s)})$

$$774\ \text{kJ}\cdot\text{mol}^{-1} = \Delta H^\circ_f(\text{Rb, g}) + 79\ \text{kJ}\cdot\text{mol}^{-1} + 402\ \text{kJ}\cdot\text{mol}^{-1} - 328\ \text{kJ}\cdot\text{mol}^{-1} + 558\ \text{kJ}\cdot\text{mol}^{-1}$$

$$\Delta H^\circ_f(\text{Rb, g}) = 63\ \text{kJ}\cdot\text{mol}^{-1}$$

6.73 (a)

break:	3 mol C=C bonds	$3(612)\ \text{kJ}\cdot\text{mol}^{-1}$
form:	6 mol C=C bonds	$-6(518)\ \text{kJ}\cdot\text{mol}^{-1}$
	Total	$-1272\ \text{kJ}\cdot\text{mol}^{-1}$

(b) break:	4 mol C—H bonds	4(412) kJ·mol^{-1}
	4 mol Cl—Cl bonds	4(242) kJ·mol^{-1}
form:	4 mol C—Cl bonds	−4(338) kJ·mol^{-1}
	4 mol H—Cl bonds	−4(431) kJ·mol^{-1}
	Total	−460 kJ·mol^{-1}

(c) The number and types of bonds on both sides of the equations are equal, so we expect the enthalpy of the reaction to be essentially 0.

6.75

(a) break:	1 mol N—N triple bonds	(1 mol)(944 kJ·mol^{-1})
	3 mol F—F bonds	(3 mol)(158 kJ·mol^{-1})
form:	6 mol N—F bonds	(6 mol)(−195 kJ·mol^{-1})
	Total	+248 kJ
(b) break:	1 mol C=C bonds	(1 mol)(612 kJ·mol^{-1})
	1 mol O—H bonds	(1 mol)(463 kJ·mol^{-1})
form:	1 mol C—C bonds	−(1 mol)(348 kJ·mol^{-1})
	1 mol C—O bonds	−(1 mol)(360 kJ·mol^{-1})
	1 mol C—H bonds	−(1 mol)(412 kJ·mol^{-1})
	Total	−45 kJ
(c) break:	1 mol C—H bonds	(1 mol)(412 kJ·mol^{-1})
	1 mol Cl—Cl bonds	(1 mol)(242 kJ·mol^{-1})
form:	1 mol C—Cl bonds	−(1 mol)(338 kJ·mol^{-1})
	1 mol H—Cl bonds	−(1 mol)(431 kJ·mol^{-1})
	Total	−115 kJ

6.77 The value that we want is given simply by the difference between three isolated C=C bonds and three isolated C—C single bonds, versus six resonance-stabilized bonds:

3 C=C bonds + 3 C—C bonds = 3(348 kJ) + 3(612 kJ) = 2880 kJ

6 resonance-stabilized bonds = 6(518 kJ) = 3108 kJ

As can be seen, the six resonance-stabilized bonds are more stable by ca. 228 kJ.

6.79 (a) The enthalpy of vaporization is the enthalpy change associated with the conversion $C_6H_6(l) \longrightarrow C_6H_6(g)$ at constant pressure. The value at 298.2 K will be given by

$$\Delta H^\circ_{\text{vaporization at 298 K}} = \Delta H^\circ_f(C_6H_6, g) - \Delta H^\circ_f(C_6H_6, l)$$
$$= 82.93\ kJ\cdot mol^{-1} - (49.0\ kJ\cdot mol^{-1})$$
$$= 33.93\ kJ\cdot mol^{-1}$$

(b) In order to take into account the difference in temperature, we need to use the heat capacities of the reactants and products in order to raise the temperature of the system to 353.2 K. We can rewrite the reactions as follows, to emphasize temperature, and then combine them according to Hess's law:

$$C_6H_6(l)_{\text{at 298 K}} \longrightarrow C_6H_6(g)_{\text{at 298 K}} \quad \Delta H^\circ = 33.93\ kJ$$

$$C_6H_6(l)_{\text{at 298 K}} \longrightarrow C_6H_6(l)_{\text{at 353.2 K}} \quad \Delta H^\circ = (1\ mol)(353.2\ K - 298.2\ K)(136.1\ J\cdot mol^{-1}\cdot K^{-1}) = 7.48\ kJ$$

$$C_6H_6(g)_{\text{at 298 K}} \longrightarrow C_6H_6(g)_{\text{at 353.2 K}} \quad \Delta H^\circ = (1\ mol)(353.2\ K - 298.2\ K)(81.67\ J\cdot mol^{-1}\cdot K^{-1}) = 4.49\ kJ$$

To add these together to get the overall equation at 353.2 K, we must reverse the second equation:

$$C_6H_6(l)_{\text{at 298 K}} \longrightarrow C_6H_6(g)_{\text{at 298 K}} \quad \Delta H^\circ = 33.93\ kJ$$
$$C_6H_6(l)_{\text{at 353.2 K}} \longrightarrow C_6H_6(l)_{\text{at 298 K}} \quad \Delta H^\circ = -7.48\ kJ$$
$$C_6H_6(g)_{\text{at 298 K}} \longrightarrow C_6H_6(g)_{\text{at 353.2 K}} \quad \Delta H^\circ = 4.49\ kJ$$

$$C_6H_6(l)_{\text{at 353.2 K}} \longrightarrow C_6H_6(g)_{\text{at 353.2 K}} \quad \Delta H^\circ = 30.94\ kJ$$

(c) The value in the table is 30.8 $kJ\cdot mol^{-1}$ for the enthalpy of vaporization of benzene. The value is close to that calculated as corrected by heat capacities. At least part of the error can be attributed to the fact that heat capacities are not strictly constant with temperature.

6.81 This process involves five separate steps: (1) raising the temperature of the ice from −30.27°C to 0.00°C, (2) melting the ice at 0.00°C, (3) raising the temperature of the liquid water from 0.00°C to 100.00°C, (4) vaporizing the water at 100.00°C, and (5) raising the temperature of the water vapor from 100.00°C to 150.35°C.

Step 1: $\Delta H = (27.96\ g)(2.03\ J\cdot(°C)^{-1}\cdot g^{-1})(0.00°C - (-30.27°C)) = 1.72\ kJ$

Step 2: $\Delta H = \left(\dfrac{27.96\ g}{18.02\ g\cdot mol^{-1}}\right)(6.01\ kJ\cdot mol^{-1}) = 9.32\ kJ$

Step 3: $\Delta H = (27.96\ g)(4.18\ J\cdot(°C)^{-1}\cdot g^{-1})(100.00°C - 0.00°C) = 11.7\ kJ$

Step 4: $\Delta H = \left(\dfrac{27.96\ g}{18.02\ g\cdot mol^{-1}}\right)(40.7\ kJ\cdot mol^{-1}) = 63.1\ kJ$

Step 5: $\Delta H = (27.96\ g)(2.01\ J\cdot(°C)^{-1}\cdot g^{-1})(150.35°C - 100.00°C) = 2.83\ kJ$

The total heat required = 1.72 kJ + 9.32 kJ + 11.7 kJ + 63.1 kJ + 2.83 kJ
= 88.7 kJ

6.83 Appendix 2A provides us with the heat of formation of $I_2(g)$ at 298K (+62.44 $kJ \cdot mol^{-1}$) and the heat capacities of $I_2(g)$ (36.90 $J \cdot K^{-1} \cdot mol^{-1}$) and $I_2(s)$ (54.44 $J \cdot K^{-1} \cdot mol^{-1}$). We can calculate the $\Delta H_{sub}{}^0$ at 298K:

$I_2(s) \longrightarrow I_2(g) \qquad \Delta H_{sub}{}^0 = +62.44\ kJ \cdot mol^{-1}$

We can calculate the enthalpy of fusion from the relationship

$\Delta H_{sub}{}^0 = \Delta H_{fus}{}^0 + \Delta H_{vap}{}^0$

but these values need to be at the same temperature.

To correct the value for the fact that we want all the numbers for 298K, we need to alter the heat of vaporization, using the heat capacities for liquid and gaseous iodine.

$I_2(l)$ at 184.3°C $\longrightarrow$ $I_2(g)$ at 184.3°C $\qquad \Delta H_{vap}{}^0 = +41.96\ kJ \cdot mol^{-1}$

From Section 6.22, we find the following relationship

$\Delta H_{r,2}{}^0 = \Delta H_{r,1}{}^0 + \Delta C_{P,m}{}^0 (T_2 - T_1)$

$\Delta H_{vap,\ 298K}{}^0 = \Delta H_{vap,\ 475.5K}{}^0 + (C_{P,m}{}^0(I_2, g) - C_{P,m}{}^0(I_2, l))(T_2 - T_1)$

$\Delta H_{vap,\ 298K}{}^0 = +41.96\ kJ \cdot mol^{-1}$
$\qquad + (36.90\ J \cdot K^{-1} \cdot mol^{-1} - 80.7\ J \cdot K^{-1} \cdot mol^{-1})(298K - 475.5K)$
$\qquad = +49.73\ kJ \cdot mol^{-1}$

So, at 298K:

$+62.44\ kJ \cdot mol^{-1} = \Delta H_{fus}{}^0 + 49.73\ kJ \cdot mol^{-1}$

$\Delta H_{fus}{}^0 = +12.71\ kJ \cdot mol^{-1}$

6.85 (a) n = 1.0 $\quad T$ = 300 $\quad V_f/V_i$ = 1 to 10

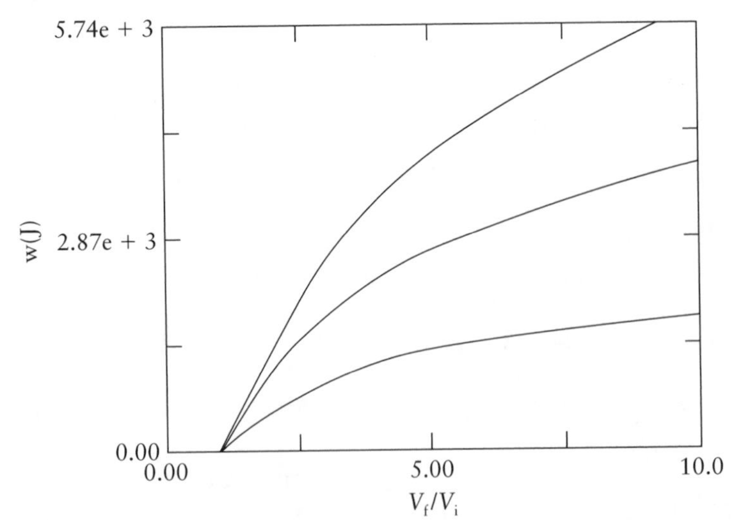

(b) The amount of work done is greater at the higher temperature. This can be seen from the equation:

$$w = -nRT \ln\frac{V_{\text{final}}}{V_{\text{initial}}}$$

The amount of work done is directly proportional to the temperature at which the expansion takes place.

(c) The comparison requested is the comparison of the terms

$$\ln\frac{V_{\text{final}}}{V_{\text{initial}}}$$

for the two processes. Even though in both cases the gas expands by 4 L, the relative amount of work done is different. We can get a numerical comparison by taking the ratio of this term for the two conditions:

$$\frac{\ln\left(\frac{9.00\text{ L}}{5.00\text{ L}}\right)}{\ln\left(\frac{5.00\text{ L}}{1.00\text{ L}}\right)} = \frac{0.588}{1.61} = 0.365$$

The second expansion by 4.00 L produces only about one third the amount of work that the first expansion did.

6.87 First, we need to calculate how much energy from the sunshine will be hitting the surface of the ethanol, so we convert the rate to $\text{kJ}\cdot\text{cm}^{-2}\cdot\text{s}^{-1}$:

$$1\text{ kJ}\cdot\text{m}^{-2}\cdot\text{s}^{-1}\left(\frac{1\text{ m}}{100\text{ cm}}\right)^2 = 1 \times 10^{-4}\text{ kJ}\cdot\text{cm}^{-2}\cdot\text{s}^{-1}$$

$$(1 \times 10^{-4}\text{ kJ}\cdot\text{cm}^{-2}\cdot\text{s}^{-1})(50.0\text{ cm}^2)\left(10\text{ min} \times \frac{60\text{ s}}{\text{min}}\right) = 3\text{ kJ}$$

The enthalpy of vaporization of ethanol is 43.5 $\text{kJ}\cdot\text{mol}^{-1}$ (see Table 6.2). We will assume that the enthalpy of vaporization is approximately the same at ambient conditions as it would be at the boiling point of ethanol.

$$\left(\frac{3\text{ kJ}}{43.5\text{ kJ}\cdot\text{mol}^{-1}}\right)(46.07\text{ g}\cdot\text{mol}^{-1}) = 3\text{ g}$$

6.89 (a) $C_6H_5NH_2(l) + \frac{31}{4}\,O_2(g) \longrightarrow 6\,CO_2(g) + \frac{7}{2}\,H_2O(l) + \frac{1}{2}\,N_2(g)$

(b) $$m_{CO_2} = \left(\frac{0.1754\text{ g aniline}}{93.12\text{ g}\cdot\text{mol}^{-1}\text{ aniline}}\right)\left(\frac{6\text{ mol }CO_2}{1\text{ mol aniline}}\right)(28.01\text{ g}\cdot\text{mol}^{-1}\text{ }CO_2)$$
$$= 0.4873\text{ g }CO_2(g)$$

$$m_{H_2O} = \left(\frac{0.1754\text{ g aniline}}{93.12\text{ g}\cdot\text{mol}^{-1}\text{ aniline}}\right)\left(\frac{3.5\text{ mol }H_2O}{1\text{ mol aniline}}\right)(18.02\text{ g}\cdot\text{mol}^{-1}\text{ }H_2O)$$
$$= 0.1188\text{ g }H_2O(l)$$

$$m_{N_2} = \left(\frac{0.1754\text{ g aniline}}{93.12\text{ g}\cdot\text{mol}^{-1}\text{ aniline}}\right)\left(\frac{0.5\text{ mol }N_2}{1\text{ mol aniline}}\right)(28.02\text{ g}\cdot\text{mol}^{-1}\text{ }N_2)$$
$$= 0.026\ 39\text{ g }N_2(g)$$

(c) $n_{O_2} = \left(\dfrac{0.1754 \text{ g aniline}}{93.12 \text{ g}\cdot\text{mol}^{-1} \text{ aniline}}\right)\left(\dfrac{\frac{31}{4} \text{ mol } O_2}{1 \text{ mol aniline}}\right)$

$= 0.014\ 60 \text{ g } O_2(g)$

$$P = \frac{nRT}{V} = \frac{(0.014\ 60 \text{ mol } O_2)(0.082\ 06 \text{ L}\cdot\text{atm}\cdot\text{K}^{-1}\cdot\text{mol}^{-1})(296\text{K})}{0.355 \text{ L}}$$

$= 0.999 \text{ atm}$

6.91 (a) The reaction enthalpy is obtained by Hess's law:

$\Delta H°_r = \Delta H°_f(\text{CO, g}) - \Delta H°_f(\text{H}_2\text{O, g})$

$\Delta H°_r = (1)(-110.53 \text{ kJ}\cdot\text{mol}^{-1}) - (1)(-241.82 \text{ kJ}\cdot\text{mol}^{-1})$

$\Delta H°_r = +131.29 \text{ kJ}\cdot\text{mol}^{-1}$

endothermic

(b) The number of moles of H_2 produced is obtained from the ideal gas law:

$$n = \frac{PV}{RT} = \frac{\left(\dfrac{500 \text{ Torr}}{760 \text{ Torr}\cdot\text{atm}^{-1}}\right)(200 \text{ L})}{(0.082\ 06 \text{ L}\cdot\text{atm}\cdot\text{K}^{-1}\cdot\text{mol}^{-1})(338 \text{ K})} = 4.74 \text{ mol}$$

The enthalpy change accompanying the production of this amount of hydrogen will be given by

$\Delta H = (4.74 \text{ mol})(131.29 \text{ kJ}\cdot\text{mol}^{-1}) = 623 \text{ kJ}$

6.93 (a) The enthalpy of combustion of $H_2(g)$ is the same as the enthalpy of formation of liquid water: $\Delta H°_f(\text{H}_2\text{O, l}) = -285.83 \text{ kJ}\cdot\text{mol}^{-1}$, whereas the enthalpy of combustion of octane is $-5471 \text{ kJ}\cdot\text{mol}^{-1}$. On a per gram basis, the heat released per gram of H_2 is

$-285.83 \text{ kJ}\cdot\text{mol}^{-1} \div 2.0158 \text{ g}\cdot\text{mol}^{-1} = -142 \text{ kJ}\cdot\text{g}^{-1}$

whereas that of octane is

$-5471 \text{ kJ}\cdot\text{mol}^{-1} \div 114.23 \text{ g}\cdot\text{mol}^{-1} = -47.89 \text{ kJ}\cdot\text{g}^{-1}$

Hydrogen produces more heat on a per gram basis than does the organic liquid, and it would be significantly cleaner to burn.

(b) The use of hydrogen as a fuel is limited by its storage problems. Normally, hydrogen is a gas, and to use it as a fuel requires carrying a large quantity of hydrogen under pressure. Much research is aimed at creating safe but also economical hydrogen storage tanks (certain metals such as Pd absorb hydrogen). Safety is a major issue because $H_2(g)$ can react explosively with air. This hazard is a serious concern for automobiles involved in accidents.

6.95 (a) First we must balance the chemical reaction:

$C_6H_6(l) + \frac{15}{2} O_2(g) \longrightarrow 6\ CO_2(g) + 3\ H_2O(g)$

For 1 mol $C_6H_6(l)$ burned, the change in the number of moles of gas is
$(9.00 - 7.50)\text{ mol} = +2.50\text{ mol} = \Delta n$

$$w = -P\Delta V = -P\left(\frac{\Delta nRT}{P}\right) = -\Delta nRT$$

$$w = -(+2.50\text{ mol})(8.314\text{ J}\cdot\text{K}^{-1}\cdot\text{mol}^{-1})(298\text{ K}) = -6.19 \times 10^3\text{ J} = -6.19\text{ kJ}$$

(b) $\Delta H_c = 6(-393.51\text{ kJ}\cdot\text{mol}^{-1}) + 3(-241.82\text{ kJ}\cdot\text{mol}^{-1}) - (+49.0\text{ kJ}\cdot\text{mol}^{-1})$
$= -3135.5\text{ kJ}$

(c) $\Delta U° = \Delta H° + w = (-3135.5 - 6.19)\text{ kJ} = -3141.7\text{ kJ}$

6.97 (a) The heat given off by the reaction, which was absorbed by the calorimeter, is given by
$\Delta H = -(525.0\text{ J}\cdot(°\text{C})^{-1})(20.0°\text{C} - 18.6°\text{C}) = -0.74\text{ kJ}$
This, however, is not all the heat produced, as the 100.0 mL of solution resulting from mixing also absorbed some heat. If we assume that the volume of NaOH and HNO_3 are negligible compared to the volume of water present and that the density of the solution is $1.00\text{ g}\cdot\text{mL}^{-1}$, then the change in heat of the solution is given by
$\Delta H = -(20.0°\text{C} - 18.6°\text{C})(4.18\text{ J}\cdot(°\text{C})^{-1}\cdot\text{g}^{-1})(100.0\text{ g}) = -0.59\text{ kJ}$
The total heat given off will be $-0.74\text{ kJ} + (-0.59\text{ kJ}) = -1.33\text{ kJ}$
(b) This heat is for the reaction of $(0.500\text{ M})(0.0500\text{ L}) = 0.0250\text{ mol } HNO_3$, so the amount of heat produced per mole of HNO_3 will be given by

$$\frac{-1.33\text{ kJ}}{0.0250\text{ mol}} = -53.2\text{ kJ}\cdot\text{mol}^{-1}$$

6.99 (a) $\xrightarrow{H_2}$ $\xrightarrow{H_2}$ $\xrightarrow{H_2}$

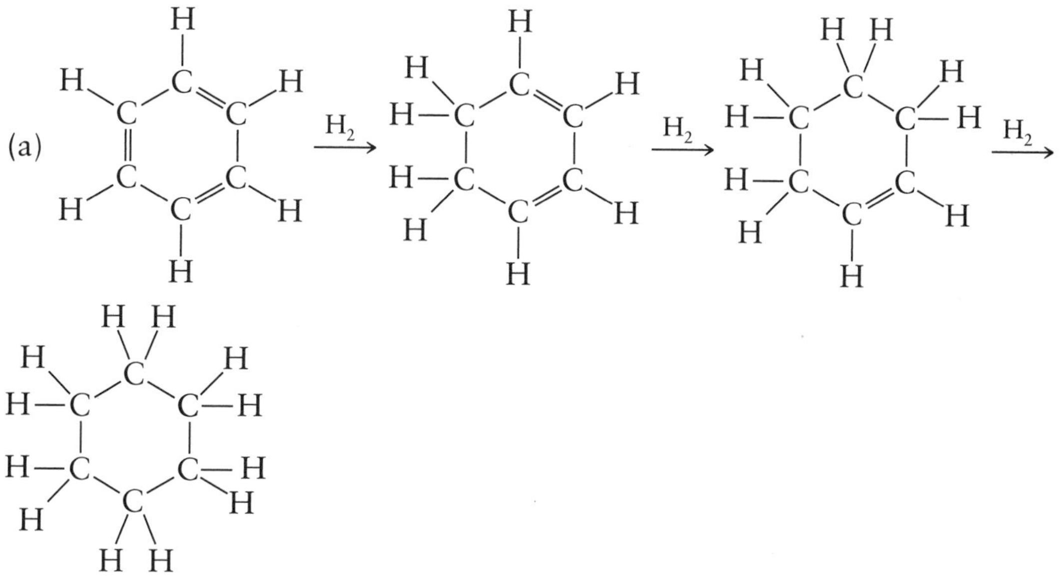

(b) From bond enthalpies, each step is identical, as the number and types of bonds broken and formed are the same:

break:	1 mol C═C bonds	612 kJ
	1 mol H—H bonds	436 kJ
form:	1 mol C—C bonds	−348 kJ
	2 mol C—H bonds	2(−412 kJ)
Total:		−124 kJ

The total energy change should be equal to the sum of the three steps or $3(-124\ \text{kJ}) = -372\ \text{kJ}$.

(c) The Hess's law calculation using standard enthalpies of formation is easily performed on the composite reaction:

$$C_6H_6(l) + 3\ H_2(g) \longrightarrow C_6H_{12}(l)$$

$$\begin{aligned}\Delta H^\circ_r &= \Sigma\ \Delta H^\circ_f\ (\text{products}) - \Sigma\ \Delta H^\circ_f\ (\text{reactants})\\ &= \Delta H^\circ_f\ (\text{cyclohexane}) - \Delta H^\circ_f\ (\text{benzene})\\ &= -156.4\ \text{kJ}\cdot\text{mol}^{-1} - (+49.0\ \text{kJ}\cdot\text{mol}^{-1})\\ &= -205.4\ \text{kJ}\cdot\text{mol}^{-1}\end{aligned}$$

(d) The hydrogenation of benzene is much less exothermic than predicted by bond enthalpy estimations. Part of this difference can be due to the inherent inaccuracy of using average values, but the difference is so large that this cannot be the complete explanation. As may be expected, the resonance energy of benzene makes it more stable than would be expected by treating it as a set of three isolated double and three isolated single bonds. The difference in these two values $[-205\ \text{kJ} - (-372\ \text{kJ}) = 167\ \text{kJ}]$ is a measure of how much more stable benzene is than the Kekulé structure would predict.

6.101 Equations for these various processes can be rearranged and summed according to Hess's law to give the desired lattice enthalpy of KBr(s).

Reaction	ΔH°
(1) $K(s) \longrightarrow K(g)$	+89.2 kJ
(2) $K(g) \longrightarrow K^+(g) + e^-$	+425.0 kJ
(3) $Br_2(l) \longrightarrow Br_2(g)$	+30.9
(4) $Br_2(g) \longrightarrow 2\ Br(g)$	+192.9
(5) $Br(g) + e^- \longrightarrow Br^-(g)$	−331.0
(6) $K(s) + \frac{1}{2}\ Br_2(l) \longrightarrow KBr(s)$	−394

Multiply (3) and (4) by $\frac{1}{2}$, reverse (6) and then sum:

$K(s) \longrightarrow K(g)$	$+89.2$ kJ
$K(g) \longrightarrow K^+(g) + e^-$	$+425.0$ kJ
$\frac{1}{2}\,Br_2(l) \longrightarrow \frac{1}{2}\,Br_2(g)$	$\frac{1}{2}(+30.9$ kJ$)$
$\frac{1}{2}\,Br_2(g) \longrightarrow Br(g)$	$\frac{1}{2}(+192.9$ kJ$)$
$Br(g) + e^- \longrightarrow Br^-(g)$	-331.0 kJ
$KBr(s) \longrightarrow K(s) + \frac{1}{2}\,Br_2(l)$	$+394$ kJ
$KBr(s) \longrightarrow K^+(g) + Br^-(g)$	689 $kJ \cdot mol^{-1}$

6.103 (a) The combustion reaction is

$C_{60}(s) + 60\ O_2(g) \rightarrow 60\ CO_2(g)$

The enthalpy of formation of $C_{60}(s)$ will be given by

$\Delta H°_c = 60\ \Delta H°_f(CO_2, g) - \Delta H°_f\,(C_{60}, s)$

$-25\,937\ kJ = 60\ mol \times (-393.51\ kJ \cdot mol^{-1}) - \Delta H°_f\,(C_{60}, s)$

$\Delta H°_f\,(C_{60}, s) = +2326\ kJ \cdot mol^{-1}$

(b) The bond enthalpy calculation is

$60\ C(gr) \longrightarrow 60\ C(g)$	$(60)(+717\ kJ \cdot mol^{-1})$
Form 60 mol C—C bonds	$-60\ (348\ kJ \cdot mol^{-1})$
Form 30 mol C═C bonds	$-30\ (612\ kJ \cdot mol^{-1})$
$C_{60}(g) \longrightarrow C_{60}(s)$	-233 kJ
$60\ C(gr) \longrightarrow C_{60}(s)$	$+3547$ kJ

(c) From the experimental data, the enthalpy of formation of C_{60} shows that it is *more* stable by (3547 kJ − 2326 kJ) = 1221 kJ than predicted by the isolated bond model.

(d) 1221 kJ ÷ 60 = 20 kJ per carbon atom

(e) 150 kJ ÷ 6 = 25 kJ per carbon atom

(f) Although the comparison of the stabilization of benzene with that of C_{60} should be treated with caution, it does appear that there is slightly less stabilization per carbon atom in C_{60} than in benzene. This fits with expectations, as the C_{60} molecule is forced by its geometry to be curved. This means that the overlap of the *p*-orbitals, which gives rise to the delocalization that results in resonance, will not be as favorable as in the planar benzene molecule. Another perspective on this is obtained by noting that the C atoms in C_{60} are forced to be partially sp^3 hybridized because they cannot be rigorously planar as required by sp^2 hybridization.

6.105 The balanced combustion reactions are

$$C_6H_3(NO_2)_3(s) + \tfrac{15}{4}\,O_2(g) \longrightarrow 6\,CO_2(g) + \tfrac{3}{2}\,H_2O(l) + \tfrac{3}{2}\,N_2(g)$$

$$C_6H_3(NH_2)_3(s) + \tfrac{33}{4}\,O_2(g) \longrightarrow 6\,CO_2(g) + \tfrac{9}{2}\,H_2O(l) + \tfrac{3}{2}\,N_2(g)$$

Because the fundamental structures of the two molecules are the same, we need only look at the differences between the two, which in this case are concerned with the groups attached to nitrogen

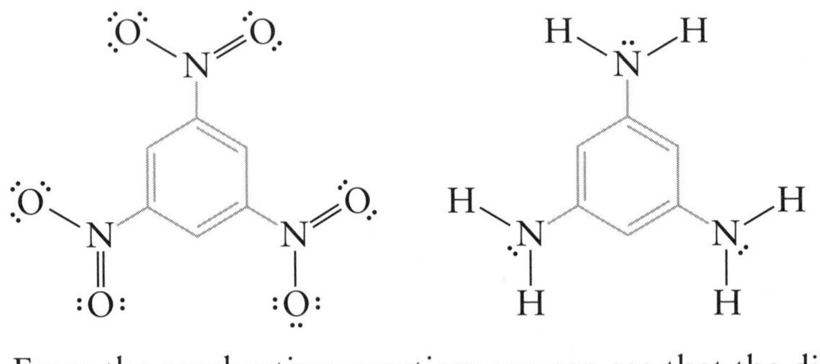

From the combustion equations we can see that the differences are (1) the consumption of $\frac{18}{4}$ more moles of $O_2(g)$ and (2) the production of six more moles of $H_2O(l)$ for the combustion of aniline. Because the ΔH°_f of $O_2(g)$ is 0, the net difference will be the production of 6 more moles of $H_2O(l)$ or $6 \times (-285.83\ \text{kJ}\cdot\text{mol}^{-1}) = -1715.0\ \text{kJ}$.

CHAPTER 7
THERMODYNAMICS: THE SECOND AND THIRD LAWS

7.1 (a) rate of entropy generation $= \dfrac{\Delta S_{surr}}{time} = -\dfrac{q_{rev}}{time \cdot T}$

$$= -\frac{\text{rate of heat generation}}{T}$$

$$= \frac{-(100\ \text{J}\cdot\text{s}^{-1})}{293\ \text{K}} = 0.341\ \text{J}\cdot\text{K}^{-1}\cdot\text{s}^{-1}$$

(b) $\Delta S_{day} = (0.341\ \text{J}\cdot\text{K}^{-1}\cdot\text{s}^{-1})(60\ \text{sec}\cdot\text{min}^{-1})(60\ \text{min}\cdot\text{hr}^{-1})(24\ \text{hr}\cdot\text{day}^{-1})$
$= 29.5\ \text{kJ}\cdot\text{K}^{-1}\cdot\text{day}^{-1}$

(c) Less, because in the equation $\Delta S = \dfrac{-\Delta H}{T}$, if T is larger, ΔS is smaller.

7.3 (a) $\Delta S = \dfrac{q_{rev}}{T} = \dfrac{65\ \text{J}}{298\ \text{K}} = 0.22\ \text{J}\cdot\text{K}^{-1}$

(b) $\Delta S = \dfrac{65\ \text{J}}{373\ \text{K}} = 0.17\ \text{J}\cdot\text{K}^{-1}$

(c) The entropy change is smaller at higher temperatures, because the matter is already more chaotic. The same amount of heat has a greater effect on entropy changes when transferred at lower temperatures.

7.5 (a) The relationship to use is $dS = \dfrac{dq}{T}$. At constant pressure, we can substitute $dq = n\,C_P\,dT$:

$$dS = \frac{n\,C_p\,dT}{T}$$

Upon integration, this gives $\Delta S = n\,Cp \ln \dfrac{T_2}{T_1}$. The answer is calculated by simply plugging in the known quantities. Remember that for an ideal monatomic gas $C_P = \frac{5}{2}R$:

$$\Delta S = (1.00\ \text{mol})(\tfrac{5}{2} \times 8.314\ \text{J}\cdot\text{K}^{-1}\cdot\text{mol}^{-1}) \ln \frac{431.0\ \text{K}}{310.8\ \text{K}} = 6.80\ \text{J}\cdot\text{K}^{-1}$$

(b) A similar analysis using C_V gives $\Delta S = n\, C_V \ln \dfrac{T_2}{T_1}$, where C_V for a monatomic ideal gas is $\frac{3}{2} R$:

$$\Delta S = (1.00 \text{ mol})(\tfrac{3}{2} \times 8.314 \text{ J}\cdot\text{K}^{-1}\cdot\text{mol}^{-1}) \ln \frac{431.0 \text{ K}}{310.8 \text{ K}} = 4.08 \text{ J}\cdot\text{K}^{-1}$$

7.7 Because the process is isothermal and reversible, the relationship $\text{d}S = \dfrac{\text{d}q}{T}$ can be used. Because the process is isothermal, $\Delta U = 0$ and hence $q = -w$, where $w = -P\text{d}V$. Making this substitution, we obtain

$$\text{d}S = \frac{P\,\text{d}V}{T} = \frac{nRT}{TV}\,\text{d}V = \frac{nR}{V}\,\text{d}V$$

$$\therefore\ \Delta S = nR \ln \frac{V_2}{V_1}$$

Substituting the known quantities, we obtain

$$\Delta S = (7.29 \text{ mol})(8.314 \text{ J}\cdot\text{K}^{-1}\cdot\text{mol}^{-1}) \ln \frac{10.589 \text{ L}}{2.475 \text{ L}}$$

$$= 88.1 \text{ J}\cdot\text{K}^{-1}$$

7.9 (a) $\Delta S° = \dfrac{q}{T} = \dfrac{\Delta H°}{T} = \dfrac{1.00 \text{ mol} \times (-6.01 \text{ kJ}\cdot\text{mol}^{-1})}{273.2 \text{ K}} = -22.0 \text{ J}\cdot\text{K}^{-1}$

(b) $\Delta S = \dfrac{q}{T} = \dfrac{\Delta H}{T} = \dfrac{\dfrac{50.0 \text{ g}}{46.07 \text{ g}\cdot\text{mol}^{-1}} \times 43.5 \text{ kJ}\cdot\text{mol}^{-1}}{351.5 \text{ K}} = +134 \text{ J}\cdot\text{K}^{-1}$

7.11 (a) The boiling point of a liquid may be obtained from the relationship $\Delta S_{\text{vap}} = \dfrac{\Delta H_{\text{vap}}}{T_{\text{B}}}$, or $T_{\text{B}} = \dfrac{\Delta H_{\text{vap}}}{\Delta S_{\text{vap}}}$. This relationship should be rigorously true if we have the actual enthalpy and entropy of vaporization. The data in the Appendix, however, are for 298 K. Thus, calculation of $\Delta H°_{\text{vap}}$ or $\Delta S°_{\text{vap}}$, using the enthalpy and entropy differences between the gas and liquid forms at 298 K, give a good approximation of these quantities but the values are not exact. For ethanal(l) $\longrightarrow$ ethanal(g), the data in the appendix give

$$\Delta H_{\text{vap}} \cong -166.19 \text{ kJ}\cdot\text{mol}^{-1} - (-192.30) \text{ kJ}\cdot\text{mol}^{-1} = 26.11 \text{ kJ}\cdot\text{mol}^{-1}$$

$$\Delta S_{\text{vap}} \cong 250.3 \text{ J}\cdot\text{K}^{-1}\cdot\text{mol}^{-1} - 160.2 \text{ J}\cdot\text{K}^{-1}\cdot\text{mol}^{-1} = 90.1 \text{ J}\cdot\text{K}^{-1}\cdot\text{mol}^{-1}$$

$$T_{\text{B}} = \frac{26.11 \times 10^3 \text{ J}\cdot\text{mol}^{-1}}{90.1 \text{ J}\cdot\text{K}^{-1}\cdot\text{mol}^{-1}} = 290 \text{ K}$$

(b) The boiling point of ethanal is 20.8°C or 293.9 K.

(c) These numbers are in very good agreement.

(d) Differences arise partly because the enthalpy and entropy of vaporization are slightly different from the values calculated at 298 K, but the boiling point of ethanal is not 298 K.

7.13 Trouton's rule indicates that the entropy of vaporization for a number of organic liquids is approximately 85 $\text{J}\cdot\text{K}^{-1}\cdot\text{mol}^{-1}$. Using this information and the relationship $T_B = \dfrac{\Delta H^\circ_{vap}}{\Delta S^\circ_{vap}} = \dfrac{21.51 \times 10^3\ \text{J}\cdot\text{mol}^{-1}}{85\ \text{J}\cdot\text{K}^{-1}\cdot\text{mol}^{-1}} = 253\ \text{K}$. The experimental boiling point of dimethyl ether is 248 K, which is in reasonably close agreement, given the nature of the approximation.

7.15 (a) The value can be estimated from

$$\Delta H^\circ_{vap} = T\Delta S^\circ_{vap}$$

$$\Delta H^\circ_{vap} = (353\ \text{K})(85\ \text{J}\cdot\text{mol}^{-1}\cdot\text{K}^{-1})$$

$$= +30\ \text{kJ}\cdot\text{mol}^{-1}$$

(b) $\Delta S^\circ_{surr} = -\dfrac{\Delta H^\circ_{system}}{T}$

$$\Delta S_{surr} = -\left(\frac{10\ \text{g}}{78.11\ \text{g}\cdot\text{mol}^{-1}}\right)\left(\frac{30\ \text{kJ}\cdot\text{mol}^{-1}}{353\ \text{K}}\right) = -11\ \text{J}\cdot\text{K}^{-1}$$

7.17 COF_2. COF_2 and BF_3 are both trigonal planar molecules, but it would be possible for the molecule to be disordered with the fluorine and oxygen atoms occupying the same locations. Because all the groups attached to boron are identical, such disorder is not possible.

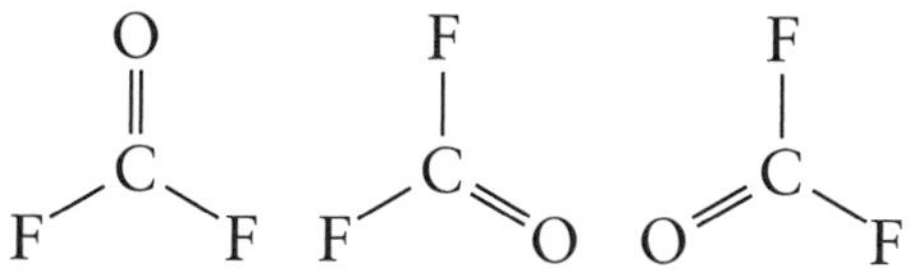

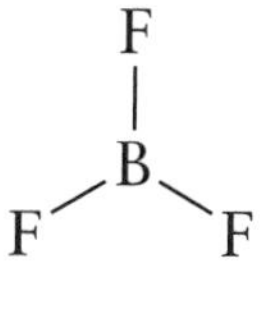

O=C(F)F   FC(F)=O   O=C(F)F

F–B(F)F

7.19 There are six orientations of an SO_2F_2 molecule as shown below:

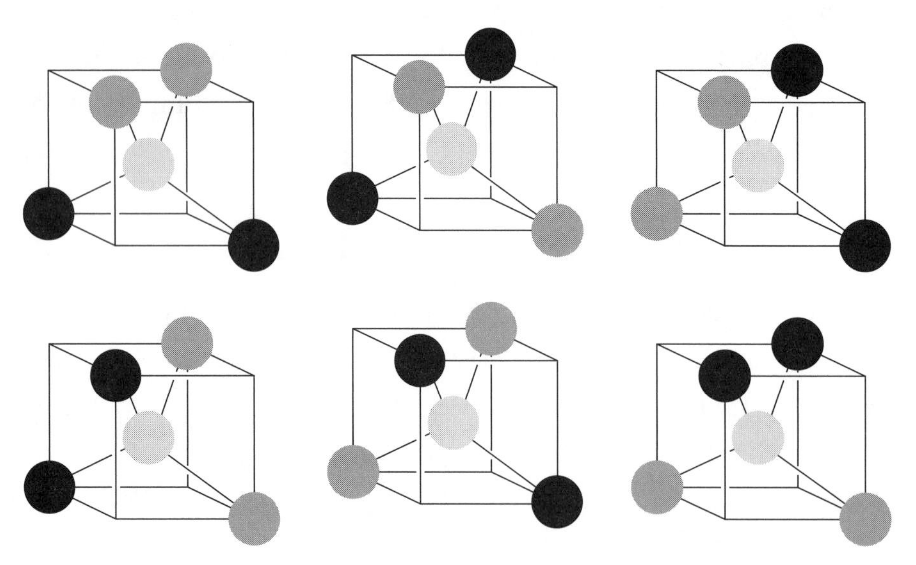

The Boltzmann expression for one mole of SO_2F_2 molecules having six possible orientations is

$S = k \ln 6^{6.02 \times 10^{23}} = (1.38 \times 10^{-23}\ \text{J}\cdot\text{K}^{-1}) \ln 6^{6.02 \times 10^{23}}$

$S = 14.9\ \text{J}\cdot\text{K}^{-1}$

7.21 (a) HBr, because Br is more massive and contains more elementary particles than F in HF; (b) NH_3, because it has greater complexity, being a molecule rather than a single atom; (c) I_2(l) because molecules in liquids are more randomly oriented than molecules in solids; (d) 1.0 mol Ar(g) at 1.00 atm, because it will occupy a larger volume than 1.0 mol of Ar(g) at 2.00 atm.

7.23 It is easy to order H_2O in its various phases because entropy will increase when going from a solid to a liquid to a gas. The main question concerns where to place C(s) in this order, and that will essentially become a question of whether C(s) should have more or less entropy than H_2O(s), because we would automatically expect C(s) to have less entropy than any liquid. Because water is a molecular substance held together in the solid phase by weak hydrogen bonds, and in C(s), which we will take to be the standard state from graphite, the carbon is more rigidly held in place and will have less entropy. C(s) < H_2O(s) < H_2O(l) < H_2O(g)

7.25 (a) In the standard state, bromine is a liquid and iodine is a solid. If the two substances had the same state (i.e., both were gases or both liquids) we would expect iodine to have the higher entropy due to its larger mass and consequently larger number of fundamental particles. However, because the compounds are in different states, we would expect the liquid to have a higher entropy than the solid.

(b) When we consider the two structures, it is clear that pentene will have more flexibility in its framework than cyclopentane, which will be comparatively rigid. Therefore, we predict pentene to have a higher entropy.

(c) Ethene (or ethylene) is a gas and polyethylene is a solid, so we automatically expect ethene to have a higher entropy. Also, for the same mass, a sample of ethene will be composed of many small molecules, whereas polyethylene will be made up of fewer but larger molecules.

7.27 The Cl_2 molecule has not only translational energy but also rotational and vibrational energy. The disordering due to this energy contributes to the overall energy disorder. The molecule also has twice as many atoms. However, matter is more organized in the molecule, so that the molar entropy of the molecule is not twice that of the atom.

7.29 (a) The number of moles of gas decreases, so we would expect entropy to decrease.

(b) The conversion of any substance from a solid to a gas should increase entropy due to the increased disorder in the gaseous state.

(c) Entropy should decrease because it decreases as temperature is lowered. At higher temperatures there is more molecular motion, which increases disorder.

7.31 (a) Entropy should decrease because the number of moles of gas is less on the product side of the reaction.

(b) Entropy should increase because the dissolution of the solid copper phosphate will increase the randomness of the copper and phosphate ions.

(c) Entropy should decrease as the total number of moles decreases.

7.33 (a) $H_2(g) + \frac{1}{2}O_2(g) \longrightarrow H_2O(l)$

$$\Delta S^\circ_f = \Sigma S^\circ_{\text{products}} - \Sigma S^\circ_{\text{reactants}}$$

$$= 69.91\ \text{J}\cdot\text{K}^{-1}\cdot\text{mol}^{-1} - [130.68\ \text{J}\cdot\text{K}^{-1}\cdot\text{mol}^{-1} + \tfrac{1}{2}(205.14\ \text{J}\cdot\text{K}^{-1}\cdot\text{mol}^{-1})]$$

$$= -163.34\ \text{J}\cdot\text{K}^{-1}\cdot\text{mol}^{-1}$$

The entropy change is negative because the number of moles of gas has decreased by 1.5. Note that the absolute entropies of the elements are not 0, and that the entropy change for the reaction in which a compound is formed from the elements is also not 0.

(b) $CO(g) + \frac{1}{2}O_2(g) \longrightarrow CO_2(g)$

$$\Delta S^\circ_r = 213.74\ \text{J}\cdot\text{K}^{-1}\cdot\text{mol}^{-1} - [197.67\ \text{J}\cdot\text{K}^{-1}\cdot\text{mol}^{-1} + \tfrac{1}{2}(205.14\ \text{J}\cdot\text{K}^{-1}\cdot\text{mol}^{-1})]$$

$$= -86.5\ \text{J}\cdot\text{K}^{-1}\cdot\text{mol}^{-1}$$

The entropy change is negative because the number of moles of gas has decreased by 0.5.

(c) $CaCO_3(s) \longrightarrow CaO(s) + CO_2(g)$

$$\Delta S°_r = 39.75\ J\cdot K^{-1}\cdot mol^{-1} + 213.74\ J\cdot K^{-1}\cdot mol^{-1} - [+92.9\ J\cdot K^{-1}\cdot mol^{-1}]$$
$$= +160.6\ J\cdot K^{-1}\cdot mol^{-1}$$

The entropy change is positive because the number of moles of gas has increased by 1.

(d) $4\ KClO_3(s) \longrightarrow 3\ KClO_4(s) + KCl(s)$

$$\Delta S°_r = 3(151.0\ J\cdot K^{-1}\cdot mol^{-1}) + 82.59\ J\cdot K^{-1}\cdot mol^{-1} - [4(143.1\ J\cdot K^{-1}\cdot mol^{-1})]$$
$$= -36.8\ J\cdot K^{-1}\cdot mol^{-1}$$

It is not immediately obvious, but the four moles of solid products are more ordered than the four moles of solid reactants.

7.35 (a) $\Delta S_{surr} = \dfrac{-\Delta H}{T} = \dfrac{-(-1 \times 10^{-3}\ J)}{2 \times 10^{-7}\ K} = 5 \times 10^{3}\ J\cdot K^{-1}$

(b) $\Delta S_{surr} = \dfrac{-(-1\ J)}{310\ K} = 3 \times 10^{-3}\ J\cdot K^{-1}$

(c) $\Delta S_{surr} = \dfrac{-(-20\ J)}{3.5\ K} = 5.7\ J\cdot K^{-1}$

7.37 (a) The change in entropy will be given by

$$\Delta S_{surr} = \frac{-\Delta H_{system}}{T} = \frac{-1.00\ mol \times 8.2 \times 10^{3}\ J\cdot mol^{-1}}{111.7\ K} = -73\ J\cdot K^{-1}$$

$$\Delta S_{system} = \frac{\Delta H_{system}}{T} = \frac{1.00\ mol \times 8.2 \times 10^{3}\ J\cdot mol^{-1}}{111.7\ K} = +73\ J\cdot K^{-1}$$

(b) $$\Delta S_{surr} = \frac{-\Delta H_{system}}{T} = \frac{-1.00\ mol \times 4.60 \times 10^{3}\ J\cdot mol^{-1}}{158.7\ K} = -29.0\ J\cdot K^{-1}$$

$$\Delta S_{system} = \frac{\Delta H_{system}}{T} = \frac{1.00\ mol \times 4.60 \times 10^{3}\ J\cdot mol^{-1}}{158.7\ K} = +29.0\ J\cdot K^{-1}$$

(c) $$\Delta S_{surr} = \frac{-\Delta H_{system}}{T} = \frac{-(1.00\ mol \times -4.60 \times 10^{3}\ J\cdot mol^{-1})}{158.7\ K} = +29.0\ J\cdot K^{-1}$$

$$\Delta S_{system} = \frac{\Delta H_{system}}{T} = \frac{1.00\ mol \times -4.60 \times 10^{3}\ J\cdot mol^{-1}}{158.7\ K} = -29.0\ J\cdot K^{-1}$$

7.39 (a) The total entropy change is given by $\Delta S_{tot} = \Delta S_{surr} + \Delta S$. ΔS for an isothermal, reversible process is calculated from $\Delta S = \dfrac{q_{rev}}{T} = \dfrac{-w_{rev}}{T} = nR \ln \dfrac{V_2}{V_1}$. To do the calculation we need the value of n, which is obtained by use of the ideal gas law: $(4.95\ \text{atm})(1.67\ \text{L}) = n(0.082\ 06\ \text{L}\cdot\text{atm}\cdot\text{K}^{-1}\cdot\text{mol}^{-1})(323\ \text{K})$; $n = 0.312$ mol.

$\Delta S = (0.312\ \text{mol})(8.314\ \text{J}\cdot\text{K}^{-1}\cdot\text{mol}^{-1}) \ln \dfrac{7.33\ \text{L}}{1.67\ \text{L}} = +3.84\ \text{J}\cdot\text{K}^{-1}$. Because the process is reversible, $\Delta S_{tot} = 0$, so $\Delta S_{surr} = -\Delta S = -3.84\ \text{J}\cdot\text{K}^{-1}$.

(b) For the irreversible process, ΔS is the same, $+3.84\ \text{J}\cdot\text{K}^{-1}$. No work is done in free expansion (see Section 6.6) so $w = 0$. Because $\Delta U = 0$, it follows that $q = 0$. Therefore, no heat is transferred into the surroundings, and their entropy is unchanged: $\Delta S_{surr} = 0$. The total change in entropy is therefore $\Delta S_{tot} = +3.84\ \text{J}\cdot\text{K}^{-1}$.

7.41 The change in entropy of the surroundings will be obtained from $\Delta S_{surr} = \dfrac{-\Delta H_{system}}{T}$. $\Delta S_{surr} = \dfrac{-1.96\ \text{kJ}}{298\ \text{K}} = -6.58\ \text{J}\cdot\text{K}^{-1}$. The total change in entropy will be negative, so we predict the process will be nonspontaneous.

7.43 Exothermic reactions tend to be spontaneous because the result is an increase in the entropy of the surroundings. Using the mathematical relationship $\Delta G_r = \Delta H_r - T\Delta S_r$, it is clear that if ΔH_r is large compared to ΔS_r, then the reaction will generally be spontaneous.

7.45 (a) $\Delta H°_r = 3(-824.2\ \text{kJ}\cdot\text{mol}^{-1}) - [2(-1118.4\ \text{kJ}\cdot\text{mol}^{-1})] = -235.8\ \text{kJ}\cdot\text{mol}^{-1}$

$\Delta S°_r = 3(87.40\ \text{J}\cdot\text{K}^{-1}\cdot\text{mol}^{-1}) - [2(146.4\ \text{J}\cdot\text{K}^{-1}\cdot\text{mol}^{-1}) + \tfrac{1}{2}(205.14\ \text{J}\cdot\text{K}^{-1}\cdot\text{mol}^{-1})]$

$= -133.17\ \text{J}\cdot\text{K}^{-1}\cdot\text{mol}^{-1}$

$\Delta G°_r = 3(-742.2\ \text{kJ}\cdot\text{mol}^{-1}) - [2(-1015.4\ \text{kJ}\cdot\text{mol}^{-1})] = -195.8\ \text{kJ}\cdot\text{mol}^{-1}$

ΔG_r may also be calculated from $\Delta H°_r$ and $\Delta S°_r$ (the numbers calculated differ slightly from the two methods due to rounding differences):

$\Delta G°_r = \Delta H°_r - T\Delta S°_r$

$= -235.8\ \text{kJ}\cdot\text{mol}^{-1} - (298\ \text{K})(-133.17\ \text{J}\cdot\text{K}^{-1}\cdot\text{mol}^{-1})/(100\ \text{J}\cdot\text{kJ}^{-1})$

$= -196.1\ \text{kJ}\cdot\text{mol}^{-1}$

(b) $\Delta H°_r = -1208.09\ \text{kJ}\cdot\text{mol}^{-1} - [-1219.6\ \text{kJ}\cdot\text{mol}^{-1}] = 11.5\ \text{kJ}\cdot\text{mol}^{-1}$

$\Delta S°_r = -80.8\ \text{J}\cdot\text{K}^{-1}\cdot\text{mol}^{-1} - [68.87\ \text{J}\cdot\text{K}^{-1}\cdot\text{mol}^{-1}] = -149.7\ \text{J}\cdot\text{K}^{-1}\cdot\text{mol}^{-1}$

$\Delta G°_r = -1111.15\ \text{kJ}\cdot\text{mol}^{-1} - [-1167.3\ \text{kJ}\cdot\text{mol}^{-1}] = +56.2\ \text{kJ}\cdot\text{mol}^{-1}$

or

$$\Delta G^\circ_r = \Delta H^\circ_r - T\Delta S^\circ_r$$
$$= 11.5\ \text{kJ}\cdot\text{mol}^{-1} - (298\ \text{K})(-149.7\ \text{J}\cdot\text{K}^{-1}\cdot\text{mol}^{-1})/(1000\ \text{J}\cdot\text{kJ}^{-1})$$
$$= 56.1\ \text{kJ}\cdot\text{mol}^{-1}$$

(c) $\Delta H^\circ_r = 9.16\ \text{kJ}\cdot\text{mol}^{-1} - [2(33.18\ \text{kJ}\cdot\text{mol}^{-1})] = -57.20\ \text{kJ}\cdot\text{mol}^{-1}$

$$\Delta S^\circ_r = 304.29\ \text{J}\cdot\text{K}^{-1}\cdot\text{mol}^{-1} - [2(240.06\ \text{J}\cdot\text{K}^{-1}\cdot\text{mol}^{-1})]$$
$$= -175.83\ \text{J}\cdot\text{K}^{-1}\cdot\text{mol}^{-1}$$
$$\Delta G^\circ_r = 97.89\ \text{kJ}\cdot\text{mol}^{-1} - [2(51.31\ \text{kJ}\cdot\text{mol}^{-1})] = -4.73\ \text{kJ}\cdot\text{mol}^{-1}$$

or

$$\Delta G^\circ_r = \Delta H^\circ_r - T\Delta S^\circ_r$$
$$= -57.2\ \text{kJ}\cdot\text{mol}^{-1} - (298\ \text{K})(-175.83\ \text{J}\cdot\text{K}^{-1}\cdot\text{mol}^{-1})/(1000\ \text{J}\cdot\text{kJ}^{-1})$$
$$= -4.80\ \text{kJ}\cdot\text{mol}^{-1}$$

7.47 (a) $\frac{1}{2}\,N_2(g) + \frac{3}{2}\,H_2(g) \longrightarrow NH_3(g)$

$$\Delta H^\circ_r = \Delta H^\circ_f(NH_3) = -46.11\ \text{kJ}\cdot\text{mol}^{-1}$$
$$\Delta S^\circ_r = S^\circ_m(NH_3, g) - [\tfrac{1}{2}S^\circ_m(N_2, g) + \tfrac{3}{2}S^\circ_m(H_2, g)]$$
$$= 192.45\ \text{J}\cdot\text{K}^{-1}\cdot\text{mol}^{-1} - [\tfrac{1}{2}(191.61\ \text{J}\cdot\text{K}^{-1}\cdot\text{mol}^{-1}) + \tfrac{3}{2}(130.68\ \text{J}\cdot\text{K}^{-1}\cdot\text{mol}^{-1})]$$
$$= -99.38\ \text{J}\cdot\text{K}^{-1}\cdot\text{mol}^{-1}$$
$$\Delta G^\circ_r = -46.11\ \text{kJ}\cdot\text{mol}^{-1} - (298\ \text{K})(-99.38\ \text{J}\cdot\text{K}^{-1}\cdot\text{mol}^{-1})/(1000\ \text{J}\cdot\text{kJ}^{-1})$$
$$= -16.49\ \text{kJ}\cdot\text{mol}^{-1}$$
$$S^\circ_m(NH_3) = 192.45\ \text{J}\cdot\text{K}^{-1}\cdot\text{mol}^{-1}$$

$\Delta S^\circ_f(NH_3)$ is negative because several gas molecules combine to form 1 NH_3 molecule.

(b) $H_2(g) + \frac{1}{2}\,O_2(g) \longrightarrow H_2O(g)$

$$\Delta H^\circ_r = \Delta H^\circ_f(H_2O, g) = -241.82\ \text{kJ}\cdot\text{mol}^{-1}$$
$$\Delta S^\circ_r = S^\circ_m(H_2O, g) - [S^\circ_m(H_2, g) + \tfrac{1}{2}S^\circ_m(O_2, g)]$$
$$= 188.83\ \text{J}\cdot\text{K}^{-1}\cdot\text{mol}^{-1} - [130.68\ \text{J}\cdot\text{K}^{-1}\cdot\text{mol}^{-1} + \tfrac{1}{2}(205.14\ \text{J}\cdot\text{K}^{-1}\cdot\text{mol}^{-1})]$$
$$= -44.42\ \text{J}\cdot\text{K}^{-1}\cdot\text{mol}^{-1}$$
$$\Delta G^\circ_r = -241.82\ \text{kJ}\cdot\text{mol}^{-1} - (298\ \text{K})(-44.42\ \text{J}\cdot\text{K}^{-1}\cdot\text{mol}^{-1})/(1000\ \text{J}\cdot\text{kJ}^{-1})$$
$$= -228.58\ \text{kJ}\cdot\text{mol}^{-1}$$
$$S^\circ_m(H_2O, g) = 188.83\ \text{J}\cdot\text{K}^{-1}\cdot\text{mol}^{-1}$$

$\Delta S^\circ_f(H_2O, g)$ is a negative number because there is a reduction in the number of gas molecules in the reaction when S°_m is positive.

(c) $C(s)$, graphite $+ \frac{1}{2} O_2(g) \longrightarrow CO(g)$

$\Delta H°_r = \Delta H°_f(CO, g) = -110.53\ kJ \cdot mol^{-1}$

$$\Delta S°_r = S°_m(CO, g) - [S°_m(C, s) + \tfrac{1}{2}S°_m(O_2, g)]$$
$$= 197.67\ J \cdot K^{-1} \cdot mol^{-1} - [5.740\ J \cdot K^{-1} \cdot mol^{-1} + \tfrac{1}{2}(205.14\ J \cdot K^{-1} \cdot mol^{-1})]$$
$$= 89.36\ J \cdot K^{-1} \cdot mol^{-1}$$

$$\Delta G°_r = -110.53\ kJ \cdot mol^{-1} - (298\ K)(89.36\ J \cdot K^{-1} \cdot mol^{-1})/(1000\ J \cdot kJ^{-1})$$
$$= -137.2\ kJ \cdot mol^{-1}$$

$S°_m(CO, g) = 197.67\ J \cdot K^{-1} \cdot mol^{-1}$

The $S°_m(CO, g)$ is larger than $\Delta S°_f(CO, g)$ because in the formation reaction the number of moles of gas is reduced.

(d) $\frac{1}{2} N_2(g) + O_2(g) \longrightarrow NO_2(g)$

$\Delta H°_r = \Delta H°_f(NO_2) = 33.18\ kJ \cdot mol^{-1}$

$$\Delta S°_r = S°_m(NO_2, g) - [\tfrac{1}{2}S°_m(N_2, g) + S°_m(O_2, g)]$$
$$= 240.06\ J \cdot K^{-1} \cdot mol^{-1} - [\tfrac{1}{2}(191.61\ J \cdot K^{-1} \cdot mol^{-1}) + 205.14\ J \cdot K^{-1} \cdot mol^{-1}]$$
$$= -60.89\ J \cdot K^{-1} \cdot mol^{-1}$$

$$\Delta G°_r = 33.18\ kJ \cdot mol^{-1} - (298\ K)(-60.89\ J \cdot K^{-1} \cdot mol^{-1})/(1000\ J \cdot kJ^{-1})$$
$$= 51.33\ kJ \cdot mol^{-1}$$

$S°_m(NO_2, g) = 240.06\ J \cdot K^{-1} \cdot mol^{-1}$

The $\Delta S°_f(NO_2, g)$ is somewhat negative due to the reduction in the number of gas molecules during the reaction.

For all of these, the important point to gain is that the $S°_m$ value of a compound is not the same as the $\Delta S°_f$ for the formation of that compound. $\Delta S°_f$ is often negative because one is bringing together a number of elements to form that compound.

7.49 Use the relationship $\Delta G°_r = \Sigma\ \Delta G°_f(\text{products}) - \Sigma\ \Delta G°_r(\text{reactants})$:

(a) $\Delta G°_r = 2\Delta G°_f(SO_3, g) - [2\Delta G°_f(SO_2, g)]$
$$= 2(-371.06\ kJ \cdot mol^{-1}) - [2(-300.19\ kJ \cdot mol^{-1})]$$
$$= -141.74\ kJ \cdot mol^{-1}$$

The reaction is spontaneous.

(b) $\Delta G°_r = \Delta G°_f(CaO, s) + \Delta G°_f(CO_2, g) - \Delta G°_f(CaCO_3, s)$
$$= (-604.03\ kJ \cdot mol^{-1}) + (-394.36\ kJ \cdot mol^{-1}) - (-1128.8\ kJ \cdot mol^{-1})$$
$$= +130.41\ kJ \cdot mol^{-1}$$

The reaction is not spontaneous.

(c) $\Delta G^\circ_r = 16\Delta G^\circ_f(CO_2, g) + 18\Delta G^\circ_f(H_2O, l) - [2\Delta G^\circ_f(C_8H_{18}, l)]$

$= 16(-394.36\ kJ\cdot mol^{-1}) + 18(-237.13\ kJ\cdot mol^{-1}) - [2(6.4\ kJ\cdot mol^{-1})]$

$= -10\ 590.9\ kJ\cdot mol^{-1}$

The reaction is spontaneous.

7.51 The standard free energies of formation of the compounds are: (a) $PCl_5(g)$, $-305.0\ kJ\cdot mol^{-1}$; (b) HCN(g), $+124.7\ kJ\cdot mol^{-1}$; (c) NO(g), $+86.55\ kJ\cdot mol^{-1}$; (d) $SO_2(g)$, $-300.19\ kJ\cdot mol^{-1}$. Those compounds with a positive free energy of formation are unstable with respect to the elements. Thus (a) and (d) are thermodynamically stable.

7.53 To answer this question, we need to calculate ΔG° and ΔS° for the reaction

$4\ KClO_3(s) \longrightarrow 3\ KClO_4(s) + KCl(s)$

From the data in Appendix 2A, these values can be calculated:

$\Delta H^\circ_r = 3\Delta H^\circ_f(KClO_4, s) + \Delta H^\circ_f(KCl, s) - [4\Delta H^\circ_f(KClO_3, s)]$

$= 3(-432.75\ kJ\cdot mol^{-1}) + (-436.75) - [4(-397.73\ kJ\cdot mol^{-1})]$

$= -144.08\ kJ\cdot mol^{-1}$

$\Delta S^\circ_r = 3S^\circ_m(KClO_4, s) + S^\circ_m(KCl, s) - [4S^\circ_m(KClO_3, s)]$

$= 3(151.0\ J\cdot K^{-1}\cdot mol^{-1}) + 82.59\ J\cdot K^{-1}\cdot mol^{-1} - [4(143.1\ J\cdot K^{-1}\cdot mol^{-1})]$

$= -36.8\ J\cdot K^{-1}\cdot mol^{-1}$

$\Delta G^\circ_r = 3\Delta G^\circ_f(KClO_4, s) + \Delta G^\circ_f(KCl, s) - [4\Delta G^\circ_f(KClO_3, s)]$

$= 3(-303.09\ kJ\cdot mol^{-1}) + (-409.14\ kJ\cdot mol^{-1}) - [4(-296.25\ kJ\cdot mol^{-1})]$

$= -133.41\ kJ\cdot mol^{-1}$

The standard free energy of the reaction is negative, which means there is a thermodynamic tendency for the reaction to occur. To determine the effect of temperature, we need to look at the entropy change in the reaction. This is obtained from the relationship $\Delta G^\circ_r = \Delta H^\circ_r - T\Delta S^\circ_r$. Because ΔS° is a negative number, the reaction will be less favorable at higher temperatures. The reaction will be favored by lower temperatures.

7.55 To understand what happens to ΔG°_r as temperature is raised, we use the relationship $\Delta G^\circ_r = \Delta H^\circ_r - T\Delta S^\circ_r$. From this it is clear that the free energy of the reaction becomes less favorable (more positive) as temperature increases, only if ΔS°_r is a negative number. Therefore, we need only to find out whether the standard entropy of formation of the compound is a negative number. This is calculated for each compound as follows:

(a) $P(s) + \frac{5}{2}\,Cl_2(g) \longrightarrow PCl_5(g)$

$$\Delta S^\circ_r = S^\circ_m(PCl_5, g) - [S^\circ_m(P, s) + \tfrac{5}{2}S^\circ_m(Cl_2, g)]$$
$$= 364.6\ J\cdot K^{-1}\cdot mol^{-1} - [41.09\ J\cdot K^{-1}\cdot mol^{-1} + \tfrac{5}{2}(223.07\ J\cdot K^{-1}\cdot mol^{-1})]$$
$$= -234.2\ J\cdot K^{-1}\cdot mol^{-1}$$

The compound is less stable at higher temperatures.

(b) $C(s)$, graphite $+ \frac{1}{2}\,N_2(g) + \frac{1}{2}\,H_2(g) \longrightarrow HCN(g)$

$$\Delta S^\circ_r = S^\circ_m(HCN, g) - [S^\circ_m(C, s) + \tfrac{1}{2}S^\circ_m(N_2, g) + \tfrac{1}{2}S^\circ_m(H_2, g)]$$
$$= 201.78\ J\cdot K^{-1}\cdot mol^{-1} - [5.740\ J\cdot K^{-1}\cdot mol^{-1} + \tfrac{1}{2}(191.61\ J\cdot K^{-1}\cdot mol^{-1}) + \tfrac{1}{2}(130.68\ J\cdot K^{-1}\cdot mol^{-1})]$$
$$= +34.90\ J\cdot K^{-1}\cdot mol^{-1}$$

$HCN(g)$ is more stable at higher T.

(c) $\frac{1}{2}\,N_2(g) + \frac{1}{2}\,O_2(g) \longrightarrow NO(g)$

$$\Delta S^\circ_r = S^\circ_m(NO, g) - [\tfrac{1}{2}S^\circ_m(N_2, g) + \tfrac{1}{2}S^\circ_m(O_2, g)]$$
$$= 210.76\ J\cdot K^{-1}\cdot mol^{-1} - [\tfrac{1}{2}(191.61\ J\cdot K^{-1}\cdot mol^{-1}) + \tfrac{1}{2}(205.14\ J\cdot K^{-1}\cdot mol^{-1})]$$
$$= +12.38\ J\cdot K^{-1}\cdot mol^{-1}$$

$NO(g)$ is more stable as T increases.

(d) $S(s) + O_2(g) \longrightarrow SO_2(g)$

$$\Delta S^\circ_r = S^\circ(SO_2, g) - [S^\circ(S, s) + S^\circ(O_2, g)]$$
$$= 248.22\ J\cdot K^{-1}\cdot mol^{-1} - [31.80\ J\cdot K^{-1}\cdot mol^{-1} + 205.14\ J\cdot K^{-1}\cdot mol^{-1}]$$
$$= +11.28\ J\cdot K^{-1}\cdot mol^{-1}$$

$SO_2(g)$ is more stable as T increases.

7.57 (a) $2\,H_2O_2(l) \longrightarrow 2\,H_2O(l) + O_2(g)$

$$\Delta S^\circ_r = 2S^\circ_m(H_2O, l) + S^\circ_m(O_2, g) - 2S^\circ_m(H_2O_2, l)$$
$$= 2(69.91\ J\cdot K^{-1}\cdot mol^{-1}) + 205.14\ J\cdot K^{-1}\cdot mol^{-1} - 2(109.6\ J\cdot K^{-1}\cdot mol^{-1})$$
$$= +125.8\ J\cdot K^{-1}\cdot mol^{-1}$$

$$\Delta H^\circ_r = 2\Delta H^\circ_f(H_2O, l) - 2\Delta H^\circ_f(H_2O_2, l)$$
$$= 2(-285.83\ kJ\cdot mol^{-1}) - 2(-187.78\ kJ\cdot mol^{-1})$$
$$= -196.10\ kJ\cdot mol^{-1}$$

$$\Delta G^\circ_r = 2\Delta G^\circ_f(H_2O, l) - 2\Delta G^\circ_f(H_2O_2, l)$$
$$= 2(-237.13\ kJ\cdot mol^{-1}) - 2(-120.35\ kJ\cdot mol^{-1})$$
$$= -233.56\ kJ\cdot mol^{-1}$$

ΔG°_r can also be calculated from ΔS°_r and ΔH°_r using the relationship:

$$\Delta G^\circ_r = \Delta H^\circ_r - T\Delta S^\circ_r$$
$$= -196.1\ \text{kJ}\cdot\text{mol}^{-1} - (298\ \text{K})(+125.8\ \text{J}\cdot\text{K}^{-1}\cdot\text{mol}^{-1})/1000\ \text{J}\cdot\text{kJ}^{-1})$$
$$= -233.6\ \text{kJ}\cdot\text{mol}^{-1}$$

(b) $2\ F_2(g) + 2\ H_2O(l) \longrightarrow 4\ HF\ (aq) + O_2(g)$

$$\Delta S^\circ_r = 4S^\circ_m(\text{HF, aq}) + S^\circ_m(\text{O}_2, \text{g}) - [2S^\circ_m(\text{F}_2, \text{g}) + 2S^\circ_m(\text{H}_2\text{O, l})]$$
$$= 4\ (-88.7\ \text{J}\cdot\text{K}^{-1}\cdot\text{mol}^{-1}) + 205.14\ \text{J}\cdot\text{K}^{-1}\cdot\text{mol}^{-1} - [2(202.78\ \text{J}\cdot\text{K}^{-1}\cdot\text{mol}^{-1}) + 2(69.91\ \text{J}\cdot\text{K}^{-1}\cdot\text{mol}^{-1})]$$
$$= +14.6\ \text{J}\cdot\text{K}^{-1}\cdot\text{mol}^{-1}$$
$$\Delta H^\circ_r = 4\Delta H^\circ_f(\text{HF, aq}) - [2\Delta H^\circ_f(\text{H}_2\text{O, l})]$$
$$= 4(-330.08\ \text{kJ}\cdot\text{mol}^{-1}) - [2(-285.83\ \text{kJ}\cdot\text{mol}^{-1})]$$
$$= -748.66\ \text{kJ}\cdot\text{mol}^{-1}$$
$$\Delta G^\circ_r = 4 \times \Delta G^\circ_f(\text{HF, aq}) - [2 \times \Delta G^\circ_f(\text{H}_2\text{O, l})]$$
$$= 4(-296.82\ \text{kJ}\cdot\text{mol}^{-1}) - [2(-237.13\ \text{kJ}\cdot\text{mol}^{-1})]$$
$$= -713.02\ \text{kJ}\cdot\text{mol}^{-1}$$

ΔG°_r can also be calculated from ΔS°_r and ΔH°_r using the relationship:

$$\Delta G^\circ_r = \Delta H^\circ_r - T\Delta S^\circ_r$$
$$= -748.66\ \text{kJ}\cdot\text{mol}^{-1} - (298\ \text{K})(14.6\ \text{J}\cdot\text{K}^{-1}\cdot\text{mol}^{-1})/(1000\ \text{J}\cdot\text{kJ}^{-1})$$
$$= -753.01\ \text{kJ}\cdot\text{mol}^{-1}$$

7.59 In order to find ΔG°_r at a temperature other than 298 K, we must first calculate ΔH°_r and ΔS°_r and then use the relationship $\Delta G^\circ_r = \Delta H^\circ_r + T\Delta S^\circ_r$ to calculate ΔG°_r.

(a) $\Delta H^\circ_r = 2\Delta H^\circ_f(\text{BF}_3, \text{g}) + 3\Delta H^\circ_f(\text{H}_2\text{O, l}) - [\Delta H^\circ_f(\text{B}_2\text{O}_3, \text{s}) + 6\Delta H^\circ_f(\text{HF, g})]$

$$= 2(-1137.0\ \text{kJ}\cdot\text{mol}^{-1}) + 3(-285.83\ \text{kJ}\cdot\text{mol}^{-1}) - [(-1272.8\ \text{kJ}\cdot\text{mol}^{-1}) + 6(-271.1\ \text{kJ}\cdot\text{mol}^{-1})]$$
$$= -232.1\ \text{kJ}\cdot\text{mol}^{-1}$$
$$\Delta S^\circ_r = 2S^\circ_m(\text{BF}_3, \text{g}) + 3S^\circ_m(\text{H}_2\text{O, l}) - [S^\circ_m(\text{B}_2\text{O}_3, \text{S}) + 6S^\circ_m(\text{HF, g})]$$
$$= 2(254.12\ \text{J}\cdot\text{K}^{-1}\cdot\text{mol}^{-1}) + 3(69.91\ \text{J}\cdot\text{K}^{-1}\cdot\text{mol}^{-1}) - [53.97\ \text{J}\cdot\text{K}^{-1}\cdot\text{mol}^{-1} + 6(173.78\ \text{J}\cdot\text{K}^{-1}\cdot\text{mol}^{-1})]$$
$$= -378.68\ \text{J}\cdot\text{K}^{-1}\cdot\text{mol}^{-1}$$
$$\Delta G^\circ_r = -232.1\ \text{J}\cdot\text{K}^{-1}\cdot\text{mol}^{-1} - (353\ \text{K})(-378.68\ \text{J}\cdot\text{K}^{-1}\cdot\text{mol}^{-1})/(1000\ \text{J}\cdot\text{kJ}^{-1})$$
$$= -98.42\ \text{kJ}\cdot\text{mol}^{-1}$$

In order to determine the range over which the reaction will be spontaneous, we consider the relative signs of ΔH°_r and ΔS°_r and their effect on ΔG°_r. Because ΔH°_r is negative and ΔS°_r is also negative, we expect the reaction to be spontaneous at low temperatures, where the term $T\Delta S°_r$ will be less than $\Delta H°_r$. To find the temperature of the cutoff, we calculate the temperature at which $\Delta G°_r = 0$. For this reaction, that temperature is

$\Delta G°_r = 0 = -232.1\ \text{kJ}\cdot\text{mol}^{-1} - (T)(-378.68\ \text{J}\cdot\text{K}^{-1}\cdot\text{mol}^{-1})/(1000\ \text{J}\cdot\text{kJ}^{-1})$

$T = 612.9\ \text{K}$

The reaction should be spontaneous below 612.9 K.

(b) $\Delta H°_r = \Delta H°_f(\text{CaCl}_2, \text{aq}) + \Delta H°_f(\text{C}_2\text{H}_2, \text{g}) - [\Delta H°_f(\text{CaC}_2, \text{s}) + 2\Delta H°_f(\text{HCl, aq})]$

$= (-877.1\ \text{kJ}\cdot\text{mol}^{-1}) + 226.73\ \text{kJ}\cdot\text{mol}^{-1} - [(-59.8\ \text{kJ}\cdot\text{mol}^{-1}) + 2(-167.16\ \text{kJ}\cdot\text{mol}^{-1})]$

$= -256.3\ \text{kJ}\cdot\text{mol}^{-1}$

$\Delta S°_r = S°_m(\text{CaCl}_2, \text{aq}) + S°_m(\text{C}_2\text{H}_2, \text{g}) - [S°_m(\text{CaC}_2, \text{s}) + 2S°_m(\text{HCl, aq})]$

$= 59.8\ \text{J}\cdot\text{K}^{-1}\cdot\text{mol}^{-1} + 200.94\ \text{J}\cdot\text{K}^{-1}\cdot\text{mol}^{-1} - [69.96\ \text{J}\cdot\text{K}^{-1}\cdot\text{mol}^{-1} + 2(56.5\ \text{J}\cdot\text{K}^{-1}\cdot\text{mol}^{-1})]$

$= +77.8\ \text{J}\cdot\text{K}^{-1}\cdot\text{mol}^{-1}$

$\Delta G°_r = -256.2\ \text{kJ}\cdot\text{mol}^{-1} - (353\ \text{K})(+77.8\ \text{J}\cdot\text{K}^{-1}\cdot\text{mol}^{-1})/(1000\ \text{J}\cdot\text{kJ}^{-1})$

$= -283.7\ \text{kJ}\cdot\text{mol}^{-1}$

Because $\Delta H°_r$ is negative and $\Delta S°_r$ is positive, the reaction will be spontaneous at all temperatures.

(c) $\Delta H°_r = \Delta H°_f(\text{C (s), diamond}) = +1.895\ \text{kJ}\cdot\text{mol}^{-1}$

$\Delta S°_r = S°_m(\text{C (s), diamond}) - S°_m(\text{C (s), graphite})$

$= +2.377\ \text{J}\cdot\text{K}^{-1}\cdot\text{mol}^{-1} - 5.740\ \text{J}\cdot\text{K}^{-1}\cdot\text{mol}^{-1}$

$= -3.363\ \text{J}\cdot\text{K}^{-1}$

$\Delta G°_r = +1.895\ \text{kJ}\cdot\text{mol}^{-1} - (353\ \text{K})(-3.363\ \text{J}\cdot\text{K}^{-1}\cdot\text{mol}^{-1})/(1000\ \text{J}\cdot\text{kJ}^{-1})$

$= +3.082\ \text{kJ}\cdot\text{mol}^{-1}$

Because $\Delta H°_r$ is positive and $\Delta S°_r$ is negative, the reaction will be nonspontaneous at all temperatures. Note: This calculation is for atmospheric pressure. Diamond can be produced from graphite at elevated pressures and high temperatures.

7.61 (a) 1-propanol (C_3H_8O) and 2-propanone (C_3H_6O) have similar numbers of electrons so that we would expect the molar entropies to be similar. Because 1-propanol exhibits hydrogen bonding, however, we might expect the liquid phase to be more ordered than for 2-propanone. This is observed. The standard molar entropy for 2-propanone is 200 $\text{J}\cdot\text{K}^{-1}\cdot\text{mol}^{-1}$ while that of 1-propanol is 193 $\text{J}\cdot\text{K}^{-1}\cdot\text{mol}^{-1}$. (b) In the gas phase, hydrogen bonding will not be important because the molecules are too far apart, so the standard molar entropies should be more similar.

7.63 For the cis compound there will be 12 different orientations:

```
  X           X           X           X
  |           |           |           |
Y   Y       Y   Y       Y   Y       Y   X
 >M<         >M<         >M<         >M<
Y   X       X   Y       X   Y       Y   Y
  |           |           |           |
  Y           Y           Y           Y

  Y           Y           Y           Y
  |           |           |           |
Y   Y       Y   Y       X   Y       Y   X
 >M<         >M<         >M<         >M<
Y   X       X   Y       Y   Y       Y   Y
  |           |           |           |
  X           X           X           X

  Y           Y           Y           Y
  |           |           |           |
Y   X       X   X       X   Y       Y   Y
 >M<         >M<         >M<         >M<
Y   X       Y   Y       X   Y       X   X
  |           |           |           |
  Y           Y           Y           Y
```

For the trans compound there will only be 3 different orientations.

```
  X           Y           Y
  |           |           |
Y   Y       Y   X       X   Y
 >M<         >M<         >M<
Y   Y       X   Y       Y   X
  |           |           |
  X           Y           Y
```

Comparing the Boltzmann entropy calculations for the cis and trans forms:

cis:

$S = k \ln 12^{6.02 \times 10^{23}} = (1.38 \times 10^{-23}\ \mathrm{J \cdot K^{-1}}) \ln 12^{6.02 \times 10^{23}}$

$S = 20.6\ \mathrm{J \cdot K^{-1}}$

trans:

$S = k \ln 3^{6.02 \times 10^{23}} = (1.38 \times 10^{-23}\ \mathrm{J \cdot K^{-1}}) \ln 3^{6.02 \times 10^{23}}$

$S = 9.13\ \mathrm{J \cdot K^{-1}}$

The cis form should have the higher residual entropy.

7.65 According to Trouton's rule, the entropy of vaporization of an organic liquid is a constant of approximately 85 $\mathrm{J \cdot mol^{-1} \cdot K^{-1}}$. The relationship between entropy of fusion, enthalpy of fusion, and melting point is given by $\Delta S°_{fus} = \dfrac{\Delta H°_{fus}}{T_{fus}}$.

For Pb: $\Delta S°_{fus} = \dfrac{5100\ \mathrm{J}}{600\ \mathrm{K}} = 8.50\ \mathrm{J \cdot K^{-1}}$

For Hg: $\Delta S°_{fus} = \dfrac{2290\ \mathrm{J}}{234\ \mathrm{K}} = 9.79 \cdot \mathrm{K^{-1}}$

For Na: $\Delta S°_{fus} = \dfrac{2640\ \mathrm{J}}{371\ \mathrm{K}} = 7.12\ \mathrm{J \cdot K^{-1}}$

These numbers are reasonably close but clearly much smaller than the value associated with Trouton's rule.

7.67 This is best answered by considering the reaction that interconverts the two compounds

$4\ Fe_3O_4(s) + O_2(g) \longrightarrow 6\ Fe_2O_3(s)$

We calculate $\Delta G°_r$ using data from Appendix 2A:

$\Delta G°_r = 6\Delta G°_f(Fe_2O_3, s) - [4\Delta G°_f(Fe_3O_4, s)]$

$\Delta G°_r = 6(-742.2\ kJ\cdot mol^{-1}) - [4(-1015.4\ kJ\cdot mol^{-1})]$

$= -391.6\ kJ\cdot mol^{-1}$

Because $\Delta G°_r$ is negative, the process is spontaneous at 25°C. Therefore, Fe_2O_3 is thermodynamically more stable.

7.69 (a) The simplest way to calculate $\Delta G°_r$ is to use the $\Delta G°_f$ values from Appendix 2A, but because we want to calculate the temperature when $\Delta G°_r = 0$, we need to calculate $\Delta H°_r$ and $\Delta S°_r$.

$\Delta G°_r = \Delta G°_f(CO, g) - [\Delta G°_f(H_2O, g)]$

$= (-137.17\ kJ\cdot mol^{-1}) - [-228.57\ kJ\cdot mol^{-1}]$

$= +91.40\ kJ\cdot mol^{-1}$

$\Delta H°_r = \Delta H°_f(CO, g) - [\Delta H°_f(H_2O, g)]$

$= -110.53\ kJ\cdot mol^{-1} - [-241.82\ kJ\cdot mol^{-1}]$

$= +131.29\ kJ\cdot mol^{-1}$

$\Delta S°_r = S°_m(CO, g) + S°_m(H_2, g) - [S°_m(H_2O, g) + S°_m(C\ (s, graphite)]$

$= 197.67\ J\cdot K^{-1}\cdot mol^{-1} + 130.68\ J\cdot K^{-1}\cdot mol^{-1} - [188.83\ J\cdot K^{-1}\cdot mol^{-1} + 5.740\ J\cdot K^{-1}\cdot mol^{-1}]$

$= +133.78\ J\cdot K^{-1}\cdot mol^{-1}$

$\Delta G°_r = \Delta H°_r - T\Delta S°_r$

$= +131.29\ kJ\cdot mol^{-1} - (298.2\ K)(133.78\ J\cdot K^{-1}\cdot mol^{-1})/(1000\ J\cdot kJ^{-1})$

$= +91.40\ kJ\cdot mol^{-1}$

(b) $\Delta G°_r = \Delta H°_r - T\Delta S°_r$

$= +131.29\ kJ\cdot mol^{-1} - (T)(133.78\ J\cdot K^{-1}\cdot mol^{-1})/(1000\ J\cdot kJ^{-1}) = 0$

$T = 981.39\ K$

Even though the calculation will allow us to give T to five significant figures, the fact that ΔH_r and ΔS_r vary somewhat with temperature make reporting that number of figures unreasonable. A better answer, given the limitations of the data, is 980 K.

7.71 We can calculate the free energy changes associated with the conversions:

(a) $2\ FeS(s) \longrightarrow Fe(s) + FeS_2(s)$

(b) $FeS_2(s) \longrightarrow S(s) + FeS(s)$

For (a), $\Delta G^\circ_r = -166.9\ kJ\cdot mol^{-1} - 2(-100.4\ kJ\cdot mol^{-1}) = +33.9\ kJ\cdot mol^{-1}$

This process is predicted to be nonspontaneous.

For (b), $\Delta G^\circ_r = -100.4\ kJ\cdot mol^{-1} - (-166.9\ kJ\cdot mol^{-1}) = +66.5\ kJ\cdot mol^{-1}$

This process is predicted to be nonspontaneous.

(c) The presence of added sulfur or iron should not affect these answers.

7.73 (a) Because the enthalpy change for dissolution is positive, the entropy change of the surroundings must be a negative number $\left(\Delta S^\circ_{surr} = -\dfrac{\Delta H^\circ_{system}}{T}\right)$. Because spontaneous processes are accompanied by an increase in entropy, the change in enthalpy does not favor the dissolution process. (b) In order for the process to be spontaneous (because it occurs readily, we know it is spontaneous), the entropy change of the system must be positive. (c) Locational disorder is dominant. (d) Because the surroundings participate in the solution process only as a source of heat, the entropy change of the surroundings is primarily a result of the dispersal of thermal motion. (e) The driving force for the dissolution is the dispersal of matter, resulting in an overall positive ΔS.

7.75 The values are calculated simply from the Hess's law relationship that the sums of the various energy quantities for the products minus the similar sum for the reactants will give the overall change in the state function desired:

C(s), graphite $\longrightarrow$ C(s), diamond

$\Delta H^\circ = \Delta H^\circ_f(\text{C(s), diamond}) = +1.895\ kJ\cdot mol^{-1}$

$$\begin{aligned}\Delta S^\circ &= \Delta S^\circ_m(\text{C(s), diamond}) - \Delta S^\circ_m(\text{C(s), graphite}) \\ &= 2.377\ J\cdot K^{-1}\cdot mol^{-1} - 5.740\ J\cdot K^{-1}\cdot mol^{-1} \\ &= -3.363\ J\cdot K^{-1}\cdot mol^{-1}\end{aligned}$$

$\Delta G^\circ = \Delta G^\circ_f(\text{C(s), diamond}) = +2.900\ kJ\cdot mol^{-1}$

Notice that the values used for the entropy calculation are absolute entropies, not ΔS° values and that the S°_m value for C(s), graphite, is not 0. Graphite has a delocalized structure similar to that of benzene, whereas in diamond, all the carbon atoms are bonded to four other carbon atoms in a very rigid lattice. It is not surprising that the change in entropy upon going from graphite to diamond would decrease, because we would expect graphite to have a higher molar entropy than diamond. Similarly, one can compare the bond formation and breaking that accompanies a change from graphite to diamond. Given the actual numbers, the change is clearly small however. In graphite, the carbon atom is bonded to three other carbon atoms with delocalized bonds (we can approximate this very roughly, using the values of 518 $kJ \cdot mol^{-1}$ of delocalized C—C bonds, as given in Table 6.7). In diamond, the carbon atom is bonded to four other carbon atoms by single C—C bonds (approximately by 348 $kJ \cdot mol^{-1}$). Even though these approximations overestimate the effect of delocalization on the C—C bond strength in graphite, the trend is expected—the three delocalized bonds in C(s) graphite are actually slightly more exothermic than the four C—C single bonds in diamond, making the standard enthalpy change a positive number. (Notice that if one considered the C—C bonds in graphite to be localized, the opposite prediction would have been made.) The standard free energy change for this reaction is positive as follows, from a positive standard enthalpy change and a negative standard entropy change for the reaction.

7.77 The entries all correspond to aqueous ions. The fact that they are negative is due to the reference point that has been established. Because ions cannot actually be separated and measured independently, a reference point that defines $S^{\circ}_{m}(H^{+}, aq) = 0$ has been established. This definition is then used to calculate the standard entropies for the other ions. The fact that they are negative will arise in part because the solvated ion $M(H_2O)_x^{n+}$ will be more ordered than the isolated ion and solvent molecules ($M^{n+} + x\ H_2O$).

7.79 In this reaction, we expect the change in enthalpy to be close to 0 (it should not be exactly 0) because the number and types of bonds that are broken are equal to the number and types of bonds that are formed. The free energy change should therefore be dominated by the entropy change of the reaction. We would expect the entropy change of the reaction to be positive, because the mixed complex $InCl_2Br$ would be expected to have a larger disorder than a molecule composed of all the same types of halide ions, InX_3. Consequently, we predict a positive value for ΔS°_{r} and a negative value of ΔG°_{r}.

7.81 (a) In order to calculate the free energy at different temperatures, we need to know ΔH° and ΔS° for the process: $H_2O(l) \longrightarrow H_2O(g)$

$$\Delta H^{\circ}_{r} = \Delta H^{\circ}_{f}(H_2O, g) - \Delta H^{\circ}_{f}(H_2O, l)$$
$$= (-241.82\ kJ \cdot mol^{-1}) - [-285.83\ kJ \cdot mol^{-1}]$$
$$= 44.01\ kJ \cdot mol^{-1}$$
$$\Delta S^{\circ}_{r} = S^{\circ}_{m}(H_2O, g) - S^{\circ}_{m}(H_2O, l)$$
$$= 188.83\ J \cdot K^{-1} \cdot mol^{-1} - [69.91\ J \cdot K^{-1} \cdot mol^{-1}]$$
$$= 118.92\ J \cdot K^{-1} \cdot mol^{-1}$$

$$\Delta G^\circ_r = \Delta H^\circ_r - T\Delta S^\circ_r$$
$$= 44.01\ \text{kJ}\cdot\text{mol}^{-1} - T(118.92\ \text{J}\cdot\text{K}^{-1}\cdot\text{mol}^{-1})/(1000\ \text{J}\cdot\text{kJ}^{-1})$$

T(K)	ΔG°_r(kJ)
298	8.57 kJ
373	−0.35 kJ
423	−6.29 kJ

The reaction goes from being nonspontaneous near room temperature to being spontaneous above 100°C.

(b) The value at 100°C should be exactly 0, because this is the normal boiling point of water.

(c) The discrepancy arises because the enthalpy and entropy values calculated from the tables are not rigorously constant with temperature. Better values would be obtained using the actual enthalpy and entropy of vaporization measured at the boiling point.

7.83 The dehydrogenation of cyclohexane to benzene follows the following equation:

$$C_6H_{12}(l) \longrightarrow C_6H_6(l) + 3\ H_2(g)$$

We can confirm that this process is nonspontaneous by calculating the ΔG°_r for the process, using data in Appendix 2A:

$$\Delta G^\circ_r = \Delta G^\circ_f(C_6H_6, l) - \Delta G^\circ_f(C_6H_{12}, l)$$
$$= 124.3\ \text{kJ}\cdot\text{mol}^{-1} - 26.7\ \text{kJ}\cdot\text{mol}^{-1}$$
$$= +97.6\ \text{kJ}\cdot\text{mol}^{-1}$$

The reaction of ethene with hydrogen can be examined similarly:

$$C_2H_2(g) + H_2(g) \longrightarrow C_2H_6(g)$$
$$\Delta G^\circ_r = \Delta G^\circ_f(C_2H_6, g) - \Delta G^\circ_f(C_2H_2, g)$$
$$= (-32.82\ \text{kJ}\cdot\text{mol}^{-1}) - 68.15\ \text{kJ}\cdot\text{mol}^{-1}$$
$$= -100.97\ \text{kJ}\cdot\text{mol}^{-1}$$

We can now combine these two reactions so that $C_2H_2(g)$ accepts the hydrogen that is formed in the dehydrogenation reaction:

$$C_6H_{12}(l) \longrightarrow C_6H_6(l) + 3\ H_2(g) \qquad \Delta G^\circ_r = +97.6\ \text{kJ}\cdot\text{mol}^{-1}$$
$$+\ 3[C_2H_2(g) + H_2(g) \longrightarrow C_2H_6(g)] \qquad \Delta G^\circ_r = 3(-100.97\ \text{kJ}\cdot\text{mol}^{-1})$$

$$C_6H_{12}(l) + 3\ C_2H_2(g) \longrightarrow C_6H_6(l) + 3\ C_2H_6(g) \qquad \Delta G^\circ_r = -205.13\ \text{kJ}\cdot\text{mol}^{-1}$$

We can see that by combining these two reactions, the overall process becomes spontaneous. Essentially, we are using the energy of the favorable reaction to drive the nonfavorable process.

7.85 (a)

(1) cis-2-butene: $H_3C(H)C{=}C(H)CH_3$ with both CH_3 groups on the same side

(2) trans-2-butene: $H_3C(H)C{=}C(CH_3)H$ with the CH_3 groups on opposite sides

(3) 2-methylpropene: $(H_3C)_2C{=}CH_2$

cis-2-butene *trans*-2-butene 2-methylpropene

(b) For the three reactions, the calculation of $\Delta G°$, $\Delta H°$, and $\Delta S°$ are as follows:

$\mathbf{1} \rightleftharpoons \mathbf{2}$

$$\Delta G°_r = \Delta G°_f(\mathbf{2}) - \Delta G°_f(\mathbf{1})$$
$$= 62.97\ \text{kJ}\cdot\text{mol}^{-1} - 65.86\ \text{kJ}\cdot\text{mol}^{-1}$$
$$= -2.89\ \text{kJ}\cdot\text{mol}^{-1}$$
$$\Delta H°_r = \Delta H°_f(\mathbf{2}) - \Delta H°_f(\mathbf{1})$$
$$= (-11.17\ \text{kJ}\cdot\text{mol}^{-1}) - (-6.99\ \text{kJ}\cdot\text{mol}^{-1})$$
$$= -4.18\ \text{kJ}\cdot\text{mol}^{-1}$$
$$\Delta G°_r = \Delta H°_r - T\Delta S°_r$$
$$-2.89\ \text{kJ}\cdot\text{mol}^{-1} = -4.18\ \text{kJ}\cdot\text{mol}^{-1} - (298\ \text{K})(\Delta S°_r)/(1000\ \text{J}\cdot\text{kJ}^{-1})$$
$$\Delta S°_r = -4.33\ \text{J}\cdot\text{K}^{-1}\cdot\text{mol}^{-1}$$

$\mathbf{1} \rightleftharpoons \mathbf{3}$

$$\Delta G°_r = \Delta G°_f(\mathbf{3}) - \Delta G°_f(\mathbf{1})$$
$$= 58.07\ \text{kJ}\cdot\text{mol}^{-1} - 65.86\ \text{kJ}\cdot\text{mol}^{-1}$$
$$= -7.79\ \text{kJ}\cdot\text{mol}^{-1}$$
$$\Delta H°_f = \Delta H°_f(\mathbf{3}) - \Delta H°_f(\mathbf{1})$$
$$= (-16.90\ \text{kJ}\cdot\text{mol}^{-1}) - (-6.99\ \text{kJ}\cdot\text{mol}^{-1})$$
$$= -9.91\ \text{kJ}\cdot\text{mol}^{-1}$$
$$\Delta G°_r = \Delta H°_r - T\Delta S°_r$$
$$-7.79\ \text{kJ}\cdot\text{mol}^{-1} = -9.91\ \text{kJ}\cdot\text{mol}^{-1} - (298\ \text{K})(\Delta S°_r)/(1000\ \text{J}\cdot\text{kJ}^{-1})$$
$$\Delta S°_r = -7.11\ \text{J}\cdot\text{K}^{-1}\cdot\text{mol}^{-1}$$

$\mathbf{2} \rightleftharpoons \mathbf{3}$

$$\Delta G°_r = \Delta G°_f(\mathbf{3}) - \Delta G°_f(\mathbf{2})$$
$$= 58.07\ \text{kJ}\cdot\text{mol}^{-1} - 62.97\ \text{kJ}\cdot\text{mol}^{-1}$$
$$= -4.90\ \text{kJ}\cdot\text{mol}^{-1}$$
$$\Delta H°_r = \Delta H°_f(\mathbf{3}) - \Delta H°_f(\mathbf{2})$$
$$= (-16.90\ \text{kJ}\cdot\text{mol}^{-1}) - (-11.17\ \text{kJ}\cdot\text{mol}^{-1})$$
$$= -5.73\ \text{kJ}\cdot\text{mol}^{-1}$$
$$\Delta G°_r = \Delta H°_r - T\Delta S°_r$$
$$-4.90\ \text{kJ}\cdot\text{mol}^{-1} = -5.73\ \text{kJ}\cdot\text{mol}^{-1} - (298\ \text{K})(\Delta S°_r)/(1000\ \text{J}\cdot\text{kJ}^{-1})$$
$$\Delta S°_r = -2.78\ \text{J}\cdot\text{K}^{-1}\cdot\text{mol}^{-1}$$

(b) The most stable of the three compounds is 2-methylpropene.

(c) Because $\Delta S°$ is also equal to the difference in the $S°_m$ values for the compounds, we can examine those values to place the three compounds in order of their relative absolute entropies. The ordering is $S°_m(\mathbf{1}) > S°_m(\mathbf{2}) > S°_m(\mathbf{3})$.

7.87 We need to calculate $\Delta H°_r$ and $\Delta S°_r$ for each process.

(a) $HCOOH(l) \longrightarrow H_2(g) + CO_2(g)$

$$\Delta H°_r = \Delta H°_f(CO_2, g) - \Delta H°_f(HCOOH, l)$$
$$= (-393.51 \text{ kJ·mol}^{-1}) - (-424.72 \text{ kJ·mol}^{-1})$$
$$= +31.21 \text{ kJ·mol}^{-1}$$
$$\Delta S°_r = S°_m(H_2, g) + S°(CO_2, g) - [S°_m(HCOOH, l)]$$
$$= 130.68 \text{ J·K}^{-1}\text{·mol}^{-1} + 213.74 \text{ J·K}^{-1}\text{·mol}^{-1} - [128.95 \text{ J·K}^{-1}\text{·mol}^{-1}]$$
$$= +215.47 \text{ J·K}^{-1}\text{·mol}^{-1}$$

The reaction will become spontaneous above the temperature at which $\Delta G°_r = 0$:

$$0 = 31.21 \text{ kJ·mol}^{-1} - T(215.47 \text{ J·K}^{-1}\text{·mol}^{-1})/(1000 \text{ J·kJ}^{-1})$$
$$T = 145 \text{ K}$$

(b) $CH_3COOH(l) \longrightarrow CH_4(g) + CO_2(g)$

$$\Delta H°_r = \Delta H°_f(CH_4, g) + \Delta H°_f(CO_2, g) - \Delta H°_f(CH_3COOH, l)$$
$$= (-74.81 \text{ kJ·mol}^{-1}) + (-393.51 \text{ kJ·mol}^{-1}) - (-484.5 \text{ kJ·mol}^{-1})$$
$$= +16.18 \text{ kJ·mol}^{-1}$$
$$\Delta S°_r = S°_m(CH_4, g) + S°(CO_2, g) - [S°_m(CH_3COOH, l)]$$
$$= 186.26 \text{ J·K}^{-1}\text{·mol}^{-1} + 213.74 \text{ J·K}^{-1}\text{·mol}^{-1} - [159.8 \text{ J·K}^{-1}\text{·mol}^{-1}]$$
$$= +240.2 \text{ J·K}^{-1}\text{·mol}^{-1}$$

The reaction will become spontaneous above the temperature at which $\Delta G°_r = 0$:

$$0 = 16.18 \text{ kJ·mol}^{-1} - T(240.2 \text{ J·K}^{-1}\text{·mol}^{-1})/(1000 \text{ J·kJ}^{-1})$$
$$T = 67 \text{ K}$$

(c) $C_6H_5COOH(s) \longrightarrow C_6H_6(l) + CO_2(g)$

$$\Delta H°_r = \Delta H°_f(C_6H_6, l) + \Delta H°_f(CO_2, g) - [\Delta H°_f(C_6H_5COOH, s)]$$
$$= +49.0 \text{ kJ·mol}^{-1} + (-393.51 \text{ kJ·mol}^{-1}) - [-385.1 \text{ kJ·mol}^{-1}]$$
$$= +40.6 \text{ kJ·mol}^{-1}$$
$$\Delta S°_r = S°_m(C_6H_6, l) + S°(CO_2, g) - [S°_m(C_6H_5COOH, s)]$$
$$= +173.3 \text{ J·K}^{-1}\text{·mol}^{-1} + 213.74 \text{ J·K}^{-1}\text{·mol}^{-1} - (167.6 \text{ J·K}^{-1}\text{·mol}^{-1})$$
$$= +219.4 \text{ J·K}^{-1}\text{·mol}^{-1}$$

The reaction will become spontaneous above the temperature at which $\Delta G°_r = 0$:

$$0 = 40.58 \text{ kJ·mol}^{-1} - T(219.4 \text{ J·K}^{-1}\text{·mol}^{-1})/(1000 \text{ J·kJ}^{-1})$$
$$T = 185 \text{ K}$$

It is clear from these calculations that all of these carboxylic acids are thermodynamically unstable with respect to decomposition to produce $CO_2(g)$. A consideration of the parameters shows that this is driven by the entropy increase in the production of the gas, because the enthalpy of the reaction in all cases is endothermic.

CHAPTER 8
PHYSICAL EQUILIBRIA

8.1 In a 1.0 L vessel at 20°C, there will be 17.5 Torr of water vapor. The ideal gas law can be used to calculate the mass of water present:

$$PV = nRT$$

let m = mass of water

$$\left(\frac{17.5 \text{ Torr}}{760 \text{ Torr}\cdot\text{atm}^{-1}}\right)(1.0 \text{ L}) = \left(\frac{m}{18.02 \text{ g}\cdot\text{mol}^{-1}}\right)(0.082\ 06 \text{ L}\cdot\text{atm}\cdot\text{K}^{-1}\cdot\text{mol}^{-1})(293 \text{ K})$$

$$m = \frac{(17.5 \text{ Torr})(1.0 \text{ L})(18.02 \text{ g}\cdot\text{mol}^{-1})}{(760 \text{ Torr}\cdot\text{atm}^{-1})(0.082\ 06 \text{ L}\cdot\text{atm}\cdot\text{K}^{-1}\cdot\text{mol}^{-1})(293 \text{ K})}$$

$$m = 0.017 \text{ g}$$

8.3 The volume of the bathroom (neglecting internal fixtures) is $4.0 \times 3.0 \times 3.0 \text{ m}^3 = 36 \text{ m}^3$ or $36 \text{ m}^3 \times 1000 \text{ L}\cdot\text{m}^{-3} = 36\ 000 \text{ L}$. The ideal gas law can be used to calculate the mass of water present:

$$PV = nRT$$

let m = mass of water

$$\left(\frac{7.4 \text{ kPa}}{101.325 \text{ kPa}\cdot\text{atm}^{-1}}\right)(36\ 000 \text{ L})$$

$$= \left(\frac{m}{18.02 \text{ g}\cdot\text{mol}^{-1}}\right)(0.082\ 06 \text{ L}\cdot\text{atm}\cdot\text{K}^{-1}\cdot\text{mol}^{-1})(313 \text{ K})$$

$$m = \frac{(7.4 \text{ kPa})(36\ 000 \text{ L})(18.02 \text{ g}\cdot\text{mol}^{-1})}{(101.325 \text{ kPa}\cdot\text{atm}^{-1})(0.082\ 06 \text{ L}\cdot\text{atm}\cdot\text{K}^{-1}\cdot\text{mol}^{-1})(313 \text{ K})}$$

$$m = 1.8 \times 10^3 \text{ g}$$

8.5 (a) 99.2°C; (b) 99.7°C

8.7 (a) The quantities $\Delta H°_{vap}$ and $\Delta S°_{vap}$ can be calculated using the relationship

$$\ln P = -\frac{\Delta H°_{vap}}{R}\cdot\frac{1}{T} + \frac{\Delta S°_{vap}}{R}$$

Because we have two temperatures with corresponding vapor pressures, we can set up two equations with two unknowns and solve for $\Delta H°_{vap}$ and $\Delta S°_{vap}$. If the equation is used as is, P must be expressed in atm, which is the standard reference state. Remember that the value used for P is really activity, which for pressure is P

divided by the reference state of 1 atm, so that the quantity inside the ln term is dimensionless.

$$8.314\ \text{J}\cdot\text{K}^{-1}\cdot\text{mol}^{-1} \times \ln\frac{58\ \text{Torr}}{760\ \text{Torr}} = -\frac{\Delta H^\circ_{\text{vap}}}{250.4\ \text{K}} + \Delta S^\circ_{\text{vap}}$$

$$8.314\ \text{J}\cdot\text{K}^{-1}\cdot\text{mol}^{-1} \times \ln\frac{512\ \text{Torr}}{760\ \text{Torr}} = -\frac{\Delta H^\circ_{\text{vap}}}{298.2\ \text{K}} + \Delta S^\circ_{\text{vap}}$$

which give, upon combining terms,

$$-21.39\ \text{J}\cdot\text{K}^{-1}\cdot\text{mol}^{-1} = -0.003\ 994\ \text{K}^{-1} \times \Delta H^\circ_{\text{vap}} + \Delta S^\circ_{\text{vap}}$$

$$-3.284\ \text{J}\cdot\text{K}^{-1}\cdot\text{mol}^{-1} = -0.003\ 353\ \text{K}^{-1} \times \Delta H^\circ_{\text{vap}} + \Delta S^\circ_{\text{vap}}$$

Subtracting one equation from the other will eliminate the $\Delta S^\circ_{\text{vap}}$ term and allow us to solve for $\Delta H^\circ_{\text{vap}}$:

$$-18.11\ \text{J}\cdot\text{K}^{-1}\cdot\text{mol}^{-1} = -0.000\ 641 \times \Delta H^\circ_{\text{vap}}$$

$$\Delta H^\circ_{\text{vap}} = +28.3\ \text{kJ}\cdot\text{mol}^{-1}$$

(b) We can then use $\Delta H^\circ_{\text{vap}}$ to calculate $\Delta S^\circ_{\text{vap}}$ using either of the two equations:

$$-21.39\ \text{J}\cdot\text{K}^{-1}\cdot\text{mol}^{-1} = -0.003\ 994\ \text{K}^{-1} \times (+28\ 200\ \text{J}\cdot\text{mol}^{-1}) + \Delta S^\circ_{\text{vap}}$$

$$\Delta S^\circ_{\text{vap}} = 91.2\ \text{J}\cdot\text{K}^{-1}\cdot\text{mol}^{-1}$$

$$-3.284\ \text{J}\cdot\text{K}^{-1}\cdot\text{mol}^{-1} = -0.003\ 353\ \text{K}^{-1} \times (+28\ 200\ \text{J}\cdot\text{mol}^{-1}) + \Delta S^\circ_{\text{vap}}$$

$$\Delta S^\circ_{\text{vap}} = 91.3\ \text{J}\cdot\text{K}^{-1}\cdot\text{mol}^{-1}$$

(c) The $\Delta G^\circ_{\text{vap}}$ is calculated using $\Delta G^\circ_{\text{r}} = \Delta H^\circ_{\text{r}} - T\Delta S^\circ_{\text{r}}$

$$\Delta G^\circ_{\text{r}} = +28.3\ \text{kJ}\cdot\text{mol}^{-1} - (298\ \text{K})(91.2\ \text{J}\cdot\text{K}^{-1}\cdot\text{mol}^{-1})/(1000\ \text{J}\cdot\text{kJ}^{-1})$$

$$\Delta G^\circ_{\text{r}} = +1.1\ \text{kJ}\cdot\text{mol}^{-1}$$

(d) The boiling point can be calculated using one of several methods. The easiest to use is the one developed in the last chapter:

$$\Delta G^\circ_{\text{vap}} = \Delta H^\circ_{\text{vap}} - T_{\text{B}}\,\Delta S^\circ_{\text{vap}} = 0$$

$$\Delta H^\circ_{\text{vap}} = T_{\text{B}}\,\Delta S^\circ_{\text{vap}} \text{ or } T_{\text{B}} = \frac{\Delta H^\circ_{\text{vap}}}{\Delta S^\circ_{\text{vap}}}$$

$$T_{\text{B}} = \frac{28.2\ \text{kJ}\cdot\text{mol}^{-1} \times 1000\ \text{J}\cdot\text{kJ}^{-1}}{91.2\ \text{J}\cdot\text{K}^{-1}\cdot\text{mol}^{-1}} = 309\ \text{K or } 36^\circ\text{C}$$

Alternatively, we could use the relationship $\ln\frac{P_2}{P_1} = -\frac{\Delta H^\circ_{\text{vap}}}{R}\left[\frac{1}{T_2} - \frac{1}{T_1}\right]$. Here we would substitute, in one of the known vapor pressure points, the value of the enthalpy of vaporization and the condition that $P = 1$ atm at the normal boiling point.

8.9 (a) The quantities $\Delta H^\circ_{\text{vap}}$ and $\Delta S^\circ_{\text{vap}}$ can be calculated using the relationship

$$\ln P = -\frac{\Delta H^\circ_{\text{vap}}}{R}\cdot\frac{1}{T} + \frac{\Delta S^\circ_{\text{vap}}}{R}$$

Because we have two temperatures with corresponding vapor pressures (we know that the vapor pressure = 1 atm at the boiling point), we can set up two equations with two unknowns and solve for ΔH°_{vap} and ΔS°_{vap}. If the equation is used as is, P must be expressed in atm, which is the standard reference state. Remember that the value used for P is really activity, which for pressure is P divided by the reference state of 1 atm, so that the quantity inside the ln term is dimensionless.

$$8.314\ \text{J}\cdot\text{K}^{-1}\cdot\text{mol}^{-1} \times \ln 1 = -\frac{\Delta H^\circ_{vap}}{292.7\ \text{K}} + \Delta S^\circ_{vap}$$

$$8.314\ \text{J}\cdot\text{K}^{-1}\cdot\text{mol}^{-1} \times \ln \frac{359\ \text{Torr}}{760\ \text{Torr}} = -\frac{\Delta H^\circ_{vap}}{273.2\ \text{K}} + \Delta S^\circ_{vap}$$

which give, upon combining terms,

$$0\ \text{J}\cdot\text{K}^{-1}\cdot\text{mol}^{-1} = -0.003\,416\ \text{K}^{-1} \times \Delta H^\circ_{vap} + \Delta S^\circ_{vap}$$

$$-6.235\ \text{J}\cdot\text{K}^{-1}\cdot\text{mol}^{-1} = -0.003\,660\ \text{K}^{-1} \times \Delta H^\circ_{vap} + \Delta S^\circ_{vap}$$

Subtracting one equation from the other will eliminate the ΔS°_{vap} term and allow us to solve for ΔH°_{vap}:

$$+6.235\ \text{J}\cdot\text{K}^{-1}\cdot\text{mol}^{-1} = +0.000\,244\ \text{K}^{-1} \times \Delta H^\circ_{vap}$$

$$\Delta H^\circ_{vap} = +25.6\ \text{kJ}\cdot\text{mol}^{-1}$$

(b) We can then use ΔH°_{vap} to calculate ΔS°_{vap} using either of the two equations:

$$0 = -0.003\,416\ \text{K}^{-1} \times (+25\,600\ \text{J}\cdot\text{mol}^{-1}) + \Delta S^\circ_{vap}$$

$$\Delta S^\circ_{vap} = 87.4\ \text{J}\cdot\text{K}^{-1}\cdot\text{mol}^{-1}$$

$$-6.235\ \text{J}\cdot\text{K}^{-1}\cdot\text{mol}^{-1} = -0.003\,660\ \text{K}^{-1} \times (+25\,600\ \text{J}\cdot\text{mol}^{-1}) + \Delta S^\circ_{vap}$$

$$\Delta S^\circ_{vap} = 87.5\ \text{J}\cdot\text{K}^{-1}\cdot\text{mol}^{-1}$$

(c) The vapor pressure at another temperature is calculated using

$$\ln \frac{P_2}{P_1} = -\frac{\Delta H^\circ_{vap}}{R}\left[\frac{1}{T_2} - \frac{1}{T_1}\right]$$

We need to insert the calculated value of the enthalpy of vaporization and one of the known vapor pressure points:

$$\ln \frac{P_{\text{at } 8.5^\circ\text{C}}}{1\ \text{atm}} = -\frac{25\,600\ \text{J}\cdot\text{mol}^{-1}}{8.314\ \text{J}\cdot\text{K}^{-1}\cdot\text{mol}^{-1}}\left[\frac{1}{291.7\ \text{K}} - \frac{1}{292.7\ \text{K}}\right]$$

$$P_{\text{at } 8.5^\circ\text{C}} = 1.0\ \text{atm or } 7.6 \times 10^2\ \text{Torr}$$

8.11 (a) The quantities ΔH°_{vap} and ΔS°_{vap} can be calculated using the relationship

$$\ln P = \frac{\Delta H^\circ_{vap}}{R}\cdot\frac{1}{T} + \frac{\Delta S^\circ_{vap}}{R}$$

Because we have two temperatures with corresponding vapor pressures, we can set up two equations with two unknowns and solve for ΔH°_{vap} and ΔS°_{vap}. If the equation is used as is, P must be expressed in atm, which is the standard reference state. Remember that the value used for P is really activity which, for pressure, is

P divided by the reference state of 1 atm, so that the quantity inside the ln term is dimensionless.

$$8.314\ \text{J}\cdot\text{K}^{-1}\cdot\text{mol}^{-1} \times \ln\frac{35\ \text{Torr}}{760\ \text{Torr}} = -\frac{\Delta H^\circ_{\text{vap}}}{161.2\ \text{K}} + \Delta S^\circ_{\text{vap}}$$

$$8.314\ \text{J}\cdot\text{K}^{-1}\cdot\text{mol}^{-1} \times \ln\frac{253\ \text{Torr}}{760\ \text{Torr}} = -\frac{\Delta H^\circ_{\text{vap}}}{189.6\ \text{K}} + \Delta S^\circ_{\text{vap}}$$

which give, upon combining terms,

$$-25.59\ \text{J}\cdot\text{K}^{-1}\cdot\text{mol}^{-1} = -0.006\ 203\ \text{K}^{-1} \times \Delta H^\circ_{\text{vap}} + \Delta S^\circ_{\text{vap}}$$

$$-9.145\ \text{J}\cdot\text{K}^{-1}\cdot\text{mol}^{-1} = -0.005\ 274\ \text{K}^{-1} \times \Delta H^\circ_{\text{vap}} + \Delta S^\circ_{\text{vap}}$$

Subtracting one equation from the other will eliminate the $\Delta S^\circ_{\text{vap}}$ term and allow us to solve for $\Delta H^\circ_{\text{vap}}$:

$$-16.45\ \text{J}\cdot\text{K}^{-1}\cdot\text{mol}^{-1} = -0.000\ 929 \times \Delta H^\circ_{\text{vap}}$$

$$\Delta H^\circ_{\text{vap}} = +17.7\ \text{kJ}\cdot\text{mol}^{-1}$$

(b) We can then use $\Delta H^\circ_{\text{vap}}$ to calculate $\Delta S^\circ_{\text{vap}}$ using either of the two equations:

$$-25.59\ \text{J}\cdot\text{K}^{-1}\cdot\text{mol}^{-1} = -0.006\ 203\ \text{K}^{-1} \times (+17\ 700\ \text{J}\cdot\text{mol}^{-1}) + \Delta S^\circ_{\text{vap}}$$

$$\Delta S^\circ_{\text{vap}} = 84\ \text{J}\cdot\text{K}^{-1}\cdot\text{mol}^{-1}$$

$$-9.145\ \text{J}\cdot\text{K}^{-1}\cdot\text{mol}^{-1} = -0.005\ 274\ \text{K}^{-1} \times (+17\ 700\ \text{J}\cdot\text{mol}^{-1}) + \Delta S^\circ_{\text{vap}}$$

$$\Delta S^\circ_{\text{vap}} = 84\ \text{J}\cdot\text{K}^{-1}\cdot\text{mol}^{-1}$$

(c) The $\Delta G^\circ_{\text{vap}}$ is calculated using $\Delta G^\circ_{\text{r}} = \Delta H^\circ_{\text{r}} - T\Delta S^\circ_{\text{r}}$

$$\Delta G^\circ_{\text{r}} = +17.7\ \text{kJ}\cdot\text{mol}^{-1} - (298\ \text{K})(84\ \text{J}\cdot\text{K}^{-1}\cdot\text{mol}^{-1})/(1000\ \text{J}\cdot\text{kJ}^{-1})$$

$$\Delta G^\circ_{\text{r}} = -7.3\ \text{kJ}\cdot\text{mol}^{-1}$$

Notice that the standard $\Delta G^\circ_{\text{r}}$ is negative, so that the vaporization of arsine is spontaneous, which is as expected; under those conditions arsine is a gas at room temperature.

(d) The boiling point can be calculated from one of several methods. The easiest to use is that developed in the last chapter:

$$\Delta G^\circ_{\text{vap}} = \Delta H^\circ_{\text{vap}} - T_B\Delta S^\circ_{\text{vap}} = 0$$

$$\Delta H^\circ_{\text{vap}} = T_{\text{B}}\ \Delta S^\circ_{\text{vap}} \text{ or } T_{\text{B}} = \frac{\Delta H^\circ_{\text{vap}}}{\Delta S^\circ_{\text{vap}}}$$

$$T_{\text{B}} = \frac{17.7\ \text{kJ}\cdot\text{mol}^{-1} \times 1000\ \text{J}\cdot\text{kJ}^{-1}}{84\ \text{J}\cdot\text{K}^{-1}\cdot\text{mol}^{-1}} = 211\ \text{K}$$

Alternatively, we could use the relationship $\ln\frac{P_2}{P_1} = -\frac{\Delta H^\circ_{\text{vap}}}{R}\left[\frac{1}{T_2} - \frac{1}{T_1}\right]$. Here, we would substitute, in one of the known vapor pressure points, the value of the enthalpy of vaporization and the condition that $P = 1$ atm at the normal boiling point.

8.13 (a) The quantities $\Delta H°_{vap}$ and $\Delta S°_{vap}$ can be calculated, using the relationship

$$\ln P = -\frac{\Delta H°_{vap}}{R}\cdot\frac{1}{T} + \frac{\Delta S°_{vap}}{R}$$

Because we have two temperatures with corresponding vapor pressures (we know that the vapor pressure = 1 atm at the boiling point), we can set up two equations with two unknowns and solve for $\Delta H°_{vap}$ and $\Delta S°_{vap}$. If the equation is used as is, P must be expressed in atm which is the standard reference state. Remember that the value used for P is really activity which, for pressure, is P divided by the reference state of 1 atm, so that the quantity inside the ln term is dimensionless.

$$8.314\ \text{J}\cdot\text{K}^{-1}\cdot\text{mol}^{-1} \times \ln 1 = -\frac{\Delta H°_{vap}}{315.58\ \text{K}} + \Delta S°_{vap}$$

$$8.314\ \text{J}\cdot\text{K}^{-1}\cdot\text{mol}^{-1} \times \ln\frac{140\ \text{Torr}}{760\ \text{Torr}} = -\frac{\Delta H°_{vap}}{273.2\ \text{K}} + \Delta S°_{vap}$$

which give, upon combining terms,

$0\ \text{J}\cdot\text{K}^{-1}\cdot\text{mol}^{-1} = -0.003\ 169\ \text{K}^{-1} \times \Delta H°_{vap} + \Delta S°_{vap}$

$-14.06\ \text{J}\cdot\text{K}^{-1}\cdot\text{mol}^{-1} = -0.003\ 660\ \text{K}^{-1} \times \Delta H°_{vap} + \Delta S°_{vap}$

Subtracting one equation from the other will eliminate the $\Delta S°_{vap}$ term and allow us to solve for $\Delta H°_{vap}$.

$-14.06\ \text{J}\cdot\text{K}^{-1}\cdot\text{mol}^{-1} = -0.000\ 491\ \text{K}^{-1} \times \Delta H°_{vap}$

$\Delta H°_{vap} = +28.6\ \text{kJ}\cdot\text{mol}^{-1}$

(b) We can then use $\Delta H°_{vap}$ to calculate $\Delta S°_{vap}$ using either of the two equations:

$0 = -0.003\ 169\ \text{K}^{-1} \times (+28\ 600\ \text{J}\cdot\text{mol}^{-1}) + \Delta S°_{vap}$

$\Delta S°_{vap} = 90.6\ \text{J}\cdot\text{K}^{-1}\cdot\text{mol}^{-1}$

$-14.06\ \text{J}\cdot\text{K}^{-1}\cdot\text{mol}^{-1} = -0.003\ 660\ \text{K}^{-1} \times (+28\ 600\ \text{J}\cdot\text{mol}^{-1}) + \Delta S°_{vap}$

$\Delta S°_{vap} = 90.6\ \text{J}\cdot\text{K}^{-1}\cdot\text{mol}^{-1}$

(c) The vapor pressure at another temperature is calculated using

$$\ln\frac{P_2}{P_1} = -\frac{\Delta H°_{vap}}{R}\left[\frac{1}{T_2} - \frac{1}{T_1}\right]$$

We need to insert the calculated value of the enthalpy of vaporization and one of the known vapor pressure points:

$$\ln\frac{P_{\text{at }25.0°\text{C}}}{1\ \text{atm}} = -\frac{28\ 600\ \text{J}\cdot\text{mol}^{-1}}{8.314\ \text{J}\cdot\text{K}^{-1}\cdot\text{mol}^{-1}}\left[\frac{1}{298.2\ \text{K}} - \frac{1}{315.58\ \text{K}}\right]$$

$P_{\text{at }25.0°\text{C}} = 0.53\ \text{atm or } 4.0 \times 10^2\ \text{Torr}$

8.15 Table 6.2 contains the enthalpy of vaporization and the boiling point of methanol (at which the vapor pressure = 1 atm). Using this data and the equation

$$\ln \frac{P_2}{P_1} = -\frac{\Delta H^\circ_{vap}}{R}\left[\frac{1}{T_2} - \frac{1}{T_1}\right]$$

$$\ln \frac{P_{25.0°C}}{1} = -\frac{35\ 300\ J\cdot mol^{-1}}{8.314\ J\cdot K^{-1}\cdot mol^{-1}}\left[\frac{1}{298.2\ K} - \frac{1}{337.2\ K}\right]$$

$P_{25.0°C} = 0.19$ atm or 1.5×10^2 Torr

8.17 The relationship of choice is

$$\ln \frac{P_2}{P_1} = -\frac{\Delta H^\circ_{vap}}{R}\left[\frac{1}{T_2} - \frac{1}{T_1}\right]$$

because the Table contains the enthalpy of vaporization and the normal boiling point (vapor pressure = 1 atm or 760 Torr).

$$\ln \frac{P_2}{1} = -\frac{5.93 \times 10^4\ J\cdot mol^{-1}}{8.314\ J\cdot K^{-1}\cdot mol^{-1}}\left[\frac{1}{448\ K} - \frac{1}{629.7\ K}\right]$$

$P_2 = 0.010$ atm or 7.6 Torr

8.19 (a) vapor; (b) liquid; (c) vapor; (d) vapor

8.21 (a) 2.4 K; (b) ~10 atm; (c) 5.5 K; (d) no

8.23 (a) At the lower pressure triple point, liquid helium-I and -II are in equilibrium with helium gas; at the higher pressure triple point, liquid helium-I and -II are in equilibrium with solid helium. (b) helium-I

8.25 The pressure increase would bring CO_2 into the solid region.

8.27 (a) KCl is an ionic solid, so water would be the best choice; (b) CCl_4 is nonpolar, so the best choice is benzene; (c) CH_3COOH is polar, so water is the better choice.

8.29 (a) hydrophilic, because NH_2 is polar, and has a lone pair and H atoms that can participate in hydrogen bonding to water molecules; (b) hydrophobic, because the CH_3 group is not very polar; (c) hydrophobic, because the Br group is not very polar; (d) hydrophilic, because the carboxylic acid group has lone pairs on oxygen and an acidic proton that can participate in hydrogen bonding to water molecules.

8.31 (a) The solubility of $O_2(g)$ in water is $1.3 \times 10^{-3}\ mol \cdot L^{-1} \cdot atm^{-1}$.

$$\text{solubility at 50 kPa} = \frac{50\ kPa}{101.325\ kPa \cdot atm^{-1}} \times 1.3 \times 10^{-3}\ mol \cdot L^{-1} \cdot atm^{-1}$$
$$= 6.4 \times 10^{-4}\ mol \cdot L^{-1}$$

(b) The solubility of $CO_2(g)$ in water is $2.3 \times 10^{-2}\ mol \cdot L^{-1} \cdot atm^{-1}$.

$$\text{solubility at 500 Torr} = \frac{500\ Torr}{760\ Torr \cdot atm^{-1}} \times 2.3 \times 10^{-2}\ mol \cdot L^{-1} \cdot atm^{-1}$$
$$= 1.5 \times 10^{-2}\ mol \cdot L^{-1}$$

(c) solubility $= 0.10\ atm \times 2.3 \times 10^{-2}\ mol \cdot L^{-1} \cdot atm^{-1} = 2.3 \times 10^{-3}\ mol \cdot L^{-1}$.

8.33 The question essentially asks how much CO_2 is dissolved in the solution. This we can calculate using the solubility from Table 8.5. This produces an answer in moles, which can be converted into volume by using the ideal gas equation:

$$PV = nRT \text{ or } V = \frac{nRT}{P}$$

$$V_{CO_2} = (3.00\ atm)(0.355\ L)(2.3 \times 10^{-2}\ mol \cdot L^{-1} \cdot atm^{-1}) \left(\frac{0.082\ 06\ L \cdot atm \cdot K^{-1} \cdot mol^{-1}}{1\ atm}\right)(293\ K)$$
$$= 0.59\ L$$

8.35 (a) By Henry's law, the concentration of CO_2 in solution will double. (b) No change in the equilibrium will occur; the partial pressure of CO_2 is unchanged and the concentration is unchanged.

8.37 (a) $4\ mg \cdot L^{-1} \times 1000\ mL \cdot L^{-1} \times 1.00\ g \cdot mL^{-1} \times 1\ kg/1000\ g$
$= 4\ mg \cdot L^{-1}$ or 4 ppm

(b) The solubility of O_2 (g) in water is $1.3 \times 10^{-3}\ mol \cdot L^{-1} \cdot atm^{-1}$ which can be converted to parts per million as follows:

In 1.00 L (corresponding to 1 kg) of solution there will be 1.3×10^{-3} mol O_2

$1.3 \times 10^{-3}\ mol \cdot kg^{-1} \cdot atm^{-1}\ O_2 \times 32.00\ g \cdot mol^{-1}\ O_2 \times 10^3\ mg \cdot g^{-1}$
$= 42\ mg \cdot kg^{-1} \cdot atm^{-1}$ or $42\ ppm \cdot atm^{-1}$

$4\ mg \div 41\ mg \cdot kg^{-1} \cdot atm^{-1} = 0.1\ atm$

(c) $P = \dfrac{0.1\ atm}{0.21} = 0.5\ atm$

8.39 Ion hydration enthalpies for ions of the same charge within a group of elements become less negative moving down the group. This trend parallels increasing ion size. Hydration energy is an ion-dipole interaction, so it should be higher for small, high charge density ions.

8.41 (a) Because it is exothermic, the enthalpy change must be negative; (b) $Li_2SO_4(s) \rightleftharpoons 2\ Li^+(aq) + SO_4^{2-}(aq) + \text{heat}$; (c) Given that $\Delta H_L + \Delta H_{\text{hydration}} = \Delta H$ of solution, the enthalpy of hydration should be larger. If the lattice energy were greater, the overall process would be endothermic.

8.43 To answer this question we must first determine the molar enthalpies of solution and multiply this by the number of moles of solid dissolved to get the actual amount of heat released.

(a) $\Delta H°_{\text{hydration}} = +3.9$

$$\Delta H = \frac{10.0\ \text{g NaCl}}{58.44\ \text{g}\cdot\text{mol}^{-1}} \times (+3.9\ \text{kJ}\cdot\text{mol}^{-1}) = 0.67\ \text{kJ or } 6.7 \times 10^2\ \text{J}$$

(b) $\Delta H°_{\text{hydration}} = -0.6$

$$\Delta H = \frac{10.0\ \text{g NaBr}}{102.90\ \text{g}\cdot\text{mol}^{-1}} \times (-0.6\ \text{kJ}\cdot\text{mol}^{-1}) = 0.06\ \text{kJ or } -6 \times 10\ \text{J}$$

(c) $\Delta H°_{\text{hydration}} = -329\ \text{kJ}\cdot\text{mol}^{-1}$

$$\Delta H = \frac{10.0\ \text{g AlCl}_3}{133.33\ \text{g}\cdot\text{mol}^{-1}} \times (-329\ \text{kJ}\cdot\text{mol}^{-1}) = -24.7\ \text{kJ}$$

(d) $\Delta H°_{\text{hydration}} = 25.7\ \text{k}\cdot\text{mol}^{-1}$

$$\Delta H = \frac{10.0\ \text{g NH}_4\text{NO}_3}{80.04\ \text{g}\cdot\text{mol}^{-1}} \times (+25.7\ \text{kJ}\cdot\text{mol}^{-1}) = +3.21\ \text{kJ}$$

8.45 From the tables we find that the $\Delta H°_L = 2153\ \text{kJ}\cdot\text{mol}^{-1}$, and hydration enthalpy of $Sr^{2+} = -1524\ \text{kJ}\cdot\text{mol}^{-1}$, hydration enthalpy of $Cl^- = -340\ \text{kJ}\cdot\text{mol}^{-1}$.

$$\begin{aligned}\Delta H_{\text{hydration}}(\text{SrCl}_2) &= 1\ \text{mol} \times \Delta H_{\text{hydration}}(\text{Sr}^{2+}) + 2\ \text{mol} \times \Delta H_{\text{hydration}}(\text{Cl}^-) \\ &= 1\ \text{mol} \times (-1524\ \text{kJ}\cdot\text{mol}^{-1}) + 2\ \text{mol} \times (-340\ \text{kJ}\cdot\text{mol}^{-1}) \\ &= -2204\ \text{kJ}\end{aligned}$$

This corresponds to 1 mol $SrCl_2$ so it can also be written as $-2204\ \text{kJ}\cdot\text{mol}^{-1}$.

$$\begin{aligned}\Delta H°_{\text{solution}} &= \Delta H°_L + \Delta H_{\text{hydration}} \\ &= 2153\ \text{kJ}\cdot\text{mol}^{-1} + (-2204\ \text{kJ}\cdot\text{mol}^{-1}) \\ &= -51\ \text{kJ}\cdot\text{mol}^{-1}\end{aligned}$$

8.47 All the enthalpies of solution are positive. Those of the alkali metal chlorides increase as the cation becomes larger and less strongly hydrated by water. All the alkali metal chlorides are soluble in water, but AgCl is not. AgCl has a relatively large positive enthalpy of solution. When dissolving is highly endothermic, the small increase in disorder due to solution formation may not be enough to compensate for the decrease in disorder for the surroundings, and a solution does not form. This is the case for AgCl.

8.49 (a) $$m_{\text{NaCl}} = \frac{\left(\dfrac{10.0\text{ g NaCl}}{58.44\text{ g}\cdot\text{mol}^{-1}}\right)}{0.250\text{ kg}} = 0.684\ m$$

(b) $$m_{\text{NaOH}} = \frac{\left(\dfrac{\text{mass NaOH}}{40.00\text{ g}\cdot\text{mol}^{-1}}\right)}{0.345\text{ kg}} = 0.22\ m$$

mass NaOH = 3.0 g

(c) $$m_{\text{urea}} = \frac{\left(\dfrac{1.54\text{ g urea}}{60.06\text{ g}\cdot\text{mol}^{-1}\text{ urea}}\right)}{(515\text{ cm}^3)(1.00\text{ g}\cdot\text{cm}^{-3})\left(\dfrac{1\text{ kg}}{1000\text{ g}}\right)} = 0.0498\ m$$

8.51 (a) 1 kg of 5.00% K_3PO_4 will contain 50.0 g K_3PO_4 and 950.0 g H_2O.

$$\frac{\left(\dfrac{50.0\text{ g K}_3\text{PO}_4}{212.27\text{ g}\cdot\text{mol}^{-1}\text{ K}_3\text{PO}_4}\right)}{0.950\text{ kg}} = 0.248\ m$$

(b) The mass of 1.00 L of solution will be 1043 g, which will contain 1043 g × 0.0500 = 52.2 g K_3PO_4.

$$\frac{\left(\dfrac{52.2\text{ g K}_3\text{PO}_4}{212.27\text{ g}\cdot\text{mol}^{-1}\text{ K}_3\text{PO}_4}\right)}{1.00\text{ L}} = 0.246\text{ M}$$

8.53 (a) If x_{MgCl_2} is 0.0175, then there are 0.0175 mol $MgCl_2$ for every 0.9825 mol H_2O. The mass of water will be 18.02 $\text{g}\cdot\text{mol}^{-1}$ × 0.9825 mol = 17.70 g or 0.017 70 kg.

$$m_{\text{Cl}^-} = \frac{\left(\dfrac{2\text{ Cl}^-}{\text{MgCl}_2}\right)(0.0175\text{ mol MgCl}_2)}{0.017\,70\text{ kg solvent}} = 1.98\ m$$

(b) $$m_{\text{Cl}^-} = \frac{\left(\dfrac{7.12\text{ g NaOH}}{40.00\text{ g}\cdot\text{mol}^{-1}\text{ NaOH}}\right)}{0.325\text{ kg solvent}} = 0.548\ m$$

(c) 1.000 L of 15.00 M HCl(aq) will contain 15.00 mol with a mass of 15.00 × 36.46 $\text{g}\cdot\text{mol}^{-1}$ = 546.9 g. The density of the 1.000 L of solution is 1.0745 $\text{g}\cdot\text{cm}^{-3}$ so the total mass in the solution is 1074.5 g. This leaves 1074.5 g − 546.9 g = 527.6 g as water.

$$\frac{15.00 \text{ mol HCl}}{0.5276 \text{ kg solvent}} = 28.43\ m$$

8.55 (a) Molar mass of $CaCl_2 \cdot 6\ H_2O = 219.08\ g \cdot mol^{-1}$, which consists of 110.98 g $CaCl_2$ and 108.10 g of water.

$$m_{CaCl_2} = \frac{x \text{ mol } CaCl_2 \cdot 6\ H_2O}{0.500 \text{ kg} + x\ (6 \times 0.018\ 02 \text{ kg } H_2O)}$$

Note: 18.02 g H_2O = 0.018 02 kg H_2O = 1.000 mol H_2O

x = number of moles of $CaCl_2 \cdot 6\ H_2O$ needed to prepare a solution of molality m_{CaCl_2}, in which each mole of $CaCl_2 \cdot 6\ H_2O$ produces 6(0.018 02 kg) of water as solvent (assuming we begin with 0.250 kg H_2O).

For a 0.175 m solution of $CaCl_2 \cdot 6\ H_2O$,

$$0.175\ m = \frac{x}{0.500 \text{ kg} + x(6)(0.018\ 02 \text{ kg } H_2O)}$$

$x = 0.0875 \text{ mol} + x(0.0189 \text{ mol})$

$x - 0.0189\ x = 0.0875 \text{ mol}$

$0.981\ x = 0.0875 \text{ mol}$

$x = 0.0892 \text{ mol } CaCl_2 \cdot 6\ H_2O$

$$\therefore (0.0892 \text{ mol } CaCl_2 \cdot 6\ H_2O) \times \left(\frac{219.08 \text{ g } CaCl_2 \cdot 6\ H_2O}{1 \text{ mol } CaCl_2 \cdot 6\ H_2O}\right) =$$

19.5 g $CaCl_2 \cdot 6\ H_2O$

(b) Molar mass of $NiSO_4 \cdot 6\ H_2O = 262.86\ g \cdot mol^{-1}$, which consists of 154.77 g $NiSO_4$ and 108.09 g H_2O.

$$m_{NiSO_4} = \frac{x \text{ mol } NiSO_4 \cdot 6\ H_2O}{0.500 \text{ kg} + x\ (6 \times 0.018\ 02 \text{ kg } H_2O)}$$

where x = number of moles of $NiSO_4 \cdot 6\ H_2O$ needed to prepare a solution of molality m_{NiSO_4}, in which each mole of $NiSO_4 \cdot 6\ H_2O$ produces 6(0.018 02 kg) of water as solvent. Assuming we begin with 0.500 kg H_2O, for a 0.33 m solution of $NiSO_4 \cdot 6\ H_2O$,

$$0.33\ m = \frac{x}{0.500 \text{ kg} + x(6)(0.018\ 02 \text{ kg } H_2O)}$$

$x = 0.16 \text{ mol} + x(0.0357 \text{ mol})$

$x - 0.0357\ x = 0.16 \text{ mol}$

$0.964\ x = 0.16 \text{ mol}$

$x = 0.17 \text{ mol } NiSO_4 \cdot 6\ H_2O$

$$\therefore (0.17 \text{ mol } NiSO_4 \cdot 6\ H_2O) \times \left(\frac{262.86 \text{ g } NiSO_4 \cdot 6\ H_2O}{1 \text{ mol } NiSO_4 \cdot 6\ H_2O}\right) = 45 \text{ g } NiSO_4 \cdot 6\ H_2O$$

8.57 (a) $P = x_{\text{solvent}} \times P_{\text{pure solvent}}$

At 100°C, the normal boiling point of water, the vapor pressure of water is 1.00 atm. If the mole fraction of sucrose is 0.100, then the mole fraction of water is 0.900:

$P = 0.900 \times 1.000 \text{ atm} = 0.900 \text{ atm or } 684 \text{ Torr}$

(b) First, the molality must be converted to mole fraction. If the molality is 0.100 $\text{mol}\cdot\text{kg}^{-1}$, then there will be 0.100 mol sucrose per 1000 g of water.

$$x_{H_2O} = \frac{n_{H_2O}}{n_{H_2O} + n_{\text{sucrose}}} = \frac{\dfrac{1000 \text{ g}}{18.02 \text{ g}\cdot\text{mol}^{-1}}}{\dfrac{1000 \text{ g}}{18.02 \text{ g}\cdot\text{mol}^{-1}} + 0.100 \text{ mol}} = 0.998$$

$P = 0.998 \times 1.000 \text{ atm} = 0.998 \text{ atm or } 758 \text{ Torr}$

8.59 (a) The vapor pressure of water at 0°C is 4.58 Torr. The concentration of the solution must be converted to mole fraction in order to perform the calculation. A solution that is 2.50% ethylene glycol by mass will contain 2.50 g of ethylene glycol per 97.50 g of water.

$$x_{H_2O} = \frac{n_{H_2O}}{n_{H_2O} + n_{\text{ethylene glycoal}}} = \frac{\dfrac{97.50 \text{ g}}{18.02 \text{ g}\cdot\text{mol}^{-1}}}{\dfrac{97.50 \text{ g}}{18.02 \text{ g}\cdot\text{mol}^{-1}} + \dfrac{2.50 \text{ g}}{62.07 \text{ g}\cdot\text{mol}^{-1}}} = 0.993$$

$P = x_{\text{solvent}} \times P_{\text{pure solvent}}$

$P = 0.993 \times 4.58 \text{ Torr} = 4.55 \text{ Torr}$

(b) The vapor pressure of water is 355.26 Torr at 80°C. The concentration given in $\text{mol}\cdot\text{kg}^{-1}$ must be converted to mole fraction.

$$x_{H_2O} = \frac{n_{H_2O}}{n_{H_2O} + n_{Na^+} + n_{OH^-}} = \frac{\dfrac{1000 \text{ g}}{18.02 \text{ g}\cdot\text{mol}^{-1}}}{\dfrac{1000 \text{ g}}{18.02 \text{ g}\cdot\text{mol}^{-1}} + 2 \times 0.155 \text{ mol}} = 0.9944$$

$P = 0.9972 \times 355.26 \text{ Torr} = 354.3 \text{ Torr}$

(c) At 10°C, the vapor pressure of water is 9.21 Torr. The concentration must be expressed in terms of mole fraction:

$$x_{H_2O} = \frac{n_{H_2O}}{n_{H_2O} + n_{\text{urea}}} = \frac{\dfrac{100 \text{ g}}{18.02 \text{ g}\cdot\text{mol}^{-1}}}{\dfrac{100 \text{ g}}{18.02 \text{ g}\cdot\text{mol}^{-1}} + \dfrac{5.95 \text{ g}}{60.06 \text{ g}\cdot\text{mol}^{-1}}} = 0.982$$

$P = 0.982 \times 9.21 \text{ Torr} = 9.04 \text{ Torr}$

The change in the vapor pressure will therefore be $9.21 - 9.04 = 0.17$ Torr

8.61 (a) From the relationship $P = x_{\text{solvent}} \times P_{\text{pure solvent}}$ we can calculate the mole fraction of the solvent:

$94.8 \text{ Torr} = x_{\text{solvent}} \times 100.0 \text{ Torr}$

$x_{\text{solvent}} = 0.948$

The mole fraction of the unknown compound will be $1.000 - 0.948 = 0.052$.

(b) The molar mass can be calculated by using the definition of mole fraction for either the solvent or solute. In this case, the math is slightly easier if the definition of mole fraction of the solvent is used:

$$x_{\text{solvent}} = \frac{n_{\text{solvent}}}{n_{\text{unknown}} + n_{\text{solvent}}}$$

$$0.948 = \frac{\dfrac{100 \text{ g}}{78.11 \text{ g}\cdot\text{mol}^{-1}}}{\dfrac{100 \text{ g}}{78.11 \text{ g}\cdot\text{mol}^{-1}} + \dfrac{8.05 \text{ g}}{M_{\text{unknown}}}}$$

$$M_{\text{unknown}} = \frac{8.05 \text{ g}}{\left[\left(\dfrac{100 \text{ g}}{78.11 \text{ g}\cdot\text{mol}^{-1}}\right)\Big/0.948\right] - \left(\dfrac{100 \text{ g}}{78.11 \text{ g}\cdot\text{mol}^{-1}}\right)} = 115 \text{ g}\cdot\text{mol}^{-1}$$

8.63 (a) $\Delta T_{\text{b}} = ik_{\text{b}}m$

Because sucrose is a nonelectrolyte, $i = 1$.

$\Delta T_{\text{b}} = 0.51 \text{ K}\cdot\text{kg}\cdot\text{mol}^{-1} \times 0.10 \text{ mol}\cdot\text{kg}^{-1} = 0.051 \text{ K or } 0.051°\text{C}$

The boiling point will be 100.000°C + 0.051°C = 100.051°C.

(b) $\Delta T_{\text{b}} = ik_{\text{b}}m$

For NaCl, $i = 2$.

$\Delta T_{\text{b}} = 2 \times 0.51 \text{ K}\cdot\text{kg}\cdot\text{mol}^{-1} \times 0.22 \text{ mol}\cdot\text{kg}^{-1} = 0.22 \text{ K or } 0.22°\text{C}$

The boiling point will be 100.00°C + 0.22°C = 100.22°C.

(c) $\Delta T_{\text{b}} = ik_{\text{b}}m$

$$\Delta T_{\text{b}} = 2 \times 0.51 \text{ K}\cdot\text{kg}\cdot\text{mol}^{-1} \times \frac{\left(\dfrac{0.230 \text{ g}}{25.94 \text{ g}\cdot\text{mol}^{-1}}\right)}{0.100 \text{ kg}} = 0.090 \text{ K or } 0.090°\text{C}$$

The boiling point will be 100.000°C + 0.090°C or 100.090°C.

8.65 (a) Pure water has a vapor pressure of 760.00 Torr at 100°C. The mole fraction of the solution can be determined from $P = x_{\text{solvent}} \times P_{\text{pure solvent}}$.

$751 \text{ Torr} = x_{\text{solvent}} \times 760.00 \text{ Torr}$

$x_{\text{solvent}} = 0.988$

The mole fraction needs to be converted to molality:

$$x_{\text{solvent}} = 0.988 = \frac{n_{H_2O}}{n_{H_2O} + n_{\text{solute}}}$$

Because the absolute amount of solution is not important, we can assume that the total number of moles = 1.00.

$$0.988 = \frac{n_{H_2O}}{1.00 \text{ mol}}$$

$n_{H_2O} = 0.988 \text{ mol}$; $n_{\text{solute}} = 0.012 \text{ mol}$

$$\text{molality} = \frac{0.012 \text{ mol}}{\left(\frac{0.988 \text{ mol} \times 18.02 \text{ g}\cdot\text{mol}^{-1}}{1000 \text{ g}\cdot\text{kg}^{-1}}\right)} = 0.67 \text{ mol}\cdot\text{kg}^{-1}$$

Knowing the mole fraction, one can calculate ΔT_b:

$\Delta T_b = k_b m$

$\Delta T_b = 0.51 \text{ K}\cdot\text{mol}\cdot\text{kg}^{-1} \times 0.67 \text{ mol}\cdot\text{kg}^{-1} = 0.34 \text{ K or } 0.34°\text{C}$

Boiling point = 100.00°C + 0.34°C = 100.34°C

(b) The procedure is the same as in (a). Pure benzene has a vapor pressure of 760.00 Torr at 80.1°C. The mole fraction of the solution can be determined from

$P = x_{\text{solvent}} \times P_{\text{pure solvent}}$

$740 \text{ Torr} = x_{\text{solvent}} \times 760.00 \text{ Torr}$

$x_{\text{solvent}} = 0.974$

The mole fraction needs to be converted to molality:

$$x_{\text{solvent}} = 0.974 = \frac{n_{\text{benzene}}}{n_{\text{benzene}} + n_{\text{solute}}}$$

Because the absolute amount of solution is not important, we can assume that the total number of moles = 1.00:

$$0.974 = \frac{n_{\text{benzene}}}{1.00 \text{ mol}}$$

$n_{\text{benzene}} = 0.974 \text{ mol}$; $n_{\text{solute}} = 0.026 \text{ mol}$

$$\text{molality} = \frac{0.026 \text{ mol}}{\left(\frac{0.974 \text{ mol} \times 78.11 \text{ g}\cdot\text{mol}^{-1}}{1000 \text{ g}\cdot\text{kg}^{-1}}\right)} = 0.34 \text{ mol}\cdot\text{kg}^{-1}$$

Knowing the mole fraction, one can calculate ΔT_b:

$\Delta T_b = k_b m$

$\Delta T_b = 2.53 \text{ K}\cdot\text{kg}\cdot\text{mol}^{-1} \times 0.34 \text{ mol}\cdot\text{kg}^{-1} = 0.86 \text{ K or } 0.86 \text{ °C}$

Boiling point = 80.1°C + 0.86°C = 81.0°C

8.67 $\Delta T_b = 61.51°\text{C} - 61.20°\text{C} = 0.31°\text{C or } 0.31 \text{ K}$

$\Delta T_b = k_b \times \text{molality}$

$0.31 \text{ K} = 4.95 \text{ K}\cdot\text{kg}\cdot\text{mol}^{-1} \times \text{molality}$

$$0.31\ \text{K} = 4.95\ \text{K}\cdot\text{kg}\cdot\text{mol}^{-1} \times \frac{\left(\dfrac{1.05\ \text{g}}{M_{\text{unknown}}}\right)}{0.100\ \text{kg}}$$

$$\frac{0.100\ \text{kg} \times 0.31\ \text{K}}{4.95\ \text{K}\cdot\text{kg}\cdot\text{mol}^{-1}} = \frac{1.05\ \text{g}}{M_{\text{unknown}}}$$

$$M_{\text{unknown}} = \frac{1.05\ \text{g} \times 4.95\ \text{K}\cdot\text{kg}\cdot\text{mol}^{-1}}{0.100\ \text{kg} \times 0.31\ \text{K}} = 1.7 \times 10^{2}\ \text{g}\cdot\text{mol}^{-1}$$

8.69 (a) $\Delta T_f = ik_f m$

Because sucrose is a nonelectrolyte, $i = 1$.

$\Delta T_f = 1.86\ \text{K}\cdot\text{kg}\cdot\text{mol}^{-1} \times 0.10\ \text{mol}\cdot\text{kg}^{-1} = 0.19\ \text{K}$ or 0.19°C

The freezing point will be 0.000°C − 0.19°C = −0.19°C.

(b) $\Delta T_f = ik_f m$

For NaCl, $i = 2$

$\Delta T_f = 2 \times 1.86\ \text{K}\cdot\text{kg}\cdot\text{mol}^{-1} \times 0.22\ \text{mol}\cdot\text{kg}^{-1} = 0.82\ \text{K}$ or 0.82°C

The freezing point will be 0.00°C − 0.82°C = −0.82°C.

(c) $\Delta T_f = ik_f m$

$i = 2$ for LiF

$$\Delta T_f = 2 \times 1.86\ \text{K}\cdot\text{kg}\cdot\text{mol}^{-1} \times \frac{\left(\dfrac{0.120\ \text{g}}{25.94\ \text{g}\cdot\text{mol}^{-1}}\right)}{0.100\ \text{kg}} = 0.172\ \text{K or } 0.172°\text{C}$$

The freezing point will be 0.000°C − 0.172°C = −0.172°C.

8.71 $\Delta T_f = 179.8°\text{C} - 176.9°\text{C} = 2.9°\text{C}$ or 2.9 K

$\Delta T_f = k_f m$

$2.9\ \text{K} = (39.7\ \text{K}\cdot\text{kg}\cdot\text{mol}^{-1})m$

$$2.9\ \text{K} = (39.7\ \text{K}\cdot\text{kg}\cdot\text{mol}^{-1})\frac{\left(\dfrac{1.14\ \text{g}}{M_{\text{unknown}}}\right)}{0.100\ \text{kg}}$$

$$\frac{0.100\ \text{kg} \times 2.9\ \text{K}}{39.7\ \text{K}\cdot\text{kg}\cdot\text{mol}^{-1}} = \frac{1.14\ \text{g}}{M_{\text{unknown}}}$$

$$M_{\text{unknown}} = \frac{1.14\ \text{g} \times 39.7\ \text{K}\cdot\text{kg}\cdot\text{mol}^{-1}}{0.100\ \text{kg} \times 2.9\ \text{K}} = 1.6 \times 10^{2}\ \text{g}\cdot\text{mol}^{-1}$$

8.73 (a) First, calculate the molality using the change in boiling point, and then use that value to calculate the change in freezing point.

$\Delta T_b = k_b m$

$\Delta T_b = 82.0°\text{C} - 80.1°\text{C} = 1.9°\text{C}$ or 1.9 K

$1.9\ \text{K} = 2.53\ \text{K}\cdot\text{kg}\cdot\text{mol}^{-1} \times \text{molality}$

$\text{molality} = 0.75\ \text{mol}\cdot\text{kg}^{-1}$

$\Delta T_f = k_f m$

$\Delta T_f = 5.12\ \text{K}\cdot\text{kg}\cdot\text{mol}^{-1} \times 0.75\ \text{mol}\cdot\text{kg}^{-1}$

$\Delta T_f = 3.84\ \text{K or } 3.84°\text{C}$

The freezing point will be 5.5°C − 3.84°C = 1.7°C

(b) $\Delta T_f = k_f m$

Because the freezing point of water = 0.00°C, the freezing point of the solution equals the freezing point depression.

$3.04\ \text{K} = 1.86\ \text{K}\cdot\text{kg}\cdot\text{mol}^{-1} \times \text{molality}$

$\text{molality} = 1.63\ \text{mol}\cdot\text{kg}^{-1}$

(c) $\Delta T_f = k_f m$

$1.94\ \text{K} = (1.86\ \text{K}\cdot\text{kg}\cdot\text{mol}^{-1})m$

$$1.94\ \text{K} = 1.86\ \text{K}\cdot\text{kg}\cdot\text{mol}^{-1} \times \frac{n_{\text{solute}}}{\text{kg (solvent)}}$$

$$1.94\ \text{K} = 1.86\ \text{K}\cdot\text{kg}\cdot\text{mol}^{-1} \times \frac{n_{\text{solute}}}{0.200\ \text{kg}}$$

$n_{\text{solute}} = 0.209\ \text{mol}$

8.75 (a) A 1.00% aqueous solution of NaCl will contain 1.00 g of NaCl for 99.0 g of water. To use the freezing point depression equation, we need the molality of the solution:

$$\text{molality} = \frac{\left(\dfrac{1.00\ \text{g}}{58.44\ \text{g}\cdot\text{mol}^{-1}}\right)}{0.0990\ \text{kg}} = 0.173\ \text{mol}\cdot\text{kg}^{-1}$$

$\Delta T_f = i k_f m$

$\Delta T_f = i\,(1.86\ \text{K}\cdot\text{kg}\cdot\text{mol}^{-1})(0.173\ \text{mol}\cdot\text{kg}^{-1}) = 0.593\ \text{K}$

$i = 1.84$

(b) molality of all solute species (undissociated NaCl(aq) plus Na^+(aq) + Cl^-(aq)) $= 1.84 \times 0.173\ \text{mol}\cdot\text{kg}^{-1} = 0.318\ \text{mol}\cdot\text{kg}^{-1}$

(c) If all the NaCl had dissociated, the total molality in solution would have been $0.346\ \text{mol}\cdot\text{kg}^{-1}$, giving an i value equal to 2. If no dissociation had taken place, the molality in solution would have equaled $0.173\ \text{mol}\cdot\text{kg}^{-1}$.

$$\begin{array}{lll} \text{NaCl(aq)} & \rightleftharpoons & Na^+\text{(aq)} + Cl^-\text{(aq)} \\ 0.173\ \text{mol}\cdot\text{kg}^{-1} - x & & x \qquad\quad x \end{array}$$

$0.173\ \text{mol}\cdot\text{kg}^{-1} - x + x + x = 0.318\ \text{mol}\cdot\text{kg}^{-1}$

$0.173\ \text{mol}\cdot\text{kg}^{-1} + x = 0.318\ \text{mol}\cdot\text{kg}^{-1}$

$x = 0.145\ \text{mol}\cdot\text{kg}^{-1}$

$$\% \text{ dissociation} = \frac{0.145 \text{ mol·kg}^{-1}}{0.173 \text{ mol·kg}^{-1}} \times 100 = 83.8\%$$

8.77 For an electrolyte that dissociates into two ions, the van't Hoff i factor will be 1 plus the degree of dissociation, in this case 0.075. This can be readily seen for the general case MX. Let A = initial concentration of MX (if none is dissociated) and let Y = the concentration of MX that subsequently dissociates:

$$\begin{array}{ccc} MX(aq) \longrightarrow & M^{n+}(aq) + & X^{n-}(aq) \\ A - Y & Y & Y \end{array}$$

The total concentration of solute species is (A − Y) + Y + Y = A + Y

The value of i will then be equal to A + Y or 1.075.

The freezing point change is then easy to calculate:

$$\begin{aligned} \Delta T_f &= ik_f m \\ &= 1.075 \times 1.86 \text{ K·kg·mol}^{-1} \times 0.10 \text{ mol·kg}^{-1} \\ &= 0.20 \end{aligned}$$

Freezing point of the solution will be 0.00°C − 0.20°C = −0.20°C.

8.79 (a) $\Pi = iRT \times \text{molarity}$

$$\begin{aligned} &= 1 \times 0.082\ 06 \text{ L·atm·K}^{-1}\text{·mol}^{-1} \times 293 \text{ K} \times 0.010 \text{ mol·L}^{-1} \\ &= 0.24 \text{ atm or } 1.8 \times 10^2 \text{ Torr} \end{aligned}$$

(b) Because HCl is a strong acid, it should dissociate into two ions, H^+ and Cl^-, so $i = 2$.

$$\begin{aligned} \Pi &= 2 \times 0.082\ 06 \text{ L·atm·K}^{-1}\text{·mol}^{-1} \times 293 \text{ K} \times 1.0 \text{ mol·L}^{-1} \\ &= 48 \text{ atm} \end{aligned}$$

(c) $CaCl_2$ should dissociate into 3 ions in solution, therefore $i = 3$.

$$\begin{aligned} \Pi &= 3 \times 0.082\ 06 \text{ L·atm·K}^{-1}\text{·mol}^{-1} \times 293 \text{ K} \times 0.010 \text{ mol·L}^{-1} \\ &= 0.72 \text{ atm or } 5.5 \times 10^2 \text{ Torr} \end{aligned}$$

8.81 The polypeptide is a nonelectrolyte, so $i = 1$.

$\Pi = iRT \times \text{molarity}$

$$\Pi = \frac{3.74 \text{ Torr}}{760 \text{ Torr·atm}^{-1}} = 1 \times 0.082\ 06 \text{ L·atm·K}^{-1}\text{·mol}^{-1} \times 300 \text{ K} \times \frac{\left(\frac{0.40 \text{ g}}{M_{\text{unknown}}}\right)}{1.0 \text{ L}}$$

$$\begin{aligned} M_{\text{unknown}} &= \frac{0.082\ 06 \text{ L·atm·K}^{-1}\text{·mol}^{-1} \times 300 \text{ K} \times 0.40 \text{ g} \times 760 \text{ Torr·atm}^{-1}}{3.74 \text{ Torr} \times 1.0 \text{ L}} \\ &= 2.0 \times 10^3 \text{ g·mol}^{-1} \end{aligned}$$

8.83 We assume the polymer to be a nonelectrolyte, so $i = 1$.

$\Pi = iRT \times \text{molarity}$

$$\Pi = \frac{5.4\ \text{Torr}}{760\ \text{Torr}\cdot\text{atm}^{-1}} = 1 \times 0.082\ 06\ \text{L}\cdot\text{atm}\cdot\text{K}^{-1}\cdot\text{mol}^{-1} \times 293\ \text{K} \times \frac{\left(\dfrac{0.10\ \text{g}}{M_{\text{unknown}}}\right)}{0.100\ \text{L}}$$

$$M_{\text{unknown}} = \frac{0.082\ 06\ \text{L}\cdot\text{atm}\cdot\text{K}^{-1}\cdot\text{mol}^{-1} \times 293\ \text{K} \times 0.10\ \text{g} \times 760\ \text{Torr}\cdot\text{atm}^{-1}}{5.4\ \text{Torr} \times 0.100\ \text{L}}$$

$$= 3.4 \times 10^{3}\ \text{g}\cdot\text{mol}^{-1}$$

8.85 (a) $C_{12}H_{22}O_{11}$ should be a nonelectrolyte, so $i = 1$.

$\Pi = iRT \times \text{molarity}$

$= 1 \times 0.082\ 06\ \text{L}\cdot\text{atm}\cdot\text{K}^{-1}\cdot\text{mol}^{-1} \times 293\ \text{K} \times 0.050\ \text{mol}\cdot\text{L}^{-1}$

$= 1.2\ \text{atm}$

(b) NaCl dissociates to give 2 ions in solution, so $i = 2$.

$\Pi = iRT \times \text{molarity}$

$= 2 \times 0.082\ 06\ \text{L}\cdot\text{atm}\cdot\text{K}^{-1}\cdot\text{mol}^{-1} \times 293\ \text{K} \times 0.0010\ \text{mol}\cdot\text{L}^{-1}$

$= 0.048\ \text{atm or } 36\ \text{Torr}$

(c) AgCN dissociates in solution to give two ions (Ag^{+} and CN^{-}), so $i = 2$.

$\Pi = iRT \times \text{molarity}$

We must assume that the AgCN does not significantly affect either the volume or density of the solution, which is reasonable given the very small amount of it that dissolves.

$$\Pi = 2 \times 0.082\ 06\ \text{L}\cdot\text{atm}\cdot\text{K}^{-1}\cdot\text{mol}^{-1} \times 293\ \text{K} \times \frac{\left(\dfrac{2.3 \times 10^{-5}\ \text{g}}{133.89\ \text{g}\cdot\text{mol}^{-1}}\right)}{\left(\dfrac{100\ \text{g H}_2\text{O}}{1\ \text{g}\cdot\text{cm}^{-3}\ \text{H}_2\text{O} \times 1000\ \text{cm}^{3}\cdot\text{L}^{-1}}\right)}$$

$$= 8.3 \times 10^{-5}\ \text{atm}$$

8.87 $\Pi = n\dfrac{RT}{V} = \dfrac{m}{M}\dfrac{RT}{V}$

where m is the mass of unknown compound.

$$M = \frac{mRT}{V\Pi}$$

$$M = \frac{(0.166\ \text{g})(0.082\ 06\ \text{L}\cdot\text{atm}\cdot\text{K}^{-1}\cdot\text{mol}^{-1})(293\ \text{K})}{(0.010\ \text{L})\left(\dfrac{1.2\ \text{Torr}}{760\ \text{Torr}\cdot\text{atm}^{-1}}\right)} = 2.5 \times 10^{5}\ \text{g}\cdot\text{mol}^{-1}$$

8.89 (a) To determine the vapor pressure of the solution, we need to know the mole fraction of each component.

$$x_{\text{benzene}} = \frac{1.50\text{ mol}}{1.50\text{ mol} + 0.50\text{ mol}} = 0.75$$

$x_{\text{toluene}} = 1 - x_{\text{benzene}} = 0.25$

$P_{\text{total}} = (0.75 \times 94.6\text{ Torr}) + (0.25 \times 29.1\text{ Torr}) = 78.2\text{ Torr}$

The vapor phase composition will be given by

$$x_{\text{benzene in vapor phase}} = \frac{P_{\text{benzene}}}{P_{\text{total}}} = \frac{0.75 \times 94.6\text{ Torr}}{78.2\text{ Torr}} = 0.91$$

$x_{\text{toluene in vapor phase}} = 1 - 0.91 = 0.09$

The vapor is richer in the more volatile benzene, as expected.

(b) The procedure is the same as in (a) but the number of moles of each component must be calculated first:

$$n_{\text{benzene}} = \frac{15.0\text{ g}}{78.11\text{ g}\cdot\text{mol}^{-1}} = 0.192$$

$$n_{\text{toluene}} = \frac{65.3\text{ g}}{92.14\text{ g}\cdot\text{mol}^{-1}} = 0.709$$

$$x_{\text{benzene}} = \frac{0.192\text{ mol}}{0.192\text{ mol} + 0.709\text{ mol}} = 0.213$$

$x_{\text{toluene}} = 1 - x_{\text{benzene}} = 0.787$

$P_{\text{total}} = (0.213 \times 94.6\text{ Torr}) + (0.787 \times 29.1\text{ Torr}) = 43.0\text{ Torr}$

The vapor phase composition will be given by

$$x_{\text{benzene in vapor phase}} = \frac{P_{\text{benzene}}}{P_{\text{total}}} = \frac{0.213 \times 94.6\text{ Torr}}{43.0\text{ Torr}} = 0.469$$

$x_{\text{toluene in vapor phase}} = 1 - 0.469 = 0.531$

8.91 To calculate this quantity, we must first find the mole fraction of each that will be present in the mixture. This value is obtained from the relationship

$$P_{\text{total}} = [x_{\text{1,1-dichloroethane}} \times P_{\text{pure 1,1-dichloroethane}}] + [x_{\text{1,1-dichlorotetrafluoroethane}} \times P_{\text{pure 1,1-dichlorotetrafluoroethane}}]$$

$157\text{ Torr} = [x_{\text{1,1-dichloroethane}} \times 228\text{ Torr}] + [x_{\text{1,1-dichlorotetrafluoroethane}} \times 79\text{ Torr}]$

$157\text{ Torr} = [x_{\text{1,1-dichloroethane}} \times 228\text{ Torr}] + [(1 - x_{\text{1,1-dichloroethane}}) \times 79\text{ Torr}]$

$157\text{ Torr} = 79\text{ Torr} + [(x_{\text{1,1-dichloroethane}} \times 228\text{ Torr}) - (x_{\text{1,1-dichloroethane}} \times 79\text{ Torr})]$

$78\text{ Torr} = x_{\text{1,1-dichloroethane}} \times 149\text{ Torr}$

$x_{\text{1,1-dichloroethane}} = 0.52$

$x_{\text{1,1-dichlorotetrafluoroethane}} = 1 - 0.52 = 0.48$

To calculate the number of grams of 1,1-dichloroethane, we use the definition of mole fraction. Mathematically, it is simpler to use the $x_{\text{1,1-dichlorotetrafluoroethane}}$ definition:

$$n_{\text{1,1-dichlorotetrafluoroethane}} = \frac{100.0 \text{ g}}{170.92 \text{ g}\cdot\text{mol}^{-1}} = 0.5851 \text{ mol}$$

$$x_{\text{1,1-dichlorotetrafluoroethane}} = \frac{0.5851 \text{ mol}}{0.5851 \text{ mol} + \dfrac{m}{98.95 \text{ g}\cdot\text{mol}^{-1}}} = 0.48$$

$$0.5851 \text{ mol} = 0.48 \times \left[0.5851 \text{ mol} + \frac{m}{98.95 \text{ g}\cdot\text{mol}^{-1}}\right]$$

$$0.5851 \text{ mol} - (0.48 \times 0.5851 \text{ mol}) = 0.48 \times \left[\frac{m}{98.95 \text{ g}\cdot\text{mol}^{-1}}\right]$$

$$0.3042 \text{ mol} = 0.004\,851 \text{ mol}\cdot\text{g}^{-1} \times m$$

$$m = 63 \text{ g}$$

8.93 Raoult's Law applies to the vapor pressure of the mixture, so positive deviation means that the vapor pressure is higher than expected for an ideal solution. Negative deviation means that the vapor pressure is lower than expected for an ideal solution. Negative deviation will occur when the interactions between the different molecules are somewhat stronger than the interactions between molecules of the same kind. (a) For methanol and ethanol, we expect the types of intermolecular attractions in the mixture to be similar to those in the component liquids, so that an ideal solution is predicted. (b) For HF and H_2O, the possibility of intermolecular hydrogen bonding between water and HF would suggest that negative deviation would be observed, which is the case. HF and H_2O form an azeotrope that boils at 111°C, a temperature higher than the boiling pont of either HF (19.4°C) or water. (c) Because hexane is nonpolar and water is polar with hydrogen bonding, we would expect a mixture of these two to exhibit positive deviation (the interactions between the different molecules would be weaker than the intermolecular forces between like molecules). Hexane and water do form an azeotrope that boils at 61.6°C, a temperature below the boiling point of either hexane or water.

8.95 (a) stronger; (b) low; (c) high; (d) weaker; (e) weak, low; (f) low; (g) strong, high

8.97 (a, b) Viscosity and surface tension decrease with increasing temperature; at high temperatures the molecules readily move away from their neighbors because of increased kinetic energy. (c) Evaporation rate and vapor pressure increase with increasing temperature because the kinetic energy of the molecules increases with temperature, and the molecules are more likely to escape into the gas phase.

8.99 If the external pressure is lowered, water boils at a lower temperature. The boiling temperature is the temperature at which the vapor pressure equals the external pressure. Vapor pressure increases with temperature. If the external pressure is reduced, the vapor pressure of the water will reach that value at a lower temperature; thus, lowering the external pressure lowers the boiling temperature.

8.101 (a) No. Solid helium will melt before converting to a gas at 5 K. (b) Yes. Raising the temperature of rhombic sulfur at 1 atm will result in a phase change to monoclinic sulfur at 96°C. (c) Rhombic sulfur is the form observed. (d) At 100°C and 1 atm, the monoclinic form is more stable. (e) No. The phase diagram shows no area where solid diamond and gaseous carbon are adjacent.

8.103 (a) $\dfrac{25.0 \text{ Torr}}{31.83 \text{ Torr}} \times 100 = 78.5\%$; (b) At 25°C the vapor pressure of water is only 23.76 Torr, so some of the water vapor in the air would condense as dew or fog.

8.105 Assume 100 g of solution:

$$16.0\% \text{ C}_{12}\text{H}_{22}\text{O}_{11}(\text{aq}) \text{ is } \equiv \frac{16.0 \text{ g C}_{12}\text{H}_{22}\text{O}_{11}}{100 \text{ g soln}}$$

(a)
$$\text{molality} = \left(\frac{16.0 \text{ g C}_{12}\text{H}_{22}\text{O}_{11}}{100 \text{ g soln}}\right)\left(\frac{1.0635 \text{ g soln}}{1 \text{ mL soln}}\right)\left(\frac{1 \text{ mL}}{10^{-3} \text{ L}}\right)\left(\frac{1 \text{ mol C}_{12}\text{H}_{22}\text{O}_{11}}{342.30 \text{ g C}_{12}\text{H}_{22}\text{O}_{11}}\right)$$
$$= 0.497 \text{ mol}\cdot\text{L}^{-1}$$

(b) The solution contains 16.0 g $C_{12}H_{22}O_{11}$ in 84.0 g H_2O:

$$n_{\text{C}_{12}\text{H}_{22}\text{O}_{11}} = (16.0 \text{ g C}_{12}\text{H}_{22}\text{O}_{11})\left(\frac{1 \text{ mol C}_{12}\text{H}_{22}\text{O}_{11}}{342.30 \text{ g C}_{12}\text{H}_{22}\text{O}_{11}}\right)$$
$$= 0.0467 \text{ C}_{12}\text{H}_{22}\text{O}_{11}$$

$$n_{\text{H}_2\text{O}} = (84.0 \text{ g H}_2\text{O})\left(\frac{1 \text{ mol H}_2\text{O}}{18.02 \text{ g H}_2\text{O}}\right)$$
$$= 4.66 \text{ mol H}_2\text{O}$$

$$x_{\text{H}_2\text{O}} = \frac{4.66 \text{ mol H}_2\text{O}}{4.66 \text{ mol H}_2\text{O} + 0.0467 \text{ mol C}_{12}\text{H}_{22}\text{O}_{11}} = 0.990$$

$$P_{\text{H}_2\text{O}} = x_{\text{H}_2\text{O}}P_{\text{pure H}_2\text{O}} = 0.990 \times 17.54 \text{ Torr} = 17.4 \text{ Torr}$$

(c)
$$\Delta T_\text{b} = ik_\text{b}m = (0.51 \text{ K}\cdot\text{kg}\cdot\text{mol}^{-1})\left(\frac{0.0468 \text{ mol C}_{12}\text{H}_{22}\text{O}_{11}}{0.0840 \text{ kg H}_2\text{O}}\right)$$
$$= 0.28 \text{ K or } 0.28°\text{C}$$

$$\text{b.p.} = 100.00°\text{C} + 0.28°\text{C} = 100.28°\text{C}$$

8.107 (a) At 30°C, the vapor pressure of pure water = 31.83 Torr. According to Raoult's law, $P = x_{\text{solvent}} \cdot P_{\text{pure}}$. To calculate the x_{solvent}: 0.50 *m* NaCl gives 0.50 moles Na^{+} and 0.50 moles Cl^{-} per kg solvent. The number of moles of solvent is 1000 kg ÷ 18.02 $g \cdot mol^{-1}$ = 55.49 mol. The total number of moles = 0.50 mol + 0.50 mol + 55.49 mol = 56.49 mol.

$$x_{\text{solvent}} = \frac{55.49\text{ mol}}{56.49\text{ mol}} = 0.9823$$

$P = x_{\text{solvent}} \cdot P_{\text{pure}} = 0.9823 \times 31.83\text{ Torr} = 31.26\text{ Torr}$

(b) At 100°C, $P_{\text{pure}} = 760$ Torr: $P = x_{\text{solvent}} \cdot P_{\text{pure}} = 0.9823 \times 760\text{ Torr} = 747$ Torr

At 0°C, $P_{\text{pure}} = 4.58$ Torr; $P = x_{\text{solvent}} \cdot P_{\text{pure}} = 0.9823 \times 4.58\text{ Torr} = 4.50\text{ Torr}$

8.109 (a) $\Delta T_{\text{f}} = 1.20°\text{C} - 5.5°\text{C} = -4.3°\text{C}$

$$\Delta T_{\text{f}} = ik_f m = ik_f \frac{\left(\frac{m_{\text{solute}}}{M_{\text{solute}}}\right)}{\text{kg solvent}}$$

$$M_{\text{solute}} = \frac{ik_f m_{\text{solute}}}{(\text{kg solvent})(\Delta T_{\text{f}})} = \frac{(1)(5.12\text{ K}\cdot\text{kg}\cdot\text{mol}^{-1})(10.0\text{ g})}{(0.0800\text{ kg})(4.3°\text{C})}$$

$$= 1.5 \times 10^{2}\text{ g}\cdot\text{mol}^{-1}$$

(b) The empirical formula mass = 73.4 $g \cdot mol^{-1}$; the experimental formula mass is roughly double this value, so the molecular formula is $C_6H_4Cl_2$.

(c) molar mass = 146.99 $g \cdot mol^{-1}$

8.111 (a) The elemental analysis yields the following:

element	% by mass	mol in 100 g	ratio to smallest moles (0.542 for N)
C	59.0	4.91	9.04
O	26.2	1.64	3.02
H	7.10	7.04	13.0
N	7.60	0.542	1

The empirical formula is $C_9H_{13}O_3N$ (formula mass ~ 183 $g \cdot mol^{-1}$).

(b) and (c)

$$\Delta T_{\text{f}} = ik_f m = ik_f \frac{\left(\frac{\text{mass}_{\text{solute}}}{M_{\text{solute}}}\right)}{\text{kg solvent}}$$

$$M_{\text{solute}} = \frac{ik_f \text{mass}_{\text{solute}}}{(\text{kg solvent})(\Delta T_f)} = \frac{(1)(5.12\ \text{K}\cdot\text{kg}\cdot\text{mol}^{-1})(0.64\ \text{g})}{(0.036\ \text{kg})(0.50°\text{C})}$$
$$= 1.8 \times 10^2\ \text{g}\cdot\text{mol}^{-1}$$

The molecular formula is the same as the empirical formula, $C_9H_{13}O_3N$. Molar mass of epinephrine = 183.20 g·mol^{-1}.

8.113 Hypotonic solutions will cause a net flow of solvent into the cells to equalize the osmotic pressure. The cells will burst and die (hemolysis). Hypertonic solutions will cause a net flow of solvent out of the cells to equalize the osmotic pressure. The cells will shrink and die.

8.115 (a) To determine the vapor pressure of the solution, we need to know the mole fraction of each component.

$$x_{\text{diethyl ether}} = \frac{0.75\ \text{mol}}{0.75\ \text{mol} + 0.50\ \text{mol}} = 0.60$$

$$x_{\text{ethyl methyl ether}} = 1 - x_{\text{diethyl ether}} = 0.40$$

$$P_{\text{total}} = (0.60 \times 185\ \text{Torr}) + (0.40 \times 554\ \text{Torr}) = 3.3 \times 10^2\ \text{Torr}$$

The vapor phase composition will be given by

$$x_{\text{diethyl ether in vapor phase}} = \frac{P_{\text{diethyl ether}}}{P_{\text{total}}} = \frac{0.60 \times 185\ \text{Torr}}{3.3 \times 10^2\ \text{Torr}} = 0.34$$

$$x_{\text{ethyl methyl ether in vapor phase}} = 1 - 0.34 = 0.66$$

The vapor is richer in the more volatile ethyl methyl ether, as expected.

(b) The procedure is the same as in (a) but the number of moles of each component must be calculated first:

$$n_{\text{diethyl ether}} = \frac{25.0\ \text{g}}{74.12\ \text{g}\cdot\text{mol}^{-1}} = 0.337$$

$$n_{\text{ethyl methyl ether}} = \frac{35.0\ \text{g}}{60.10\ \text{g}\cdot\text{mol}^{-1}} = 0.582$$

$$x_{\text{diethyl ether}} = \frac{0.337\ \text{mol}}{0.337\ \text{mol} + 0.582\ \text{mol}} = 0.367$$

$$x_{\text{ethyl methyl ether}} = 1 - x_{\text{diethyl ether}} = 0.633$$

$$P_{\text{total}} = (0.367 \times 185\ \text{Torr}) + (0.633 \times 554\ \text{Torr}) = 419\ \text{Torr}$$

The vapor phase composition will be given by

$$x_{\text{diethyl ether in vapor phase}} = \frac{P_{\text{diethyl ether}}}{P_{\text{total}}} = \frac{0.367 \times 185\ \text{Torr}}{419\ \text{Torr}} = 0.162$$

$$x_{\text{ethyl methyl ether in vapor phase}} = 1 - 0.162 = 0.838$$

8.117 (a) The partial pressure remains the same. Some of the ethanol will condense to return the ethanol(l) $\rightleftharpoons$ ethanol(g) reaction to the equilibrium point. (b) The total pressure will be the sum of the pressures due to the ethanol vapor and to the air. The air will not condense, so its pressure should follow the ideal gas law. If the total pressure initially is 750 Torr and 58.9 Torr is due to ethanol vapor, the 750 Torr − 58.9 Torr = 691 Torr will be due to the air. If the volume is halved at constant temperature, then the pressure due to the air will be doubled, or 2 × 691 Torr = 1.38×10^3 Torr. The total pressure will then be 1.38×10^3 Torr + 58.9 Torr = 1.44×10^3 Torr.

8.119 (a) Given $d = 1.00\ \text{g}\cdot\text{cm}^{-3} = 1.00\ \text{g}\cdot\text{mL}^{-1}$,

$$\therefore \frac{1.54\ \text{g}}{100\ \text{mL solution}} = 15.4\ \text{g}\cdot\text{L}^{-1}$$

$$15.4\ \text{g}\cdot\text{L}^{-1} \times \frac{1\ \text{mol Li}_2\text{CO}_3}{73.89\ \text{g Li}_2\text{CO}_3} = 0.208\ \text{mol}\cdot\text{L}^{-1}$$

$$\Pi = iRT \times \text{molarity}$$

$$= 3 \times 0.082\,06\ \text{L}\cdot\text{atm}\cdot\text{K}^{-1}\cdot\text{mol}^{-1} \times 273\ \text{K} \times 0.208\ \text{mol}\cdot\text{L}^{-1} = 14.0\ \text{atm}$$

(b) $P_{\text{solution}} = x_{\text{solvent}} \times P_{\text{pure solvent}}$

$751\ \text{Torr} = x_{\text{solvent}} \times 760\ \text{Torr}$

$x_{\text{solvent}} = 0.988$

$x_{\text{solute}} = 1 - 0.988 = 0.012$

Consider a solution of 0.012 mol solute in 0.988 mol H_2O. Assume that the volume of solution equals the volume of H_2O. Then

$$V = 0.988\ \text{mol H}_2\text{O} \times \frac{18.02\ \text{g H}_2\text{O}}{1.00\ \text{mol H}_2\text{O}} \times \frac{1.00\ \text{mL H}_2\text{O}}{1.00\ \text{g H}_2\text{O}} = 17.8\ \text{mL} = 0.0178\ \text{L}$$

$$\text{molarity of solution} = \frac{0.012\ \text{mol}}{0.0178\ \text{L}} = 0.67\ \text{mol}\cdot\text{L}^{-1}$$

Assuming that $i = 1$,

$$\Pi = 1 \times 0.082\,06\ \text{L}\cdot\text{atm}\cdot\text{K}^{-1}\cdot\text{mol}^{-1} \times 373\ \text{K} \times 0.67\ \text{mol}\cdot\text{L}^{-1} = 21\ \text{atm}$$

(c) $\Delta T = ik_b \times \text{molality}$, assume $i = 1$, then

$$\text{molality} = \frac{\Delta T_b}{k_b} = \frac{1\ \text{K}}{0.51\ \text{K}\cdot\text{kg}\cdot\text{mol}^{-1}} = 2.0\ \text{mol}\cdot\text{kg}^{-1}$$

Given that $d_{\text{solution}} = d_{\text{water}} = 1.00\ \text{g}\cdot\text{cm}^{-3} = 1.00\ \text{g}\cdot\text{mL}^{-1}$, we have

$$2.0\ \text{mol}\cdot\text{kg}^{-1} \times \frac{1\ \text{kg}}{1000\ \text{g}} \times \frac{1.00\ \text{g}}{1.00\ \text{mL}} \times \frac{1000\ \text{mL}}{1.00\ \text{L}} = 2.0\ \text{mol}\cdot\text{L}^{-1}$$

Then $\Pi = iRT \times \text{molarity}$

$$= 1 \times 0.082\,06\ \text{L}\cdot\text{atm}\cdot\text{K}^{-1}\cdot\text{mol}^{-1} \times 374\ \text{K} \times 2.0\ \text{mol}\cdot\text{L}^{-1}$$

$$= 61\ \text{atm}$$

8.121 (a) Because the osmotic pressure is calculated from $\Pi = iRT \times \text{molarity}$ (both compounds are nonelectrolytes, so $i = 1$), the solution with the higher concentration will have the higher osmotic pressure, which in this case is the solution of urea, $CO(NH_2)_2$.

(b) The more concentrated solution, $CO(NH_2)_2$, will become more dilute with the passage of H_2O molecules through the membrane. (c) Pressure must be applied to the more concentrated solution, $CO(NH_2)_2$, to equilibrate the water flow. (d) $\Pi = iRT \times \Delta M$ (where ΔM is the difference in molar concentration between the two solutions):

$$\Pi = 1 \times 0.082\ 06\ \text{L}\cdot\text{atm}\cdot\text{K}^{-1}\cdot\text{mol}^{-1} \times 298\ \text{K} \times 0.030\ \text{mol}\cdot\text{L}^{-1} = 0.73\ \text{atm}$$

8.123 (a) The data in Table 6.2 can be used to obtain the heat of vaporization and boiling point of benzene.

$$C_6H_6(l) \rightleftharpoons C_6H_6(g)$$

$$\Delta H°_{vap} = 30.8\ \text{kJ}\cdot\text{mol}^{-1}$$

The boiling point is 353.2 K or 80.0°C, at which the vapor pressure = 1 atm. To derive the general equation, we start with the expression that $\Delta G°_{vap} = -RT \ln P$, where P is the vapor pressure of the solvent. Because $\Delta G°_{vap} = \Delta H°_{vap} - T\Delta S°_{vap}$, this is the relationship to use to determine the temperature dependence of $\ln P$:

$$\Delta H°_{vap} - T\Delta S°_{vap} = -RT \ln P$$

This equation can be rearranged to give:

$$\ln P = -\frac{\Delta H°_{vap}}{R} \times \frac{1}{T} + \frac{\Delta S°_{vap}}{R}$$

Because we do not have enough data to calculate $\Delta S°_{vap}$ from the table or appendix, we can use the alternate form of the equation, which relates the vapor pressure at two points to the corresponding temperatures. This equation is obtained by subtracting one specific point from another as shown:

$$\ln P_2 = -\frac{\Delta H°_{vap}}{R} \times \frac{1}{T_2} + \frac{\Delta S°_{vap}}{R}$$

$$-\left[\ln P_1 = -\frac{\Delta H°_{vap}}{R} \times \frac{1}{T_1} + \frac{\Delta S°_{vap}}{R}\right]$$

$$\ln\frac{P_2}{P_1} = -\frac{\Delta H°_{vap}}{R}\left[\frac{1}{T_2} - \frac{1}{T_1}\right]$$

Because we know $\Delta H°_{vap}$ and one point, we can introduce those values:

$$\ln\frac{P}{1} = -\frac{30\ 800\ \text{J}\cdot\text{mol}^{-1}}{8.314\ \text{J}\cdot\text{K}^{-1}\cdot\text{mol}^{-1}}\left[\frac{1}{T} - \frac{1}{353.2\ \text{K}}\right]$$

$$\ln P = -\frac{3.70 \times 10^3\ \text{K}}{T} + 10.5$$

(b) The appropriate quantities to plot to get a straight line are ln P versus $\frac{1}{T}$.

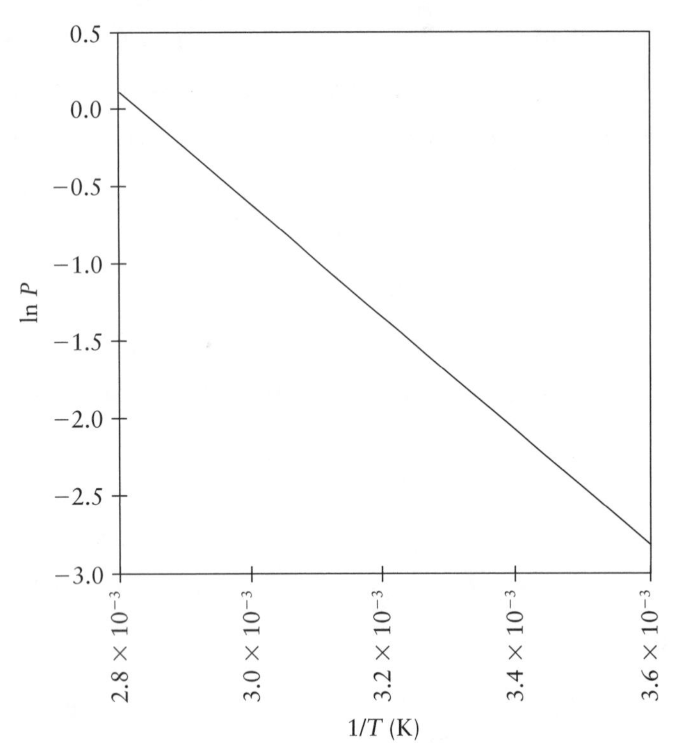

(c) The solution will boil when the vapor pressure equals the external pressure. Simply substitute the numbers, once the equation has been derived:

$$\ln P = -\frac{3.70 \times 10^3 \text{ K}}{T} + 10.5$$

$$\ln 0.655 = -\frac{3.70 \times 10^3 \text{ K}}{T} + 10.5$$

$T = 339$ K or 66°C

(d) The equation of the line is written in such a form that the constant 10.5 is equal to $\frac{\Delta S°_{\text{vap}}}{R}$.

$$\frac{\Delta S°_{\text{vap}}}{R} = 10.5$$

$\Delta S°_{\text{vap}} = 10.5 \times 8.314 \text{ J}\cdot\text{K}^{-1}\cdot\text{mol}^{-1} = 87.3 \text{ J}\cdot\text{K}^{-1}\cdot\text{mol}^{-1}$

Because $\Delta S°_{\text{vap}} = S°_{\text{m}}(\text{benzene, g}) - S°_{\text{m}}(\text{benzene, l})$

we can now calculate the standard molar entropy of benzene(g)

$87.3 \text{ J}\cdot\text{K}^{-1}\cdot\text{mol}^{-1} = S°_{\text{m}}(\text{benzene, g}) - 173.3 \text{ J}\cdot\text{K}^{-1}\cdot\text{mol}^{-1}$

$S°_{\text{m}}(\text{benzene, g}) = 260.6 \text{ J}\cdot\text{K}^{-1}\cdot\text{mol}^{-1}$

8.125 The equation to give vapor pressure as a function of temperature is

$$\ln P = -\frac{\Delta H°_{\text{vap}}}{R} \times \frac{1}{T} + \frac{\Delta S°_{\text{vap}}}{R}$$

$$\ln P = -\frac{57\,814\ \text{J}\cdot\text{mol}^{-1}}{8.314\ \text{J}\cdot\text{K}^{-1}\cdot\text{mol}^{-1}} \times \frac{1}{T} + \frac{124\ \text{J}\cdot\text{mol}^{-1}}{8.314\ \text{J}\cdot\text{K}^{-1}\cdot\text{mol}^{-1}}$$

$$\ln P = -\frac{6954\ \text{K}}{T} + 14.9$$

In order for the substance to boil, we need to reduce the external pressure on the sample to its vapor pressure at that temperature. Thus, the problem becomes simply a matter of calculating the vapor pressure at 80°C:

$$\ln P = -\frac{6954\ \text{K}}{353\ \text{K}} + 14.9 = -4.80$$

$P = 0.008\ 23$ atm or 6.26 Torr

The pressure needs to be reduced to 6.26 Torr.

8.127 The critical pressures and temperatures of the compounds are as follows:

Compound	T_C (°C)	P_C (atm)
Methane, CH_4	−82.1	45.8
Methyl amine, CH_3NH_2	156.9	40.2
Ammonia, NH_3	132.5	112.5
Tetrafluoromethane, CF_4	−45.7	41.4

In order to access the supercritical fluid state, we must have conditions in excess of the critical temperature and pressure. Given the rating of the autoclave, ammonia would not be suitable because one could not access the supercritical state due to the pressure limitation. Methylamine would not be suitable for a room temperature extraction because its T_C is too high. Either methane or tetrafluoromethane would be suitable for this application.

8.129 (a) If there is enough diethyl ether present, then the pressure will be due to the vapor pressure of diethyl ether at that temperature. If there is insufficient diethyl ether present, then it will all convert to gas and the pressure will be determined from the ideal gas law. We can calculate the amount of diethyl ether necessary to achieve the vapor pressure from the ideal gas equation, using the vapor pressure as the pressure of the gas.

$$\left(\frac{57\ \text{Torr}}{760\ \text{Torr}\cdot\text{atm}^{-1}}\right)(1.00\ \text{L}) = \left(\frac{m_{\text{diethyl ether}}}{74.12\ \text{g}\cdot\text{mol}^{-1}\ \text{diethyl ether}}\right)(0.082\ 06\ \text{L}\cdot\text{atm}\cdot\text{K}^{-1}\cdot\text{mol}^{-1})(228\ \text{K})$$

$$m_{\text{diethyl ether}} = 0.297\ \text{g}$$

Because there are 1.50 g of diethyl ether, the vapor pressure will be achieved, so the answer to (a) is 57 Torr.

(b)

$$\left(\frac{535 \text{ Torr}}{760 \text{ Torr}\cdot\text{atm}^{-1}}\right)(1.00 \text{ L}) = \left(\frac{m_{\text{diethyl ether}}}{74.12 \text{ g}\cdot\text{mol}^{-1} \text{ diethyl ether}}\right)(0.082\ 06 \text{ L}\cdot\text{atm}\cdot\text{K}^{-1}\cdot\text{mol}^{-1})(298 \text{ K})$$

$$m_{\text{diethyl ether}} = 2.13 \text{ g}$$

In this case, there is not enough diethyl ether to achieve equilibrium so all of the ether will vaporize. The pressure will be

$$(P)(1.00 \text{ L}) = \left(\frac{1.50 \text{ g diethyl ether}}{74.12 \text{ g}\cdot\text{mol}^{-1} \text{ diethyl ether}}\right)(0.082\ 06 \text{ L}\cdot\text{atm}\cdot\text{K}^{-1}\cdot\text{mol}^{-1})(298 \text{ K})$$

$P = 0.495$ atm or 376 Torr

(c) The system is at the same temperature as in (a) but the volume is now doubled. This volume will accommodate twice as much diethyl ether or $2 \times 0.297 \text{ g} = 0.594$ g. There is still sufficient diethyl ether for this to occur so that the pressure will again be equal to the vapor pressure of diethyl ether, or 57 Torr.

(d) If flask B is cooled by liquid nitrogen, its temperature will approach −196°C. The vapor pressure of ether at that temperature is negligible and the ether will solidify in flask B. As the ether condenses in flask B, the liquid ether in flask A will continue to vaporize and will do so until all the ether has condensed in flask B. At the end of this, there will be no ether left in flask A and the pressure in the apparatus will be 0.

8.131

$$0.50 = \frac{\chi_{\text{pentane, liquid}} P^{\circ}_{\text{pentane}}}{\chi_{\text{pentane, liquid}} P^{\circ}_{\text{pentane}} + (1 - \chi_{\text{pentane, liquid}}) P^{\circ}_{\text{hexane}}}$$

$$\chi_{\text{pentane, gas}} = 0.50 = \frac{\chi_{\text{pentane, liquid}} P^{\circ}_{\text{pentane}}}{\chi_{\text{pentane, liquid}} P^{\circ}_{\text{pentane}} + \chi_{\text{hexane, liquid}} P^{\circ}_{\text{hexane}}}$$

$$0.50 = \frac{(\chi_{\text{pentane, liquid}})(512 \text{ Torr})}{(\chi_{\text{pentane, liquid}})(512 \text{ Torr}) + (1 - \chi_{\text{pentane, liquid}})(151 \text{ Torr})}$$

$$\chi_{\text{pentane, liquid}} = 0.228;\ \chi_{\text{hexane, liquid}} = 0.772$$

8.133 (a) The vapor pressures over the two solutions are different. The volatile component, ethanol, will transfer from one solution to the other until the vapor pressures (and therefore concentrations) are equal. The vapor pressure is lower over the more concentrated solution so ethanol will condense on that side of the apparatus. As this takes place, there will be a net transfer of ethanol from the less concentrated side to the more concentrated side of the apparatus.

(b) The vapor pressure will be determined by the concentration of the solution (the solutions in both sides of the apparatus will have the same concentration) once the system reaches equilibrium.

For the 0.15 m solution, there will be 0.15 mol sucrose (342.29 $g \cdot mol^{-1}$) for 1000 g ethanol (46.07 $g \cdot mol^{-1}$). This corresponds to 51 g of sucrose for a total solution mass of 1051 g. In 15.0 g of solution there will be 0.73 g sucrose and 14.27 g of ethanol.

Similarly, for the 0.050 m solution, there will be 0.050 mol sucrose for 1000 g ethanol, corresponding to 17 g sucrose and a total solution mass of 1017 g. In 15.0 g of solution there will be 0.25 g sucrose and 14.75 g ethanol.

At equilibrium, the concentrations must be equal. In order to calculate the vapor pressure, we need to know the concentration in terms of mole fractions. The concentrations will be made equal by transferring some ethanol from one solution to the other. This can be expressed mathematically by the following relationship:

$$\frac{\left(\dfrac{0.73\ \text{g}}{342.29\ \text{g}\cdot\text{mol}^{-1}}\right)}{\left(\dfrac{14.27 + x\ \text{g}}{46.07\ \text{g}\cdot\text{mol}^{-1}}\right) + \left(\dfrac{0.73\ \text{g}}{342.29\ \text{g}\cdot\text{mol}^{-1}}\right)} = \frac{\left(\dfrac{0.25\ \text{g}}{342.29\ \text{g}\cdot\text{mol}^{-1}}\right)}{\left(\dfrac{14.75 - x\ \text{g}}{46.07\ \text{g}\cdot\text{mol}^{-1}}\right) + \left(\dfrac{0.25\ \text{g}}{342.29\ \text{g}\cdot\text{mol}^{-1}}\right)}$$

Solving this equation gives $x = 7.3$ g. The mole fraction of sucrose for the solution will be 0.0045, and the mole fraction of ethanol will be $1 - 0.0045 = 0.9955$. The vapor pressure will be $P = \chi_{\text{ethanol}} P^{\circ}_{\text{ethanol}} = (0.9955)(60\ \text{Torr}) \cong 60$ Torr. A 0.0045 mole fraction corresponds to a 0.098 m solution of sucrose.

8.135 Compound (a). The stationary phase is more polar than the liquid phase, so the more polar compound of (a) and (b) should be attracted more strongly to the stationary phase and should remain on the column longer. Because compound (a) has two carboxylic acid units (COOH), it will be more polar and will have the greater value of k.

CHAPTER 9
CHEMICAL EQUILIBRIA

9.1 (a) False. Equilibrium is dynamic. At equilibrium, the concentrations of reactants and products will not change, but the reaction will continue to proceed in both directions.

(b) False. Equilibrium reactions are affected by the presence of both products and reactants.

(c) False. The value of the equilibrium constant is not affected by the amounts of reactants or products added as long as the temperature is constant.

(d) True.

9.3

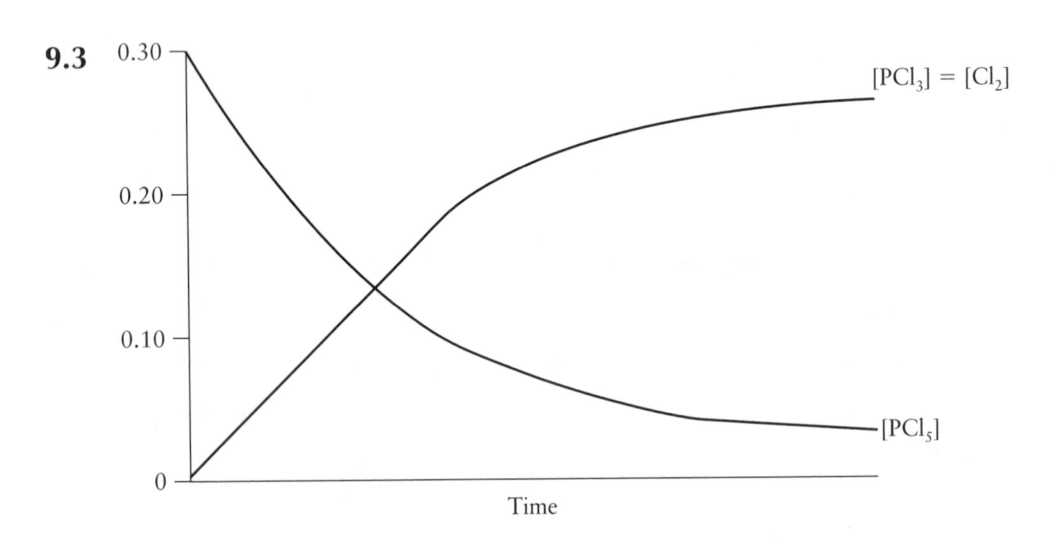

9.5 The calculations for any dilution or concentration process are obtained from the relationship: $G_m = G°_m + RT \ln Q$, where Q in this case represents the concentration or pressure that is different from the standard state value of 1 bar or $mol \cdot L^{-1}$. We never know the absolute values of G, but we can calculate the change in free energy from one state to another. Thus, we can write this relationship for conditions 1 and 2, and then find the difference between them:

$$G_{m,1} = G°_m + RT \ln P_1$$

$$G_{m,2} = G°_m + RT \ln P_2$$

$$G_{m,2} - G_{m,1} = G°_m + RT \ln P_2 - (G°_m + RT \ln P_1)$$

$$\Delta G_m = RT \ln \frac{P_2}{P_1}$$

A similar expression can be written for concentrations in place of pressures. Note that the nature of the compound is irrelevant—an identical free energy change is observed regardless of what type of gas is compressed or expanded, or what type of solution (solute or solvent) is diluted or concentrated.

(a) $\Delta G_m = (8.314\ \text{J}\cdot\text{K}^{-1}\cdot\text{mol}^{-1})(298\ \text{K})\left(\ln\dfrac{25.00}{0.500}\right)/(1000\ \text{J}\cdot\text{kJ}^{-1})$

$= 9.69\ \text{kJ}\cdot\text{mol}^{-1}$

(b) $\Delta G_m = (8.314\ \text{J}\cdot\text{K}^{-1}\cdot\text{mol}^{-1})(298\ \text{K})\left(\ln\dfrac{5.00}{10.00}\right)/(1000\ \text{J}\cdot\text{kJ}^{-1})$

$= -1.72\ \text{kJ}\cdot\text{mol}^{-1}$

(c) $\Delta G_m = (8.314\ \text{J}\cdot\text{K}^{-1}\cdot\text{mol}^{-1})(298\ \text{K})\left(\ln\dfrac{10.00}{20.00}\right)/(1000\ \text{J}\cdot\text{kJ}^{-1})$

$= -1.72\ \text{kJ}\cdot\text{mol}^{-1}$

(d) $\Delta G_m = (8.314\ \text{J}\cdot\text{K}^{-1}\cdot\text{mol}^{-1})(298\ \text{K})\left(\ln\dfrac{10.00}{20.00}\right)/(1000\ \text{J}\cdot\text{kJ}^{-1})$

$= -1.72\ \text{kJ}\cdot\text{mol}^{-1}$

9.7 See explanation for Exercise 9.5.

(a) $\Delta G_m = (8.314\ \text{J}\cdot\text{K}^{-1}\cdot\text{mol}^{-1})(298\ \text{K})\left(\ln\dfrac{0.350}{0.450}\right)/(1000\ \text{J}\cdot\text{kJ}^{-1})$

$= -0.623\ \text{kJ}\cdot\text{mol}^{-1}$

(b) $\Delta G_m = (8.314\ \text{J}\cdot\text{K}^{-1}\cdot\text{mol}^{-1})(298\ \text{K})\left(\ln\dfrac{0.700}{0.900}\right)/1000\ \text{J}\cdot\text{kJ}^{-1})$

$= -0.623\ \text{kJ}\cdot\text{mol}^{-1}$

(c) $\Delta G_m = (8.314\ \text{J}\cdot\text{K}^{-1}\cdot\text{mol}^{-1})(298\ \text{K})\left(\ln\dfrac{0.350}{0.175}\right)/(1000\ \text{J}\cdot\text{kJ}^{-1})$

$= 1.72\ \text{kJ}\cdot\text{mol}^{-1}$

(d) $\Delta G_m = (8.314\ \text{J}\cdot\text{K}^{-1}\cdot\text{mol}^{-1})(298\ \text{K})\left(\ln\dfrac{0.350}{0.175}\right)/(1000\ \text{J}\cdot\text{kJ}^{-1})$

$= 1.72\ \text{kJ}\cdot\text{mol}^{-1}$

9.9 To answer these questions, we will first calculate $\Delta G°$ for each reaction and then use that value in the expression $\Delta G° = -RT \ln K$.

(a) $2\ H_2(g) + O_2(g) \longrightarrow 2\ H_2O(g)$

$\Delta G°_r = 2(\Delta G°_f(H_2O, g) = 2(-228.57\ \text{kJ}\cdot\text{mol}^{-1}) = -457.14\ \text{kJ}$

$\Delta G°_r = -RT \ln K$ or

$$\ln K = -\frac{\Delta G°_p}{RT}$$

$$\ln K = -\frac{-457\ 140\ \text{J}}{(8.314\ \text{J}\cdot\text{K}^{-1}\cdot\text{mol}^{-1})(298\ \text{K})} = +184.51$$

$K = 1 \times 10^{80}$

(b) $2\ CO(g) + O_2(g) \longrightarrow 2\ CO_2(g)$

$$\begin{aligned}\Delta G^\circ_r &= 2 \times \Delta G^\circ_f(CO_2, g) - [2 \times \Delta G^\circ_f(CO, g)] \\ &= 2(-394.36\ \text{kJ}\cdot\text{mol}^{-1}) - [2(-137.17\ \text{kJ}\cdot\text{mol}^{-1})] \\ &= -514.38\ \text{kJ}\end{aligned}$$

$$\ln K = -\frac{-514\ 380\ \text{J}}{(8.314\ \text{J}\cdot\text{K}^{-1}\cdot\text{mol}^{-1})(298\ \text{K})} = +208$$

$K = 1 \times 10^{90}$

(c) $CaCO_3(s) \longrightarrow CaO(s) + CO_2(g)$

$$\begin{aligned}\Delta G^\circ_r &= \Delta G^\circ_f(CaO, s) + \Delta G^\circ_f(CO_2, g) - [\Delta G^\circ_f(CaCO_3\ (s)] \\ &= (-604.03\ \text{kJ}\cdot\text{mol}^{-1}) + (-394.36\ \text{kJ}\cdot\text{mol}^{-1}) - [-1128.8\ \text{kJ}\cdot\text{mol}^{-1}] \\ &= +130.4\ \text{kJ}\end{aligned}$$

$$\ln K = -\frac{+130\ 400\ \text{J}}{(8.314\ \text{J}\cdot\text{K}^{-1}\cdot\text{mol}^{-1})(298\ \text{K})} = -52.6$$

$K = 1 \times 10^{-23}$

9.11 (a) $\Delta G^\circ_r = -RT \ln K = -(8.314\ \text{J}\cdot\text{K}^{-1}\cdot\text{mol}^{-1})(1200\ \text{K})\ln 6.8 = -19\ \text{kJ}\cdot\text{mol}^{-1}$

(b) $\Delta G^\circ_r = -RT \ln K = -(8.314\ \text{J}\cdot\text{K}^{-1}\cdot\text{mol}^{-1})(298\ \text{K})\ln 1.1 \times 10^{-12}$

$= +68\ \text{kJ}\cdot\text{mol}^{-1}$

9.13 First we must calculate K for the reaction, which can be done using data from Appendix 2A:

$\Delta G^\circ_r = 2 \times \Delta G^\circ_f(NO, g) = 2 \times 86.55\ \text{kJ}\cdot\text{mol}^{-1} = 173.1\ \text{kJ}\cdot\text{mol}^{-1}$

$\Delta G^\circ = -RT \ln K$

$$\ln K = -\frac{\Delta G^0}{RT}$$

$$\ln K = -\frac{+173\ 100\ \text{J}\cdot\text{mol}^{-1}}{(8.314\ \text{J}\cdot\text{K}^{-1}\cdot\text{mol}^{-1})(298\ \text{K})} = -69.9$$

$K = 4 \times 10^{-31}$

Because $Q > K$, the reaction will tend to proceed to produce reactants.

9.15 The free energy at a specific set of conditions is given by

$\Delta G_r = \Delta G^\circ_r + RT \ln Q$

$\Delta G_r = -RT \ln K + RT \ln Q$

$$\Delta G_r = -RT\ln K + RT\ln\frac{[I]^2}{[I_2]}$$

$$= -(8.314\ J\cdot K^{-1}\cdot mol^{-1})(1200\ K)\ln 0.068 + (8.314\ J\cdot K^{-1}\cdot mol^{-1})(1200\ K)\ln\frac{(0.0084)^2}{(0.026)}$$

$$= -32\ kJ\cdot mol^{-1}$$

Because ΔG_r is negative, the reaction will be spontaneous to produce I.

9.17 The free energy at a specific set of conditions is given by

$$\Delta G_r = \Delta G°_r + RT\ln Q$$

$$\Delta G_r = -RT\ln K + RT\ln Q$$

$$\Delta G_r = -RT\ln K + RT\ln\frac{[NH_3]^2}{[N_2][H_2]^3}$$

$$= -(8.314\ J\cdot K^{-1}\cdot mol^{-1})(400\ K)\ln 41 + (8.314\ J\cdot K^{-1}\cdot mol^{-1})(400\ K)\ln\frac{(15)^2}{(5.7)(1.9)^3}$$

$$= -6.5\ kJ\cdot mol^{-1}$$

Because ΔG_r is negative, the reaction will proceed to form products.

9.19 (a) $K_C = \dfrac{[COCl][Cl]}{[CO][Cl_2]}$; (b) $K_C = \dfrac{[HBr]^2}{[H_2][Br_2]}$; (c) $K_C = \dfrac{[SO_2]^2[H_2O]^2}{[H_2S]^2[O_2]^3}$

9.21 (a) $K = \dfrac{P_{NO}{}^2 P_{Cl_2}}{P_{NOCl}{}^2} = 1.8 \times 10^{-2}$

$$K = \left(\frac{T}{12.027\ K}\right)^{\Delta n} K_C$$

$$K_C = \left(\frac{12.027\ K}{T}\right)^{\Delta n} K = \left(\frac{12.027\ K}{500\ K}\right)^{(3-2)} 1.8 \times 10^{-2} = 4.3 \times 10^{-4}$$

(b) $K = P_{CO_2} = 167$

$$K_C = \left(\frac{12.027\ K}{T}\right)^{\Delta n} K = \left(\frac{12.027\ K}{1073\ K}\right)^{(1)} 167 = 1.87$$

9.23 (a) Because the volume is the same, the number of moles of O_2 is larger in the second experiment. (b) Because K_C is a constant and the denominator is larger in the second case, the numerator must also be larger; so the concentration of O_2 is larger in the second case. (c) Although $[O_2]^3/[O_3]^2$ is the same, $[O_2]/[O_3]$ will be different, a result seen by solving for K_C in each case. (d) Because K_C is a constant, $[O_2]^3/[O_3]^2$ is the same. (e) Because $[O_2]^3/[O_3]^2$ is the same, its reciprocal must be the same.

9.25 For the reaction written as $N_2(g) + 3\ H_2(g) \rightleftharpoons 2\ NH_3(g)$ Eq. 1

$$K = \frac{P_{NH_3}{}^2}{P_{N_2}P_{H_2}{}^3} = 41$$

(a) For the reaction written as $2\ NH_3(g) \rightleftharpoons N_2(g) + 3\ H_2(g)$ Eq. 2

$$K = \frac{P_{N_2}P_{H_2}{}^3}{P_{NH_3}{}^2}$$

This is $\dfrac{1}{K_{Eq.\ 1}} = \dfrac{1}{41} = 0.024$

(b) For the reaction written as $\frac{1}{2}\ N_2(g) + \frac{3}{2}\ H_2(g) \rightleftharpoons NH_3(g)$ Eq. 3

$$K_{Eq.\ 3} = \frac{P_{NH_3}}{P_{N_2}{}^{1/2}P_{H_2}{}^{3/2}} = \sqrt{K_{Eq.\ 1}} = \sqrt{41} = 6.4$$

Note that Eq. 3 = $\frac{1}{2}$ Eq. 1 and thus $K_{Eq.\ 3} = K_{Eq.\ 1}{}^{1/2}$.

(c) For the reaction written as $2\ N_2(g) + 6\ H_2(g) \rightleftharpoons 4\ NH_3(g)$ Eq. 4

$$K_{Eq.\ 4} = \frac{P_{NH_3}{}^4}{P_{N_2}{}^2P_{H_2}{}^6} = K_{Eq.\ 1}{}^2 = 41^2 = 1.7 \times 10^3$$

Note that Eq. 4 = 2 Eq. 1 and thus $K_{Eq.\ 3} = K_{Eq.\ 1}{}^2$.

9.27 $K_C = \dfrac{[HI]^2}{[H_2][I_2]}$

for condition 1, $K_C = \dfrac{(0.0137)^2}{(6.47 \times 10^{-3})(0.594 \times 10^{-3})} = 48.8$

for condition 2, $K_C = \dfrac{(0.0169)^2}{(3.84 \times 10^{-3})(1.52 \times 10^{-3})} = 48.9$

for condition 3, $K_C = \dfrac{(0.0100)^2}{(1.43 \times 10^{-3})(1.43 \times 10^{-3})} = 48.9$

9.29 (a) $\dfrac{1}{P_{O_2}{}^3}$

(b) $P_{H_2O}{}^7$

(c) $\dfrac{P_{NO}P_{NO_2}}{P_{N_2O_3}}$

9.31 $H_2(g) + I_2(g) \rightleftharpoons 2\ HI(g) \qquad K_C = 160$

$$K_C = \frac{[HI]^2}{[H_2][I_2]}$$

$$160 = \frac{(2.21 \times 10^{-3})^2}{[H_2](1.46 \times 10^{-3})}$$

$$[H_2] = \frac{(2.21 \times 10^{-3})^2}{(160)(1.46 \times 10^{-3})}$$

$$[H_2] = 2.1 \times 10^{-5}$$

9.33 $K = \frac{P_{PCl_3}P_{Cl_2}}{P_{PCl_5}}$

$$25 = \frac{P_{PCl_3}(2.35)}{1.45}$$

$$P_{PCl_3} = \frac{(25)(1.45)}{2.35} = 15$$

$$P_{PCl_3} = 15 \text{ bar}$$

9.35 (a) $K_C = \frac{[HI]^2}{[H_2][I_2]} = 160$

$$Q_C = \frac{[2.4 \times 10^{-3}]^2}{[4.8 \times 10^{-3}][2.4 \times 10^{-3}]} = 0.50$$

(b) $Q_C \neq K_C$, $\therefore$ the system is not at equilibrium

(c) Because $Q_C < K_C$, more products will be formed.

9.37 (a) $K_C = \frac{[SO_3]^2}{[SO_2]^2[O_2]} = 1.7 \times 10^6$

$$Q_C = \frac{\left[\frac{1.0 \times 10^{-4}}{0.500}\right]^2}{\left[\frac{1.20 \times 10^{-3}}{0.500}\right]^2\left[\frac{5.0 \times 10^{-4}}{0.500}\right]} = 6.9$$

(b) Because $Q_C < K_C$, more products will tend to form, which will result in the formation of more SO_3.

9.39 $\frac{1.90 \text{ g HI}}{127.91 \text{ g}\cdot\text{mol}^{-1}} = 0.0148 \text{ mol HI}$

$$2\,HI(g) \rightleftharpoons H_2(g) + I_2$$

$$0.0172 \text{ mol} - 2x \qquad x \qquad x$$

$$0.0148 \text{ mol} = 0.0172 \text{ mol} - 2x$$

$$x = 0.0012 \text{ mol}$$

$$K_C = \frac{\left[\frac{0.0012}{2.00}\right]\left[\frac{0.0012}{2.00}\right]}{\left[\frac{0.0148}{2.00}\right]^2} = \frac{\left[\frac{0.0012}{2.00}\right]^2}{\left[\frac{0.0148}{2.00}\right]^2} = 0.0066 \text{ or } 6.6 \times 10^{-3}$$

9.41 $\dfrac{25.0\text{ g }NH_4(NH_2CO_2)}{78.07\text{ g}\cdot\text{mol}^{-1}\ NH_4(NH_2CO_2)} = 0.320\text{ mol }NH_4(NH_2CO_2)$

$\dfrac{0.0174\text{ g }CO_2}{44.01\text{ g}\cdot\text{mol}^{-1}CO_2} = 3.95 \times 10^{-4}\text{ mol }CO_2$

2 mol NH_3 are formed per mol of CO_2, so mol NH_3 $= 2 \times 3.95 \times 10^{-4}$
$= 7.90 \times 10^{-4}$

$$K_C = [NH_3]^2[CO_2] = \left(\frac{7.90 \times 10^{-4}}{0.250}\right)^2\left(\frac{3.95 \times 10^{-4}}{0.250}\right) = 1.58 \times 10^{-8}$$

9.43 For each set of conditions, we need to calculate the final concentrations of reactants present. Because the stoichiometry is one mole $CH_3COOC_2H_5$ produced per mole of CH_3COOH and per mole of C_2H_5OH consumed, and the amount of water produced equals the amount of $CH_3COOC_2H_5$ produced, the values are easy to obtain:

	$[CH_3COOH]$	$[C_2H_5OH]$	$[CH_3COOC_2H_5]$	$[H_2O]$
final	1.00 − 0.171 = 0.829	0.180 − 0.171 = 0.009	0.171	0.171
final	1.00 − 0.667 = 0.333	1.00 − 0.667 = 0.333	0.667	0.667
final	1.00 − 0.966 = 0.034	8.00 − 0.966 = 7.03	0.966	0.966

$$K_C = \frac{[CH_3COOC_2H_5][H_2O]}{[CH_3COOH][C_2H_5OH]}$$

$$\#1\ K_C = \frac{[0.171][0.171]}{[0.829][0.009]} = 4$$

$$\#2\ K_C = \frac{[0.667][0.667]}{[0.333][0.333]} = 4.01$$

$$\#3\ K_C = \frac{[0.966][0.966]}{[0.034][7.03]} = 3.9$$

These numbers are identical within the experimental error.

9.45 (a) The balanced equation is: $Cl_2(g) \rightleftharpoons 2\ Cl(g)$

The initial concentration of $Cl_2(g)$ is $\dfrac{0.0020\text{ mol }Cl_2}{2.0\text{ L}} = 0.0010\text{ mol}\cdot\text{L}^{-1}$

Concentration ($\text{mol}\cdot\text{L}^{-1}$)	$Cl_2(g)$	$\rightleftharpoons$	$2\ Cl(g)$
initial	0.0010		0
change	$-x$		$+2x$
equilibrium	$0.0010 - x$		$+2x$

$$K_C = \frac{[Cl]^2}{[Cl_2]} = \frac{(2\,x)^2}{(0.0010 - x)} = 1.2 \times 10^{-7}$$

$$4\,x^2 = (1.2 \times 10^{-7})(0.0010 - x)$$

$$4\,x^2 + (1.2 \times 10^{-7})x - (1.2 \times 10^{-10}) = 0$$

$$x = \frac{-(1.2 \times 10^{-7}) \pm \sqrt{(1.2 \times 10^{-7})^2 - 4(4)(-1.2 \times 10^{-10})}}{2 \cdot 4}$$

$$x = \frac{-(1.2 \times 10^{-7}) \pm 4.4 \times 10^{-5}}{8}$$

$$x = -5.5 \times 10^{-6} \text{ or } +5.5 \times 10^{-6}$$

The negative answer is not meaningful, so we choose $x = +5.5 \times 10^{-6}\ \text{mol}\cdot\text{L}^{-1}$. The concentration of Cl_2 is essentially unchanged because $0.0010 - 5.5 \times 10^{-6} \cong 0.0010$. The concentration of Cl atoms is $2 \times (5.5 \times 10^{-6}) = 1.1 \times 10^{-5}\ \text{mol}\cdot\text{L}^{-1}$. The percentage decomposition of Cl_2 is given by

$$\frac{5.5 \times 10^{-6}}{0.0010} \times 100 = 0.55\%$$

(b) The balanced equation is: $F_2(g) \rightleftharpoons 2\,F(g)$

The problem is worked in an identical fashion to (a) but the equilibrium constant is now 1.2×10^{-4}.

The initial concentration of $F_2(g)$ is $\frac{0.0020\ \text{mol}\ F_2}{2.0\ \text{L}} = 0.0010\ \text{mol}\cdot\text{L}^{-1}$

Concentration ($\text{mol}\cdot\text{L}^{-1}$)	$F_2(g)$	$\rightleftharpoons$	$2\,F(g)$
initial	0.0010		0
change	$-x$		$+2\,x$
equilibrium	$0.0010 - x$		$+2\,x$

$$K_C = \frac{[F]^2}{[F_2]} = \frac{(2\,x)^2}{(0.0010 - x)} = 1.2 \times 10^{-4}$$

$$4\,x^2 = (1.2 \times 10^{-4})(0.0010 - x)$$

$$4\,x^2 + (1.2 \times 10^{-4})x - (1.2 \times 10^{-7}) = 0$$

$$x = \frac{-(1.2 \times 10^{-4}) \pm \sqrt{(1.2 \times 10^{-4})^2 - 4(4)(-1.2 \times 10^{-7})}}{2 \cdot 4}$$

$$x = \frac{-(1.2 \times 10^{-4}) \pm 1.4 \times 10^{-3}}{8}$$

$$x = -1.9 \times 10^{-4} \text{ or } +1.6 \times 10^{-4}$$

The negative answer is not meaningful, so we choose $x = +1.6 \times 10^{-4}\ \text{mol}\cdot\text{L}^{-1}$. The concentration of F_2 is $0.0010 - 1.6 \times 10^{-4} = 8 \times 10^{-4}\ \text{mol}\cdot\text{L}^{-1}$. The concentration of F atoms is $2 \times (1.6 \times 10^{-4}) = 3.2 \times 10^{-4}\ \text{mol}\cdot\text{L}^{-1}$. The percentage decomposition of F_2 is given by

$$\frac{1.6 \times 10^{-4}}{0.0010} \times 100 = 16\%$$

(c) Cl_2 is more stable. This can be seen even without the aid of the calculation from the larger equilibrium constant for the dissociation for F_2 compared to Cl_2.

9.47

Concentration ($\text{mol}\cdot\text{L}^{-1}$)	$2\,HBr(g)$	$\rightleftharpoons$	$H_2(g)$ +	$Br_2(g)$
initial	1.2×10^{-3}		0	0
change	$-2x$		$+x$	$+x$
final	$1.2 \times 10^{-3} - 2x$		$+x$	$+x$

$$K_C = \frac{[H_2][Br_2]}{[HBr]^2}$$

$$7.7 \times 10^{-11} = \frac{(x)(x)}{(1.2 \times 10^{-3} - 2x)^2} = \frac{x^2}{(1.2 \times 10^{-3} - 2x)^2}$$

$$\sqrt{7.7 \times 10^{-11}} = \sqrt{\frac{x^2}{(1.2 \times 10^{-3} - 2x)^2}}$$

$$\frac{x}{(1.2 \times 10^{-3} - 2x)} = 8.8 \times 10^{-6}$$

$$x = (8.8 \times 10^{-6})(1.2 \times 10^{-3} - 2x)$$

$$x + 2(8.8 \times 10^{-6})x = (8.8 \times 10^{-6})(1.2 \times 10^{-3})$$

$$x \cong 1.1 \times 10^{-8}$$

$[Br_2] = [H_2] = 1.1 \times 10^{-8}\ \text{mol}\cdot\text{L}^{-1}$; the concentration of HBr is essentially unaffected by the formation of Br_2 and H_2.

The percentage decomposition is given by

$$\frac{2\,(1.1 \times 10^{-8}\ \text{mol}\cdot\text{L}^{-1})}{1.2 \times 10^{-3}\ \text{mol}\cdot\text{L}^{-1}} \times 100 = 1.8 \times 10^{-3}\%$$

9.49 (a) Concentration of PCl_5 initially $= \dfrac{\left(\dfrac{1.0\ \text{g}\ PCl_5}{208.22\ \text{g}\cdot\text{mol}^{-1}\ PCl_5}\right)}{0.250\ \text{L}} = 0.019\ \text{mol}\cdot\text{L}^{-1}$

Concentration ($\text{mol}\cdot\text{L}^{-1}$)	$PCl_5(g)$	$\rightleftharpoons$	$PCl_3(g)$ +	Cl_2
initial	0.019		0	0
change	$-x$		$+x$	$+x$
final	$0.019 - x$		$+x$	$+x$

$$K_C = \frac{[PCl_2][Cl_2]}{[PCl_5]} = \frac{(x)(x)}{(0.019 - x)} = \frac{x^2}{(0.019 - x)}$$

$$\frac{x^2}{(0.019 - x)} = 1.1 \times 10^{-2}$$

$$x^2 = (1.1 \times 10^{-2})(0.019 - x)$$

$$x^2 + (1.1 \times 10^{-2})x - 2.1 \times 10^{-4} = 0$$

$$x = \frac{-(1.1 \times 10^{-2}) \pm \sqrt{(1.1 \times 10^{-2})^2 - (4)(1)(-2.1 \times 10^{-4})}}{2 \cdot 1}$$

$$x = \frac{-(1.1 \times 10^{-2}) \pm 0.031}{2 \cdot 1} = +0.010 \text{ or } -0.021$$

The negative root is not meaningful, so we choose $x = 0.010\ \text{mol} \cdot \text{L}^{-1}$.

$[PCl_3] = [Cl_2] = 0.010\ \text{mol} \cdot \text{L}^{-1}$; $[PCl_5] = 0.009\ \text{mol} \cdot \text{L}^{-1}$.

(b) The percentage decomposition is given by

$$\frac{0.010}{0.019} \times 100 = 53\%$$

9.51 Starting concentration of $NH_3 = \frac{0.400\ \text{mol}}{2.00\ \text{L}} = 0.200\ \text{mol} \cdot \text{L}^{-1}$

Concentration $(\text{mol} \cdot \text{L}^{-1})$	$NH_4HS(s)$ $\rightleftharpoons$	$NH_3(g)$ +	$H_2S(g)$
initial	—	0.200	0
change	—	$+x$	$+x$
final	—	$0.200 + x$	$+x$

$$K_C = [NH_3][H_2S] = (0.200 + x)(x)$$

$$1.6 \times 10^{-4} = (0.200 + x)(x)$$

$$x^2 + 0.200x - 1.6 \times 10^{-4} = 0$$

$$x = \frac{-(+0.200) \pm \sqrt{(+0.200)^2 - (4)(1)(-1.6 \times 10^{-4})}}{2 \cdot 1}$$

$$x = \frac{-0.200 \pm 0.2016}{2 \cdot 1} = +0.0008 \text{ or } -0.2008$$

The negative root is not meaningful, so we choose $x = 8 \times 10^{-4}\ \text{mol} \cdot \text{L}^{-1}$ (note that in order to get this number we have had to ignore our normal significant figure conventions).

$[NH_3] = +0.200\ \text{mol} \cdot \text{L}^{-1} + 8 \times 10^{-4}\ \text{mol} \cdot \text{L}^{-1} = 0.200\ \text{mol} \cdot \text{L}^{-1}$

$[H_2S] = 8 \times 10^{-4}\ \text{mol} \cdot \text{L}^{-1}$

Alternatively, we could have assumed that $x \ll 0.2$, the $0.200x = 1.6 \times 10^{-4}$, $x = 8.0 \times 10^{-4}$.

9.53 The initial concentrations of PCl_5 and PCl_3 are calculated as follows:

$$[PCl_5] = \frac{0.200 \text{ mol}}{4.00 \text{ L}} = 0.0500 \text{ mol}\cdot\text{L}^{-1};\ [PCl_3] = \frac{0.600 \text{ mol}}{4.00 \text{ L}} = 0.150 \text{ mol}\cdot\text{L}^{-1}$$

Concentrations ($\text{mol}\cdot\text{L}^{-1}$)	PCl_5	$\rightleftharpoons$	$PCl_3(g)$	+	$Cl_2(g)$
initial	0.0500		0.150		0
change	$-x$		$+x$		$+x$
final	$0.0500 - x$		$0.150 + x$		$+x$

$$K_C = \frac{[PCl_3][Cl_2]}{[PCl_5]} = \frac{(0.150 + x)(x)}{(0.0500 - x)} = 33.3$$

$$x^2 + 0.150x = (33.3)(0.0500 - x)$$

$$x^2 + 33.45x - 1.665 = 0$$

$$x = \frac{-33.45 \pm \sqrt{(33.45)^2 - (4)(1)(-1.665)}}{(2)(1)} = \frac{-33.4 \pm 33.6}{2} = 0.0497$$

The negative root has no physical meaning and so it can be discarded.

$[PCl_5] = 0.0500 \text{ mol}\cdot\text{L}^{-1} - 0.0497 \text{ mol}\cdot\text{L}^{-1} = 3 \times 10^{-4} \text{ mol}\cdot\text{L}^{-1}$

$[PCl_3] = 0.150 \text{ mol}\cdot\text{L}^{-1} + 0.0497 \text{ mol}\cdot\text{L}^{-1} = 0.200 \text{ mol}\cdot\text{L}^{-1}$

$[Cl_2] = 0.0497 \text{ mol}\cdot\text{L}^{-1}$

9.55 The initial concentrations of N_2 and O_2 are equal at $0.114 \text{ mol}\cdot\text{L}^{-1}$ because the vessel has a volume of 1.00 L.

Concentrations ($\text{mol}\cdot\text{L}^{-1}$)	$N_2(g)$	+	$O_2(g)$	$\rightleftharpoons$	$2\,NO(g)$
initial	0.114		0.114		0
change	$-x$		$-x$		$+2\,x$
final	$0.114 - x$		$0.114 - x$		$+2\,x$

$$K_C = \frac{[NO]^2}{[N_2][O_2]} = \frac{(2\,x)^2}{(0.114 - x)(0.114 - x)} = \frac{(2\,x)^2}{(0.114 - x)^2}$$

$$1.00 \times 10^{-5} = \frac{(2\,x)^2}{(0.114 - x)^2}$$

$$\sqrt{1.00 \times 10^{-5}} = \sqrt{\frac{(2\,x)^2}{(0.114 - x)^2}}$$

$$3.16 \times 10^{-3} = \frac{(2\,x)}{(0.114 - x)}$$

$$2\,x = (3.16 \times 10^{-3})(0.114 - x)$$

$$2.003\,16\,x = 3.60 \times 10^{-4}$$

$$x = 1.8 \times 10^{-4}$$

$[NO] = 2\,x = 2 \times 1.8 \times 10^{-4} = 3.6 \times 10^{-4}$; the concentrations of N_2 and O_2 remain essentially unchanged at $0.114 \text{ mol}\cdot\text{L}^{-1}$.

9.57 The initial concentrations of H_2 and I_2 are:

$$[H_2] = \frac{0.400\text{ mol}}{3.00\text{ L}} = 0.133\text{ mol}\cdot\text{L}^{-1};\ [I_2] = \frac{1.60\text{ mol}}{3.00\text{ L}} = 0.533\text{ mol}\cdot\text{L}^{-1}$$

Concentrations (mol·L^{-1})	$H_2(g)$	+	$I_2(g)$	$\rightleftharpoons$	2 HI(g)
initial	0.133		0.533		0
change	$-x$		$-x$		$+2x$
final	$0.133 - x$		$0.533 - x$		$+2x$

At equilibrium, 60.0% of the H_2 had reacted, so 40.0% of the H_2 remains:

$(0.400)(0.133\text{ mol}\cdot\text{L}^{-1}) = 0.133\text{ mol}\cdot\text{L}^{-1} - x$

$x = 0.133\text{ mol}\cdot\text{L}^{-1} - (0.400)(0.133\text{ mol}\cdot\text{L}^{-1})$

$x = 0.080\text{ mol}\cdot\text{L}^{-1}$

at equilibrium: $[H_2] = 0.133\text{ mol}\cdot\text{L}^{-1} - 0.080\text{ mol}\cdot\text{L}^{-1} = 0.053\text{ mol}\cdot\text{L}^{-1}$

$[I_2] = 0.533\text{ mol}\cdot\text{L}^{-1} - 0.080\text{ mol}\cdot\text{L}^{-1} = 0.453\text{ mol}\cdot\text{L}^{-1}$

$[HI] = 2 \times 0.080\text{ mol}\cdot\text{L}^{-1} = 0.16\text{ mol}\cdot\text{L}^{-1}$

$$K_C = \frac{[HI]^2}{[H_2][I_2]} = \frac{0.16^2}{(0.053)(0.453)} = 1.1$$

9.59 Initial concentrations of CO and O_2 are given by

$$[CO] = \frac{\left(\dfrac{0.28\text{ g CO}}{28.01\text{ g}\cdot\text{mol}^{-1}\text{ CO}}\right)}{2.0\text{ L}} = 5.0 \times 10^{-3}\text{ mol}\cdot\text{L}^{-1}$$

$$[O_2] = \frac{\left(\dfrac{0.032\text{ g }O_2}{32.00\text{ g}\cdot\text{mol}^{-1}\text{ }O_2}\right)}{2.0\text{ L}} = 5.0 \times 10^{-4}\text{ mol}\cdot\text{L}^{-1}$$

Concentration (mol·L^{-1})	2 CO(g)	+	$O_2(g)$	$\rightleftharpoons$	2 $CO_2(g)$
initial	5.0×10^{-3}		5.0×10^{-4}		0
change	$-2x$		$-x$		$+2x$
final	$5.0 \times 10^{-3} - 2x$		$5.0 \times 10^{-4} - x$		$+2x$

$$K_C = \frac{[CO_2]^2}{[CO]^2[O_2]}$$

$$= \frac{(2x)^2}{(5.0 \times 10^{-3} - 2x)^2(5.0 \times 10^{-4} - x)}$$

$$= \frac{4x^2}{(4x^2 - 0.020x + 2.5 \times 10^{-5})(5.0 \times 10^{-4} - x)}$$

$$0.66 = \frac{4x^2}{-4x^3 + 0.022x^2 - 3.5 \times 10^{-5}x + 1.25 \times 10^{-8}}$$

$$4x^2 = (0.66)(-4x^3 + 0.022x^2 - 3.5 \times 10^{-5}x + 1.25 \times 10^{-8})$$

$$6.06\,x^2 = -4\,x^3 + 0.022\,x^2 - 3.5 \times 10^{-5}x + 1.25 \times 10^{-8}$$

$$0 = -4\,x^3 - 6.04\,x^2 - 3.5 \times 10^{-5}x + 1.25 \times 10^{-8}$$

$$x = 4.3 \times 10^{-5}$$

$$[CO_2] = 8.6 \times 10^{-5}\ \text{mol}\cdot\text{L}^{-1};\ [CO] = 4.9 \times 10^{-3}\ \text{mol}\cdot\text{L}^{-1}$$

$$[O_2] = 4.6 \times 10^{-4}\ \text{mol}\cdot\text{L}^{-1}$$

9.61 Concentrations

$(\text{mol}\cdot\text{L}^{-1})$	CH_3COOH +	$C_2H_5OH \rightleftharpoons$	$CH_3COOC_2H_5$ +	H_2O
initial	0.32	6.30	0	0
change	$-x$	$-x$	$+x$	$+x$
final	$0.32 - x$	$6.30 - x$	$+x$	$+x$

$$K_C = \frac{[CH_3COOC_2H_5][H_2O]}{[CH_3COOH][C_2H_5OH]} = \frac{(x)(x)}{(0.32 - x)(6.30 - x)} = \frac{x^2}{x^2 - 6.62\,x + 2.02}$$

$$4.0 = \frac{x^2}{x^2 - 6.62\,x + 2.02}$$

$$4.0\,x^2 - 26.48\,x + 8.08 = x^2$$

$$3.0\,x^2 - 26.48\,x + 8.08 = 0$$

$$x = \frac{-(-26.48) \pm \sqrt{(-26.48)^2 - (4)(3.0)(8.08)}}{(2)(3.0)} = \frac{+26.48 \pm 24.58}{6.0}$$

$x = 8.51$ or 0.317

The root 8.51 is meaningless because it is larger than the concentration of acetic acid and ethanol, so the value 0.317 is chosen. The equilibrium concentration of the product ester is, therefore, $0.317\ \text{mol}\cdot\text{L}^{-1}$. The numbers can be confirmed by placing them into the equilibrium expression:

$$K_C = \frac{[CH_3COOC_2H_5][H_2O]}{[CH_3COOH][C_2H_5OH]} = \frac{(0.317)(0.317)}{(0.320 - 0.317)(6.300 - 0.317)} = 5.6$$

Note: This number does not appear to agree well with the given value of $K_C = 4.0$. If 0.316 is used, the agreement is better, giving a quotient of 4.1. If 0.315 is used, the quotient is 3.3. The better answer is thus $0.316\ \text{mol}\cdot\text{L}^{-1}$. The discrepancy is caused by rounding errors in places that are really beyond the accuracy of the measurement. Given that K_C is only given to two significant figures, the best report of the concentration of ester would be $0.32\ \text{mol}\cdot\text{L}^{-1}$, even though this value will not satisfy the equilibrium expression as well as $0.316\ \text{mol}\cdot\text{L}^{-1}$.

9.63 $K_C = \dfrac{[BrCl]^2}{[Cl_2][Br_2]}$

$$0.031 = \frac{0.145^2}{(0.495)[Br_2]}$$

$$[Br_2] = \frac{0.145^2}{(0.495)(0.031)} = 9.4\ mol \cdot L^{-1}$$

9.65 We can calculate changes according to the reaction stoichiometry:

Amount (mol)	$CO(g)$	+ $3\ H_2(g)$ $\rightleftharpoons$	$CH_4(g)$	+ $H_2O(g)$
initial	2.00	3.00	0	0
change	$-x$	$-3x$	$+x$	$+x$
final	$2.00 - x$	$3.00 - 3x$	0.478	$+x$

According to the stoichiometry, 0.478 mol = x; therefore, at equilibrium, there are 2.00 mol − 0.478 mol = 1.52 mol CO, 3.00 − 3(0.478 mol) = 1.57 mol H_2, and 0.478 mol H_2O. To employ the equilibrium expression, we need either concentrations or pressures; because K_C is given, we will choose to express these as concentrations. This calculation is easy because $V = 10.0$ L:

$[CO] = 0.152\ mol \cdot L^{-1}$; $[H_2] = 0.157\ mol \cdot L^{-1}$; $[CH_4] = 0.0478\ mol \cdot L^{-1}$; $[H_2O] = 0.0478\ mol \cdot L^{-1}$

$$K_C = \frac{[CH_4][H_2O]}{[CO][H_2]^3} = \frac{(0.0478)(0.0478)}{(0.152)(0.157)^3} = 3.88$$

9.67 First, we calculate the initial concentrations of each species:

$$[SO_2] = [NO] = \frac{0.100\ mol}{5.00\ L} = 0.0200\ mol \cdot L^{-1};$$

$$[NO_2] = \frac{0.200\ mol}{5.00\ L} = 0.0400\ mol \cdot L^{-1};\ [SO_3] = \frac{0.150\ mol}{5.00\ L} = 0.0300\ mol \cdot L^{-1}$$

We can use these values to calculate Q in order to see which direction the reactions will go:

$Q = \dfrac{(0.0200)(0.0300)}{(0.0200)(0.0400)} = 0.75$. Because $Q < K_C$, the reaction will proceed to produce more products.

Concentration ($mol \cdot L^{-1}$)	$SO_2(g)$	+ $NO_2(g)$ $\rightleftharpoons$	$NO(g)$	+ $SO_3(g)$
initial	0.0200	0.0400	0.0200	0.0300
change	$-x$	$-x$	$+x$	$+x$
final	$0.0200 - x$	$0.0400 - x$	$0.0200 + x$	$0.0300 + x$

$$K_C = \frac{[NO][SO_3]}{[SO_2][NO_2]}$$

$$85.0 = \frac{(0.0200 + x)(0.0300 + x)}{(0.0200 - x)(0.0400 - x)}$$

$$= \frac{x^2 + 0.0500\,x + 0.000\,600}{x^2 - 0.0600\,x + 0.000\,800}$$

$$85.0(x^2 - 0.0600\,x + 0.000\,800) = x^2 + 0.0500\,x + 0.000\,600$$

$$85.0\,x^2 - 5.10\,x + 0.0680 = x^2 + 0.0500\,x + 0.000\,600$$

$$84.0\,x^2 - 5.15\,x + 0.0674 = 0$$

$$x = \frac{-(-5.15) \pm \sqrt{(-5.15)^2 - (4)(84.0)(0.0674)}}{(2)(84.0)} = \frac{+5.15 \pm 1.97}{168}$$

$x = +0.0424$ or $+0.0189$

The root 0.0424 is not meaningful because it is larger than the concentration of NO_2. The root of choice is therefore 0.0189.

At equilibrium:

$[SO_2] = 0.0200\ mol\cdot L^{-1} - 0.0189\ mol\cdot L^{-1} = 0.0011\ mol\cdot L^{-1}$

$[NO_2] = 0.0400\ mol\cdot L^{-1} - 0.0189\ mol\cdot L^{-1} = 0.0211\ mol\cdot L^{-1}$

$[NO] = 0.0200\ mol\cdot L^{-1} + 0.0189\ mol\cdot L^{-1} = 0.0389\ mol\cdot L^{-1}$

$[SO_3] = 0.0300\ mol\cdot L^{-1} + 0.0189\ mol\cdot L^{-1} = 0.0489\ mol\cdot L^{-1}$

To check, we can put these numbers back into the equilibrium constant expression:

$$K_C = \frac{[NO][SO_3]}{[SO_2][NO_2]}$$

$$\frac{(0.0389)(0.0489)}{(0.0011)(0.0211)} = 82.0$$

Compared to $K_C = 85.0$, this is reasonably good agreement given the nature of the calculation. We can check to see, by trial and error, if a better answer could be obtained. Because the K_C value is low for the concentrations we calculated, we can choose to alter x slightly so that this ratio becomes larger. If we let $x = 0.0190$, the concentrations of NO and SO_3 are increased to 0.0390 and 0.0490, and the concentrations of SO_2 and NO_2 are decreased to 0.0010 and 0.0200 (the stoichiometry of the reaction is maintained by calculating the concentrations in this fashion). Then the quotient becomes 91.0, which is further from the value for K_C than the original answer. So, although the agreement is not the best with the numbers we obtained, it is the best possible, given the limitation on the number of significant figures we are allowed to use in the calculation.

9.69 (a) The initial concentrations of O_2 and N_2O are $0.0560\ mol\cdot L^{-1}$ and $0.0200\ mol\cdot L^{-1}$; the final concentration of $NO_2 = 0.0200\ mol\cdot L^{-1}$ because $V = 1.00$ L.

Concentrations ($mol \cdot L^{-1}$)	$2\ N_2O(g)$	+	$3\ O_2(g)$	$\rightleftharpoons$	$4\ NO_2(g)$
initial	0.0200		0.0560		0
change	$-2\,x$		$-3\,x$		$+4\,x$
final	$0.0200 - 2\,x$		$0.0560 - 3\,x$		0.0200

Thus $4\,x = 0.0200\ mol \cdot L^{-1}$, or $x = 0.005\ 00\ mol \cdot L^{-1}$.

At equilibrium:

$[N_2O] = 0.0200\ mol \cdot L^{-1} - 2\ (0.005\ 00\ mol \cdot L^{-1}) = 0.0100\ mol \cdot L^{-1}$

$[O_2] = 0.0560\ mol \cdot L^{-1} - 3\ (0.005\ 00\ mol \cdot L^{-1}) = 0.0410\ mol \cdot L^{-1}$

$[NO_2] = 0.0200\ mol \cdot L^{-1}$

(b) $K_C = \dfrac{(0.0200)^4}{(0.0100)^2(0.0410)^3} = 23.2$

9.71 (a) The initial concentrations are:

$[PCl_5] = \dfrac{1.50\ mol}{0.500\ L} = 3.00\ mol \cdot L^{-1}$; $[PCl_3] = \dfrac{3.00\ mol}{0.500\ L} = 6.00\ mol \cdot L^{-1}$;

$[Cl_2] = \dfrac{0.500\ mol}{0.500\ L} = 1.00\ mol \cdot L^{-1}$

First calculate Q:

$Q = \dfrac{[PCl_5]}{[PCl_3][Cl_2]} = \dfrac{3.00}{(6.00)(1.00)} = 0.500$

Because $Q \neq K$, the reaction is not at equilibrium.

(b) Because $Q < K_C$, the reaction will proceed to produce products.

(c)

Concentrations ($mol \cdot L^{-1}$)	$PCl_3(g)$	+	$Cl_2(g)$	$\rightleftharpoons$	$PCl_5(g)$
initial	6.00		1.00		3.00
change	$-x$		$-x$		$+x$
final	$6.00 - x$		$1.00 - x$		$3.00 + x$

$K_C = \dfrac{[PCl_5]}{[PCl_3][Cl_2]}$

$0.56 = \dfrac{3.00 + x}{(6.00 - x)(1.00 - x)} = \dfrac{3.00 + x}{x^2 - 7\,x + 6.00}$

$(0.56)(x^2 - 7\,x + 6.00) = 3.00 + x$

$0.56\,x^2 - 3.92\,x + 3.36 = 3.00 + x$

$0.56\,x^2 - 4.92\,x + 0.36 = 0$

$x = \dfrac{-(-4.92) \pm \sqrt{(-4.92)^2 - (4)(0.56)(0.36)}}{(2)(0.56)} = \dfrac{+4.92 \pm 4.84}{1.12}$

$x = 9.2$ or 0.07

Because the root 9.2 is larger than the amount of PCl_3 or Cl_2 available, it is physically meaningless and can be discarded. Thus, $x = 0.071\ mol \cdot L^{-1}$, giving

$[PCl_5] = 3.00\ mol \cdot L^{-1} + 0.07\ mol \cdot L^{-1} = 3.07\ mol \cdot L^{-1}$

$[PCl_3] = 6.00\ mol \cdot L^{-1} - 0.07\ mol \cdot L^{-1} = 5.93\ mol \cdot L^{-1}$

$[Cl_2] = 1.00\ mol \cdot L^{-1} - 0.07\ mol \cdot L^{-1} = 0.93\ mol \cdot L^{-1}$

The number can be checked by substituting them back into the equilibrium constant expression:

$$K_C = \frac{[PCl_5]}{[PCl_3][Cl_2]}$$

$$\frac{(3.07)}{(5.93)(0.93)} \stackrel{?}{=} 0.56$$

$$0.56 \stackrel{\checkmark}{=} 0.56$$

9.73

Pressures (bar)	$2\ HCl(g)$ $\rightleftharpoons$	$H_2(g)$ +	$Cl_2(g)$
initial	0.22	0	0
change	$-2x$	$+x$	$+x$
final	$0.22 - 2x$	$+x$	$+x$

$$K = \frac{P_{H_2} P_{Cl_2}}{P_{HCl}^{\ 2}}$$

$$3.2 \times 10^{-34} = \frac{(x)(x)}{(0.22 - 2x)^2}$$

Because the equilibrium constant is small, assume that $x \ll 0.22$

$$3.2 \times 10^{-34} = \frac{x^2}{(0.22)^2}$$

$$x^2 = (3.2 \times 10^{-34})(0.22)^2$$

$$x = \sqrt{(3.2 \times 10^{-34})(0.22)^2}$$

$$x = \pm 3.9 \times 10^{-18}$$

The negative root is not physically meaningful and can be discarded. x is small compared to 0.22, so the initial assumption was valid. The pressures at equilibrium are

$P_{HCl} = 0.22$ bar; $P_{H_2} = P_{Cl_2} = 3.9 \times 10^{-18}$ bar

The values can be checked by substituting them into the equilibrium expression:

$$\frac{(3.9 \times 10^{-18})(3.9 \times 10^{-18})}{(0.22)^2} \stackrel{?}{=} 3.2 \times 10^{-34}$$

$$3.1 \times 10^{-34} \stackrel{\checkmark}{=} 3.2 \times 10^{-34}$$

$P_{HCl} = 0.22$ bar; $P_{H_2} = P_{Cl_2} = 3.9 \times 10^{-18}$ bar
The numbers agree very well for a calculation of this type.

9.75 (a) To determine on which side of the equilibrium position the conditions lie, we will calculate Q:

$$[CO] = \frac{0.342 \text{ mol}}{3.00 \text{ L}} = 0.114 \text{ mol} \cdot \text{L}^{-1}; [H_2] = \frac{0.215 \text{ mol}}{3.00 \text{ L}} = 0.0717 \text{ mol} \cdot \text{L}^{-1};$$

$$[CH_3OH] = \frac{0.125 \text{ mol}}{3.00 \text{ L}} = 0.0417 \text{ mol} \cdot \text{L}^{-1}$$

$$Q = \frac{[CH_3OH]}{[CO][H_2]^2} = \frac{0.0417}{(0.114)(0.0717)^2} = 71.1 \times 10^3$$

Because $Q > K_C$, the reaction will proceed to produce more of the reactants, which means that the concentration of methanol will decrease.

(b)

Concentrations ($\text{mol} \cdot \text{L}^{-1}$)	$CO(g)$	+	$2\ H_2(g)$	$\rightleftharpoons$	$CH_3OH(g)$
initial	0.114		0.0717		0.0417
change	$+x$		$+2\,x$		$-x$
final	$0.0114 + x$		$0.0717 + 2\,x$		$0.0417 - x$

$$K_C = \frac{0.0417 - x}{(0.0114 + x)(0.0717 + 2\,x)^2} = 1.1 \times 10^{-2}$$

$$0.0417 - x = (1.1 \times 10^{-2})(0.114 + x)(0.0717 + 2\,x)^2$$
$$= (1.2 \times 10^{-3} + 1.1 \times 10^{-2}x)(4\,x^2 + 0.287x + 5.14 \times 10^{-3})$$
$$= 4.4 \times 10^{-2}x^3 + 8.0 \times 10^{-3}x^2 + 4.0 \times 10^{-4}x + 6.2 \times 10^{-6}$$
$$0 = 4.4 \times 10^{-2}x^3 + 8.0 \times 10^{-3}x^2 + 1.00x - 0.0417$$

This equation can be solved approximately, simply by inspection: it is clear that the x term will be very much larger than the x^3 and the x^2 terms, because their coefficients are very small compared to 1.00. This leads to a prediction that $x = 0.0417 \text{ mol} \cdot \text{L}^{-1}$ to within the accuracy of the data. Essentially all of the CH_3OH will react, so that $[CO] = 0.114 \text{ mol} \cdot \text{L}^{-1} + 0.0417 \text{ mol} \cdot \text{L}^{-1} = 0.156 \text{ mol} \cdot \text{L}^{-1}$; $[H_2] = 0.0717 \text{ mol} \cdot \text{L}^{-1} + 2\ (0.0417 \text{ mol} \cdot \text{L}^{-1}) = 0.155 \text{ mol} \cdot \text{L}^{-1}$. The mathematical situation is odd in that clearly a $[CH_3OH] = 0$ will not satisfy the equilibrium constant. Knowing that the methanol concentration is very small compared to the CO and H_2 concentrations, we can now back-calculate to get a concentration value that will satisfy the equilibrium expression:

$$K_C = \frac{y}{(0.156)(0.155)^2} = 1.1 \times 10^{-2}$$

$y = (1.1 \times 10^{-2})(0.156)(0.155)^2 = 2.6 \times 10^{-4}$

Alternatively, the cubic equation can be solved with the aid of a graphing calculator like the one supplied on the CD accompanying this book.

9.77 (a) According to Le Chatelier's principle, an increase in the partial pressure of CO_2 will result in creation of reactants, which will decrease the H_2 partial pressure.

(b) According to Le Chatelier's principle, if the CO pressure is reduced, the reaction will shift to form more CO, which will decrease the pressure of CO_2.

(c) According to Le Chatelier's principle, if the concentration of CO is increased, the reaction will proceed to form more products, which will result in a higher pressure of H_2.

(d) The equilibrium constant for the reaction is unchanged, because it is unaffected by any change in concentration.

9.79 (a) According to Le Chatelier's principle, increasing the concentration of NO will cause the reaction to form reactants in order to reduce the concentration of NO; the amount of water will decrease.

(b) For the same reason as in (a), the amount of O_2 will increase.

(c) According to Le Chatelier's principle, removing water will cause the reaction to shift toward products, resulting in the formation of more NO.

(d) According to Le Chatelier's principle, removing a reactant will cause the reaction to shift in the direction to replace the removed substance; the amount of NH_3 should increase.

(e) According to Le Chatelier's principle, adding ammonia will shift the reaction to the right, but the equilibrium constant, which is a constant, will not be affected.

(f) According to Le Chatelier's principle, removing NO will cause the formation of more products; the amount of NH_3 will decrease.

(g) According to Le Chatelier's principle, adding reactants will promote the formation of products; the amount of oxygen will decrease.

9.81 As per Le Chatelier's principle, whether increasing the pressure on a reaction will affect the distribution of species within an equilibrium mixture of gases depends largely upon the difference in the number of moles of gases between the reactant and product sides of the equation. If there is a net increase in the amount of gas, then applying pressure will shift the reaction toward reactants in order to remove the stress applied by increasing the pressure. Similarly, if there is a net decrease in the amount of gas, applying pressure will cause the formation of products. If the number of moles of gas is the same on the product and reactant side, then changing the pressure will have little or no effect on the equilibrium distribution of species present. Using this information, we can apply it to the specific reactions given. The answers are: (a) reactants; (b) reactants; (c) reactants; (d) no change (there is the same number of moles of gas on both sides of the equation); (e) reactants

9.83 (a) If the pressure of NO (a product) is increased, the reaction will shift to form more reactants; the pressure of NH_3 should increase.
(b) If the pressure of NH_3 (a reactant) is decreased, then the reaction will shift to form more reactants; the pressure of O_2 should increase.

9.85 If a reaction is exothermic, raising the temperature will tend to shift the reaction toward reactants, whereas if the reaction is endothermic, a shift toward products will be observed. For the specific examples given, (a) and (b) are endothermic (the values for (b) can be calculated, but we know that it requires energy to break an X—X bond, so those processes will all be endothermic) and raising the temperature should favor the formation of products; (c) and (d) are exothermic and raising the temperature should favor the formation of reactants.

9.87 Even though numbers are given, we do not need to do a calculation to answer this qualitative question. Because the equilibrium constant for the formation of ammonia is smaller at the higher temperature, raising the temperature will favor the formation of reactants. Less ammonia will be present at higher temperature, assuming no other changes occur to the system (i.e., the volume does not change, no reactants or products are added or removed from the container, etc.).

9.89 To answer this question we must calculate Q:

$$Q = \frac{[NH_3]^2}{[N_2][H_2]^3} = \frac{(0.500)^2}{(3.00)(2.00)^3} = 0.0104$$

Because $Q \neq K$, the system is not at equilibrium and because $Q < K$, the reaction will proceed to produce more products.

9.91 Because we want the equilibrium constant at two temperatures, we will need to calculate $\Delta H°_r$ and $\Delta S°_r$ for each reaction:
(a) $NH_4Cl \rightleftharpoons NH_3(g) + HCl(g)$

$\Delta H°_r = \Delta H°_f(NH_3, g) + \Delta H°_f(HCl, g) - \Delta H°_f(NH_4Cl, s)$

$\Delta H°_r = (-46.11\ kJ\cdot mol^{-1}) + (-92.31\ kJ\cdot mol^{-1}) - (-314.43\ kJ\cdot mol^{-1})$

$\Delta H°_r = 176.01\ kJ\cdot mol^{-1}$

$\Delta S°_r = S°\ (NH_3, g) + S°\ (HCl, g) - S°\ (NH_4Cl, s)$

$\Delta S°_r = 192.45\ J\cdot K^{-1}\cdot mol^{-1} + 186.91\ J\cdot K^{-1}\cdot mol^{-1} - 94.6\ J\cdot K^{-1}\cdot mol^{-1}$

$\Delta S°_r = 284.8\ J\cdot K^{-1}\cdot mol^{-1}$

$\Delta G°_r = \Delta H°_r - T\Delta S°_r$

At 298 K:

$$\Delta G°_{r(298\ K)} = 176.01\ kJ - (298\ K)(284.8\ J\cdot K^{-1})/(1000\ J\cdot kJ^{-1})$$
$$= 91.14\ kJ\cdot mol^{-1}$$

$\Delta G°_{r(298\ K)} = -RT \ln K$

$$\ln K = -\frac{\Delta G°_{r(298\ K)}}{RT}$$
$$= -\frac{91\ 140\ J}{(8.314\ J\cdot K^{-1})(298\ K)} = -36.8$$

$K = 1 \times 10^{-16}$

At 423 K:

$$\Delta G°_{r(423\ K)} = 176.01\ kJ - (423\ K)(284.8\ J\cdot K^{-1})/(1000\ J\cdot kJ^{-1})$$
$$= 55.54\ kJ\cdot mol^{-1}$$

$\Delta G°_{r(423\ K)} = -RT \ln K$

$$\ln K = -\frac{\Delta G°_{r(423\ K)}}{RT}$$
$$= -\frac{55\ 540\ J}{(8.314\ J\cdot K^{-1})(423\ K)} = -15.8$$

$K = 1 \times 10^{-7}$

(b) $H_2(g) + D_2O(l) \rightleftharpoons D_2(g) + H_2O(l)$

$\Delta H°_r = \Delta H°_f(H_2O, l) - [\Delta H°_f(D_2O, l)]$

$\Delta H°_r = (-285.83\ kJ\cdot mol^{-1}) - [-294.60\ kJ\cdot mol^{-1}]$

$\Delta H°_r = 8.77\ kJ\cdot mol^{-1}$

$\Delta S°_r = S°(D_2, g) + S°(H_2O, l) - [S°(H_2, g) + S°(D_2O, l)]$

$$\Delta S°_r = 144.96\ J\cdot K^{-1}\cdot mol^{-1} + 69.91\ J\cdot K^{-1}\cdot mol^{-1}$$
$$- [130.68\ J\cdot K^{-1}\cdot mol^{-1} + 75.94\ J\cdot K^{-1}\cdot mol^{-1}]$$

$\Delta S°_r = 8.25\ J\cdot K^{-1}\cdot mol^{-1}$

At 298 K:

$$\Delta G°_{r(298\ K)} = 8.77\ kJ\cdot mol^{-1} - (298\ K)(8.25\ J\cdot K^{-1}\cdot mol^{-1})/(1000\ J\cdot kJ^{-1})$$
$$= 6.31\ kJ\cdot mol^{-1}$$

$$\Delta G^\circ_{r(298\ K)} = -RT \ln K$$

$$\ln K = -\frac{\Delta G^\circ_{r(298\ K)}}{RT}$$

$$= -\frac{6310\ J}{(8.314\ J\cdot K^{-1})(298\ K)} = -2.55$$

$$K = 7.8 \times 10^{-2}$$

At 423 K:

$$\Delta G^\circ_{r(423\ K)} = 8.77\ kJ\cdot mol^{-1} - (423\ K)(8.25\ J\cdot K^{-1}\cdot mol^{-1})/(1000\ J\cdot kJ^{-1})$$

$$= 5.28\ kJ\cdot mol^{-1}$$

$$\ln K = -\frac{5280\ J}{(8.314\ J\cdot K^{-1})(423\ K)} = -1.50$$

$$K = 0.22$$

9.93 (a) The initial pressure of NOCl is calculated as follows:

$$P = \frac{nRT}{V} = \frac{\left(\frac{m}{M}\right)RT}{V}$$

$$= \frac{\left(\frac{30.1\ g}{65.46\ g\cdot mol^{-1}}\right)(0.082\ 06\ L\cdot atm\cdot K^{-1}\cdot mol^{-1})(500\ K)\left(\frac{1.013\ 25\ bar}{1\ atm}\right)}{0.200\ L}$$

$$= 95.6\ bar$$

Pressures (bar)	$2\ NOCl(g)$	$\rightleftharpoons$	$2\ NO(g)$	+	$Cl_2(g)$
initial	95.6		0		0
change	$-2x$		$+2x$		$+x$
final	$95.6 - 2x$		$+2x$		$+x$

$$K = \frac{P_{NO}^{\ 2}P_{Cl_2}}{P_{NOCl}^{\ 2}}$$

$$1.13 \times 10^{-3} = \frac{(2x)^2(x)}{(95.6 - 2x)^2}$$

Assume that $x \ll 95.6$:

$$1.13 \times 10^{-3} = \frac{4x^3}{(95.6)^2}$$

$$x^3 = (1.13 \times 10^{-3})(95.6)^2$$

$$x = \sqrt[3]{(1.13 \times 10^{-3})(95.6)^2} = 2.18$$

Because x is not very much less than 95.6, the assumption was not good and the expression should be solved more rigorously:

$$1.13 \times 10^{-3} = \frac{4x^3}{(95.6 - 2x)^2}$$

$$= \frac{4x^3}{4x^2 - 382.4x + 9139}$$

$$(1.13 \times 10^{-3})(4x^2 - 382.4x + 9139) = 4x^3$$

$$4x^3 - 4.52 \times 10^{-3}x^2 + 0.432x - 10.3 = 0$$

x can be obtained from a graphing calculator or by the trial and error method to give 1.345 bar.

$P_{NOCl} = 95.6 \text{ bar} - 2(1.345 \text{ bar}) = 92.9 \text{ bar}$

$P_{NO} = 2 \times 1.345 \text{ bar} = 2.69 \text{ bar}$

$P_{Cl_2} = 1.34 \text{ bar}$

These numbers can be checked by substituting them into the equilibrium expression:

$$\frac{(2.69)^2(1.34)}{(92.9)^2} \stackrel{?}{=} 1.13 \times 10^{-3}$$

$$1.12 \times 10^{-3} \stackrel{\checkmark}{=} 1.13 \times 10^{-3}$$

This is good agreement.

(b) $\frac{95.6 \text{ bar} - 92.9 \text{ bar}}{95.6 \text{ bar}} \times 100 = 2.8\%$

9.95 (a) According to Le Chatelier's principle, adding a product should cause a shift in the equilibrium toward the reactants side of the equation.

(b) Because there are equal numbers of moles of gas on both sides of the equation, there will be little or no effect upon compressing the system.

(c) If the amount of CO_2 is increased, this will cause the reaction to shift toward the formation of products.

(d) Because the reaction is endothermic, raising the temperature will favor the formation of products.

(e) If the amount of $C_6H_{12}O_6$ is removed, this will cause the reaction to shift toward the formation of products.

(f) Because water is a liquid, it is by definition present at unit concentration, so changing the amount of water will not affect the reaction. As long as the glucose solution is dilute, its concentration can be considered unchanged.

(g) Decreasing the concentration of a reactant will favor the production of more reactants.

9.97 (a) In order to solve this problem, we will manipulate the equations with the known K's so that we can combine them to give the desired overall reaction:

First, reverse equation (1) and multiply it by $\frac{1}{2}$:

$$H_2(g) + \tfrac{1}{2}\,O_2(g) \rightleftharpoons H_2O(g) \qquad K_4 = \frac{1}{(1.6 \times 10^{-11})^{1/2}} \tag{4}$$

Multiply equation (2) by $\frac{1}{2}$ also:

$$CO_2(g) \rightleftharpoons CO(g) + \tfrac{1}{2}\,O_2(g) \qquad K_5 = (1.3 \times 10^{-10})^{1/2} \tag{5}$$

Adding equations (4) and (5) gives the desired reaction. The resultant equilibrium constant will be the product of the K's for (4) and (5):

$$CO_2(g) + H_2(g) \rightleftharpoons H_2O(g) + CO(g) \qquad K_5 = \left(\frac{1.3 \times 10^{-10}}{1.6 \times 10^{-11}}\right)^{1/2} = 2.9 \tag{3}$$

(b) To obtain the K value for a net equation from two (or more) others, the K's are multiplied, but ΔG°_r's are added:

$$2\,H_2O(g) \rightleftharpoons 2\,H_2(g) + O_2(g) \tag{1}$$

$$\begin{aligned}\Delta G^\circ_r &= -RT \ln K \\ &= -(8.314\ J{\cdot}K^{-1}{\cdot}mol^{-1})(1565\ K) \ln 1.6 \times 10^{-11} \\ &= +3.2 \times 10^2\ kJ{\cdot}mol^{-1}\end{aligned}$$

$$2\,CO_2(g) \rightleftharpoons 2\,CO(g) + O_2(g) \tag{2}$$

$$\begin{aligned}\Delta G^\circ_r &= -RT \ln K \\ &= -(8.314\ J{\cdot}K^{-1}{\cdot}mol^{-1})(1565\ K) \ln 1.3 \times 10^{-10} \\ &= +3.0 \times 10^2\ kJ{\cdot}mol^{-1}\end{aligned}$$

The corresponding values for (4) and (5) are

$$\Delta G^\circ_{r(4)} = -\tfrac{1}{2}(3.2 \times 10^2\ kJ) = -1.6 \times 10^2\ kJ$$

$$\Delta G^\circ_{r(5)} = \tfrac{1}{2}(3.0 \times 10^2\ kJ) = 1.5 \times 10^2\ kJ$$

Summing these two values will give

$$\Delta G^\circ_{r(3)} = -1.6 \times 10^2\ kJ + 1.5 \times 10^2\ kJ = -10\ kJ{\cdot}mol^{-1}$$

$$\ln K = -\frac{\Delta G^\circ_r}{RT} = -\frac{-10\ 000\ J{\cdot}mol^{-1}}{(8.314\ J{\cdot}K^{-1}{\cdot}mol^{-1})(1565\ K)} = +0.8$$

$$K = 2$$

This value is in reasonable agreement with the one obtained in (a), given the problems in significant figures and due to rounding errors.

9.99

Pressure	$PCl_5(g)$ $\rightleftharpoons$	$PCl_3(g)$ +	$Cl_2(g)$
initial	n		
change	$-n\alpha$	$+n\alpha$	$+n\alpha$
final	$n(1-\alpha)$	$+n\alpha$	$+n\alpha$

$$K = \frac{(n\alpha)(n\alpha)}{n(1-\alpha)} = \frac{n^2\alpha^2}{n(1-\alpha)} = \frac{n\alpha^2}{(1-\alpha)}$$

$$P = n(1-\alpha) + n\alpha + n\alpha = n(1+\alpha)$$

$$n = \frac{P}{1 + \alpha}$$

$$K = \frac{n\alpha^2}{(1 - \alpha)} = \left(\frac{P}{1 + \alpha}\right)\left(\frac{\alpha^2}{1 - \alpha}\right) = \frac{P\alpha^2}{1 - \alpha^2}$$

$$(1 - \alpha^2)K = P\alpha^2$$

$$K - K\alpha^2 = P\alpha^2$$

$$K = P\alpha^2 + K\alpha^2$$

$$\alpha^2 = \frac{K}{P + K}$$

$$\alpha = \sqrt{\frac{K}{P + K}}$$

(a) For the specific conditions $K = 4.96$ and $P = 0.50$ bar,

$$\alpha = \sqrt{\frac{4.96}{0.50 + 4.96}} = 0.953$$

(b) For the specific conditions $K = 4.96$ and $P = 1.00$ bar,

$$\alpha = \sqrt{\frac{4.96}{1.00 + 4.96}} = 0.912$$

9.101 (a) If $K = 1.00$, then $\Delta G°$ must be equal to 0 ($\Delta G° = -RT \ln K$).

(b) This can be calculated by determining the values for $\Delta H°$ and $\Delta S°$ at 25°C.

$$\Delta H° = -393.51\ \text{kJ}\cdot\text{mol}^{-1} - [(-110.53\ \text{kJ}\cdot\text{mol}^{-1}) + (-241.82\ \text{kJ}\cdot\text{mol}^{-1})]$$
$$= -41.16\ \text{kJ}\cdot\text{mol}^{-1}$$

$$\Delta S° = 130.68\ \text{J}\cdot\text{K}^{-1}\cdot\text{mol}^{-1} + 213.74\ \text{J}\cdot\text{K}^{-1}\cdot\text{mol}^{-1} - [197.67\ \text{J}\cdot\text{K}^{-1}\cdot\text{mol}^{-1} + 188.83\ \text{J}\cdot\text{K}^{-1}\cdot\text{mol}^{-1}]$$
$$= -42.08\ \text{J}\cdot\text{K}^{-1}\cdot\text{mol}^{-1}$$

$$\Delta G = (-41.16\ \text{kJ}\cdot\text{mol}^{-1})(1000\ \text{J}\cdot\text{kJ}^{-1}) - T(-42.08\ \text{J}\cdot\text{K}^{-1}\cdot\text{mol}^{-1}) = 0$$

$$T = 978\ K \text{ (or 705°C)}$$

(c)

	$CO(g)$	+	$H_2O(g)$	$\rightleftharpoons$	$CO_2(g)$	+	$H_2(g)$
	10.00 bar		10.00 bar		5.00 bar		5.00 bar
change	$-x$		$-x$		$+x$		$+x$
net	$10.00 - x$		$10.00 - x$		$5.00 + x$		$5.00 + x$

$x = 2.50$ bar

All pressures are equal to 7.50 bar.

(d) First, check Q to determine the direction of the reaction:

$$Q = \frac{(10.00)(5.00)}{(6.00)(4.00)} = 2.08$$

Because Q is greater than 1, the reaction will shift to produce reactants.

	$CO(g)$ +	$H_2O(g)$ $\rightleftharpoons$	$CO_2(g)$ +	$H_2(g)$
	6.00 bar	4.00 bar	10.00 bar	5.00 bar
change	$+x$	$+x$	$-x$	$-x$
net	$6.00 + x$	$4.00 + x$	$10.00 - x$	$5.00 - x$

$$\frac{(10.00 - x)(5.00 - x)}{(6.00 + x)(4.00 + x)} = 1$$

$(10.00 - x)(5.00 - x) = (6.00 + x)(4.00 + x)$

$x^2 - 15.00\,x + 50.0 = x^2 + 10.00\,x + 24.0$

$25.00\,x = 26.0$

$x = 1.04$ bar

$P_{CO(g)} = 7.04$ bar; $P_{H_2O(g)} = 5.04$ bar; $P_{CO_2(g)} = 8.96$ bar; $P_{H_2(g)} = 3.96$ bar

9.103 (a) These values are easily calculated from the relationship $\Delta G° = -RT \ln K$. For the atomic species, the free energy of the reaction will be $\frac{1}{2}$ of this value because the equilibrium reactions are for the formation of two moles of halogen atoms. The results are

Halogen	Bond Dissociation Energy ($kJ \cdot mol^{-1}$)	$\Delta G°$ ($kJ \cdot mol^{-1}$)
fluorine	146	19.2
chlorine	230	47.8
bromine	181	42.8
iodine	139	5.6

(b)

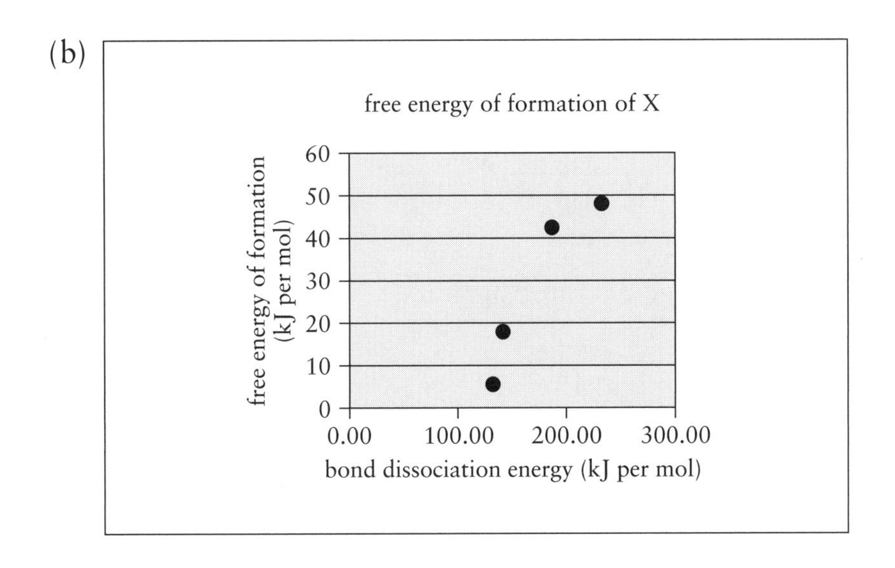

There is a correlation between the bond dissociation energy and the free energy of formation of the atomic species, but the relationship is clearly not linear.

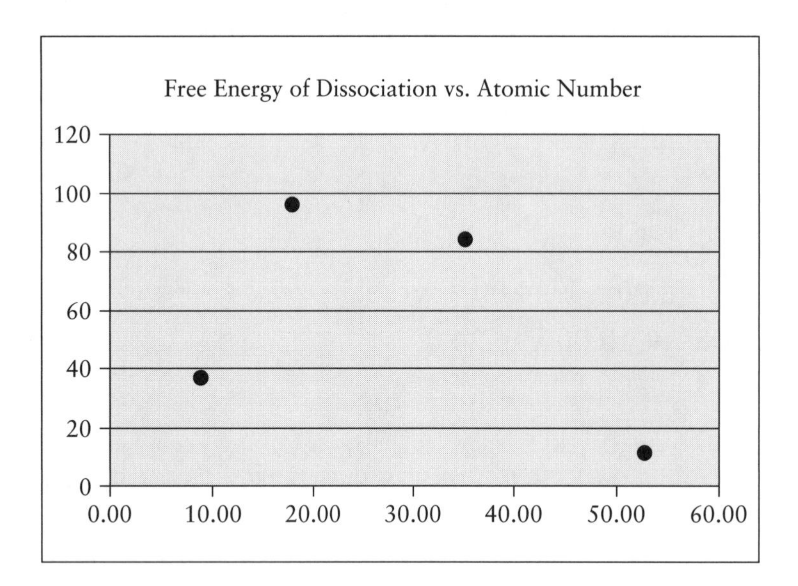

For the heavier three halogens, there is a trend to decreasing free energy of formation of the atoms as the element becomes heavier, but fluorine is anomalous. The F—F bond energy is lower than expected, owing to repulsions of the lone pairs of electrons on the adjacent F atoms because the F—F bond distance is so short.

9.105 (a) Use the relationship $\ln \frac{K_2}{K_1} = -\frac{\Delta H°}{R}\left(\frac{1}{T_2} - \frac{1}{T_1}\right)$. The value of $\Delta H°$ is first obtained by using the two points given in Table 9.1.

$$\ln \frac{1.7 \times 10^{-3}}{3.4 \times 10^{-5}} = -\frac{\Delta H°}{R}\left(\frac{1}{1200} - \frac{1}{1000}\right)$$

$$\Delta H° = 1.96 \times 10^{2}\ \text{kJ}\cdot\text{mol}^{-1}$$

We then use this value plus one of the known equilibrium constant points in the equation

$$\ln \frac{K_{298}}{3.4 \times 10^{-5}} = -\left(\frac{196\ 000\ \text{J}\cdot\text{mol}^{-1}}{8.314\ \text{J}\cdot\text{K}^{-1}\cdot\text{mol}^{-1}}\right)\left(\frac{1}{298} - \frac{1}{1000}\right)$$

$$K_{298} = 2.6 \times 10^{-29}$$

(b) Using the thermodynamic data in Appendix 2A:

$$Br_2(g) \rightleftharpoons 2\ Br(g)$$

$$\Delta G° = 2(82.40\ \text{kJ}\cdot\text{mol}^{-1}) - 3.11\ \text{kJ}\cdot\text{mol}^{-1} = 161.69\ \text{kJ}\cdot\text{mol}^{-1}$$

$$K = e^{-\Delta G°/RT} = 4.5 \times 10^{-29}$$

For equilibrium constant calculations, this is reasonably good agreement with the value obtained from part (a), especially if one considers that $\Delta H°$ will not be perfectly constant over so large a temperature range.

(c) We will use data from Appendix 2A to calculate the vapor pressure of bromine:

$Br_2(l) \rightleftharpoons Br_2(g)$

$\Delta G° = 3.11\ kJ \cdot mol^{-1}$

$K = e^{-\Delta G°/RT} = 0.285$

The vapor pressure of bromine will, therefore, be 0.285 bar or 0.289 atm. Remember that because the standard state for the thermodynamic quantities is 1 bar, the values in K will be derived in bar as well.

(d) $4.5 \times 10^{-29} = \dfrac{P_{Br(g)}{}^2}{P_{Br_2(l)}} = \dfrac{P_{Br(g)}{}^2}{0.285\ bar}$

$P_{Br(g)}{}^2 = 3.6 \times 10^{-15}$ bar or 3.6×10^{-15} atm

(e) Use the ideal gas law:

$PV = nRT$

$(0.289\ atm)V = (0.0100\ mol)(0.082\ 06\ L \cdot atm \cdot K^{-1} \cdot mol^{-1})(298K)$

$V = 0.846$ L or 846 mL

9.107 First, we calculate the equilibrium constant for the conditions given.

$K = \dfrac{(23.72)^2}{(3.11)(1.64)^3} = 41.0$, which corresponds to the reaction written as

$N_2(g) + 3\ H_2(g) \rightleftharpoons 2\ NH_3(g)$

We then set up the table of anticipated changes upon introduction of the nitrogen:

	$N_2(g)$	+ $3\ H_2(g)$	$\rightleftharpoons$ $2\ NH_3(g)$
initial	4.68 bar	1.64 bar	23.72 bar
change	$-x$	$-3x$	$+2x$
total	$4.68 - x$	$1.64 - 3x$	$23.72 + 2x$

$$41.0 = \frac{(23.72 + 2x)^2}{(4.68 - x)(1.64 - 3x)^3}$$

The equation can be solved using a graphing calculator, other computer software, or by trial and error. The solution is $x = 0.176$. The pressures of gases are

$P_{N_2} = 4.33$ bar or 4.39 atm

$P_{H_2} = 1.11$ bar or 1.12 atm

$P_{NH_3} = 24.07$ bar or 24.39 atm

9.109

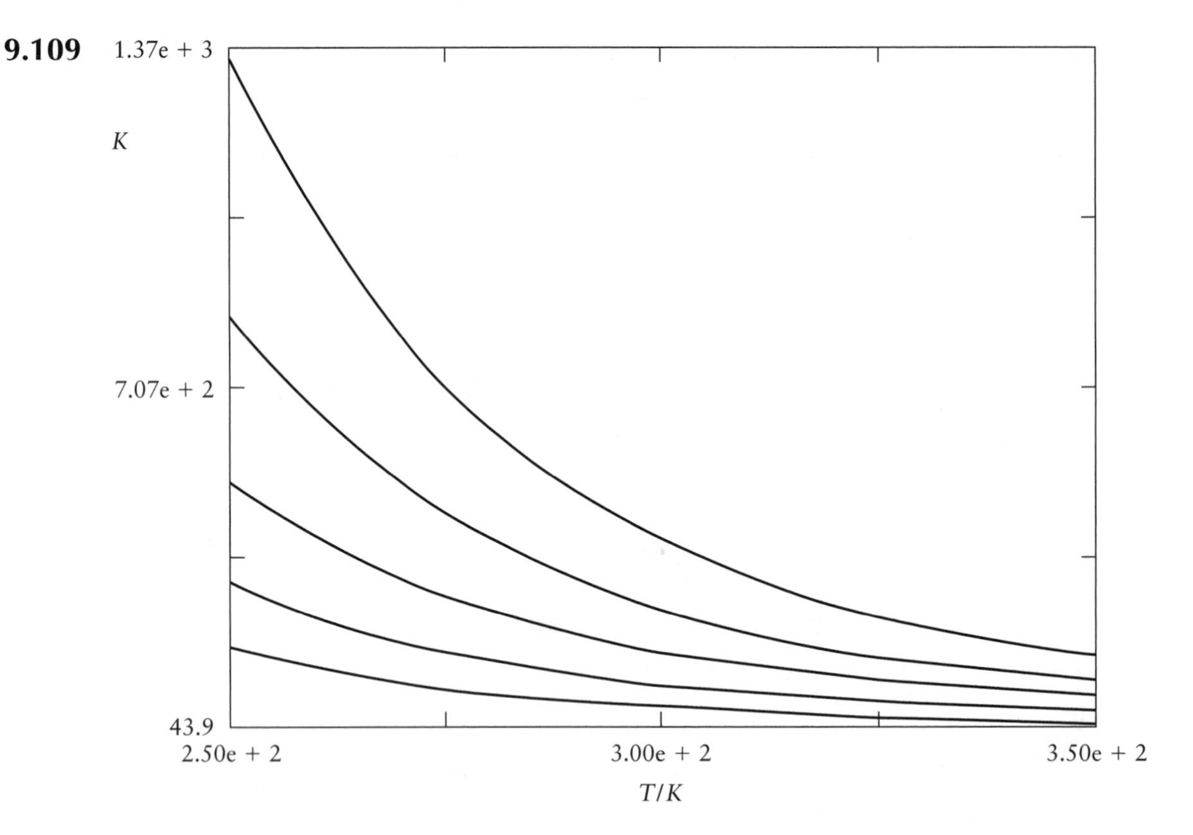

The greater change in equilibrium constant as temperature changes is seen for reactions that have the larger values of $\Delta G°$.

9.111 (a) $K_C = \dfrac{[NO_2]^2}{[N_2O_4]} = \dfrac{(2.13)^2}{0.405} = 11.2$

(b) If NO_2 is added, the equilibrium will shift to produce more N_2O_4. The amount of NO_2 will be greater than initially present, but less than the $3.13\ mol \cdot L^{-1}$ present immediately upon making the addition. K_C will not be affected.

(c)

Concentrations ($mol \cdot L^{-1}$)	N_2O_4	$\rightleftharpoons$	$2\ NO_2$
initial	0.405		3.13
change	$+x$		$-2x$
final	$0.405 + x$		$3.13 - 2x$

$$11.2 = \frac{(3.13 - 2x)^2}{0.405 + x}$$

$(11.2)(0.405 + x) = (3.13 - 2x)^2$

$11.2x + 4.536 = 4x^2 - 12.52x + 9.797$

$$4x^2 - 23.7x + 5.26 = 0$$

$$x = \frac{-(-23.7) \pm \sqrt{(-23.7)^2 - (4)(4)(5.26)}}{(2)(4)} = \frac{23.7 \pm 21.9}{8}$$

$x = 5.70$ or 0.23

At equilibrium $[N_2O_4] = 0.405\ \text{mol}\cdot\text{L}^{-1} + 0.23\ \text{mol}\cdot\text{L}^{-1} = 0.64\ \text{mol}\cdot\text{L}^{-1}$

$[NO_2] = 3.13\ \text{mol}\cdot\text{L}^{-1} - 2(0.23\ \text{mol}\cdot\text{L}^{-1}) = 2.67\ \text{mol}\cdot\text{L}^{-1}$

These concentrations are consistent with the predictions in (b).

9.113 To find the vapor pressure, we first calculate $\Delta G°$ for the conversion of the liquid to the gas at 298 K, using the free energies of formation found in the appendix:

$$\Delta G°_{H_2O(l) \rightarrow H_2O(g)} = \Delta G°_{f(H_2O(g))} - \Delta G°_{f(H_2O(l))}$$
$$= (-228.57\ \text{kJ}\cdot\text{mol}^{-1}) - [-237.13\ \text{kJ}\cdot\text{mol}^{-1}]$$
$$= 8.56\ \text{kJ}\cdot\text{mol}^{-1}$$

$$\Delta G°_{D_2O(l) \rightarrow D_2O(g)} = \Delta G°_{f(D_2O(g))} - \Delta G°_{f(D_2O(l))}$$
$$= (-234.54\ \text{kJ}\cdot\text{mol}^{-1}) - [-243.44\ \text{kJ}\cdot\text{mol}^{-1}]$$
$$= 8.90\ \text{kJ}\cdot\text{mol}^{-1}$$

The equilibrium constant for these processes is the vapor pressure of the liquid:

$K = P_{H_2O}$ or $K = P_{D_2O}$

Using $\Delta G° = -RT \ln K$, we can calculate the desired values.

For H_2O:

$$\ln K = -\frac{\Delta G°}{RT} = -\frac{8560\ \text{J}\cdot\text{mol}^{-1}}{(8.314\ \text{J}\cdot\text{K}^{-1}\cdot\text{mol}^{-1})(298\ \text{K})} = -3.45$$

$K = 0.032$ bar

$$0.032\ \text{bar} \times \frac{1\ \text{atm}}{1.013\ 25\ \text{bar}} \times \frac{760\ \text{Torr}}{1\ \text{atm}} = 24\ \text{Torr}$$

For D_2O:

$$\ln K = -\frac{\Delta G°}{RT} = -\frac{8900\ \text{J}\cdot\text{mol}^{-1}}{(8.314\ \text{J}\cdot\text{K}^{-1}\cdot\text{mol}^{-1})(298\ \text{K})} = -3.59$$

$K = 0.028$ bar

$$0.028\ \text{bar} \times \frac{1\ \text{atm}}{1.013\ 25\ \text{bar}} \times \frac{760\ \text{Torr}}{1\ \text{atm}} = 21\ \text{Torr}$$

The answer is that D has a lower zero point energy than H. This makes the D_2O—D_2O "hydrogen bond" stronger than the H_2O—H_2O hydrogen bond. Because the hydrogen bond is stronger, the intermolecular forces are stronger, the vapor pressure is lower, and the boiling point is higher.

Potential energy curves for the O—H and O—D bonds as a function of distance:

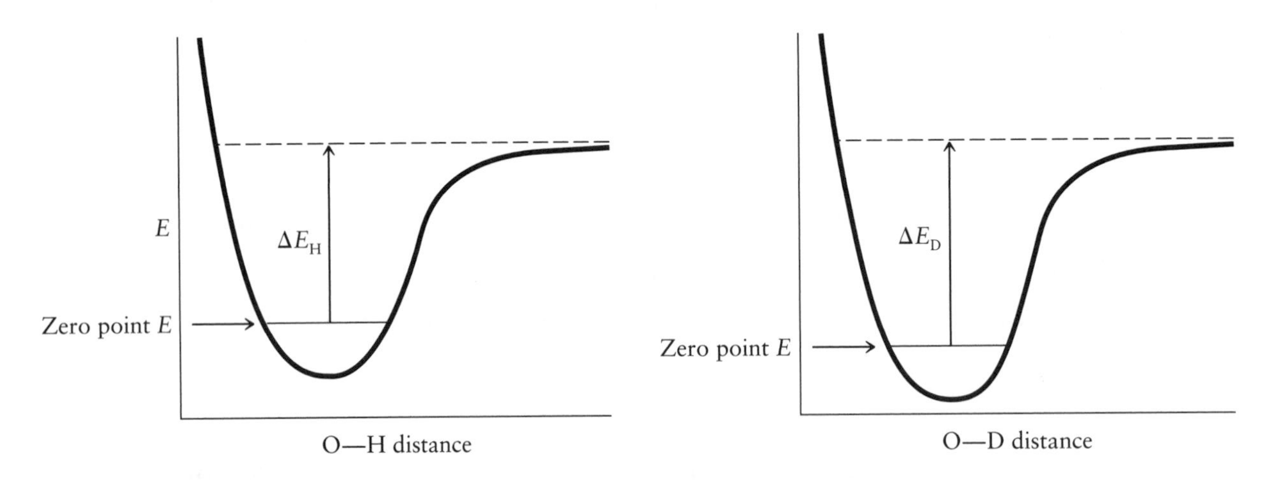

ΔE = energy required to break O—H or O—D bond.

9.115 First, we derive a general relationship that relates ΔG°_f values to nonstandard state conditions. We do this by returning to the fundamental definition of ΔG and ΔG°.

$\Delta G^\circ_f = \Sigma G^\circ_{m(products)} - \Sigma G^\circ_{m(reactants)}$

Similarly, for conditions other than standard state, we can write

$\Delta G_f = \Sigma G_{m(products)} - \Sigma G_{m(reactants)}$

Because $G_m = G^\circ_m + RT \ln Q$,

$\Delta G_f = \Sigma(G^\circ_m + RT \ln P_i)_{(products)} - \Sigma(G^\circ_m + RT \ln P_i)_{(reactants)}$

But because all reactants or products in the same system refer to the same standard state,

$\Delta G_f = \Sigma G^\circ_{m(products)} - \Sigma G^\circ_{m(reactants)} + \Delta n(RT \ln Q_{(reactants)})$

$\Delta G_f = \Delta G^\circ_f + \Delta n(RT \ln Q_{(reactants)})$

The value ΔG_f, which is the nonstandard value for the conditions of 1 atm, $1\ \text{mol}\cdot\text{L}^{-1}$, etc. becomes the new standard value.

(a) $\frac{1}{2}\ H_2(g) + \frac{1}{2}\ I_2(g) \longrightarrow HI(g)$, $\Delta n = 1\ \text{mol} - (\frac{1}{2}\ \text{mol} + \frac{1}{2}\ \text{mol}) = 0$

1 atm = 1.013 25 bar

$\Delta G_f = 1.70\ \text{kJ}\cdot\text{mol}$

$\quad + (0)(8.314\ \text{J}\cdot\text{K}^{-1}\cdot\text{mol}^{-1})(298\ \text{K})(\ln 1.013\ 25)/(1000\ \text{J}\cdot\text{kJ}^{-1})$

$= 1.70\ \text{kJ}\cdot\text{mol}$

(b) $C(s) + \frac{1}{2}\ O_2(g) \longrightarrow CO(g)$, $\Delta n = 1\ \text{mol} - \frac{1}{2}\ \text{mol} = \frac{1}{2}\ \text{mol}$

$\Delta G_f = -137.17\ \text{kJ}\cdot\text{mol}$

$\quad + (\frac{1}{2})(8.314\ \text{J}\cdot\text{K}^{-1}\cdot\text{mol}^{-1})(298\ \text{K})(\ln 1.013\ 25)/(1000\ \text{J}\cdot\text{kJ}^{-1})$

$= -137.15\ \text{kJ}\cdot\text{mol}^{-1}$

(c) $C(s) + \frac{1}{2}\ N_2(g) + \frac{1}{2}\ H_2(g) \longrightarrow HCN(g)$, $\Delta n = 1\ \text{mol} - (\frac{1}{2}\ \text{mol} + \frac{1}{2}\ \text{mol})$

$= 0$

$$1\ \text{Torr} \times \frac{1\ \text{atm}}{760\ \text{Torr}} \times \frac{1.013\ 25\ \text{bar}}{1\ \text{atm}} = 1.333 \times 10^{-3}\ \text{bar}$$

$$\begin{aligned}\Delta G_f &= 124.7\ \text{kJ}\cdot\text{mol} \\ &\quad + (0)(8.314\ \text{J}\cdot\text{K}^{-1}\cdot\text{mol}^{-1})(298\ \text{K})(\ln 1.333 \times 10^{-3})/(1000\ \text{J}\cdot\text{kJ}^{-1}) \\ &= 124.7\ \text{kJ}\cdot\text{mol}^{-1}\end{aligned}$$

(d) $C(s) + 2\ H_2(g) \longrightarrow CH_4(g)$, $\Delta n = 1\ \text{mol} - 2\ \text{mol} = -1\ \text{mol}$

$1\ \text{Pa} = 10^{-5}\ \text{bar}$

$$\begin{aligned}\Delta G_f &= -50.72\ \text{kJ}\cdot\text{mol} \\ &\quad + (-1)(8.314\ \text{J}\cdot\text{K}^{-1}\cdot\text{mol}^{-1})(298\ \text{K})(\ln 10^{-5})/(1000\ \text{J}\cdot\text{kJ}^{-1}) \\ &= -22.2\ \text{kJ}\cdot\text{mol}^{-1}\end{aligned}$$

9.117 (a)

Compound	$\Delta G^\circ_f(\text{kJ}\cdot\text{mol}^{-1})$
$CH_3Cl(g)$	−63.01
$CH_3Br(g)$	−28.16

(b) $$\begin{aligned}\Delta G^\circ_r &= (-28.16\ \text{kJ}\cdot\text{mol}^{-1}) + (-287.556\ \text{kJ}\cdot\text{mol}^{-1}) \\ &\quad - [(-63.01\ \text{kJ}\cdot\text{mol}^{-1}) + (-236.914\ \text{kJ}\cdot\text{mol}^{-1})] \\ &= -15.79\ \text{kJ}\cdot\text{mol}^{-1}\end{aligned}$$

$\Delta G^\circ_r = -RT \ln K$, $K = 584$

(c) BCl_2Br, BCl_3

(d) We can use the equilibrium expression in (b) and the equations already set up in exercise 9.112 (except that the total pressure equation 2 bar = A + B + C + D is no longer valid) to solve this problem. Because there are now six unknowns, we need six equations; we already have three from 9.112. Consistent with that exercise, we will define

$P_{BCl_3} = A$

$P_{BBr_3} = B$

$P_{BCl_2Br} = C$

$P_{BClBr_2} = D$

$P_{CH_3Cl} = E$

$P_{CH_3Br} = F$

(1) $K_1 = \dfrac{C\cdot D}{A\cdot B} = 1.5522$

(2) $K_2 = \dfrac{A\cdot D}{C^2} = 0.5897$

(3) $K_3 = \dfrac{B\cdot C}{D^2} = 1.092$

(4) $CH_3Cl(g) + BBr_3(g) \rightleftharpoons CH_3Br(g) + BClBr_2(g)$

$$K_3 = \frac{D \cdot F}{E \cdot B} = 584$$

(5) From the mass balance we know that

$$E + F = 1 \text{ bar}$$

(6) Also from the mass balance we know

$$A + B + C + D = 1 \text{ bar}$$

Solving these equations simultaneously using a computer program, we obtain

A = 0.191; B = 0.257; C = 0.291; D = 0.262; E = 0.002; F = 0.998

$P_{BCl_3} = 0.191$ bar

$P_{BBr_3} = 0.257$ bar

$P_{BCl_2Br} = 0.291$ bar

$P_{BClBr_2} = 0.262$ bar

$P_{CH_3Cl} = 0.002$ bar

$P_{CH_3Br} = 0.998$ bar

CHAPTER 10
ACIDS AND BASES

10.1 (a) $CH_3NH_3^+$ (b) $NH_2NH_3^+$ (c) H_2CO_3 (d) CO_3^{2-} (e) $C_6H_5O^-$ (f) $CH_3CO_2^-$

10.3 (a) $H_2SO_4(aq) + H_2O(l) \longrightarrow H_3O^+(aq) + HSO_4^-(aq)$

acid$_1$, base$_2$, acid$_2$, base$_1$; conjugate pairs: H_2SO_4/HSO_4^- and H_2O/H_3O^+

(b) $C_6H_5NH_3^+(aq) + H_2O(l) \rightleftharpoons H_3O^+(aq) + C_6H_5NH_2(aq)$

acid$_1$, base$_2$, acid$_2$, base$_1$; conjugate pairs: $C_6H_5NH_3^+$/$C_6H_5NH_2$ and H_2O/H_3O^+

(c) $H_2PO_4^-(aq) + H_2O(l) \rightleftharpoons H_3O^+(aq) + HPO_4^{2-}(aq)$

acid$_1$, base$_2$, acid$_2$, base$_1$; conjugate pairs: $H_2PO_4^-$/HPO_4^{2-} and H_2O/H_3O^+

(d) $HCOOH(aq) + H_2O(l) \rightleftharpoons H_3O^+(aq) + HCO_2^-(aq)$

acid$_1$, base$_2$, acid$_2$, base$_1$; conjugate pairs: $HCOOH$/HCO_2^- and H_2O/H_3O^+

(e) $NH_2NH_3^+(aq) + H_2O(l) \rightleftharpoons H_3O^+(aq) + NH_2NH_2(aq)$

acid$_1$, base$_2$, acid$_2$, base$_1$; conjugate pairs: $NH_2NH_3^+$/NH_2NH_2 and H_2O/H_3O^+

10.5 (a) Brønsted acid: HNO_3
Brønsted base: HPO_4^{2-}
(b) conjugate base to HNO_3: NO_3^-
conjugate acid to HPO_4^{2-}: $H_2PO_4^-$

10.7 (a) HCO_3^-, as an acid:

$$\underset{\text{acid}_1}{HCO_3^-(aq)} + \underset{\text{base}_2}{H_2O(l)} \rightleftharpoons \underset{\text{acid}_2}{H_3O^+(aq)} + \underset{\text{base}_1}{CO_3^{2-}(aq)}$$

conjugate (HCO_3^- / CO_3^{2-}); conjugate (H_2O / H_3O^+)

HCO_3^-, as a base:

$$\underset{\text{acid}_1}{H_2O(l)} + \underset{\text{base}_2}{HCO_3^-(aq)} \rightleftharpoons \underset{\text{acid}_2}{H_2CO_3(aq)} + \underset{\text{base}_1}{OH^-(aq)}$$

conjugate (H_2O / OH^-); conjugate (HCO_3^- / H_2CO_3)

(b) HPO_4^{2-}, as an acid:

$$\underset{\text{acid}_1}{HPO_4^{2-}(aq)} + \underset{\text{base}_2}{H_2O(l)} \rightleftharpoons \underset{\text{acid}_2}{H_3O^+(aq)} + \underset{\text{base}_1}{PO_4^{3-}(aq)}$$

conjugate (HPO_4^{2-} / PO_4^{3-}); conjugate (H_2O / H_3O^+)

HPO_4^{2-}, as a base:

$$\underset{\text{acid}_1}{H_2O(l)} + \underset{\text{base}_2}{HPO_4^{2-}(aq)} \rightleftharpoons \underset{\text{acid}_2}{H_2PO_4^-(aq)} + \underset{\text{base}_1}{OH^-(aq)}$$

conjugate (H_2O / OH^-); conjugate (HPO_4^{2-} / $H_2PO_4^-$)

10.9 (a) basic; (b) acidic; (c) amphoteric; (d) basic

10.11 In each case, use $K_w = [H_3O^+][OH^-] = 1.0 \times 10^{-14}$, then

$$[OH^-] = \frac{K_w}{[H_3O^+]} = \frac{1.0 \times 10^{-14}}{[H_3O^+]}$$

(a) $[OH^-] = \dfrac{1.0 \times 10^{-14}}{0.20} = 5.0 \times 10^{-14}\ \text{mol}\cdot\text{L}^{-1}$

(b) $[OH^-] = \dfrac{1.0 \times 10^{-14}}{1.0 \times 10^{-4}} = 1.0 \times 10^{-10}\ \text{mol}\cdot\text{L}^{-1}$

(c) $[OH^-] = \dfrac{1.0 \times 10^{-14}}{3.1 \times 10^{-2}} = 3.2 \times 10^{-3}\ \text{mol}\cdot\text{L}^{-1}$

10.13 (a) $K_w = 2.1 \times 10^{-14} = [H_3O^+][OH^-] = x^2$, where $x = [H_3O^+] = [OH^-]$

$x = \sqrt{2.1 \times 10^{-14}} = 1.4 \times 10^{-7}\ mol \cdot L^{-1}$

$pH = -\log[H_3O^+] = 6.80$

(b) $[OH^-] = [H_3O^+] = 1.4 \times 10^{-7}\ mol \cdot L^{-1}$

$pOH = -\log[OH^-] = 6.80$

10.15 Because $Ba(OH)_2$ is a strong base, $Ba(OH)_2\ (aq) \longrightarrow Ba^{2+}(aq) + 2\ OH^-(aq)$, 100%.

Then $[Ba(OH)_2]_0 = [Ba^{2+}]$, $[OH^-] = 2 \times [Ba(OH)_2]_0$, where $[Ba(OH)_2]_0$ = nominal concentration of $Ba(OH)_2$.

$$\text{moles of } Ba(OH)_2 = \frac{0.50\ g}{171.36\ g \cdot mol^{-1}} = 2.9 \times 10^{-3}\ mol$$

$$[Ba(OH)_2]_0 = \frac{2.9 \times 10^{-3}\ mol}{0.100\ L} = 2.9 \times 10^{-2}\ mol \cdot L^{-1} = [Ba^{2+}]$$

$$[OH^-] = 2 \times [Ba(OH)_2]_0 = 2 \times 2.9 \times 10^{-2}\ mol \cdot L^{-1} = 5.8 \times 10^{-2}\ mol \cdot L^{-1}$$

$$[H_3O^+] = \frac{K_w}{[OH^-]} = \frac{1.0 \times 10^{-14}}{5.8 \times 10^{-2}} = 1.7 \times 10^{-13}\ mol \cdot L^{-1}$$

10.17 Because $pH = -\log[H_3O^+]$, $\log[H_3O^+] = -pH$. Taking the antilogs of both sides gives $[H_3O^+] = 10^{-pH}\ mol \cdot L^{-1}$

(a) $[H_3O^+] = 10^{-3.3} = 5 \times 10^{-4}\ mol \cdot L^{-1}$

(b) $[H_3O^+] = 10^{-6.7}\ mol \cdot L^{-1} = 2 \times 10^{-7}\ mol \cdot L^{-1}$

(c) $[H_3O^+] = 10^{-4.4}\ mol \cdot L^{-1} = 4 \times 10^{-5}\ mol \cdot L^{-1}$

(d) $[H_3O^+] = 10^{-5.3}\ mol \cdot L^{-1} = 5 \times 10^{-6}\ mol \cdot L^{-1}$

10.19 (a) $[HNO_3] = [H_3O^+] = 1.46\ mol \cdot L^{-1}$

$pH = -\log(1.46) = -0.164, \quad pOH = 14.00 - (-0.164) = 14.16$

(b) $[HCl] = [H_3O^+] = 0.11\ mol \cdot L^{-1}$

$pH = -\log(0.11) = 0.96, \quad pOH = 14.00 - 0.96 = 13.04$

(c) $[OH^-] = 2 \times [Ba(OH)_2] = 2 \times 0.0092\ \text{M} = 0.018\ mol \cdot L^{-1}$

$pOH = -\log(0.018) = 1.74, \quad pH = 14.00 - 1.74 = 12.26$

(d) $[KOH]_0 = [OH^-]$

$$[OH^-] = \left(\frac{5.50\ mL}{500\ mL}\right) \times (0.175\ mol \cdot L^{-1}) = 1.92 \times 10^{-3}\ mol \cdot L^{-1}$$

$pOH = -\log(1.92 \times 10^{-3}) = 2.72, \quad pH = 14.00 - 2.72 = 11.28$

(e) $[NaOH]_0 = [OH^-]$

$$\text{number of moles of NaOH} = \frac{0.0279\ g}{40.00\ g\cdot mol^{-1}} = 6.98 \times 10^{-4}\ mol$$

$$[NaOH]_0 = \frac{6.98 \times 10^{-4}\ mol}{0.350\ L} = 1.99 \times 10^{-3}\ mol\cdot L^{-1} = [OH^-]$$

$$pOH = -\log(1.99 \times 10^{-3}) = 2.70,\quad pH = 14.00 - 2.70 = 11.30$$

(f) $[HBr]_0 = [H_3O^+]$

$$[H_3O^+] = \left(\frac{75.0\ mL}{250\ mL}\right) \times (3.5 \times 10^{-4}\ mol\cdot L^{-1}) = 1.0 \times 10^{-4}\ mol\cdot L^{-1}$$

$$pH = -\log(1.0 \times 10^{-4}) = 4.00,\quad pOH = 14.00 - 4.00 = 10.00$$

10.21 $pK_{a1} = -\log K_{a1}$; therefore, after taking antilogs, $K_{a1} = 10^{-pK_{a1}}$

Acid	pK_{a1}	K_{a1}
(a) H_3PO_4	2.12	7.6×10^{-3}
(b) H_3PO_3	2.00	1.0×10^{-2}
(c) H_2SeO_3	2.46	3.5×10^{-3}
(d) $HSeO_4$	1.92	1.2×10^{-2}

(e) The larger K_{a1}, the stronger the acid; therefore $H_2SeO_3 < H_3PO_4 < H_3PO_3 < HSeO_4^-$

10.23 The weakest acid has the strongest conjugate base and vice versa.

(a) $HSeO_3^-$, hydrogen selenite ion

(b) SeO_4^{2-}, hydrogen selenate ion

(c) $HSeO_3^-$, $K_b = \dfrac{K_w}{K_a} = \dfrac{1.0 \times 10^{-14}}{3.5 \times 10^{-3}} = 2.9 \times 10^{-12}$

SeO_4^{2-}, $K_b = \dfrac{1.0 \times 10^{-14}}{0.012} = 8.3 \times 10^{-13}$

(d) $HSeO_3^-$, strongest base corresponds to highest pH.

10.25

(a) $HClO_2(aq) + H_2O(l) \rightleftharpoons H_3O^+(aq) + ClO_2^-(aq)$ $\quad K_a = \dfrac{[H_3O^+][ClO_2^-]}{[HClO_2]}$

$ClO_2^-(aq) + H_2O(l) \rightleftharpoons HClO_2(aq) + OH^-(aq)$ $\quad K_b = \dfrac{[HClO_2][OH^-]}{[ClO_2^-]}$

(b) $HCN(aq) + H_2O(l) \rightleftharpoons H_3O^+(aq) + CN^-(aq)$ $\quad K_a = \dfrac{[H_3O^+][CN^-]}{[HCN]}$

$CN^-(aq) + H_2O(l) \rightleftharpoons HCN(aq) + OH^-(aq)$ $\quad K_b = \dfrac{[HCN][OH^-]}{[CN^-]}$

(c) $C_6H_5OH(aq) + H_2O(l) \rightleftharpoons H_3O^+(aq) + C_6H_5O^-(aq)$ $\qquad K_a = \frac{[H_3O^+][C_6H_5O^-]}{[C_6H_5OH]}$

$C_6H_5O^-(aq) + H_2O(l) \rightleftharpoons C_6H_5OH(aq) + OH^-(aq)$ $\qquad K_b = \frac{[C_6H_5OH][OH^-]}{[C_6H_5O^-]}$

10.27 Decreasing pK_a will correspond to increasing acid strength because $pK_a = -\log K_a$. The pK_a values (given in parentheses) determine the following ordering: $(CH_3)_2NH_2^+$ (14.00 − 3.27 = 10.73) < $^+NH_3OH$ (14.00 − 7.97 = 6.03) < HNO_2 (3.37) < $HClO_2$ (2.00).
Remember that the pK_a for the conjugate acid of a weak base will be given by $pK_a + pK_b = 14$.

10.29 Decreasing pK_b will correspond to increasing base strength because $pK_b = -\log K_b$. The pK_b values (given in parentheses) determine the following ordering: F^- (14.00 − 3.45 = 10.55) < CH_3COO^- (14.00 − 4.75 = 9.25) < C_5H_5N (8.75) ≪ NH_3 (4.75).
Remember that the pK_b for the conjugate base of a weak acid will be given by $pK_a + pK_b = 14$.

10.31 Any acid whose conjugate base lies above water in Table 10.3 will be a strong acid; that is, the conjugate base of the acid will be a weaker base than water, and so water will accept the H^+ preferentially. Based upon this information, we obtain the following analysis: (a) $HClO_3$, strong; (b) H_2S, weak; (c) HSO_4^-, weak (Note: even though H_2SO_4 is a strong acid, HSO_4^- is a weak acid. Its conjugate base is SO_4^{2-}); (d) $CH_3NH_3^+$, weak acid; (e) HCO_3^-, weak; (f) HNO_3, strong; (g) CH_4, weak.

10.33 For oxoacids, the greater the number of highly electronegative O atoms attached to the central atom, the stronger the acid. This effect is related to the increased oxidation number of the central atom as the number of O atoms increases. Therefore, HIO_3 is the stronger acid, with the lower pK_a.

10.35 (a) HCl is the stronger acid, because its bond strength is much weaker than the bond in HF, and bond strength is the dominant factor in determining the strength of binary acids.
(b) $HClO_2$ is stronger; there is one more O atom attached to the Cl atom in $HClO_2$ than in HClO. The additional O in $HClO_2$ helps to pull the electron of

the H atom out of the H—O bond. The oxidation state of Cl is higher in $HClO_2$ than in HClO.

(c) $HClO_2$ is stronger; Cl has a greater electronegativity than Br, making the H—O bond $HClO_2$ more polar than in $HBrO_2$.

(d) $HClO_4$ is stronger; Cl has a greater electronegativity than P.

(e) HNO_3 is stronger. The explanation is the same as that for part (b). HNO_3 has one more O atom.

(f) H_2CO_3 is stronger; C has greater electronegativity than Ge. See part (c).

10.37 (a) The $—CCl_3$ group that is bonded to the carboxyl group, —COOH, in trichloroacetic acid, is more electron withdrawing than the $—CH_3$ group in acetic acid. Thus, trichloroacetic acid is the stronger acid.

(b) The $—CH_3$ group in acetic acid has electron-donating properties, which means that it is less electron withdrawing than the —H attached to the carboxyl group in formic acid, HCOOH. Thus, formic acid is a slightly stronger acid than acetic acid. However, it is not nearly as strong as trichloroacetic acid. The order is $CCl_3COOH \gg HCOOH > CH_3COOH$.

10.39 The larger the K_a, the stronger the corresponding acid. 2,4,6-Trichlorophenol is the stronger acid because the chlorine atoms have a greater electron-withdrawing power than the hydrogen atoms present in the unsubstituted phenol.

10.41 The larger the pK_a of an acid, the stronger the corresponding conjugate base; hence, the order is aniline < ammonia < methylamine < ethylamine. Although we should not draw conclusions from such a small data set, we might suggest the possibility that

(1) arylamines < ammonia < alkylamines

(2) methyl < ethyl < etc.

(Arylamines are amines in which the nitrogen of the amine is attached to a benzene ring.)

10.43 (a)

Concentration ($mol\cdot L^{-1}$)	CH_3COOH	+ H_2O $\rightleftharpoons$	H_3O^-	+ CH_3CO_2
initial	0.29	—	0	0
change	$-x$	—	$+x$	$+x$
equilibrium	$0.29 - x$	—	x	x

$$K_a = 1.8 \times 10^{-5} = \frac{[H_3O^+][CH_3CO_2^-]}{[CH_3COOH]} = \frac{x^2}{0.29 - x} \approx \frac{x^2}{0.29}$$

$x = [H_3O^+] = 2.3 \times 10^{-3}\ mol \cdot L^{-1}$

$pH = -\log(1.6 \times 10^{-3}) = 2.64, \quad pOH = 14.00 - 2.64 = 11.36$

(b) The equilibrium table for (b) is similar to that for (a).

$$K_a = 3.0 \times 10^{-1} = \frac{[H_3O^+][CCl_3CO_2^-]}{[CCl_3COOH]} = \frac{x^2}{0.29 - x}$$

or $x^2 + 3.0 \times 10^{-1}x - 0.087 = 0$

$$x = \frac{-3.0 \times 10^{-1} \pm \sqrt{(3.0 \times 10^{-1})^2 - (4)(-0.087)}}{2} = 0.18, -0.48$$

The negative root is not possible and can be eliminated.

$x = [H_3O^+] = 0.18\ mol \cdot L^{-1}$

$pH = -\log(0.18) = 0.74, \quad pOH = 14.00 - 0.74 = 13.26$

(c)

Concentration ($mol \cdot L^{-1}$)	$HCOOH$ +	H_2O $\rightleftharpoons$	H_3O^+ +	HCO_2^-
initial	0.29	—	0	0
change	$-x$	—	$+x$	$+x$
equilibrium	$0.29 - x$	—	x	x

$$K_a = \frac{[H_3O^+][HCO_2^-]}{[HCOOH]} = \frac{x \cdot x}{0.29 - x} \approx \frac{x^2}{0.29} = 1.8 \times 10^{-4}$$

$x = [H_3O^+] = \sqrt{0.29 \times 1.8 \times 10^{-4}} = 7.2 \times 10^{-3}\ mol \cdot L^{-1}$

$pH = -\log(7.2 \times 10^{-3}) = 2.14, \quad pOH = 14.00 - 2.14 = 11.86$

10.45 (a)

Concentration ($mol \cdot L^{-1}$)	H_2O +	NH_3 $\rightleftharpoons$	NH_4^+ +	OH^-
initial	—	0.057	0	0
change	—	$-x$	$+x$	$+x$
equilibrium	—	$0.057 - x$	x	x

$$K_b = \frac{[NH_4^+][OH^-]}{[NH_3]} = \frac{x \cdot x}{0.057 - x} \approx \frac{x^2}{0.057} = 1.8 \times 10^{-5}$$

$x = [OH^-] = \sqrt{0.057 \times 1.8 \times 10^{-5}} = 1.0 \times 10^{-3}\ mol \cdot L^{-1}$

$pOH = -\log(1.0 \times 10^{-3}) = 3.00, \quad pH = 14.00 - 3.00 = 11.00$

$$\text{percentage protonation} = \frac{1.0 \times 10^{-3}}{0.057} \times 100\% = 1.8\%$$

(b)

Concentration ($mol \cdot L^{-1}$)	NH_2OH +	H_2O $\rightleftharpoons$	$^+NH_3OH$ +	OH^-
initial	0.162	—	0	0
change	$-x$	—	$+x$	$+x$
equilibrium	$0.162 - x$	—	x	x

$$K_b = 1.1 \times 10^{-8} = \frac{x^2}{0.162 - x} \approx \frac{x^2}{0.162}$$

$x = [OH^-] = 4.2 \times 10^{-5}\ mol \cdot L^{-1}$

$pOH = -\log(4.2 \times 10^{-5}) = 4.38, \quad pH = 14.00 - 4.38 = 9.62$

$$\text{percentage protonation} = \frac{4.2 \times 10^{-5}}{0.162} \times 100\% = 0.026\%$$

(c)

Concentration ($mol \cdot L^{-1}$)	$(CH_3)_3N$	+ H_2O $\rightleftharpoons$	$(CH_3)_3NH^+$	+ OH^-
initial	0.35	—	0	0
change	$-x$	—	$+x$	$+x$
equilibrium	$0.35 - x$	—	$+x$	$+x$

$$6.5 \times 10^{-5} = \frac{x^2}{0.35 - x}$$

Assume $x \ll 0.35$

Then $x = 4.8 \times 10^{-3}\ mol \cdot L^{-1}$

$[OH^-] = 4.8 \times 10^{-3}\ mol \cdot L^{-1}$

$pOH = -\log(4.8 \times 10^{-3}) = 2.32, \quad pH = 14.00 - 2.32 = 11.68$

$$\text{percentage protonation} = \frac{4.8 \times 10^{-3}}{0.35} \times 100\% = 1.4\%$$

(d) $pK_b = 14.00 - pK_a = 14.00 - 8.21 = 5.79, \quad K_b = 1.6 \times 10^{-6}$

codeine + $H_2O \rightleftharpoons$ codeineH$^+$ + OH^-

$$K_b = 1.6 \times 10^{-6} = \frac{x^2}{0.073 - x} \approx \frac{x^2}{0.073}$$

$x = [OH^-] = 1.1 \times 10^{-4}\ mol \cdot L^{-1}$

$pOH = -\log(1.1 \times 10^{-4}) = 3.96, \quad pH = 14.00 - 3.96 = 10.04$

$$\text{percentage protonation} = \frac{1.1 \times 10^{-4}}{0.0073} \times 100\% = 2.5\%$$

10.47 (a) $HClO_2 + H_2O \rightleftharpoons H_3O^+ + ClO_2^-$

$[H_3O^+] = [ClO_2^-] = 10^{-pH} = 10^{-1.2} = 0.06\ mol \cdot L^{-1}$

$$K_a = \frac{[H_3O^+][ClO_2^-]}{[HClO_2]} = \frac{(0.06)^2}{0.10 - 0.06} = 0.09 \text{ (1 sf)}$$

$pK_a = -\log(0.09) = 1.0$

(b) $C_3H_7NH_2 + H_2O \rightleftharpoons C_3H_7NH_3^+ + OH^-$

$pOH = 14.00 - 11.86 = 2.14$

$[C_3H_7NH_3^+] = [OH^-] = 10^{-2.14} = 7.2 \times 10^{-3}\ mol \cdot L^{-1}$

$$K_b = \frac{[C_3H_7NH_3^+][OH^-]}{[C_3H_7NH_2]} = \frac{(7.2 \times 10^{-3})^2}{0.10 - 7.2 \times 10^{-3}} = 5.6 \times 10^{-4}$$

$pK_b = -\log(5.6 \times 10^{-4}) = 3.25$

10.49 (a) $pH = 4.60$, $[H_3O^+] = 10^{-pH} = 10^{-4.60} = 2.5 \times 10^{-5}\ mol \cdot L^{-1}$

Let x = nominal concentration of HClO, then

Concentration	HClO	+	H_2O	$\rightleftharpoons$	H_3O^+	+	ClO^-
nominal	x		—		0		0
equilibrium	$x - 2.5 \times 10^{-5}$		—		2.5×10^{-5}		2.5×10^{-5}

$$K_a = 3.0 \times 10^{-8} = \frac{(2.5 \times 10^{-5})^2}{x - 2.5 \times 10^{-5}}$$

$$\text{Solve for } x;\ x = \frac{(2.5 \times 10^{-5})^2 + (2.5 \times 10^{-5})(3.0 \times 10^{-8})}{3.0 \times 10^{-8}}$$

$$= 2.1 \times 10^{-2}\ mol \cdot L^{-1} = 0.021\ mol \cdot L^{-1}$$

(b) $pOH = 14.00 - pH = 14.00 - 10.20 = 3.80$

$[OH^-] = 10^{-pOH} = 10^{-3.80} = 1.6 \times 10^{-4}$

Let x = nominal concentration of NH_2NH_2, then

Concentration	NH_2NH_2	+	H_2O	$\rightleftharpoons$	$NH_2NH_3^+$	+	OH^-
nominal	x		—		0		0
equilibrium	$x - 1.6 \times 10^{-4}$		—		1.6×10^{-4}		1.6×10^{-4}

$$K_b = 1.7 \times 10^{-6} = \frac{(1.6 \times 10^{-4})^2}{x - 1.6 \times 10^{-4}}$$

Solve for x; $x = 1.5 \times 10^{-2}\ mol \cdot L^{-1}$

10.51

Concentration ($mol \cdot L^{-1}$)	C_6H_5COOH	+	H_2O	$\rightleftharpoons$	H_3O^+	+	$C_6H_5CO_2^-$
initial	0.110		—		0		0
change	$-x$		—		$+x$		$+x$
equilibrium	$0.110 - x$		—		x		x

$x = 0.024 \times 0.110\ mol \cdot L^{-1} = [H_3O^+] = [C_6H_5CO_2^-]$

$$K_a = \frac{[H_3O^+][C_6H_5COO^-]}{[C_6H_5COOH]} = \frac{(0.024 \times 0.110)^2}{(1 - 0.024) \times 0.110} = 6.3 \times 10^{-5}$$

$pH = -\log(2.6 \times 10^{-3}) = 2.58$

10.53 H_2O + octylamine $\rightleftharpoons$ octylamineH^+ + OH^-

The change in the concentration of octylamine is $x = 0.067 \times 0.10 = 0.0067\ \text{mol}\cdot\text{L}^{-1}$. Thus the equilibrium table is

Concentration $(\text{mol}\cdot\text{L}^{-1})$	H_2O	+	octylamine	$\rightleftharpoons$	octylamineH^+	+	OH^-
initial	—		0.100		0		0
change	—		−0.0067		+0.0067		+0.0067
equilibrium	—		0.100 − 0.0067		0.0067		0.0067

The equilibrium concentrations are

$[\text{octylamine}] = 0.100 - 0.067 \times 0.10 = 0.093\ \text{mol}\cdot\text{L}^{-1}$

$[OH^-] = [\text{octylamineH}^+] = 0.0067\ \text{mol}\cdot\text{L}^{-1}$

$\text{pOH} = -\log(0.0067) = 2.17, \quad \text{pH} = 14.00 - 2.17 = 11.83$

$$K_b = \frac{[\text{octylamineH}^+][OH^-]}{[\text{octylamine}]} = \frac{(6.7 \times 10^{-3})^2}{0.093} = 4.8 \times 10^{-4}$$

10.55 (a) $\text{pH} < 7$, acidic; $NH_4^+(aq) + H_2O(l) \rightleftharpoons H_3O^+(aq) + NH_3(aq)$

(b) $\text{pH} > 7$, basic; $H_2O(l) + CO_3^{2-}(aq) \rightleftharpoons HCO_3^-(aq) + OH^-(aq)$

(c) $\text{pH} > 7$, basic; $H_2O(l) + F^-(aq) \rightleftharpoons HF(aq) + OH^-(aq)$

(d) $\text{pH} = 7$, neutral; K^+ is not an acid, Br^- is not a base

(e) $\text{pH} < 7$, acidic; $Al(H_2O)_6^{3+}(aq) + H_2O(l) \rightleftharpoons H_3O^+(aq) + Al(H_2O)_5OH^{2+}(aq)$

(f) $\text{pH} < 7$, acidic; $Cu(H_2O)_6^{2+}(aq) + H_2O(l) \rightleftharpoons H_3O^+(aq) + Cu(H_2O)_5OH^+(aq)$

10.57 (a) $K_b = \dfrac{K_w}{K_a} = \dfrac{1.00 \times 10^{-14}}{1.8 \times 10^{-5}} = 5.6 \times 10^{-10}$

Concentration $(\text{mol}\cdot\text{L}^{-1})$	$CH_3CO_2^-(aq)$	+	$H_2O(l)$	$\rightleftharpoons$	$HCH_3CO_2(aq)$	+	$OH^-(aq)$
initial	0.63		—		0		0
change	$-x$		—		$+x$		$+x$
equilibrium	$0.63 - x$		—		x		x

$$K_b = \frac{[HCH_3CO_2][OH^-]}{[CH_3CO_2^-]} = 5.6 \times 10^{-10} = \frac{x^2}{0.63 - x} \approx \frac{x^2}{0.63}$$

$x = 1.9 \times 10^{-5} = [OH^-], \quad \text{pOH} = -\log(1.9 \times 10^{-5}) = 4.72$

$\text{pH} = 14.00 - \text{pOH} = 14.00 - 4.72 = 9.28$

(b) $K_a = \dfrac{K_w}{K_b} = \dfrac{1.00 \times 10^{-14}}{1.8 \times 10^{-5}} = 5.6 \times 10^{-10}$

Concentration (mol·L^{-1})	$NH_4^+(aq)$	+ $H_2O(l)$	$\rightleftharpoons$ $H_3O^+(aq)$	+ $NH_3(aq)$
initial	0.19	—	0	0
change	$-x$	—	$+x$	$+x$
equilibrium	$0.19 - x$	—	x	x

$$K_a = \frac{[H_3O^+][NH_3]}{[NH_4Cl]} = 5.6 \times 10^{-10} = \frac{x^2}{0.19 - x} \approx \frac{x^2}{0.19}$$

$x = 1.0 \times 10^{-5}$ mol·L^{-1} = $[H_3O^+]$

pH = $-\log(1.0 \times 10^{-5}) = 5.00$

(c)

Concentration (mol·L^{-1})	$Al(H_2O)_6^{3+}(aq)$	+ $H_2O(l)$	$\rightleftharpoons$ $H_3O^+(aq)$	+ $Al(H_2O)_5OH^{2+}(aq)$
initial	0.055	—	0	0
change	$-x$	—	$+x$	$+x$
equilibrium	$0.055 - x$	—	x	x

$$K_a = \frac{[H_3O^+][Al(H_2O)_5OH^{2+}]}{[Al(H_2O)_6^{3+}]} = 1.4 \times 10^{-5} = \frac{x^2}{0.055 - x} \approx \frac{x^2}{0.055}$$

$x = 8.8 \times 10^{-4}$ mol·L^{-1} = $[H_3O^+]$

pH = $-\log(8.8 \times 10^{-4}) = 3.06$

(d)

Concentration (mol·L^{-1})	$H_2O(l)$ +	$CN^-(aq)$	$\rightleftharpoons$ $HCN(aq)$	+ $OH^-(aq)$
initial	—	0.065	0	0
change	—	$-x$	$+x$	$+x$
equilibrium	—	$0.065 - x$	x	x

$$K_b = \frac{K_w}{K_a} = \frac{1.00 \times 10^{-14}}{4.9 \times 10^{-10}} = 2.0 \times 10^{-5} = \frac{[HCN][OH^-]}{[CN^-]} = \frac{x^2}{0.065 - x} \approx \frac{x^2}{0.065}$$

$x = [OH^-] = 1.1 \times 10^{-3}$ mol·L^{-1}

pOH = $-\log(1.1 \times 10^{-3}) = 2.96$, pH = 11.04

10.59 (a) 250 mL of solution contains 5.34 g $KC_2H_3O_2$, molar mass = 98.14 g·mol^{-1}

$$(5.34 \text{ g } KC_2H_3O_2)\left(\frac{1 \text{ mol } KC_2H_3O_2}{98.14 \text{ g } KC_2H_3O_2}\right)\left(\frac{1}{0.250 \text{ L}}\right) = 0.218 \text{ M } KC_2H_3O_2$$

Concentration (mol·L^{-1})	$H_2O(l)$ +	$C_2H_3O_2^-(aq)$	$\rightleftharpoons$ $HC_2H_3O_2(aq)$	+ $OH^-(aq)$
initial	—	0.218	0	0
change	—	$-x$	$+x$	$+x$
equilibrium	—	$0.218 - x$	x	x

$$\frac{1.0 \times 10^{-14}}{1.8 \times 10^{-5}} = \frac{x^2}{0.218 - x} \approx \frac{x^2}{0.218}$$

$[OH^-] = 1.1 \times 10^{-5}\ mol \cdot L^{-1}$

$[H_3O^+] = 9.1 \times 10^{-10}\ mol \cdot L^{-1}$

$pH = -\log(9.1 \times 10^{-10}) = 9.04$

(b) 100 mL of solution contains 5.75 g NH_4Br, molar mass = 97.95 $g \cdot mol^{-1}$

$$(5.75\ g\ NH_4Br)\left(\frac{1\ mol\ NH_4Br}{97.95\ g\ NH_4Br}\right)\left(\frac{1}{0.100\ L}\right) = 0.587\ \text{M}\ NH_4Br$$

Concentration ($mol \cdot L^{-1}$)	$NH_4^+(aq)$	+ $H_2O(l)$ $\rightleftharpoons$	$NH_3(aq)$	+ $H_3O^+(aq)$
initial	0.587	—	0	0
change	$-x$	—	$+x$	$+x$
equilibrium	$0.587 - x$	—	x	x

$$\frac{1.0 \times 10^{-14}}{1.8 \times 10^{-5}} = \frac{x^2}{0.587 - x} \approx \frac{x^2}{0.587}$$

$[H_3O^+] = 1.8 \times 10^{-5}\ mol \cdot L^{-1}$

$pH = -\log(1.8 \times 10^{-5}) = 4.74$

10.61 (a) $\dfrac{0.175\ mol \cdot L^{-1}\ NaCH_3CO_2 \times 0.150\ L}{0.500\ L} = 0.0525\ mol \cdot L^{-1}$

Concentration ($mol \cdot L^{-1}$)	$H_2O(l)$ +	$CH_3CO_2^-(aq)$ $\rightleftharpoons$	$CH_3COOH(aq)$ +	$OH^-(aq)$
initial	—	0.0525	0	0
change	—	$-x$	$+x$	$+x$
equilibrium	—	$0.0525 - x$	x	x

$$K_b = \frac{K_w}{K_a} = \frac{1.00 \times 10^{-14}}{1.8 \times 10^{-5}} = 5.6 \times 10^{-10} = \frac{[CH_3COOH][OH^-]}{[CH_3CO_2^-]}$$

$$5.6 \times 10^{-10} = \frac{x^2}{0.0525 - x} \approx \frac{x^2}{0.0525}$$

$x = 5.4 \times 10^{-6}\ mol \cdot L^{-1} = [CH_3COOH]$

(b) $\left(\dfrac{4.32\ g\ NH_4Br}{400\ mL}\right)\left(\dfrac{1\ mL}{10^{-3}\ L}\right)\left(\dfrac{1\ mol\ NH_4Br}{97.95\ g\ NH_4Br}\right)$

$= 0.110\ (mol\ NH_4Br) \cdot L^{-1}$

Concentration ($mol\cdot L^{-1}$)	$NH_4^+(aq)$	+ $H_2O(l)$	$\rightleftharpoons$ $H_3O^+(aq)$	+ $NH_3(aq)$
initial	0.110	—	0	0
change	$-x$	—	$+x$	$+x$
equilibrium	$0.110 - x$	—	x	x

$$K_a = \frac{K_w}{K_b} = \frac{1.00 \times 10^{-14}}{1.8 \times 10^{-5}} = 5.6 \times 10^{-10} = \frac{[NH_3][H_3O^+]}{[NH_4^+]}$$

$$5.6 \times 10^{-10} = \frac{x^2}{0.110 - x} \approx \frac{x^2}{0.110}$$

$x = 7.8 \times 10^{-6}\ mol\cdot L^{-1} = [H_3O^+]$ and $pH = -\log(7.8 \times 10^{-6}) = 5.11$

10.63 We can use the relationship derived in the text: $[H_3O^+]^2 - [HA]_{initial}[H_3O^+] - K_w = 0$, in which HA is any strong acid.

$[H_3O^+]^2 - (6.55 \times 10^{-7})[H_3O^+] - (1.00 \times 10^{-14}) = 0$

Solving using the quadratic equation gives $[H_3O^+] = 6.70 \times 10^{-7}$, pH = 6.174. This value is slightly lower than the value calculated, based on the acid concentration alone ($pH = -\log(6.55 \times 10^{-7}) = 6.184$).

10.65 We can use the relationship derived in the text: $[H_3O^+]^2 + [B]_{initial}[H_3O^+] - K_w = 0$, in which B is any strong base.

$[H_3O^+]^2 + (9.78 \times 10^{-8})[H_3O^+] - (1.00 \times 10^{-14}) = 0$

Solving using the quadratic equation gives $[H_3O^+] = 6.24 \times 10^{-8}$, pH = 7.205. This value is higher than the value calculated, based on the base concentration alone ($pOH = -\log(9.78 \times 10^{-8}) = 7.009$).

10.67 (a) In the absence of a significant effect due to the autoprotolysis of water, the pH values of the 1.00×10^{-4} M and 1.00×10^{-6} M HBrO solutions can be calculated as described earlier.

For $1.00 \times 10^{-4}\ mol\cdot L^{-1}$:

Concentration ($mol\cdot L^{-1}$)	$HBrO(aq)$	+ $H_2O(l)$	$\rightleftharpoons$ $H_3O^+(aq)$	+ $BrO^-(aq)$
initial	1.00×10^{-4}	—	0	0
change	$-x$	—	$+x$	$+x$
final	$1.00 \times 10^{-4} - x$	—	$+x$	$+x$

$$K_a = \frac{[H_3O^+][BrO^-]}{[HBrO]}$$

$$2.0 \times 10^{-9} = \frac{(x)(x)}{1.00 \times 10^{-4} - x} = \frac{x^2}{[1.00 \times 10^{-4} - x]}$$

Assume $x \ll 1.00 \times 10^{-4}$

$x^2 = (2.0 \times 10^{-9})(1.00 \times 10^{-4})$

$x = 4.5 \times 10^{-7}$

Because $x < 1\%$ of 1.00×10^{-4}, the assumption was valid. Given this value, the pH is then calculated to be $-\log(4.5 \times 10^{-7}) = 6.35$.

For 1.00×10^{-6} mol·L^{-1}:

Concentration (mol·L^{-1})	$HBrO(aq)$ +	$H_2O(l) \rightleftharpoons$	$H_3O^+(aq)$ +	$BrO^-(aq)$
initial	1.00×10^{-6}	—	0	0
change	$-x$	—	$+x$	$+x$
final	$1.00 \times 10^{-6} - x$	—	$+x$	$+x$

$$K_a = \frac{[H_3O^+][BrO^-]}{[HBrO]}$$

$$2.0 \times 10^{-9} = \frac{(x)(x)}{1.00 \times 10^{-6} - x} = \frac{x^2}{[1.00 \times 10^{-6} - x]}$$

Assume $x \ll 1.00 \times 10^{-6}$

$x^2 = (2.0 \times 10^{-9})(1.00 \times 10^{-6})$

$x = 4.5 \times 10^{-8}$

x is 4.5% of 1.00×10^{-6}, so the assumption is less acceptable. The pH is calculated to be $-\log(4.5 \times 10^{-8}) = 7.35$. Because this predicts a basic solution, it is not reasonable.

(b) To calculate the value taking into account the autoprotolysis of water, we can use equation (22):

$x^3 + K_a x^2 - (K_w + K_a \cdot [HA]_{initial})x - K_w \cdot K_a = 0$, where $x = [H_3O^+]$.

To solve the expression, you substitute the values of $K_w = 1.00 \times 10^{-14}$, the initial concentration of acid, and $K_a = 2.0 \times 10^{-9}$ into this equation and then solve the expression either by trial and error or, preferably, using a graphing calculator such as the one found on the CD accompanying this text.

Alternatively, you can use a computer program designed to solve simultaneous equations. Because the unknowns include $[H_3O^+]$, $[OH^-]$, $[HBrO]$, and $[BrO^-]$, you will need four equations. As seen in the text, the pertinent equations are

$$K_a = \frac{[H_3O^+][BrO^-]}{[HBrO]}$$

$K_w = [H_3O^+][OH^-]$

$[H_3O^+] = [OH^-] + [BrO^-]$

$[HBrO]_{initial} = [HBrO] + [BrO^-]$

Both methods should produce the same result.

The values obtained are

$[H_3O^+] = 4.6 \times 10^{-7}\ mol\cdot L^{-1}$, pH = 6.34 (compare to 6.35 obtained in (a))

$[BrO^-] = 4.4 \times 10^{-7}\ mol\cdot L^{-1}$

$[HBrO] \cong 1.0 \times 10^{-5}\ mol\cdot L^{-1}$

$[OH^-] = 2.2 \times 10^{-8}\ mol\cdot L^{-1}$

Similarly, for $[HBrO]_{initial} = 1.00 \times 10^{-6}$:

$[H_3O^+] = 1.1 \times 10^{-7}\ mol\cdot L^{-1}$, pH = 6.96 (compare to 7.32 obtained in (a))

$[BrO^-] = 1.8 \times 10^{-8}\ mol\cdot L^{-1}$

$[HBrO] \cong 9.8 \times 10^{-7}\ mol\cdot L^{-1}$

$[OH^-] = 9.1 \times 10^{-8}\ mol\cdot L^{-1}$

Note that for the more concentrated solution, the effect of the autoprotolysis of water is very small. Notice also that the less concentrated solution is more acidic, due to the autoprotolysis of water, than would be predicted if this effect were not operating.

10.69 (a) In the absence of a significant effect due to the autoprotolysis of water, the pH values of the 8.50×10^{-5} M and 7.37×10^{-6} M HCN solutions can be calculated as described earlier.

For $8.50 \times 10^{-5}\ mol\cdot L^{-1}$:

Concentration ($mol\cdot L^{-1}$)	HCN(aq)	+ $H_2O(l) \rightleftharpoons$	$H_3O^+(aq)$ +	$CN^-(aq)$
initial	8.50×10^{-5}	—	0	0
change	$-x$	—	$+x$	$+x$
final	$8.50 \times 10^{-5} - x$	—	$+x$	$+x$

$$K_a = \frac{[H_3O^+][CN^-]}{[HCN]}$$

$$4.9 \times 10^{-10} = \frac{(x)(x)}{8.5 \times 10^{-5} - x} = \frac{x^2}{[8.5 \times 10^{-5} - x]}$$

Assume $x \ll 8.5 \times 10^{-5}$

$x^2 = (4.9 \times 10^{-10})(8.5 \times 10^{-5})$

$x = 2.0 \times 10^{-7}$

Because $x < 1\%$ of 8.50×10^{-5}, the assumption was valid. Given this value, the pH is then calculated to be $-\log(2.0 \times 10^{-7}) = 6.69$.

For $7.37 \times 10^{-6}\ mol\cdot L^{-1}$:

Concentration ($mol \cdot L^{-1}$)	HCN(aq)	+ $H_2O(l)$ ⇌	$H_3O^+(aq)$ +	$CN^-(aq)$
initial	7.37×10^{-6}	—	0	0
change	$-x$	—	$+x$	$+x$
final	$7.37 \times 10^{-6} - x$	—	$+x$	$+x$

$$K_a = \frac{[H_3O^+][CN^-]}{[HCN]}$$

$$4.9 \times 10^{-10} = \frac{(x)(x)}{7.37 \times 10^{-6} - x} = \frac{x^2}{[7.37 \times 10^{-6} - x]}$$

Assume $x \ll 7.37 \times 10^{-6}$

$x^2 = (4.9 \times 10^{-10})(7.37 \times 10^{-6})$

$x = 6.0 \times 10^{-8}$

x is $<1\%$ of 7.37×10^{-6}, so the assumption is still reasonable. The pH is then calculated to be $-\log(6.0 \times 10^{-8}) = 7.22$. This answer is not reasonable because we know HCN is an acid.

(b) To calculate the value, taking into account the autoprotolysis of water, we can use equation (21):

$x^3 + K_a x^2 - (K_w + K_a \cdot [HA]_{initial})x - K_w \cdot K_a = 0$, where $x = [H_3O^+]$.

To solve the expression, you substitute the values of $K_w = 1.00 \times 10^{-14}$, the initial concentration of acid, and $K_a = 4.9 \times 10^{-10}$ into this equation and then solve the expression either by trial and error or, preferably, using a graphing calculator such as the one found on the CD accompanying this text.

Alternatively, you can use a computer program designed to solve simultaneous equations. Because the unknowns include $[H_3O^+]$, $[OH^-]$, [HBrO], and $[BrO^-]$, you will need four equations. As seen in the text, the pertinent equations are

$$K_a = \frac{[H_3O^+][CN^-]}{[HCN]}$$

$K_w = [H_3O^+][OH^-]$

$[H_3O^+] = [OH^-] + [CN^-]$

$[HCN]_{initial} = [HCN] + [CN^-]$

Both methods should produce the same result.

For $[HCN] = 8.5 \times 10^{-5}\ mol \cdot L^{-1}$, the values obtained are

$[H_3O^+] = 2.3 \times 10^{-7}\ mol \cdot L^{-1}$, pH = 6.64 (compare to 6.69 obtained in (a))

$[CN^-] = 1.8 \times 10^{-7}\ mol \cdot L^{-1}$

$[HCN] \cong 8.5 \times 10^{-5}\ mol \cdot L^{-1}$

$[OH^-] = 4.4 \times 10^{-8}\ mol \cdot L^{-1}$

Similarly, for $[HCN]_{initial} = 7.37 \times 10^{-6}$:

$[H_3O^+] = 1.2 \times 10^{-7}\ mol \cdot L^{-1}$, pH = 6.92 (compare to 7.22 obtained in (a))

$[CN^-] = 3.1 \times 10^{-8}\ mol \cdot L^{-1}$
$[HCN] \cong 7.3 \times 10^{-6}\ mol \cdot L^{-1}$
$[OH^-] = 8.6 \times 10^{-8}\ mol \cdot L^{-1}$

Note that for the more concentrated solution, the effect of the autoprotolysis of water is smaller. Notice also that the less concentrated solution is more acidic, due to the autoprotolysis of water, than would be predicted if this effect were not operating.

10.71 (a) $H_2SO_4(aq) + H_2O(l) \rightleftharpoons H_3O^+(aq) + HSO_4^-(aq)$
$HSO_4^-(aq) + H_2O(l) \rightleftharpoons H_3O^+(aq) + SO_4^{2-}(aq)$
(b) $H_3AsO_4(aq) + H_2O(l) \rightleftharpoons H_3O^+(aq) + H_2AsO_4^-(aq)$
$H_2AsO_4^-(aq) + H_2O(l) \rightleftharpoons H_3O^+(aq) + HAsO_4^{2-}(aq)$
$HAsO_4^{2-}(aq) + H_2O(l) \rightleftharpoons H_3O^+(aq) + AsO_4^{3-}(aq)$
(c) $C_6H_4(COOH)_2(aq) + H_2O(l) \rightleftharpoons H_3O^+(aq) + C_6H_4(COOH)CO_2^-(aq)$
$C_6H_4(COOH)CO_2^-(aq) + H_2O(l) \rightleftharpoons H_3O^+(aq) + C_6H_4(CO_2)_2^{2-}(aq)$

10.73 The initial concentrations of HSO_4^- and H_3O^+ are both 0.15 $mol \cdot L^{-1}$ as a result of the complete ionization of H_2SO_4 in the first step. The second ionization is incomplete.

Concentration ($mol \cdot L^{-1}$)	HSO_4^-	+ H_2O	$\rightleftharpoons$ H_3O^+	+ SO_4^{2-}
initial	0.15	—	0.15	0
change	$-x$	—	$+x$	$+x$
equilibrium	$0.15 - x$	—	$0.15 + x$	x

$$K_{a2} = 1.2 \times 10^{-2} = \frac{[H_3O^+][SO_4^{2-}]}{[HSO_4^-]} = \frac{(0.15 + x)(x)}{0.15 - x}$$

$$x^2 + 0.162x - 1.8 \times 10^{-3} = 0$$

$$x = \frac{-0.162 + \sqrt{(0.162)^2 + (4)(1.8 \times 10^{-3})}}{2} = 0.0104\ mol \cdot L^{-1}$$

$[H_3O^+] = 0.15 + x = (0.15 + 0.0104) mol \cdot L^{-1} = 0.16\ mol \cdot L^{-1}$
$pH = -\log(0.16) = 0.80$

10.75 (a) Because $K_{a2} \ll K_{a1}$, the second ionization can be ignored.

Concentration ($mol \cdot L^{-1}$)	H_2CO_3	+ H_2O	$\rightleftharpoons$ H_3O^+	+ HCO_3^-
initial	0.010	—	0	0
change	$-x$	—	$+x$	$+x$
equilibrium	$0.010 - x$	—	x	x

$$K_{a1} = \frac{[H_3O^+][HCO_3^-]}{[H_2CO_3]} = \frac{x^2}{0.010 - x} \approx \frac{x^2}{0.010} = 4.3 \times 10^{-7}$$

$x = [H_3O^+] = 6.6 \times 10^{-5}\ mol \cdot L^{-1}$

$pH = -\log(6.6 \times 10^{-5}) = 4.18$

(b) Because $K_{a2} \ll K_{a1}$, the second ionization can be ignored.

Concentration ($mol \cdot L^{-1}$)	$(COOH)_2$	+ H_2O $\rightleftharpoons$	H_3O^+	+ $(COOH)CO_2^-$
initial	0.10	—	0	0
change	$-x$	—	$+x$	$+x$
equilibrium	$0.10 - x$	—	x	x

$$K_{a1} = 5.9 \times 10^{-2} = \frac{[H_3O^+][(COOH)CO_2^-]}{[(COOH)_2]} = \frac{x^2}{0.10 - x}$$

$x^2 + 5.9 \times 10^{-2}x - 5.9 \times 10^{-3} = 0$

$$x = \frac{-5.9 \times 10^{-2} + \sqrt{(5.9 \times 10^{-2})^2 + (4)(5.9 \times 10^{-3})}}{2} = 0.053\ mol \cdot L^{-1}$$

$pH = -\log(0.053) = 1.28$

(c) Because $K_{a2} \ll K_{a1}$, the second ionization can be ignored.

Concentration ($mol \cdot L^{-1}$)	H_2S	+ H_2O $\rightleftharpoons$	H_3O^+	+ HS^-
equilibrium	$0.20 - x$	—	x	x

$$K_{a1} = 1.3 \times 10^{-7} = \frac{[H_3O^+][HS^-]}{[H_2S]} = \frac{x^2}{0.20 - x} \approx \frac{x^2}{0.20}$$

$x = [H_3O^+] = 1.6 \times 10^{-4}\ mol \cdot L^{-1}$

$pH = -\log(1.6 \times 10^{-4}) = 3.80$

10.77 (a) The pH is given by $pH = \frac{1}{2}(pK_{a1} + pK_{a2})$. From Table 10.9, we find

$K_{a1} = 1.5 \times 10^{-2}$ $\quad pK_{a1} = 1.82$

$K_{a2} = 1.2 \times 10^{-7}$ $\quad pK_{a2} = 6.92$

$pH = \frac{1}{2}(1.82 + 6.92) = 4.37$

(b) The pH of a salt solution of a polyprotic acid is independent of the concentration of the salt, therefore pH = 4.37.

10.79 (a) The pH is given by $pH = \frac{1}{2}(pK_{a1} + pK_{a2})$. For the monosodium salt, the pertinent values are pK_{a1} and pK_{a2}:

$pH = \frac{1}{2}(3.14 + 5.95) = 4.55$

(b) For the disodium salt, the pertinent values are pK_{a2} and pK_{a3}:

$pH = \frac{1}{2}(5.95 + 6.39) = 6.17$

10.81 The equilibrium reactions of interest are

$H_2CO_3(aq) + H_2O(l) \rightleftharpoons H_3O^+(aq) + HCO_3^-(aq) \quad K_{a1} = 4.3 \times 10^{-7}$

$HCO_3^-(aq) + H_2O(l) \rightleftharpoons H_3O^+(aq) + CO_3^{2-}(aq) \quad K_{a2} = 5.6 \times 10^{-11}$

Because the second ionization constant is much smaller than the first, we can assume that the first step dominates:

Concentration ($mol \cdot L^{-1}$)	$H_2CO_3(aq)$	+ $H_2O(l)$ $\rightleftharpoons$	$H_3O^+(aq)$	+ $HCO_3^-(aq)$
initial	0.0456	—	0	0
change	$-x$	—	$+x$	$+x$
final	$0.0456 - x$	—	$+x$	$+x$

$$K_{a1} = \frac{[H_3O^+][HCO_3^-]}{[H_2CO_3]}$$

$$4.3 \times 10^{-7} = \frac{(x)(x)}{0.0456 - x} = \frac{x^2}{0.0456 - x}$$

Assume that $x \ll 0.0456$

Then $x^2 = (4.3 \times 10^{-7})(0.0456)$

$x = 1.4 \times 10^{-4}$

Because $x < 1\%$ of 0.0456, the assumption was valid.

$x = [H_3O^+] = [HCO_3^-] = 1.4 \times 10^{-4}\ mol \cdot L^{-1}$

This means that the concentration of H_2CO_3 is $0.0456\ mol \cdot L^{-1} - 0.00014\ mol \cdot L^{-1} = 0.0455\ mol \cdot L^{-1}$. We can then use the other equilibria to determine the remaining concentrations:

$$K_{a2} = \frac{[H_3O^+][CO_3^{2-}]}{[HCO_3^-]}$$

$$5.6 \times 10^{-11} = \frac{(1.4 \times 10^{-4})[CO_3^{2-}]}{(1.4 \times 10^{-4})}$$

$[CO_3^{2-}] = 5.6 \times 10^{-11}\ mol \cdot L^{-1}$

Because $5.6 \times 10^{-11} \ll 1.4 \times 10^{-4}$, the initial assumption that the first ionization would dominate is valid.

To calculate $[OH^-]$, we use the K_w relationship:

$K_w = [H_3O^+][OH^-]$

$$[OH^-] = \frac{K_w}{[H_3O^+]} = \frac{1.00 \times 10^{-14}}{1.4 \times 10^{-4}} = 7.1 \times 10^{-11}\ mol \cdot L^{-1}$$

In summary, $[H_2CO_3] = 0.0455\ mol \cdot L^{-1}$, $[H_3O^+] = [HCO_3^-] = 1.4 \times 10^{-4}\ mol \cdot L^{-1}$, $[CO_3^{2-}] = 5.6 \times 10^{-11}\ mol \cdot L^{-1}$, $[OH^-] = 7.1 \times 10^{-11}\ mol \cdot L^{-1}$.

10.83 The equilibrium reactions of interest are now the base forms of the carbonic acid equilibria, so K_b values should be calculated for the following changes:

$$CO_3^{2-}(aq) + H_2O(l) \rightleftharpoons HCO_3^-(aq) + OH^-(aq)$$

$$K_{b1} = \frac{K_w}{K_{a2}} = \frac{1.00 \times 10^{-14}}{5.6 \times 10^{-11}} = 1.8 \times 10^{-4}$$

$$HCO_3^-(aq) + H_2O(l) \rightleftharpoons H_2CO_3(aq) + OH^-(aq)$$

$$K_{b2} = \frac{K_w}{K_{a1}} = \frac{1.00 \times 10^{-14}}{4.3 \times 10^{-7}} = 2.3 \times 10^{-8}$$

Because the second hydrolysis constant is much smaller than the first, we can assume that the first step dominates:

Concentration ($mol \cdot L^{-1}$)	$CO_3^{2-}(aq)$ +	$H_2O(l)$ $\rightleftharpoons$	$HCO_3^-(aq)$ +	$OH^-(aq)$
initial	0.0456	—	0	0
change	$-x$	—	$+x$	$+x$
final	$0.0456 - x$	—	$+x$	$+x$

$$K_{b1} = \frac{[HCO_3^-][OH^-]}{[CO_3^{2-}]}$$

$$1.8 \times 10^{-4} = \frac{(x)(x)}{0.0456 - x} = \frac{x^2}{[0.0456 - x]}$$

Assume that $x \ll 0.0456$

Then $x^2 = (1.8 \times 10^{-4})(0.0456)$

$x = 2.9 \times 10^{-3}$

Because $x > 5\%$ of 0.0456, the assumption was not valid and the full expression should be solved using the quadratic equation:

$x^2 + 1.8 \times 10^{-4}x - (1.8 \times 10^{-4})(0.0456) = 0$

Solving using the quadratic equation gives $x = 0.0028\ mol \cdot L^{-1}$.

$x = [HCO_3^-] = [OH^-] = 0.0028\ mol \cdot L^{-1}$

Therefore, $[CO_3^{2-}] = 0.0456\ mol \cdot L^{-1} - 0.0028\ mol \cdot L^{-1} = 0.0428\ mol \cdot L^{-1}$

We can then use the other equilibria to determine the remaining concentrations:

$$K_{b2} = \frac{[H_2CO_3][OH^-]}{[HCO_3^-]}$$

$$2.3 \times 10^{-8} = \frac{[H_2CO_3](0.0028)}{(0.0028)}$$

$[H_2CO_3] = 2.3 \times 10^{-8}\ mol \cdot L^{-1}$

Because $2.3 \times 10^{-8} \ll 0.0028$, the initial assumption that the first hydrolysis would dominate is valid. To calculate $[H_3O^+]$, we use the K_w relationship:

$K_w = [H_3O^+][OH^-]$

$$[H_3O^+] = \frac{K_w}{[OH^-]} = \frac{1.00 \times 10^{-14}}{0.0028} = 3.6 \times 10^{-12}\ mol \cdot L^{-1}$$

In summary, $[H_2CO_3] = 2.3 \times 10^{-8}\ mol \cdot L^{-1}$, $[OH^-] = [HCO_3^-] = 0.0028\ mol \cdot L^{-1}$, $[CO_3^{2-}] = 0.0428\ mol \cdot L^{-1}$, $[H_3O^+] = 3.6 \times 10^{-12}\ mol \cdot L^{-1}$

10.85 (a) phosphorous acid: The two pK_a values are 2.00 and 6.59. Because pH = 6.30 lies between pK_{a1} and pK_{a2}, the dominant form will be the singly deprotonated HA^- ion.

(b) oxalic acid: The two pK_a values are 1.23 and 4.19. Because pH = 6.30 lies above pK_{a2}, the species present in largest concentration will be the doubly deprotonated A^{2-} ion.

(c) hydrosulfuric acid: The two pK_a values are 6.89 and 14.15. Because pH = 6.30 lies below both pK_a values, the species present in highest concentrations will be the fully protonated H_2A form.

10.87 The equilibria present in solution are

$$H_2SO_3(aq) + H_2O(l) \rightleftharpoons H_3O^+(aq) + HSO_3^-(aq) \quad K_{a1} = 1.5 \times 10^{-2}$$

$$HSO_3^-(aq) + H_2O(l) \rightleftharpoons H_3O^+(aq) + SO_3^{2-}(aq) \quad K_{a2} = 1.2 \times 10^{-7}$$

The calculation of the desired concentrations follows exactly after the method derived in Eq. 25, substituting H_2SO_3 for H_2CO_3, HSO_3^- for HCO_3^-, and SO_3^{2-} for CO_3^{2-}. First, calculate the quantity f (at pH = 5.50 $[H_3O^+] = 10^{-5.5} = 3.2 \times 10^{-6}\ mol \cdot L^{-1}$):

$$f = [H_3O^+]^2 + [H_3O^+]K_{a1} + K_{a1}K_{a2}$$
$$= (3.2 \times 10^{-6})^2 + (3.2 \times 10^{-6})(1.5 \times 10^{-2}) + (1.5 \times 10^{-2})(1.2 \times 10^{-7})$$
$$= 5.0 \times 10^{-8}$$

The fractions of the species present are then given by

$$\alpha(H_2SO_3) = \frac{[H_3O^+]}{f} = \frac{(3.2 \times 10^{-6})^2}{5.0 \times 10^{-8}} = 2.1 \times 10^{-4}$$

$$\alpha(HSO_3^-) = \frac{[H_3O^+]K_{a1}}{f} = \frac{(3.2 \times 10^{-6})(1.5 \times 10^{-2})}{5.0 \times 10^{-8}} = 0.96$$

$$\alpha(SO_3^{2-}) = \frac{K_{a1}K_{a2}}{f} = \frac{(1.5 \times 10^{-2})(1.2 \times 10^{-7})}{5.0 \times 10^{-8}} = 0.036$$

Thus, in a $0.150\ mol \cdot L^{-1}$ solution at pH 5.50, the dominant species will be HSO_3^- with a concentration of $(0.150\ mol \cdot L^{-1})(0.96) = 0.14\ mol \cdot L^{-1}$. The concentration of H_2SO_3 will be $(2.1 \times 10^{-4})(0.150\ mol \cdot L^{-1}) = 3.2 \times 10^{-5}\ mol \cdot L^{-1}$ and the concentration of SO_3^{2-} will be $(0.036)(0.150\ mol \cdot L^{-1}) = 0.0054\ mol \cdot L^{-1}$.

10.89 (a)

Concentration ($mol\cdot L^{-1}$)	$B(OH)_3$	+ $2\,H_2O$ $\rightleftharpoons$	H_3O^+	+ $B(OH)_4^-$
initial	1.0×10^{-4}	—	0	0
change	$-x$	—	$+x$	$+x$
equilibrium	$1.0 \times 10^{-4} - x$	—	x	x

$$K_a = 7.2 \times 10^{-10} = \frac{[H_3O^+][B(OH)_4^-]}{[B(OH)_3]} = \frac{x^2}{1.0 \times 10^{-4} - x} \approx \frac{x^2}{1.0 \times 10^{-4}}$$

$x = [H_3O^+] = 2.7 \times 10^{-7}\ mol\cdot L^{-1}$

$pH = -\log(2.7 \times 10^{-7}) = 6.57$

Note: this value of $[H_3O^+]$ is not much different from the value for pure water, $1.0 \times 10^{-7}\ mol\cdot L^{-1}$; therefore, it is at the lower limit of safely ignoring the contribution to $[H_3O^+]$ from the autoprotolysis of water. The exercise should be solved by simultaneously considering both equilibria.

Concentration ($mol\cdot L^{-1}$)	$B(OH)_3$	+ $2\,H_2O$ $\rightleftharpoons$	H_3O^+	+ $B(OH)_4^-$
equilibrium	$1.0 \times 10^{-4} - x$	—	x	y

Concentration ($mol\cdot L^{-1}$)	$2\,H_2O$ $\rightleftharpoons$	H_3O^+	+ OH^-
equilibrium	—	x	z

Because there are now two contributions to $[H_3O^+]$, $[H_3O^+]$ is no longer equal to $[B(OH)_4^-]$, nor is it equal to $[OH^-]$, as in pure water. To avoid a cubic equation, x will again be ignored relative to $1.0 \times 10^{-4}\ mol\cdot L^{-1}$. This approximation is justified by the approximate calculation above, and because K_a is very small relative to 1.0×10^{-4}. Let a = initial concentration of $B(OH)_3$, then

$$K_a = 7.2 \times 10^{-10} = \frac{xy}{a - x} \approx \frac{xy}{a} \text{ or } y = \frac{aK_a}{x}$$

$K_w = 1.0 \times 10^{-14} = xz$

Electroneutrality requires $x = y + z$ or $z = x - y$; hence, $K_w = xz = x(x - y)$. Substituting for y from above:

$$x \times \left(x - \frac{aK_a}{x}\right) = K_w$$

$x^2 - aK_a = K_w$

$x^2 = K_w + aK_a$

$x = \sqrt{K_w + aK_a} = \sqrt{1.0 \times 10^{-14} + 1.0 \times 10^{-4} \times 7.2 \times 10^{-10}}$

$x = 2.9 \times 10^{-7}\ mol\cdot L^{-1} = [H_3O^+]$

$pH = -\log(2.9 \times 10^{-7}) = 6.54$

This value is slightly, but measurably, different from the value 6.57 obtained by ignoring the contribution to $[H_3O^+]$ from water.

(b) In this case, the second ionization can safely be ignored; $K_{a2} \ll K_{a1}$.

Concentration (mol·L^{-1})	H_3PO_4	+ H_2O $\rightleftharpoons$	H_3O^+	+ $H_2PO_4^-$
initial	0.015	—	0	0
change	$-x$	—	$+x$	$+x$
equilibrium	$0.015 - x$	—	x	x

$$K_{a1} = 7.6 \times 10^{-3} = \frac{x^2}{0.015 - x}$$

$$x^2 + 7.6 \times 10^{-3}x - 1.14 \times 10^{-4} = 0$$

$$x = [H_3O^+] = \frac{-7.6 \times 10^{-3} + \sqrt{(7.6 \times 10^{-3})^2 + 4.56 \times 10^{-4}}}{2}$$

$$= 7.5 \times 10^{-3} \text{ mol·L}^{-1}$$

$$\text{pH} = -\log(7.5 \times 10^{-3}) = 2.12$$

(c) In this case, the second ionization can safely be ignored; $K_{a2} \ll K_{a1}$.

Concentration (mol·L^{-1})	H_2SO_3	+ H_2O $\rightleftharpoons$	H_3O^+	+ HSO_3^-
initial	0.1	—	0	0
change	$-x$	—	$+x$	$+x$
equilibrium	$0.1 - x$	—	x	x

$$K_{a1} = 1.5 \times 10^{-2} = \frac{x^2}{0.10 - x}$$

$$x^2 + 1.5 \times 10^{-2}x - 1.5 \times 10^{-3} = 0$$

$$x = [H_3O^+] = \frac{-1.5 \times 10^{-2} + \sqrt{(1.5 \times 10^{-2})^2 + 6.0 \times 10^{-3}}}{2} = 0.032 \text{ mol·L}^{-1}$$

$$\text{pH} = -\log(0.032) = 1.49$$

10.91 (a) $K_w = [H_3O^+][OH^-] = x^2 = 3.8 \times 10^7$

$[H_3O^+] = x = 1.9 \times 10^7$ mol·L^{-1}

$\text{pH} = -\log[H_3O^+] = 6.72$

(b) There are three data points available: 25°C ($K_w = 1.0 \times 10^{-14}$), 40°C ($K_w = 3.8 \times 10^{-14}$), and 37°C ($K_w = 2.1 \times 10^{-14}$)

T	$1/T$ (K^{-1})	$\ln K_w$
25°C	0.003356	−32.2362
37°C	0.003226	−31.4943
40°C	0.003195	−31.1376

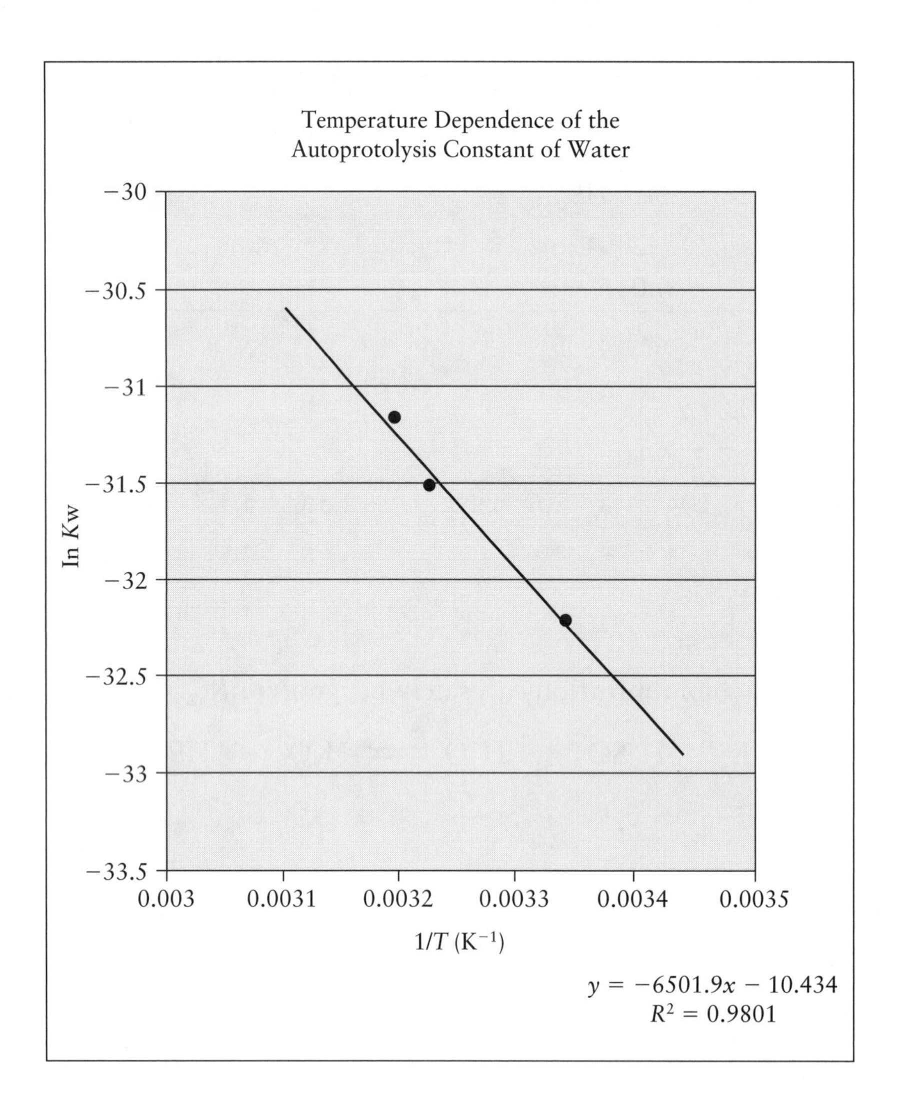

The slope is equal to $-\Delta H°/R$ and the intercept equals $\Delta S°/R$.

$\Delta H° = -(-6502\text{ K})(8.314\text{ J}\cdot\text{K}^{-1}\cdot\text{mol}^{-1}) = 54\text{ kJ}\cdot\text{mol}^{-1}$

$\Delta S° = -(-10.43)(8.314\text{ J}\cdot\text{K}^{-1}\cdot\text{mol}^{-1}) = 87\text{ J}\cdot\text{K}^{-1}\cdot\text{mol}^{-1}$

(c) The equation determined from the graph is for ln K_w. In order to write an equation for the pH dependence of pure water, we must rearrange the equation. First we note the relationship between K_w and the pH of pure water.

$$[H_3O^+] = K_w^{1/2}$$

$$\text{pH} = -\log[H_3O^+] = -\frac{1}{2}\log K_w$$

$$\ln K_w = -\frac{6501}{T} - 10.43$$

$$2.303\log K_w = -\frac{6501}{T} - 10.43$$

$$\log K_w = -\frac{2823}{T} - 4.529$$

$$pH = -\frac{1}{2}\log K_w$$

$$pH = \frac{1411}{T} + 2.264$$

10.93 First, we calculate the concentration of $[H_3O^+]$ from the pH data:

$pH = -\log[H_3O^+] = 2.8$

$[H_3O^+] = 1.6 \times 10^{-3}$

Then calculate the amount of benzoic acid present in solution:

$$\frac{[H_3O^+][C_6H_5COO^-]}{[C_6H_5COOH]} = 6.5 \times 10^{-5}$$

$$\frac{[1.6 \times 10^{-3}]^2}{[C_6H_5COOH]} = 6.5 \times 10^{-5}$$

$[C_6H_5COOH] = 0.039$

The total amount of benzoic acid that dissolves will be equal to the concentration of C_6H_5COOH plus the concentration of C_6H_5COO:

total solubility of benzoic acid $= 0.039 + 1.6 \times 10^{-3} = 0.041$ M

10.95 We wish to calculate K_a for the reaction

$HF(aq) + H_2O(l) \rightleftharpoons H_3O^+(aq) + F^-(aq)$

This equation is equivalent to

$HF(aq) \rightleftharpoons H^+(aq) + F^-(aq)$

This latter writing of the expression is simpler for the purpose of the thermodynamic calculations.

The $\Delta G°$ value for this reaction is easily calculated from the free energies given in the appendix:

$= (-278.79\ \text{kJ}\cdot\text{mol}^{-1}) - (-296.82\ \text{kJ}\cdot\text{mol}^{-1}) = 18.03\ \text{kJ}\cdot\text{mol}^{-1}$

$-RT \ln K$

$K = e^{-\Delta G°/RT}$

$K = e^{-(18030\ \text{J}\cdot\text{mol}^{-1})/[8.314\ \text{J}\cdot\text{K}^{-1}\cdot\text{mol}^{-1})(298\ \text{K})]} = 6.9 \times 10^{-4}$

10.97 (a) $D_2O + D_2O \rightleftharpoons D_3O^+ + OD^-$

(b) $K_{D_20} = [D_3O^+][OD^-] = 1.35 \times 10^{-15}$, $pK_{D_20} = -\log K_{D_20} = 14.870$

(c) $[D_3O^+] = [OD^-] = \sqrt{1.35 \times 10^{-15}} = 3.67 \times 10^{-8}\ \text{mol}\cdot\text{L}^{-1}$

(d) $pD = -\log(3.67 \times 10^{-8}) = 7.435 = pOD$

(e) $pD + pOD = pK_{D_20}(D_2O) = 14.870$

10.99 Let T = thiazole

Concentration ($mol\cdot L^{-1}$)	H_2O	+	T	$\rightleftharpoons$	TH^+	+	OH^-
initial	—		0.0010		0		0
change	—		$-x$		$+x$		$+x$
equilibrium	—		$0.0010 - x$		x		x

$x = 5.2 \times 10^{-5} \times 0.0010\ mol\cdot L^{-1} = [TH^+] \neq [OH^-]$

This is a case where the autoprotolysis of water cannot be ignored. $[OH^-]$ is not the same as [thiazolate ion].

$$K_b = \frac{[TH^+][OH^-]}{[T]}, \quad K_w = [H_3O^+][OH^-] = 1.0 \times 10^{-14}$$

Electroneutrality requires $[TH^+] + [H_3O^+] - [OH^-] = 0$,

giving $[OH^-] = 5.2 \times 10^{-8} + \dfrac{(1.0 \times 10^{-14})}{[OH^-]}$

Let $y = [OH^-]$, then $y^2 - 5.2 \times 10^{-8}y - 1.0 \times 10^{-14} = 0$

$$y = \frac{-5.2 \times 10^{-8} \pm \sqrt{(5.2 \times 10^{-8})^2 + 4.0 \times 10^{-14}}}{2}$$

$$= 1.3 \times 10^{-7}\ mol\cdot L^{-1} = [OH^-]$$

(Keeping only the positive root)

$$K_b = \frac{5.2 \times 10^{-8} \times 1.3 \times 10^{-7}}{0.0010} = 6.8 \times 10^{-12}$$

$$[H_3O^+] = \frac{1.0 \times 10^{-14}}{1.3 \times 10^{-7}} = 7.7 \times 10^{-8}\ mol\cdot L^{-1}$$

$$pH = -\log(7.7 \times 10^{-8}) = 7.11$$

10.101 $T' = 30°C = 303\ K,\ T = 20°C = 293\ K$

$$\ln\left(\frac{K'_a}{K_a}\right) = \frac{\Delta H°}{R}\left(\frac{1}{T} - \frac{1}{T'}\right) = \frac{\Delta H°}{R}\left(\frac{T' - T}{TT'}\right)$$

This is the van't Hoff equation.

$$\ln\left(\frac{1.768 \times 10^{-4}}{1.765 \times 10^{-4}}\right) = \frac{\Delta H°}{8.314\ J\cdot K^{-1}\cdot mol^{-1}}\left(\frac{303\ K - 293\ K}{293\ K \times 303\ K}\right) = 0.0017$$

$$\Delta H° = \frac{0.0017 \times 8.314\ J\cdot K^{-1}\cdot mol^{-1} \times 293\ K \times 303\ K}{10\ K} = 1.3 \times 10^2\ J\cdot mol^{-1}$$

10.103 (a) We determine the value graphically by plotting $\ln K$ versus T^{-1}. The K values are determined from pK values: $K = 10^{-pK}$.

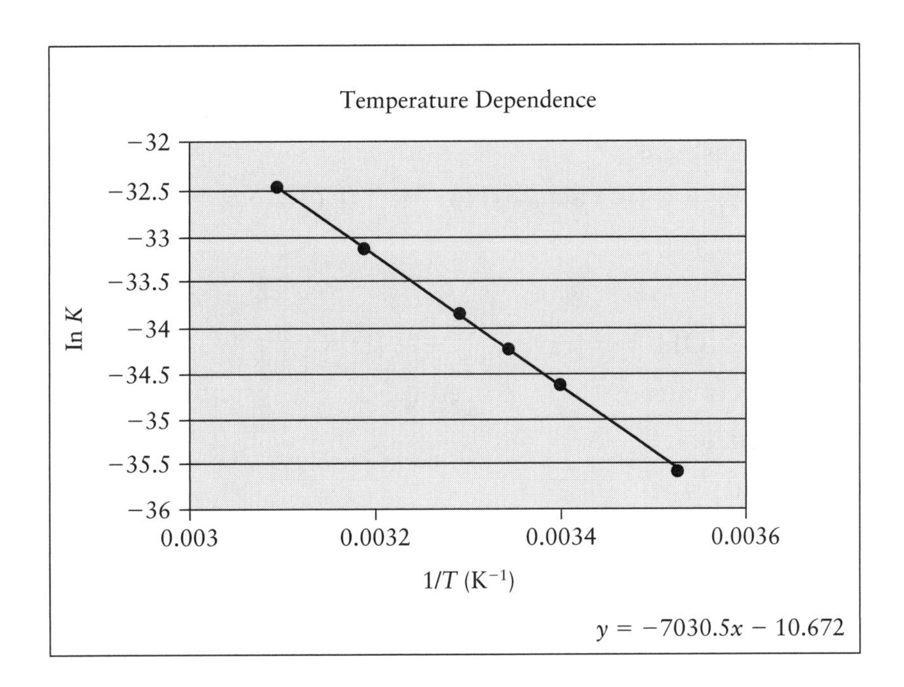

The enthalpy of ionization is obtained from the slope of the graph.

$\ln K = -7030\ T^{-1} - 10.67$

$\Delta H° = -(-7030\ \text{K})(8.314\ \text{J}\cdot\text{K}^{-1}\cdot\text{mol}^{-1}) = 58.4\ \text{kJ}\cdot\text{mol}^{-1}$

(b) The value of K for 35°C (308 K) can be determined from the equation by substituting for the temperature.

$-(7030\ \text{K})(308\ \text{K})^{-1} - 10.67$

$K_{D_2O} = 2.8 \times 10^{-15}$

$K_{D_2O} = [D_3O^+][OD^-] = 2.8 \times 10^{-15}$

$[D_3O^+] = [OD^-] = 5.3 \times 10^{-8}$

$\text{pD} = -\log[D_3O^+] = -\log(5.3 \times 10^{-8}) = 7.28$

10.105 Because the parent acid and base (phosphoric acid and ammonia) are weak, the equilibria here are very complex. The best way to approach the problem is to set up a system of simultaneous equations and use an appropriate computer program or graphing calculator to solve the set. There are eight solute species so eight independent relationships are needed.

Solute species involved:

H_3PO_4	$H_2PO_4^-$	HPO_4^{2-}	PO_4^{3-}
NH_4^+	NH_3	H_3O^+	OH^-

The equilibrium relationships that can be used:

$H_3PO_4(aq) + H_2O(l) \rightleftharpoons H_2PO_4^-(aq) + H_3O^+(aq)$ $\quad K_1 = 7.6 \times 10^{-3}$

$H_2PO_4^-(aq) + H_3O^+(aq) \rightleftharpoons HOP_4^{2-}(aq) + H_3O^+(aq)$ $\quad K_2 = 2.6 \times 10^{-7}$

$HPO_4^{2-}(aq) + H_3O^+(aq) \rightleftharpoons PO_4^{3-}(aq) + H_3O^+(aq)$ $\quad K_3 = 2.1 \times 10^{-13}$

$NH_3(aq) + + H_2O(l) \rightleftharpoons NH_4^+(aq) + OH^-(aq)$ $\quad K_4 = 1.8 \times 10^{-5}$

$2\ H_2O(l) \rightleftharpoons H_3O^+(aq) + OH^-(aq)$ $\quad K_5 = 1.0 \times 10^{-14}$

Mass balance relationships:

$[H_3PO_4] + [H_2PO_4^-] + [HPO_4^{2-}] + [PO_4^{3-}] = 0.150$

$[NH_3] + [NH_4^+] = 0.150$

Charge balance:

$[NH_4^+] + [H_3O^+] + [OH^-] + [H_2PO_4^-] + 2\ [HPO_4^{2-}] + 3\ [PO_4^{3-}]$

This gives the eight equations that we need to solve a system of eight unknowns. The values so obtained are

$[H_3PO_4] = 8.5 \times 10^{-4}$

$[H_2PO_4^-] = 0.148$

$[HPO_4^{2-}] = 8.9 \times 10^{-4}$

$[PO_4^{3-}] = 4.3 \times 10^{-12}$

$[NH_4^+] \cong = 0.150$

$[NH_3] = 1.9 \times 10^{-6}$

$[H_3O^+] = 4.3 \times 10^{-5}$

$[OH^-] = 2.3 \times 10^{-10}$

10.107 (a) and (b) Buffer regions are marked A, B, and C.

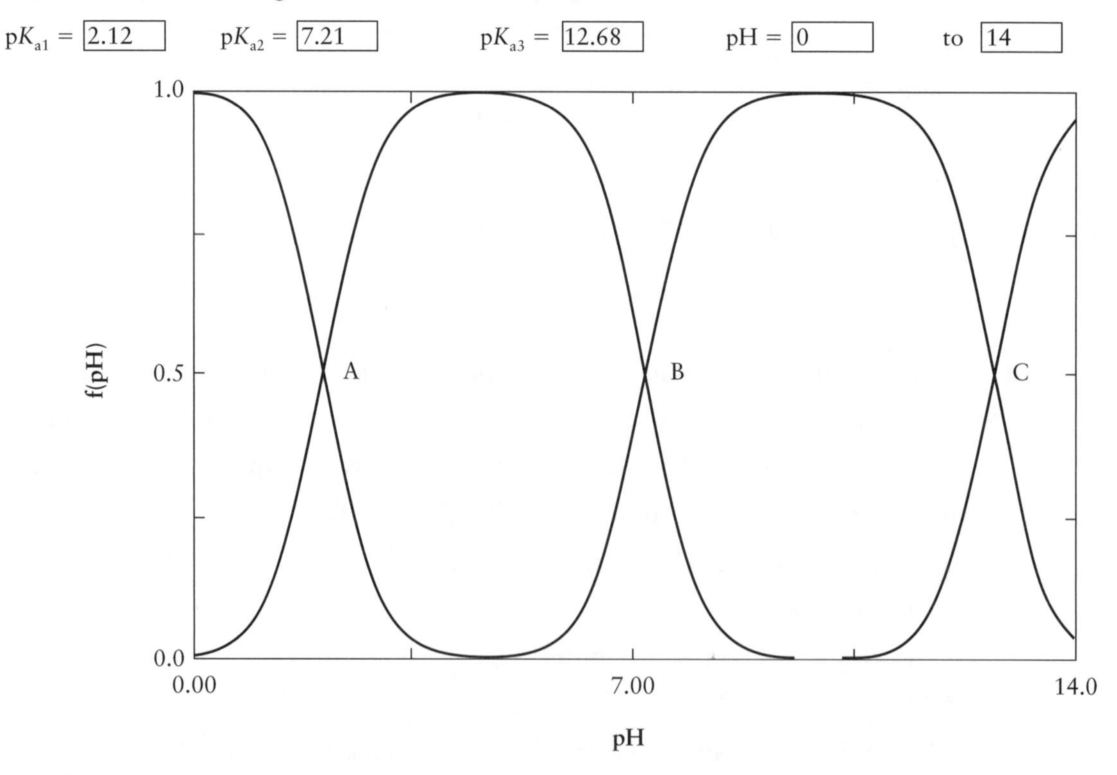

(c) Region A: H_3PO_4 and $H_2PO_4^-$

Region B: $H_2PO_4^-$ and HPO_4^{2-}

Region C: HPO_4^{2-} and PO_4^{3-}

(d)

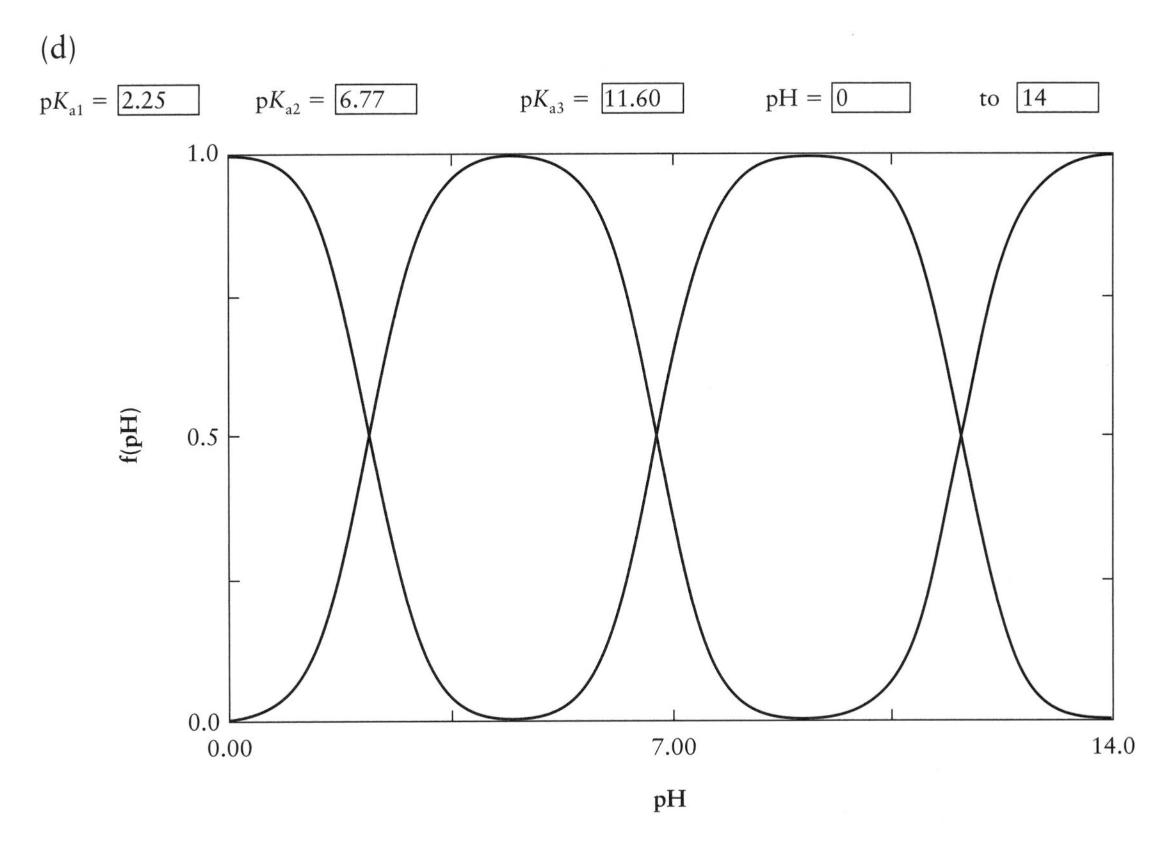

(e) The major species present are similar for both H_3PO_4 and H_3AsO_4: $H_2EO_4^-$ and HEO_4^{2-} where E = P or As. For As, there is more $HAsO_4^{2-}$ than $H_2AsO_4^-$, with a ratio of approximately 0.63 to 0.37, or 1.7 : 1. For P, the situation is reversed, with more $H_2PO_4^-$ than HPO_4^{2-} in a ratio of about 0.61 to 0.39, or 2.2 : 1.

CHAPTER 11
AQUEOUS EQUILIBRIA

11.1 (a) When solid sodium acetate is added to an acetic acid solution, the concentration of H_3O^+ decreases because the equilibrium

$HC_2H_3O_2(aq) + H_2O(l) \rightleftharpoons H_3O^+(aq) + C_2H_3O_2^-(aq)$

shifts to the left to relieve the stress imposed by the increase of $[C_2H_3O_2^-]$ (Le Chatelier's principle).

(b) When HCl is added to a benzoic acid solution, the percentage of benzoic acid that is deprotonated decreases because the equilibrium

$C_6H_5COOH(aq) + H_2O(l) \rightleftharpoons H_3O^+(aq) + C_6H_5CO_2^-(aq)$

shifts to the left to relieve the stress imposed by the increased $[H_3O^+]$ (Le Chatelier's principle).

(c) When solid NH_4Cl is added to an ammonia solution, the concentration of OH^- decreases because the equilibrium

$NH_3(aq) + H_2O(l) \rightleftharpoons NH_4^+(aq) + OH^-(aq)$

shifts to the left to relieve the stress imposed by the increased $[NH_4^+]$ (Le Chatelier's principle). Because $[OH^-]$ decreases, $[H_3O^+]$ increases and pH decreases.

11.3 (a) $K_a = \dfrac{[H_3O^+][A^-]}{[HA]}$; $pK_a = pH - \log \dfrac{[A^-]}{[HA]}$. If $[A^-] = [HA]$, then $pK_a = pH$.

$pH = pK_a = 3.08, \quad K_a = 8.3 \times 10^{-4}$

(b) Let x = [lactate ion] = $[L^{-1}]$ and $y = [H_3O^+]$

Concentration ($mol \cdot L^{-1}$)	$HL(aq)$ +	$H_2O(l) \rightleftharpoons$	$H_3O^+(aq)$ +	$L^-(aq)$
initial	$2x$	—	—	x
change	$-y$	—	$+y$	$+y$
equilibrium	$2x - y$	—	y	$y + x$

$$K_a = \frac{[H_3O^+][L^-]}{[HL]} = \frac{(y)(y + x)}{(2x - y)} \cong \frac{(y)(x)}{(2x)} = 8.3 \times 10^{-4}$$

$y = 2(8.3 \times 10^{-4}) \cong 1.7 \times 10^{-3}\ mol \cdot L^{-1} \cong [H_3O^+]$

$pH \approx 2.77$

11.5 In each case, the equilibrium involved is

$HSO_4^-(aq) + H_2O(l) \rightleftharpoons H_3O^+(aq) + SO_4^{2-}(aq)$

$HSO_4^-(aq)$ and $SO_4^{2-}(aq)$ are conjugate acid and base; therefore, the pH calculation is most easily performed with the Henderson-Hasselbalch equation:

$$pH = pK_a + \log\left(\frac{[\text{base}]}{[\text{acid}]}\right) = pK_a + \log\left(\frac{[SO_4^{2-}]}{[HSO_4^-]}\right)$$

(a) $pH = 1.92 + \log\left(\dfrac{0.40\ \text{mol}\cdot\text{L}^{-1}}{0.80\ \text{mol}\cdot\text{L}^{-1}}\right) = 1.62, \quad pOH = 14.00 - 1.62 = 12.38$

(b) $pH = 1.92 + \log\left(\dfrac{0.20\ \text{mol}\cdot\text{L}^{-1}}{0.80\ \text{mol}\cdot\text{L}^{-1}}\right) = 1.32, \quad pOH = 12.68$

(c) $pH = pK_a = 1.92, \quad pOH = 12.08$

See solution to Exercise 11.3.

11.7 $\left(\dfrac{0.356\ \text{g NaF}}{0.050\ \text{L}}\right)\left(\dfrac{1\ \text{mol NaF}}{41.99\ \text{g NaF}}\right) = 0.17\ \text{mol}\cdot\text{L}^{-1}$

Concentration ($\text{mol}\cdot\text{L}^{-1}$)	$HF(aq)$	+ $H_2O(l) \rightleftharpoons$	$H_3O^+(aq)$	+ $F^-(aq)$
initial	0.40	—	0	0.17
change	$-x$	—	$+x$	$+x$
equilibrium	$0.40 - x$	—	x	$0.17 + x$

$$K_a = \frac{[H_3O^+][F^-]}{[HF]} = \frac{(x)(0.17 + x)}{(0.40 - x)} \approx \frac{(x)(0.17)}{(0.40)} = 3.5 \times 10^{-4}$$

$x \cong 8.2 \times 10^{-4}\ \text{mol}\cdot\text{L}^{-1} \cong [H_3O^+]$

$pH = -\log[H_3O^+] = -\log(8.2 \times 10^{-4}) = 3.09$

change in $pH = 3.09 - 1.93 = 1.16$

11.9 (a) $HCN(aq) + H_2O(l) \rightleftharpoons H_3O^+(aq) + CN^-(aq)$

total volume = 100 mL = 0.100 L

moles of HCN $= 0.0200\ \text{L} \times 0.050\ \text{mol}\cdot\text{L}^{-1} = 1.0 \times 10^{-3}$ mol HCN

moles of NaCN $= 0.0800\ \text{L} \times 0.030\ \text{mol}\cdot\text{L}^{-1} = 2.4 \times 10^{-3}$ mol NaCN

$$\text{initial } [HCN]_0 = \frac{1.0 \times 10^{-3}\ \text{mol}}{0.100\ \text{L}} = 1.0 \times 10^{-2}\ \text{mol}\cdot\text{L}^{-1}$$

$$\text{initial } [CN^-]_0 = \frac{2.4 \times 10^{-3}\ \text{mol}}{0.100\ \text{L}} = 2.4 \times 10^{-2}\ \text{mol}\cdot\text{L}^{-1}$$

Concentration ($\text{mol}\cdot\text{L}^{-1}$)	$HCN(aq)$	+ $H_2O(l) \rightleftharpoons$	$H_3O^+(aq)$	+ $CN^-(aq)$
initial	1.0×10^{-2}	—	0	2.4×10^{-2}
change	$-x$	—	$+x$	$+x$
equilibrium	$1.0 \times 10^{-2} - x$	—	x	$2.4 \times 10^{-2} + x$

$$K_a = \frac{[H_3O^+][CN^-]}{[HCN]} = \frac{(x)(2.4 \times 10^{-2} + x)}{(1.0 \times 10^{-2} - x)} \approx \frac{(x)(2.4 \times 10^{-2})}{(1.0 \times 10^{-2})} = 4.9 \times 10^{-10}$$

$x \approx 2.0 \times 10^{-10}\ \text{mol}\cdot\text{L}^{-1} \approx [H_3O^+]$

$\text{pH} = -\log[H_3O^+] = -\log(2.0 \times 10^{-10}) = 9.70$

(b) The solution here is the same as for part (a), except for the initial concentrations:

$$[HCN]_0 = \frac{0.0800\ \text{L} \times 0.030\ \text{mol}\cdot\text{L}^{-1}}{0.100\ \text{L}} = 2.4 \times 10^{-2}\ \text{mol}\cdot\text{L}^{-1}$$

$$[CN^-]_0 = \frac{0.0200\ \text{L} \times 0.050\ \text{mol}\cdot\text{L}^{-1}}{0.100\ \text{L}} = 1.0 \times 10^{-2}\ \text{mol}\cdot\text{L}^{-1}$$

$$K_a = 4.9 \times 10^{-10} = \frac{(x)(1.0 \times 10^{-2})}{(2.4 \times 10^{-2})}$$

$x = [H_3O^+] = 1.2 \times 10^{-9}\ \text{mol}\cdot\text{L}^{-1}$

$\text{pH} = -\log(1.2 \times 10^{-9}) = 8.92$

(c) $[HCN]_0 = [NaCN]_0$ after mixing; therefore,

$$K_a = 4.9 \times 10^{-10} = \frac{(x)[NaCN]_0}{[HCN]_0} = x = [H_3O^+]$$

$\text{pH} = \text{p}K_a = -\log(4.9 \times 10^{-10}) = 9.31$

11.11 In a solution containing HClO(aq) and ClO^-(aq), the following equilibrium occurs:

$HClO(aq) + H_2O(l) \rightleftharpoons H_3O^+(aq) + ClO^-(aq)$

The ratio $[ClO^-]/[HClO]$ is related to pH, as given by the Henderson-Hasselbalch equation:

$$\text{pH} = \text{p}K_a + \log\left(\frac{[ClO^-]}{[HClO]}\right), \text{ or}$$

$$\log\left(\frac{[ClO^-]}{[HClO]}\right) = \text{pH} - \text{p}K_a = 6.50 - 7.53 = -1.03$$

$$\frac{[ClO^-]}{[HClO]} = 9.3 \times 10^{-2}$$

11.13 The rule of thumb we use is that the effective range of a buffer is roughly within plus or minus one pH unit of the $\text{p}K_a$ of the acid. Therefore,

(a) $\text{p}K_a = 3.08$; pH range, 2–4

(b) $\text{p}K_a = 4.19$; pH range, 3–5

(c) $\text{p}K_{a3} = 12.68$; pH range, 11.5–13.5

(d) $\text{p}K_{a2} = 7.21$; pH range, 6–8

(e) $\text{p}K_b = 7.97$, $\text{p}K_a = 6.03$; pH range, 5–7

11.15 Choose a buffer system in which the conjugate acid has a pK_a close to the desired pH. Therefore,

(a) $HClO_2$ and $NaClO_2$, pK_a = 2.00

(b) NaH_2PO_4 and Na_2HPO_4, pK_{a2} = 7.21

(c) $CH_2ClCOOH$ and $NaCH_2ClCO_2$, pK_a = 2.85

(d) Na_2HPO_4 and Na_3PO_4, pK_a = 12.68

11.17 (a) $HCO_3^-(aq) + H_2O(l) \rightleftharpoons CO_3^{2-}(aq) + H_3O^+(aq)$

$$K_{a2} = \frac{[H_3O^+][CO_3^{2-}]}{[HCO_3^-]}, \quad pK_{a2} = 10.25$$

$$pH = pK_{a2} + \log\left(\frac{[CO_3^{2-}]}{[HCO_3^-]}\right)$$

$$\log\left(\frac{[CO_3^{2-}]}{[HCO_3^-]}\right) = pH - pK_{a2} = 11.0 - 10.25 = 0.75$$

$$\frac{[CO_3^{2-}]}{[HCO_3^-]} = 5.6$$

(b) $[CO_3^{2-}] = 5.6 \times [HCO_3^-] = 5.6 \times 0.100\ mol\cdot L^{-1} = 0.56\ mol\cdot L^{-1}$

moles of CO_3^{2-} = moles of K_2CO_3 = $0.56\ mol\cdot L^{-1} \times 1\ L = 0.56\ mol$

$$\text{mass of } K_2CO_3 = 0.56\ mol \times \left(\frac{138.21\ g\ K_2CO_3}{1\ mol\ K_2CO_3}\right) = 77\ g\ K_2CO_3$$

(c) $$[HCO_3^-] = \frac{[CO_3^{2-}]}{5.6} = \frac{0.100\ mol\cdot L^{-1}}{5.6} = 1.8 \times 10^{-2}\ mol\cdot L^{-1}$$

moles of HCO_3^- = moles of $KHCO_3$ = $1.8 \times 10^{-2}\ mol\cdot L^{-1} \times 1\ L$

$= 1.8 \times 10^{-2}\ mol$

mass $KHCO_3$ = $1.8 \times 10^{-2}\ mol \times 100.12\ g\cdot mol^{-1} = 1.8\ g\ KHCO_3$

(d) $[CO_3^{2-}] = 5.6 \times [HCO_3^-]$

moles HCO_3^- = moles $KHCO_3$ = $0.100\ mol\cdot L^{-1} \times 0.100\ L = 1.00 \times 10^{-2}\ mol$

Because the final total volume is the same for both $KHCO_3$ and K_2CO_3, the number of moles of K_2CO_3 required is $5.6 \times 1.00 \times 10^{-2}\ mol = 5.6 \times 10^{-2}\ mol$. Thus,

$$\text{volume of } K_2CO_3 \text{ solution} = \frac{5.6 \times 10^{-2}\ mol}{0.200\ mol\cdot L^{-1}} = 0.28\ L = 2.8 \times 10^2\ mL$$

11.19 (a) $pH = pK_a + \log\left(\frac{[CH_3CO_2^-]}{[CH_3COOH]}\right)$ (see Exercise 11.23)

$$pH = pK_a + \log\left(\frac{0.100}{0.100}\right) = 4.75 \text{ (initial pH)}$$

final pH: (0.0100 L)(0.950 mol·L^{-1}) = 9.50 × 10^{-3} mol NaOH (strong base) produces 9.50 × 10^{-3} mol $CH_3CO_2^-$ from CH_3COOH

0.100 mol·L^{-1} × 0.100 L = 1.00 × 10^{-2} mol CH_3COOH initially

0.100 mol·L^{-1} × 0.100 L = 1.00 × 10^{-2} mol $CH_3CO_2^-$ initially

After adding NaOH:

$$[CH_3COOH] = \frac{(1.00 \times 10^{-2} - 9.5 \times 10^{-3})\ \text{mol}}{0.110\ \text{L}} = 5 \times 10^{-3}\ \text{mol·L}^{-1}$$

$$[CH_3CO_2^-] = \frac{(1.00 \times 10^{-2} + 9.5 \times 10^{-3})\ \text{mol}}{0.110\ \text{L}} = 1.77 \times 10^{-1}\ \text{mol·L}^{-1}$$

$$\text{pH} = 4.75 + \log\left(\frac{1.77 \times 10^{-1}\ \text{mol·L}^{-1}}{5 \times 10^{-3}\ \text{mol·L}^{-1}}\right) = 4.75 + 1.5 = 6.3$$

(b) (0.0200 L)(0.100 mol·L^{-1}) = 2.00 × 10^{-3} mol HNO_3 (strong acid) produces 2.00 × 10^{-3} mol CH_3COOH from $CH_3CO_2^-$.

After adding HNO_3 [see part (a) of this exercise]:

$$[CH_3COOH] = \frac{(1.00 \times 10^{-2} + 2.00 \times 10^{-3})\ \text{mol}}{0.120\ \text{L}} = 1.00 \times 10^{-1}\ \text{mol·L}^{-1}$$

$$[CH_3CO_2^-] = \frac{(1.00 \times 10^{-2} - 2.00 \times 10^{-3})\ \text{mol}}{0.120\ \text{L}} = 6.7 \times 10^{-2}\ \text{mol·L}^{-1}$$

$$\text{pH} = 4.75 + \log\left(\frac{6.7 \times 10^{-2}\ \text{mol·L}^{-1}}{1.00 \times 10^{-1}\ \text{mol·L}^{-1}}\right) = 4.75 - 0.17 = 4.58$$

ΔpH = −0.17

11.21 initial pH = −log(0.10) = 1.00

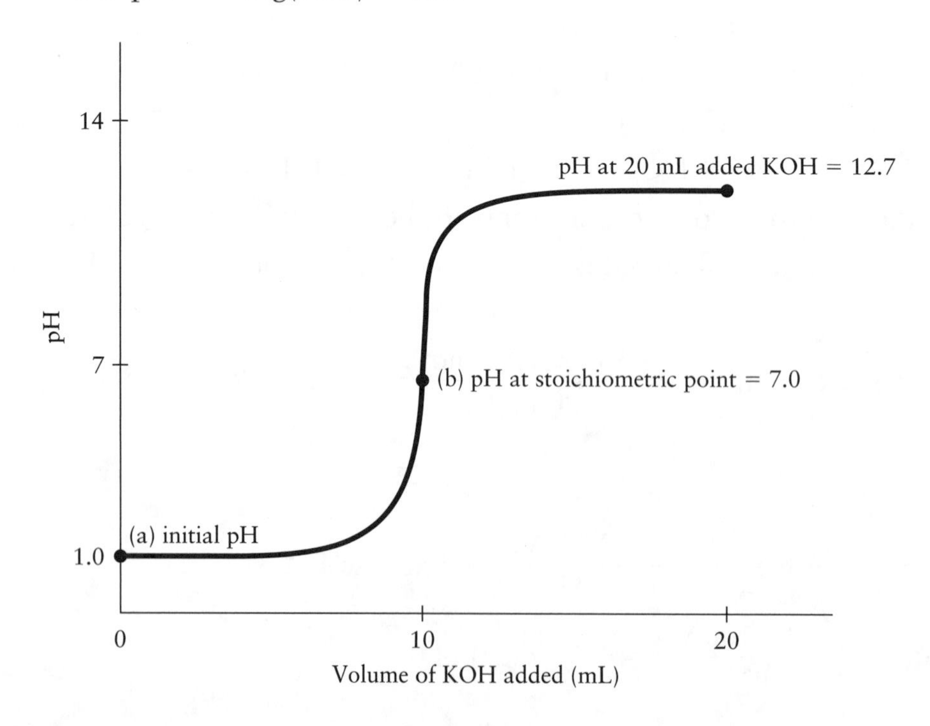

11.23 $HCl(aq) + NaOH(aq) \longrightarrow H_2O(l) + Na^+(aq) + Cl^-(aq)$

(a) $$V_{HCl} = (\tfrac{1}{2})(25.0\ mL)\left(\frac{10^{-3}\ L}{1\ mL}\right)\left(\frac{0.110\ mol\ NaOH}{1\ L}\right)\left(\frac{1\ mol\ HCl}{1\ mol\ NaOH}\right)\left(\frac{1\ L\ HCl}{0.150\ mol\ HCl}\right)$$
$$= 9.17 \times 10^{-3}\ L\ HCl$$

(b) $2 \times 9.17 \times 10^{-3}\ L = 0.0183\ L$

(c) volume $= (0.0250 + 0.0183)\ L = 0.0433\ L$

$$[Na^+] = (0.0250\ L)\left(\frac{0.110\ mol\ NaOH}{1\ L}\right)\left(\frac{1\ mol\ Na^+}{1\ mol\ NaOH}\right)\left(\frac{1}{0.0433\ L}\right)$$
$$= 0.0635\ mol \cdot L^{-1}$$

(d) number of moles of H_3O^+ (from acid) $= (0.0200\ L)\left(\frac{0.150\ mol}{1\ L}\right)$
$$= 3.00 \times 10^{-3}\ mol\ H_3O^+$$

number of moles of OH^- (from base) $= (0.0250\ L)\left(\frac{0.110\ mol\ Na^+}{1\ L}\right)$
$$= 2.75 \times 10^{-3}\ mol\ OH^-$$

excess $H_3O^+ = (3.00 - 2.75) \times 10^{-3}\ mol = 2.5 \times 10^{-4}\ mol\ H_3O^+$

$$[H_3O^+] = \frac{2.5 \times 10^{-4}\ mol}{0.0450\ L} = 5.6 \times 10^{-3}\ mol \cdot L^{-1}$$

$pH = -\log(5.6 \times 10^{-3}) = 2.25$

11.25 $HCl(aq) + NaOH(aq) \longrightarrow H_2O(l) + Na^+(aq) + Cl^-(aq)$

(a) moles of HCl $= (1.33\ g\ NaOH)\left(\frac{1\ mol\ NaOH}{40.00\ g\ NaOH}\right)\left(\frac{1\ mol\ HCl}{1\ mol\ NaOH}\right)$
$$= 3.32 \times 10^{-2}\ mol\ HCl$$

volume of HCl $= \frac{3.32 \times 10^{-2}\ mol}{0.135\ mol \cdot L^{-1}} = 0.246\ L = 246\ mL$

(b) moles of $Cl^- = 0.246\ L\left(\frac{0.135\ mol\ HCl}{1\ L}\right)\left(\frac{1\ mol\ Cl^-}{1\ mol\ HCl}\right)$
$$= 3.32 \times 10^{-2}\ mol\ Cl^-$$

$$[Cl^-] = \frac{3.32 \times 10^{-2}\ mol\ Cl^-}{(0.246 + 0.050)\ L} = 0.112\ mol \cdot L^{-1}$$

11.27 mass of pure NaOH $= (0.0406\ L\ HCl)\left(\frac{0.0695\ mol\ HCl}{1\ L\ HCl}\right)$
$$\left(\frac{1\ mol\ NaOH}{1\ mol\ HCl}\right)\left(\frac{40.00\ g\ NaOH}{1\ mol\ NaOH}\right)\left(\frac{300\ mL}{25.0\ mL}\right)$$
$= 1.35\ g$

$$\text{percent purity} = \frac{1.35\text{ g}}{1.592\text{ g}} \times 100\% = 84.8\%$$

11.29 (a) $\text{pOH} = -\log(0.110) = 0.959, \quad \text{pH} = 14.00 - 0.959 = 13.04$

(b) $\text{initial moles of OH}^- \text{ (from base)} = (0.0250\text{ L})\left(\frac{0.110\text{ mol}}{1\text{ L}}\right)$

$= 2.75 \times 10^{-3}\text{ mol OH}^-$

$\text{moles of H}_3\text{O}^+ \text{ added} = (0.0050\text{ L})\left(\frac{0.150\text{ mol}}{1\text{ L}}\right) = 7.5 \times 10^{-4}\text{ mol H}_3\text{O}^+$

$\text{excess OH}^- = (2.75 - 0.75) \times 10^{-3}\text{ mol} = 2.00 \times 10^{-3}\text{ mol OH}^-$

$$[\text{OH}^-] = \frac{2.00 \times 10^{-3}\text{ mol}}{0.030\text{ L}} = 0.067\text{ mol}\cdot\text{L}^{-1}$$

$\text{pOH} = -\log(0.067) = 1.17, \quad \text{pH} = 14.00 - 1.17 = 12.83$

(c) $\text{moles of H}_3\text{O}^+ \text{ added} = 2 \times 7.5 \times 10^{-4}\text{ mol} = 1.50 \times 10^{-3}\text{ mol H}_3\text{O}^+$

$\text{excess OH}^- = (2.75 - 1.50) \times 10^{-3}\text{ mol} = 1.25 \times 10^{-3}\text{ mol OH}^-$

$$[\text{OH}^-] = \frac{1.25 \times 10^{-3}\text{ mol}}{0.035\text{ L}} = 0.036\text{ mol}\cdot\text{L}^{-1}$$

$\text{pOH} = -\log(0.036) = 1.44, \quad \text{pH} = 14.00 - 1.44 = 12.56$

(d) $\text{pH} = 7.00$

$$V_{\text{HCl}} = (2.75 \times 10^{-3}\text{ mol NaOH})\left(\frac{1\text{ mol HCl}}{1\text{ mol NaOH}}\right)\left(\frac{1\text{ L HCl}}{0.150\text{ mol HCl}}\right)$$

$= 0.0183\text{ L}$

(e) $$[\text{H}_3\text{O}^+] = (0.0050\text{ L})\left(\frac{0.150\text{ mol}}{1\text{ L}}\right)\left(\frac{1}{(0.0250 + 0.0183 + 0.0050)\text{ L}}\right)$$

$= 0.016\text{ mol}\cdot\text{L}^{-1}$

$\text{pH} = -\log(0.016) = 1.80$

(f) $$[\text{H}_3\text{O}^+] = \left(\frac{0.010\text{ L}}{0.0533\text{ L}}\right)\left(\frac{0.150\text{ mol}}{1\text{ L}}\right) = 0.028\text{ mol}\cdot\text{L}^{-1}$$

$\text{pH} = -\log(0.028) = 1.55$

11.31 (a)

Concentration ($\text{mol}\cdot\text{L}^{-1}$)	$CH_3COOH(aq)$ +	$H_2O(l) \rightleftharpoons$	$H_3O^+(aq)$ +	$CH_3CO_2^-(aq)$
initial	0.10	—	0	0
change	$-x$	—	$+x$	$+x$
equilibrium	$0.10 - x$	—	x	x

$$K_a = \frac{[\text{H}_3\text{O}^+][\text{CH}_3\text{CO}_2^-]}{[\text{CH}_3\text{COOH}]} = 1.8 \times 10^{-5} = \frac{x^2}{0.10 - x} \approx \frac{x^2}{0.10}$$

$x^2 = 1.8 \times 10^{-6}$

$x = 1.3 \times 10^{-3}\ \text{mol}\cdot\text{L}^{-1} = [H_3O^+]$

initial pH $= -\log(1.3 \times 10^{-3}) = 2.89$

(b) moles of $CH_3COOH = (0.0250\ \text{L})(0.10\ \text{M})$

$= 2.50 \times 10^{-3}$ mol CH_3COOH

moles of NaOH $= (0.0100\ \text{L})(0.10\ \text{M}) = 1.0 \times 10^{-3}$ mol OH^-

After neutralization,

$$\frac{1.50 \times 10^{-3}\ \text{mol}\ CH_3COOH}{0.0350\ \text{L}} = 4.29 \times 10^{-2}\ \text{mol}\cdot\text{L}^{-1}\ CH_3COOH$$

$$\frac{1.0 \times 10^{-3}\ \text{mol}\ CH_3CO_2^-}{0.0350\ \text{L}} = 2.86 \times 10^{-2}\ \text{mol}\cdot\text{L}^{-1}\ CH_3CO_2^-$$

Then consider equilibrium, $K_a = \dfrac{[H_3O^+][CH_3CO_2^-]}{[CH_3COOH]}$

Concentration $(\text{mol}\cdot\text{L}^{-1})$	$CH_3COOH(aq)$	$+\ H_2O(l) \rightleftharpoons$	$H_3O^+(aq)$	$+\ CH_3CO_2^-(aq)$
initial	4.29×10^{-2}	—	0	2.86×10^{-2}
change	$-x$	—	$+x$	$+x$
equilibrium	$4.29 \times 10^{-2} - x$	—	x	$2.86 \times 10^{-2} + x$

$$1.8 \times 10^{-5} = \frac{(x)(x + 2.86 \times 10^{-2})}{(4.29 \times 10^{-2} - x)};\ \text{assume } +x \text{ and } -x \text{ negligible.}$$

$[H_3O^+] = x = 2.7 \times 10^{-5}\ \text{mol}\cdot\text{L}^{-1}$ and pH $= -\log(2.7 \times 10^{-5}) = 4.56$

(c) Because acid and base concentrations are equal, their volumes are equal at the stoichiometric point. Therefore, 25.0 mL NaOH is required to reach the stoichiometric point and 12.5 mL NaOH is required to reach the half stoichiometric point.

(d) At the half stoichiometric point, pH $= pK_a = 4.75$

(e) 25.0 mL; see part (c)

(f) The pH is that of 0.050 M $NaCH_3CO_2$.

Concentration $(\text{mol}\cdot\text{L}^{-1})$	$H_2O(l)$	$+\ CH_3CO_2^-(aq) \rightleftharpoons$	$CH_3COOH(aq)$	$+\ OH^-(aq)$
initial	—	0.050	0	0
change	—	$-x$	$+x$	$+x$
equilibrium	—	$0.050 - x$	x	x

$$K_b = \frac{K_w}{K_a} = \frac{1.00 \times 10^{-14}}{1.8 \times 10^{-5}} = 5.6 \times 10^{-10} = \frac{x^2}{0.050 - x} \approx \frac{x^2}{0.050}$$

$x^2 = 2.8 \times 10^{-11}$

$x = 5.3 \times 10^{-6}\ \text{mol·L}^{-1} = [\text{OH}^-]$

$\text{pOH} = 5.28, \quad \text{pH} = 14.00 - 5.28 = 8.72$

11.33 (a) $K_b = \dfrac{[\text{NH}_4^+][\text{OH}^-]}{[\text{NH}_3]} = 1.8 \times 10^{-5}$

Concentration (mol·L^{-1})	H_2O(l) +	NH_3(aq) ⇌	NH_4^+(aq) +	OH^-(aq)
initial	—	0.15	0	0
change	—	$-x$	$+x$	$+x$
equilibrium	—	$0.15 - x$	x	x

$$1.8 \times 10^{-5} = \frac{x^2}{0.15 - x} \approx \frac{x^2}{0.15}$$

$[\text{OH}^-] = x = 1.6 \times 10^{-3}\ \text{mol·L}^{-1}$

$\text{pOH} = 2.80, \quad \text{initial pH} = 14.00 - 2.80 = 11.20$

(b) initial moles of NH_3 = $(0.0150\ \text{L})(0.15\ \text{mol·L}^{-1}) = 2.3 \times 10^{-3}$ mol NH_3

moles of HCl = $(0.0150\ \text{L})(0.10\ \text{mol·L}^{-1}) = 1.5 \times 10^{-3}$ mol HCl

$$\frac{(2.3 \times 10^{-3} - 1.5 \times 10^{-3})\ \text{mol NH}_3}{0.0300\ \text{L}} = 2.7 \times 10^{-2}\ \text{mol·L}^{-1}\ \text{NH}_3$$

$$\frac{1.5 \times 10^{-3}\ \text{mol HCl}}{0.0300\ \text{L}} = 5.0 \times 10^{-2}\ \text{mol·L}^{-1}\ \text{HCl} \approx 5.0 \times 10^{-2}\ \text{mol·L}^{-1}\ \text{NH}_4^+$$

Then consider the equilibrium:

Concentration (mol·L^{-1})	H_2O(l) +	NH_3(aq) ⇌	NH_4^+(aq) +	OH^-(aq)
initial	—	2.7×10^{-2}	5.0×10^{-2}	0
change	—	$-x$	$+x$	$+x$
equilibrium	—	$2.7 \times 10^{-2} - x$	$5.0 \times 10^{-2} + x$	x

$$K_b = \frac{[\text{NH}_4^+][\text{OH}^-]}{[\text{NH}_3]} = 1.8 \times 10^{-5}$$

$$= \frac{(x)(5.0 \times 10^{-2} + x)}{(2.7 \times 10^{-2} - x)}$$; assume that $+x$ and $-x$ are negligible

$[\text{OH}^-] = x = 9.7 \times 10^{-6}\ \text{mol·L}^{-1}$ and pOH = 5.01

Therefore, pH = 14.00 − 5.01 = 8.99

(c) At the stoichiometric point, moles of NH_3 = moles of HCl

$$\text{volume HCl added} = \frac{(0.15\ \text{mol·L}^{-1}\ \text{NH}_3)(0.0150\ \text{L})}{0.10\ \text{mol·L}^{-1}\ \text{HCl}} = 0.0225\ \text{L HCl}$$

Therefore, halfway to the stoichiometric point, volume HCl added = 22.5/2 = 11.25 mL

(d) At half stoichiometric point, pOH = $\text{p}K_b$ and pOH = 4.75

Therefore, pH = 14.00 − 4.75 = 9.25

(e) 22.5 mL; see part (c)

(f) $NH_4^+(aq) + H_2O(l) \rightleftharpoons H_3O^+(aq) + NH_3(aq)$

The initial moles of NH_3 have now been converted to moles of NH_4^+ in a (15 + 22.5 = 37.5) mL volume:

$$[NH_4^+] = \frac{2.25 \times 10^{-3}\ \text{mol}}{0.0375\ \text{L}} = 0.060\ \text{mol}\cdot\text{L}^{-1}$$

$$K_a = \frac{K_w}{K_b} = \frac{1.00 \times 10^{-14}}{1.8 \times 10^{-5}} = 5.6 \times 10^{-10}$$

Concentration ($mol\cdot L^{-1}$)	$NH_4^+(aq)$	+ $H_2O(l)$ ⇌	$H_3O^+(aq)$	+ $NH_3(aq)$
initial	0.060	—	0	0
change	$-x$	—	$+x$	$+x$
equilibrium	$0.060 - x$	—	x	x

$$K_a = 5.6 \times 10^{-10} = \frac{x^2}{0.060 - x} \approx \frac{x^2}{0.060}$$

$x = [H_3O^+] = 5.8 \times 10^{-6}\ \text{mol}\cdot\text{L}^{-1}$

$\text{pH} = -\log(5.8 \times 10^{-6}) = 5.24$

11.35 At the stoichiometric point, the volume of solution will have doubled; therefore, the concentration of $CH_3CO_2^-$ will be 0.10 M. The equilibrium is

Concentration ($mol\cdot L^{-1}$)	$CH_3CO_2^-(aq)$	+ $H_2O(l)$ ⇌	$HCH_3CO_2(aq)$	+ $OH^-(aq)$
initial	0.10	—	0	0
change	$-x$	—	$+x$	$+x$
equilibrium	$0.10 - x$	—	x	x

$$K_b = \frac{K_w}{K_a} = \frac{1.00 \times 10^{-14}}{1.8 \times 10^{-5}} = 5.6 \times 10^{-10}$$

$$K_b = \frac{[HCH_3CO_2][OH^-]}{[CH_3CO_2^-]} = \frac{x^2}{0.10 - x} \approx \frac{x^2}{0.10} = 5.6 \times 10^{-10}$$

$x = 7.5 \times 10^{-6}\ \text{mol}\cdot\text{L}^{-1} = [OH^-]$

$\text{pOH} = -\log(7.5 \times 10^{-6}) = 5.12$, $\text{pH} = 14.00 - 5.12 = 8.88$

From Table 11.2, we see that this pH value lies within the range for phenolphthalein, so that indicator would be suitable; the others would not.

11.37 Exercise 11.31: thymol blue or phenolphthalein; Exercise 11.33: methyl red or bromocresol green.

11.39 (a) To reach the first stoichiometric point, we must add enough solution to neutralize one H^+ on the H_3AsO_4. To do this, we will require 0.0750 L × 0.137 mol·L^{-1} = 0.0103 mol OH^-. The volume of base required will be given by the number of moles of base required, divided by the concentration of base solution:

$$\frac{0.0750 \text{ L} \times 0.137 \text{ mol·L}^{-1}}{0.275 \text{ mol·L}^{-1}} = 0.0374 \text{ L or } 37.4 \text{ mL}$$

(b) and (c) To reach the second stoichiometric point will require double the amount calculated in (a), or 74.8 mL, and the third stoichiometric point will be reached with three times the amount added in (a), or 112 mL.

11.41 (a) The base HPO_3^{2-} is the fully deprotonated form of phosphorous acid H_3PO_3 (the remaining H attached to P is not acidic). It will require an equal number of moles of HNO_3 to react with HPO_3^{2-}, in order to reach the first stoichiometric point (formation of $H_2PO_3^-$). The value will be given by

$$\frac{0.0355 \text{ L} \times 0.158 \text{ mol·L}^{-1}}{0.255 \text{ mol·L}^{-1}} = 0.0220 \text{ L or } 22.0 \text{ mL}$$

(b) To reach the second stoichiometric point would require double the amount of solution calculated in (a), or 44.0 mL.

11.43 (a) This value is calculated as described in Example 10.12. First we calculate the molarity of the starting phosphorous acid solution: $\frac{0.164 \text{ g}}{81.99 \text{ g·mol}^{-1}} \Big/ 0.0500 \text{ L} =$ 0.0400 mol·L^{-1}. We then use the first acid dissociation of phosphorous acid as the dominant equilibrium. The K_{a1} is 1.0×10^{-2}. Let H_2P represent the fully-protonated phosphorus acid.

Concentration (mol·L^{-1})	$H_2P(aq)$	+ $H_2O(l)$ ⇌	$HP^-(aq)$	+ $H_3O^+(aq)$
initial	0.0400	—	0	0
change	$-x$	—	$+x$	$+x$
final	$0.0400 - x$	—	$+x$	$+x$

$$K_a = \frac{[H_3O^+][HP^-]}{[H_2P]} = 1.0 \times 10^{-2}$$

$$1.0 \times 10^{-2} = \frac{x \cdot x}{0.0400 - x} = \frac{x^2}{0.0400 - x}$$

If we assume $x \ll 0.0400$, then the equation becomes

$x^2 = (1.0 \times 10^{-2})(0.0400) = 4.00 \times 10^{-4}$

$x = 2.0 \times 10^{-2}$

Because this value is more than 10% of 0.0400, the full quadratic solution should be undertaken. The equation is

$x^2 = (1.0 \times 10^{-2})(0.0400 - x)$ or

$x^2 + (1.0 \times 10^{-2}x) - (4.0 \times 10^{-4}) = 0$

Using the quadratic formula, we obtain $x = 0.016$.

pH = 1.80

(b) First, carry out the reaction between phosphorous acid and the strong base to completion:

$H_2P(aq) + OH^-(aq) \longrightarrow HP^-(aq) + H_2O(l)$

moles of $H_2P = (0.0400\ mol \cdot L^{-1})(0.0500\ L) = 2.00 \times 10^{-3}$ mol

moles of $OH^- = (0.006\ 50\ L)(0.175\ mol \cdot L^{-1}) = 1.14 \times 10^{-3}$ mol

1.14×10^{-3} mol OH^- will react completely with 2.00×10^{-3} mol H_2P to give 1.14×10^{-3} mol HP^- with 8.6×10^{-4} moles of H_2P remaining.

$$[H_2P] = \frac{8.6 \times 10^{-4}\ mol}{0.0565\ L} = 0.0152\ mol \cdot L^{-1}$$

$$[HP^-] = \frac{1.14 \times 10^{-3}\ mol}{0.0565\ L} = 0.0202\ mol \cdot L^{-1}$$

Concentration ($mol \cdot L^{-1}$)	$H_2P(aq)$	+ $H_2O(l)$ ⇌	$HP^-(aq)$	+ $H_3O^+(aq)$
initial	0.0152	—	0.0202	0
change	$-x$	—	$+x$	$+x$
final	$0.0152 - x$	—	$0.0202 + x$	$+x$

The calculation is performed as in part (a):

$$1.0 \times 10^{-2} = \frac{(0.0202 + x)x}{0.0152 - x}$$

$x^2 + (3.0 \times 10^{-2}x) - (1.5 \times 10^{-4}) = 0$

$x = 4.4 \times 10^{-3}$

pH = 2.36

(c) The total amount of OH^- added will be 6.50 mL + 4.93 mL = 11.43 mL. The moles of $OH^- = (0.011\ 43\ L)(0.175\ mol \cdot L^{-1}) = 0.002\ 00$ mol. This amount of base will exactly neutralize one H^+ present in H_2P, so the problem becomes one of calculating the pH of a solution of NaHP. As presented in section 10.16, the pH of a solution of a salt of a polyprotic acid is given by

$pH = \frac{1}{2}(pK_{a1} + pK_{a2})$

$pH = \frac{1}{2}(2.00 + 6.59) = 4.30$

11.45 (a) The reaction of the base Na_2HPO_4 with the strong acid will be taken to completion first:

$HPO_4^{2-}(aq) + H_3O^+(aq) \longrightarrow H_2PO_4^- + H_2O(l)$

Initially, moles of HPO_4^{2-} = moles of H_3O^+ = 0.0500 L × 0.275 mol·L^{-1} = 0.0138 mol

Because this reaction proceeds with no excess base or acid, we are dealing with a solution that can be viewed as being composed of $H_2PO_4^-$. The problem then becomes one of estimating the pH of this solution, which can be done from the relationship

$\text{pH} = \frac{1}{2}(\text{p}K_{a1} + \text{p}K_{a2})$

$\text{pH} = \frac{1}{2}(2.12 + 7.21) = 4.66$

(b) This reaction proceeds as in (a), but there is more strong acid available, so the excess acid will react with $H_2PO_4^-$ to produce H_3PO_4. Addition of the first 50.0 mL of acid solution will convert all the HPO_4^{2-} into $H_2PO_4^-$. The additional 25.0 mL of the strong acid will react with $H_2PO_4^-$:

$H_2PO_4^-(aq) + H_3O^+(aq) \longrightarrow H_3PO_4(aq) + H_2O(l)$

0.0138 mol $H_2PO_4^-$ will react with 0.006 88 mol H_3O^+ to give 0.0069 mol H_3PO_4 with 0.069 mol $H_2PO_4^-$ in excess. The concentrations will be

$[H_3PO_4] = [H_2PO_4^-] = \dfrac{0.0069\text{ mol}}{0.125\text{ L}} = 0.055$. The appropriate relationship to use is then

Concentration (mol·L^{-1})	$H_3PO_4(aq)$	+ $H_2O(l)$ $\rightleftharpoons$	$H_2PO_4^-(aq)$	+ $H_3O^+(aq)$
initial	0.055	—	0.055	0
change	$-x$	—	$+x$	$+x$
final	$0.055 - x$	—	$0.055 + x$	$+x$

$$K_{a1} = \frac{[H_2PO_4^-][H_3O^+]}{[H_3PO_4]}$$

$$\frac{(0.055 + x)x}{(0.055 - x)} = 7.6 \times 10^{-3}$$

Because the equilibrium constant is not small compared to 0.055, the full quadratic solution must be calculated:

$x^2 + 0.055x = 7.6 \times 10^{-3}\,(0.055 - x)$

$x^2 + 0.063x - 4.2 \times 10^{-4} = 0$

$x = 1.6 \times 10^{-3}$

$\text{pH} = -\log(1.6 \times 10^{-3}) = 2.80$

(c) The reaction of Na_2HPO_4 with strong acid goes only halfway to completion. 0.275 mol of HPO_4^{2-} will react with (0.025 L × 0.275 mol·L^{-1}) = 6.9×10^{-3} mol HCl to produce 6.9×10^{-3} mol $H_2PO_4^-$ and leave 6.9×10^{-3} HPO_4^{2-} unreacted.

6.9×10^{-3} mol $\div$ 0.075 L = 0.092 mol·L^{-1}

Concentration (mol·L^{-1})	$H_2PO_4^-(aq)$	+ $H_2O(l)$ $\rightleftharpoons$	$H_2PO_4^{2-}(aq)$	+ $H_3O^+(aq)$
initial	0.092	—	0.092	0
change	$-x$	—	$+x$	$+x$
final	$0.092 - x$	—	$0.092 + x$	$+x$

$$K_{a2} = 6.2 \times 10^{-8} = \frac{[HPO_4^{2-}][H_3O^+]}{[H_2PO_4^-]}$$

$$\frac{[0.092 + x][H_3O^+]}{[0.092 - x]} = 6.2 \times 10^{-8}$$

assuming $x \ll$ than 0.092

$x = [H_3O^+] = 6.2 \times 10^{-8}$

$pH = -\log(6.2 \times 10^{-8}) = 7.21$

11.47 (a) The solubility equilibrium is $AgBr(s) \rightleftharpoons Ag^+(aq) + Br^-(aq)$.

$[Ag^+] = [Br^-] = 8.8 \times 10^{-7}$ mol·L^{-1} = S = solubility

$K_{sp} = [Ag^+][Br^-] = (8.8 \times 10^{-7})(8.8 \times 10^{-7}) = 7.7 \times 10^{-13}$

(b) The solubility equilibrium is $PbCrO_4(s) \rightleftharpoons Pb^{2+}(aq) + CrO_4^{2-}(aq)$

$[Pb^{2+}] = 1.3 \times 10^{-7}$ mol·L^{-1} = S, $[CrO_4^{2-}] = 1.3 \times 10^{-7}$ mol·L^{-1} = S

$K_{sp} = [Pb^{2+}][CrO_4^{2-}] = (1.3 \times 10^{-7})(1.3 \times 10^{-7}) = 1.7 \times 10^{-14}$

(c) The solubility equilibrium is $Ba(OH)_2(s) \rightleftharpoons Ba^{2+}(aq) + 2\ OH^-(aq)$

$[Ba^{2+}] = 0.11$ mol·L^{-1} = S, $[OH^-] = 0.22$ mol·L^{-1} = $2S$

$K_{sp} = [Ba^{2+}][OH^-]^2 = (0.11)(0.22)^2 = 5.3 \times 10^{-3}$

(d) The solubility equilibrium is $MgF_2(s) \rightleftharpoons Mg^{2+}(aq) + 2\ F^-(aq)$

$[Mg^{2+}] = 1.2 \times 10^{-3}$ mol·L^{-1} = S, $[F^-] = 2.4 \times 10^{-3}$ mol·L^{-1} = $2S$

$K_{sp} = [Mg^{2+}][F^-]^2 = (1.2 \times 10^{-3})(2.4 \times 10^{-3})^2 = 6.9 \times 10^{-9}$

11.49 (a) Equilibrium equation: $Ag_2S(s) \rightleftharpoons 2\ Ag^+(aq) + S^{2-}(aq)$

$K_{sp} = [Ag^+]^2[S^{2-}] = (2S)^2(S) = 4S^3 = 6.3 \times 10^{-51}$

$S = 1.2 \times 10^{-17}$ mol·L^{-1}

(b) Equilibrium equation: $CuS(s) \rightleftharpoons Cu^{2+}(aq) + S^{2-}(aq)$

$K_{sp} = [Cu^{2+}][S^{2-}] = S \times S = S^2 = 1.3 \times 10^{-36}$

$S = 1.1 \times 10^{-18}$mol·L^{-1}

(c) Equilibrium equation: $CaCO_3(s) \rightleftharpoons Ca^{2+}(aq) + CO_3^{2-}(aq)$

$K_{sp} = [Ca^{2+}][CO_3^{2-}] = S \times S = S^2 = 8.7 \times 10^{-9}$

$S = 9.3 \times 10^{-5}$ mol·L^{-1}

11.51 $Tl_2CrO_4(s) \rightleftharpoons 2\,Tl^+(aq) + CrO_4^{2-}(aq)$

$[CrO_4^{2-}] = S = 6.3 \times 10^{-5}\ mol \cdot L^{-1}$

$[Tl^+] = 2S = 2(6.3 \times 10^{-5})\ mol \cdot L^{-1}$

$K_{sp} = [Tl^+]^2[CrO_4^{2-}] = (2S)^2 \times (S)$

$K_{sp} = [2(6.3 \times 10^{-5})]^2 \times (6.3 \times 10^{-5}) = 1.0 \times 10^{-12}$

11.53 (a)

Concentration ($mol \cdot L^{-1}$)	$AgCl(s) \rightleftharpoons$	$Ag^+(aq)$	$+\ Cl^-(aq)$
initial	—	0	0.15
change	—	$+S$	$+S$
equilibrium	—	S	$S + 0.15$

$K_{sp} = [Ag^+][Cl^-] = (S) \times (S + 0.15) = 1.6 \times 10^{-10}$

Assume S in $(S + 0.15)$ is negligible, so $0.15\,S = 1.6 \times 10^{-10}$

$S = 1.1 \times 10^{-9}\ mol \cdot L^{-1} = [Ag^+]$ = molar solubility of AgCl in 0.15 M NaCl

(b)

Concentration ($mol \cdot L^{-1}$)	$Hg_2Cl_2(s) \rightleftharpoons$	$Hg_2^{2+}(aq)$	$+\ 2\,Cl^-(aq)$
initial	—	0	0.225
change	—	$+S$	$+2S$
equilibrium	—	S	$0.225 + 2S$

$K_{sp} = [Hg_2^{2+}][Cl^-]^2 = (S) \times (2S + 0.225)^2 = 1.3 \times 10^{-18}$

Assume $2S$ in $(2S + 0.225)$ is negligible, so $0.225S = 1.3 \times 10^{-18}$

$S = 5.8 \times 10^{-18}\ mol \cdot L^{-1} = [Hg_2^{2+}]$

= molar solubility of Hg_2Cl_2 in 0.225 M NaCl

(c)

Concentration ($mol \cdot L^{-1}$)	$PbCl_2(s) \rightleftharpoons$	$Pb^{2+}(aq)$	$+\ 2\,Cl^-(aq)$
initial	—	0	$2 \times 0.050 = 0.10$
change	—	$+S$	$+S$
equilibrium	—	S	$S + 0.10$

$K_{sp} = [Pb^{2+}][Cl^-]^2 = S \times (S + 0.10)^2 = 1.6 \times 10^{-5}$

S may not be negligible relative to 0.10, so the full cubic form may be required. We do it both ways:

For $S^3 + 0.20\,S^2 + (1 \times 10^{-2}S) - (1.6 \times 10^{-5}) = 0$, the solution by standard methods is $S = 1.6 \times 10^{-3}\ mol \cdot L^{-1}$.

If S had been neglected, the answer would have been the same, 1.6×10^{-3}, to within two significant figures.

(d)

Concentration ($mol \cdot L^{-1}$)	$Fe(OH)_2(s)$	$\rightleftharpoons$	$Fe^{2+}(aq)$	+	$2\ OH^-(aq)$
initial	—		2.5×10^{-3}		0
change	—		$+S$		$+2S$
equilibrium	—		$2.5 \times 10^{-3} + S$		$2S$

$K_{sp} = [Fe^{2+}][OH^-]^2 = (S + 2.5 \times 10^{-3}) \times (2S)^2 = 1.6 \times 10^{-14}$

Assume S in $(S + 2.5 \times 10^{-3})$ is negligible, so $4S^2 \times (2.5 \times 10^{-3}) = 1.6 \times 10^{-14}$

$S^2 = 1.6 \times 10^{-12}$

$S = 1.3 \times 10^{-6}\ mol \cdot L^{-1}$ = molar solubility of $Fe(OH)_2$ in 2.5×10^{-3} M $FeCl_2$

11.55 (a) $Ag^+(aq) + Cl^-(aq) \rightleftharpoons AgCl(s)$

Concentration ($mol \cdot L^{-1}$)	Ag^+	Cl^-
initial	0	1.0×10^{-5}
change	$+x$	0
equilibrium	x	1.0×10^{-5}

$K_{sp} = [Ag^+][Cl^-] = 1.6 \times 10^{-10} = (x)(1.0 \times 10^{-5})$

$x = [Ag^+] = 1.6 \times 10^{-5}\ mol \cdot L^{-1}$

(b) mass $AgNO_3$

$$= \left(\frac{1.6 \times 10^{-5}\ \text{mol AgNO}_3}{1\ \text{L}}\right)(0.100\ \text{L})\left(\frac{169.88\ \text{g AgNO}_3}{1\ \text{mol AgNO}_3}\right)\left(\frac{1\ \mu\text{g}}{10^{-6}\ \text{g}}\right)$$

$= 2.7 \times 10^2\ \mu g\ AgNO_3$

11.57 (a) $Ni^{2+}(aq) + 2\ OH^-(aq) \rightleftharpoons Ni(OH)_2(s)$

Concentration ($mol \cdot L^{-1}$)	Ni^{2+}	OH^-
initial	0.060	0
change	0	$+x$
equilibrium	0.060	x

$K_{sp} = [Ni^{2+}][OH^-]^2 = 6.5 \times 10^{-18} = (0.060)(x)^2$

$[OH^-] = x = 1.0 \times 10^{-8}\ mol \cdot L^{-1}$

$pOH = -\log(1.0 \times 10^{-8}) = 8.00, \quad pH = 14.00 - 8.00 = 6.00$

(b) A similar set up for $[Ni^{2+}] = 0.030$ M

gives $x = 1.5 \times 10^{-8}$

$pOH = -\log(1.5 \times 10^{-8}) = 7.82$

$pH = 14.00 - 7.82 = 6.18$

11.59 (a) $Ag^+(aq) + Cl^-(aq) \rightleftharpoons AgCl(s), \quad [Ag^+][Cl^-] = K_{sp}$

$$Q_{sp} = \left[\frac{(0.073\ L)(0.0040\ mol \cdot L^{-1})}{0.100\ L}\right]\left[\frac{(0.027\ L)(0.0010\ mol \cdot L^{-1})}{0.100\ L}\right]$$

$$= (2.9 \times 10^{-3})(2.7 \times 10^{-4}) = 7.9 \times 10^{-7}$$

Will precipitate, because Q_{sp} $(7.9 \times 10^{-7}) > K_{sp}$ (1.6×10^{-10})

(b) $Ca^{2+}(aq) + SO_4^{2-}(aq) \rightleftharpoons CaSO_4(s), \quad [Ca^{2+}][SO_4^{2-}] = K_{sp}$

$$Q_{sp} = \left[\frac{(0.0100\ L)(0.0030\ mol \cdot L^{-1})}{0.111\ L}\right]\left[\frac{(0.0010\ L)(1.0\ mol \cdot L^{-1})}{0.111\ L}\right]$$

$$= (2.7 \times 10^{-4})(9.0 \times 10^{-3}) = 2.4 \times 10^{-6}$$

Will not precipitate, because Q_{sp} $(2.4 \times 10^{-6}) < K_{sp}(2.4 \times 10^{-5})$

11.61 $\left(\frac{1\ mL}{20\ drops}\right) \times 1\ drop = 0.05\ mL = 0.05 \times 10^{-3}\ L = 5 \times 10^{-5}\ L$

and $(5 \times 10^{-5}\ L)(0.010\ mol \cdot L^{-1}) = 5 \times 10^{-7}$ mol NaCl $= 5 \times 10^{-7}$ mol Cl^-

(a) $Ag^+(aq) + Cl^-(aq) \rightleftharpoons AgCl(s), \quad [Ag^+][Cl^-] = K_{sp}$

$$Q_{sp} = \left[\frac{(0.010\ L)(0.0040\ mol \cdot L^{-1})}{0.010\ L}\right]\left[\frac{5 \times 10^{-7}\ mol}{0.010\ L}\right] = 2 \times 10^{-7}$$

Will precipitate, because Q_{sp} $(2 \times 10^{-7}) > K_{sp}$ (1.6×10^{-10})

(b) $Pb^{2+}(aq) + 2\ Cl^-(aq) \rightleftharpoons PbCl_2(s), \quad [Pb^{2+}][Cl^-]^2 = K_{sp}$

$$Q_{sp} = \left[\frac{(0.0100\ L)(0.0040\ mol \cdot L^{-1})}{0.010\ L}\right]\left[\frac{5 \times 10^{-7}\ mol}{0.010\ L}\right]^2 = 1 \times 10^{-11}$$

Will not precipitate, because Q_{sp} $(1 \times 10^{-11}) < K_{sp}$ (1.6×10^{-5})

11.63 (a) $K_{sp}[Ni(OH)_2] < K_{sp}[Mg(OH)_2] < K_{sp}[Ca(OH)_2]$

This is the order for the solubility products of these hydroxides. Thus, the order of precipitation is (first to last): $Ni(OH)_2$, $Mg(OH)_2$, $Ca(OH)_2$.

(b) $K_{sp}[Ni(OH)_2] = 6.5 \times 10^{-18} = [Ni^{2+}][OH^-]^2$

$$[OH^-]^2 = \frac{6.5 \times 10^{-18}}{0.0010} = 6.5 \times 10^{-15}$$

$[OH^-] = 8.1 \times 10^{-8}$

$pOH = -\log[OH^-] = 7.09 \quad pH \approx 7$

$K_{sp}[Mg(OH)_2] = 1.1 \times 10^{-11} = [Mg^{2+}][OH^-]^2$

$$[OH^-] = \sqrt{\frac{1.1 \times 10^{-11}}{0.0010}} = 1.0 \times 10^{-4}$$

$pOH = -\log(1.0 \times 10^{-4}) = 4.00 \quad pH = 14.00 - 4.00 = 10.00,\ pH \approx 10$

$K_{sp}[Ca(OH)^2] = 5.5 \times 10^{-6} = [Ca^{2+}][OH^-]^2$

$$[OH^-] = \sqrt{\frac{5.5 \times 10^{-6}}{0.0010}} = 7.4 \times 10^{-2}$$

$pOH = -\log(7.4 \times 10^{-2}) = 1.13 \quad pH = 14.00 - 1.13 = 12.87, pH \approx 13$

11.65 The K_{sp} values are

MgF_2	6.4×10^{-9}
BaF_2	1.7×10^{-6}
$MgCO_3$	1.0×10^{-5}
$BaCO_3$	8.1×10^{-9}

The difference in these numbers suggests that there is a greater solubility difference between the carbonates, and thus this anion should give a better separation. Because different numbers of ions are involved, it is instructive to convert the K_{sp} values into molar solubility. For the fluorides the reaction is

$MF_2(s) \rightleftharpoons M^{2+}(aq) + 2\ F^-(aq)$

Change $\quad +x \quad +2x$

$K_{sp} = x(2x)^2$

Solving this for MgF_2 gives 0.0012 M and for BaF_2 gives 0.0075 M.

For the carbonates:

$MCO_3(s) \rightleftharpoons M^{2+}(aq) + CO_3^{2-}(aq)$

$\quad +x \quad +x$

$K_{sp} = x^2$

Solving this for $MgCO_3$ gives 0.0032 M and for $BaCO_3$ gives 9.0×10^{-5} M. Clearly, the solubility difference is greatest between the two carbonates, and CO_3^{2-} is the better choice of anion.

11.67 $Cu(IO_3)_2$ ($K_{sp} = 1.4 \times 10^{-7}$) is more soluble than $Pb(IO_3)_2$ ($K_{sp} = 2.6 \times 10^{-13}$) so Cu^{2+} will remain in solution until essentially all the $Pb(IO_3)_2$ has precipitated. Thus, we expect very little Pb^{2+} to be left in solution by the time we reach the point at which $Cu(IO_3)_2$ begins to precipitate.

The concentration of IO_3^- at which Cu^{2+} begins to precipitate will be given by

$K_{sp} = [Cu^{2+}][IO_3^-]^2 = 1.4 \times 10^{-7} = [0.0010][IO_3^-]^2$

$[IO_3^-] = 0.012\ mol \cdot L^{-1}$

The concentration of Pb in solution when the $[IO_3^-] = 0.012\ mol \cdot L^{-1}$ is given by

$K_{sp} = [Pb^{2+}][IO_3^-]^2 = 2.6 \times 10^{-13} = [Pb^{2+}][0.012]^2$

$[Pb^{2+}] = 1.8 \times 10^{-9}\ mol \cdot L^{-1}$

11.69 (a) $pH = 7.0$; $[OH^-] = 1.0 \times 10^{-7}\ mol \cdot L^{-1}$

$Al^{3+}(aq) + 3\ OH^-(aq) \rightleftharpoons Al(OH)_3(s)$

$[Al^{3+}][OH^-]^3 = K_{sp} = 1.0 \times 10^{-33}$
$S \times (10^{-7})^3 = 1.0 \times 10^{-33}$
$S = \dfrac{1.0 \times 10^{-33}}{1 \times 10^{-21}} = 1.0 \times 10^{-12}\ \text{mol}\cdot\text{L}^{-1} = [Al^{3+}]$
= molar solubility of $Al(OH)_3$ at pH = 7.0
(b) pH = 4.5; pOH = 9.5; $[OH^-] = 3.2 \times 10^{-10}\ \text{mol}\cdot\text{L}^{-1}$
$[Al^{3+}][OH^-]^3 = K_{sp} = 1.0 \times 10^{-33}$
$S \times (3.2 \times 10^{-10})^3 = 1.0 \times 10^{-33}$
$S = \dfrac{1.0 \times 10^{-33}}{3.3 \times 10^{-29}} = 3.1 \times 10^{-5}\ \text{mol}\cdot\text{L}^{-1} = [Al^{3+}]$
= molar solubility of $Al(OH)_3$ at pH = 4.5
(c) pH = 7.0; $[OH^-] = 1.0 \times 10^{-7}\ \text{mol}\cdot\text{L}^{-1}$
$Zn^{2+}(aq) + 2\ OH^-(aq) \rightleftharpoons Zn(OH)_2(s)$
$[Zn^{2+}][OH^-]^2 = K_{sp} = 2.0 \times 10^{-17}$
$S \times (1.0 \times 10^{-7})^2 = 2.0 \times 10^{-17}$
$S = \dfrac{2.0 \times 10^{-17}}{1.0 \times 10^{-14}} = 2.0 \times 10^{-3}\ \text{mol}\cdot\text{L}^{-1} = [Zn^{2+}]$
= molar solubility of $Zn(OH)_2$ at pH = 7.0
(d) pH = 6.0; pOH = 8.0; $[OH^-] = 1.0 \times 10^{-8}\ \text{mol}\cdot\text{L}^{-1}$
$[Zn^{2+}][OH^-]^2 = 2.0 \times 10^{-17} = K_{sp}$
$S \times (1.0 \times 10^{-8})^2 = 2.0 \times 10^{-17}$
$S = \dfrac{2.0 \times 10^{-17}}{1.0 \times 10^{-16}} = 2.0 \times 10^{-1} = 0.20\ \text{mol}\cdot\text{L}^{-1} = [Zn^{2+}]$
= molar solubility of $Zn(OH)_2$ at pH = 6.0

11.71 $CaF_2(s) \rightleftharpoons Ca^{2+}(aq) + 2\ F^-(aq)$ $\quad K_{sp} = 4.0 \times 10^{-11}$
$F^-(aq) + H_2O(l) \rightleftharpoons HF(aq) + OH^-(aq)$ $\quad K_b(F^-) = 2.9 \times 10^{-11}$
(a) Multiply the second equilibrium equation by 2 and add to the first equilibrium:
$CaF_2(s) + 2\ H_2O(l) \rightleftharpoons Ca^{2+}(aq) + 2\ HF(aq) + 2\ OH^-(aq)$
$K = K_w \cdot K_b^{\,2} = (4.0 \times 10^{-11})(2.9 \times 10^{-11})^2 = 3.4 \times 10^{-32}$
(b) The calculation of K_{sp} is complicated by the fact that the anion of the salt is part of a weak base-acid pair. If we wish to solve the equation algebraically, then we need to consider which equilibrium is the dominant one at pH = 7.0, for which $[H_3O^+] = 1 \times 10^{-7}$.
To determine whether F^- or HF is the dominant species at this pH (if either), consider the base hydrolysis reaction:
$F^-(aq) + H_2O(l) \rightleftharpoons HF(aq) + OH^-(aq)$ $\quad K_b = 2.9 \times 10^{-11}$
$K_b = \dfrac{[HF][OH^-]}{[F^-]}$

$$2.9 \times 10^{-11} = \frac{[HF][1 \times 10^{-7}]}{[F^-]}$$

$$\frac{[HF]}{[F^-]} = \frac{2.9 \times 10^{-11}}{1 \times 10^{-7}} = 3 \times 10^{-4}$$

Given that the ratio of HF to F^- is on the order of 10^{-4} to 1, the F^- species is still dominant. The appropriate equation to use is thus the original one for the K_{sp} of $CaF_2(s)$:

$$CaF_2(s) \rightleftharpoons Ca^{2+}(aq) + 2\ F^-(aq) \qquad K_{sp} = 4.0 \times 10^{-11}$$

$$K_{sp} = [Ca^{2+}][F^-]^2$$

$$4.0 \times 10^{-11} = x(2x)^2 = 4x^3$$

$$x = 2.2 \times 10^{-4}$$

molar solubility $= 2.2 \times 10^{-4}\ mol \cdot L^{-1}$

(c) At pH = 3.0, $[H_3O^+] = 1 \times 10^{-3}\ mol \cdot L^{-1}$ and $[OH^-] = 1 \times 10^{-11}\ mol \cdot L^{-1}$. Under these conditions

$$K_b = \frac{[HF][OH^-]}{[F^-]}$$

$$2.9 \times 10^{-11} = \frac{[HF][1 \times 10^{-11}]}{[F^-]}$$

$$\frac{[HF]}{[F^-]} = \frac{2.9 \times 10^{-11}}{1 \times 10^{-11}} = 3$$

$$[HF] = 3\ [F^-]$$

As can be seen, at pH 3.0 the amounts of F^- and HF are comparable, so the protonation of F^- to form HF cannot be ignored. The relation $2\ [Ca^{2+}] = [F^-] + [HF]$ is required by the mass balance as imposed by the stoichiometry of the dissolution equilibrium.

$$2\ [Ca^{2+}] = [F^-] + 3\ [F^-]$$

$$2\ [Ca^{2+}] = 4\ [F^-]$$

$$[Ca^{2+}] = 2\ [F^-]$$

Using this with K_{sp} relationship:

$$(2\ [F^-])[F^-]^2 = 4.0 \times 10^{-11}$$

$$2\ [F^-]^3 = 4.0 \times 10^{-11}$$

$$[F^-] = 2 \times 10^{-4}$$

$$[Ca^{2+}] = (2)(2 \times 10^{-4}) = 4 \times 10^{-4}\ mol \cdot L^{-1}$$

The solubility is about double that at pH = 7.0.

11.73

$AgBr(s) \rightleftharpoons Ag^+(aq) + Br^-(aq)$	$K_{sp} = 7.7 \times 10^{-13}$
$Ag^+(aq) + 2\ CN^-(aq) \rightleftharpoons Ag(CN)_2^-(aq)$	$K_f = 5.6 \times 10^8$
$AgBr(s) + 2\ CN^-(aq) \rightleftharpoons Ag(CN)_2^-(aq) + Br^-(aq)$	$K = 4.3 \times 10^{-4}$

Hence, $K = \frac{[Ag(CN)_2^-][Br^-]}{[CN^-]^2} = 4.3 \times 10^{-4}$

Concentration (mol·L^{-1})	AgBr(s)	+ 2 CN$^-$(aq) ⇌	Ag(CN)$_2^-$(aq) +	Br$^-$(aq)
initial	—	0.10	0	0
change	—	$-2S$	$+S$	$+S$
equilibrium	—	$0.10 - 2S$	S	S

$$\frac{[Ag(CN)_2^-][Br^-]}{[CN^-]^2} = \frac{S^2}{(0.10 - 2S)^2} = 4.3 \times 10^{-4}$$

$$\frac{S}{0.10 - 2S} = \sqrt{4.3 \times 10^{-4}} = 2.1 \times 10^{-2}$$

$S = (2.1 \times 10^{-3}) - (4.2 \times 10^{-2}S)$

$1.042S = 2.1 \times 10^{-3}$

$S = 2.0 \times 10^{-3}$ mol·L^{-1} = molar solubility of AgBr

11.75 The two salts can be distinguished by their solubility in NH_3. The equilibria that are pertinent are

$AgCl(s) + 2\ NH_3(aq) \rightleftharpoons Ag(NH_3)_2^+(aq) + Cl^-(aq) \quad K = K_{sp} \cdot K_f = 2.6 \times 10^{-3}$

$AgI(s) + 2\ NH_3(aq) \rightleftharpoons Ag(NH_3)_2^+(aq) + I^-(aq) \quad K = K_{sp} \cdot K_f = 2.4 \times 10^{-9}$

For example, let's consider the solubility of these two salts in 1.00 M NH_3 solution:

For AgCl $K = \frac{[Ag(NH_3)_2^+][Cl^-]}{[NH_3]^2} = 2.6 \times 10^{-3}$

Concentration (mol·L^{-1})	AgCl(s)	+ 2 HN$_3$(aq) ⇌	Ag(NH$_3$)$_2^+$(aq) +	Cl$^-$(aq)
initial	—	1.00	0	0
change	—	$-2\,x$	$+x$	$+x$
final	—	$1.00 - 2x$	$+x$	$+x$

$$K = \frac{[Ag(NH_3)_2^+][Cl^-]}{[NH_3]^2} = 2.6 \times 10^{-3}$$

$$2.6 \times 10^{-3} = \frac{[x][x]}{[1.00 - 2x]^2} = \frac{x^2}{[1.00 - 2x]^2}$$

$$0.051 = \frac{x}{1.00 - 2x}$$

$x = 0.046$

0.046 mol AgCl will dissolve in 1.00 L of aqueous solution. The molar mass of AgCl is 143.32 g·mol^{-1}; this corresponds to 0.046 mol·L^{-1} × 142.32 g·mol^{-1} = 6.5 g·L^{-1}.

For AgI, the same calculation gives $x = 4.9 \times 10^{-5}$ mol·L^{-1}. The molar mass of AgI is 234.77 g·mol^{-1}, giving a solubility of 4.9×10^{-5} mol·L^{-1} $\times$ 234.77 g·mol^{-1} = 0.023 g·L^{-1}.

Thus, we could treat a 0.10 g sample of the compound with 20.0 mL of 1.00 M NH_3. The AgCl would all dissolve, whereas practically none of the AgI would. Note: AgI is also slightly yellow in color, whereas AgCl is white, so an initial distinction could be made based upon the color of the sample.

11.77 In order to use qualitative analyses, the sample must first be dissolved. This can be accomplished by digesting the sample with concentrated HNO_3 and then diluting the resulting solution. HCl or H_2SO_4 could not be used, because some of the metal compounds formed would be insoluble, whereas all of the nitrates would dissolve. Once the sample is dissolved and diluted, an aqueous solution containing chloride ions can be introduced. This should precipitate the Ag^+ as AgCl but would leave the bismuth and nickel in solution, as long as the solution was acidic. The remaining solution can then be treated with H_2S. In acidic solution, Bi_2S_3 will precipitate but NiS will not. Once the Bi_2S_3 has been precipitated, the pH of the solution can be raised by addition of base. Once this is done, NiS should precipitate.

11.79 (a)–(c) The K_b value for hydroxylamine as found in Table 10.2 is 1.1×10^{-8}. The corresponding K_a value for the hydroxylammonium ion will be $1.0 \times 10^{-14} \div 1.1 \times 10^{-8} = 9.1 \times 10^{-7}$. A table is set up to do the concentration calculation:

Concentration (mol·L^{-1})	$^{+}NH_3OH(aq)$	+ $H_2O(l)$ $\rightleftharpoons$	$NH_2OH(aq)$	+ $H_3O^+(aq)$ $K_a = 9.1 \times 10^{-7}$
initial	0.0240	—	0	0
change	$-x$	—	$+x$	$+x$
equilibrium	$0.0240 - x$	—	$+x$	$+x$

$$\frac{x^2}{0.0240 - x} = 9.1 \times 10^{-7}$$

Assuming that $x \ll 0.0240$, we calculate that $x = 1.5 \times 10^{-4}$

Similar calculations for starting concentrations of 0.0480 M and 0.0960 M yield $x = 2.1 \times 10^{-4}$ and 3.0×10^{-4}, respectively.

For the three starting concentrations, the percent deprotonation will be x divided by the starting concentration times 100.

Starting concentration	Deprotonation, %
0.0240 M	0.62
0.0480 M	0.44
0.0960 M	0.31

(d) So we can see that, even though there is more of the deprotonated form in solutions at higher concentrations, the percent deprotonation decreases with increased concentration.

(e) The ideal buffer will have equal concentrations of the acid and its conjugate base. From the relation $\text{pH} = \text{p}K_a + \log\frac{[\text{Base form}]}{[\text{Acid form}]}$, we can see that pH will be equal to $\text{p}K_a = -\log(9.1 \times 10^{-7}) = 6.0$.

11.81 The relation to use is $\text{pH} = \text{p}K_a + \log\frac{[\text{Base form}]}{[\text{Acid form}]}$. The K_a value for acetic acid is 1.8×10^{-5} and $\text{p}K_a = 4.74$. Because we are adding acid, the pH will fall upon the addition and we want the final pH to be no more than 0.20 pH units different from the initial pH, or 4.54.

$$4.54 = 4.74 + \log\frac{[\text{Base form}]}{[\text{Acid form}]}$$

$$-0.20 = \log\frac{[\text{Base form}]}{[\text{Acid form}]}$$

$$\frac{[\text{Base form}]}{[\text{Acid form}]} = 0.63$$

We want the concentration of the base form to be 0.63 times that of the acid form. We do not know the initial number of moles of base or acid forms, but we know that the two amounts were equal. Let C = initial number of moles of acetic acid and the initial number of moles of sodium acetate. The number of moles of H_3O^+ to be added (in the form of HCl(aq)) is $0.001\,00\ \text{L} \times 6.00\ \text{mol}\cdot\text{L}^{-1} = 0.006\,00\ \text{mol}$. The total final volume will be 0.1010 L.

$$\frac{\dfrac{\text{C} - 0.006\,00}{0.1010\ \text{L}}}{\dfrac{\text{C} + 0.006\,00}{0.1010\ \text{L}}} = 0.63$$

$$\frac{\text{C} - 0.006\,00}{\text{C} + 0.006\,00} = 0.63$$

$$\text{C} - 0.006\,00 = 0.63(\text{C} + 0.006\,00)$$

$$\text{C} - 0.006\,00 = 0.63\,\text{C} + 0.003\,78$$

0.37 C = 0.009 78

C = 0.026

The initial buffer solution must contain at least 0.026 mol acetic acid and 0.026 mol sodium acetate. The concentration of the initial solution will then be 0.026 mol ÷ 0.100 L = 0.260 M in both acetic acid and sodium acetate.

11.83 (a)

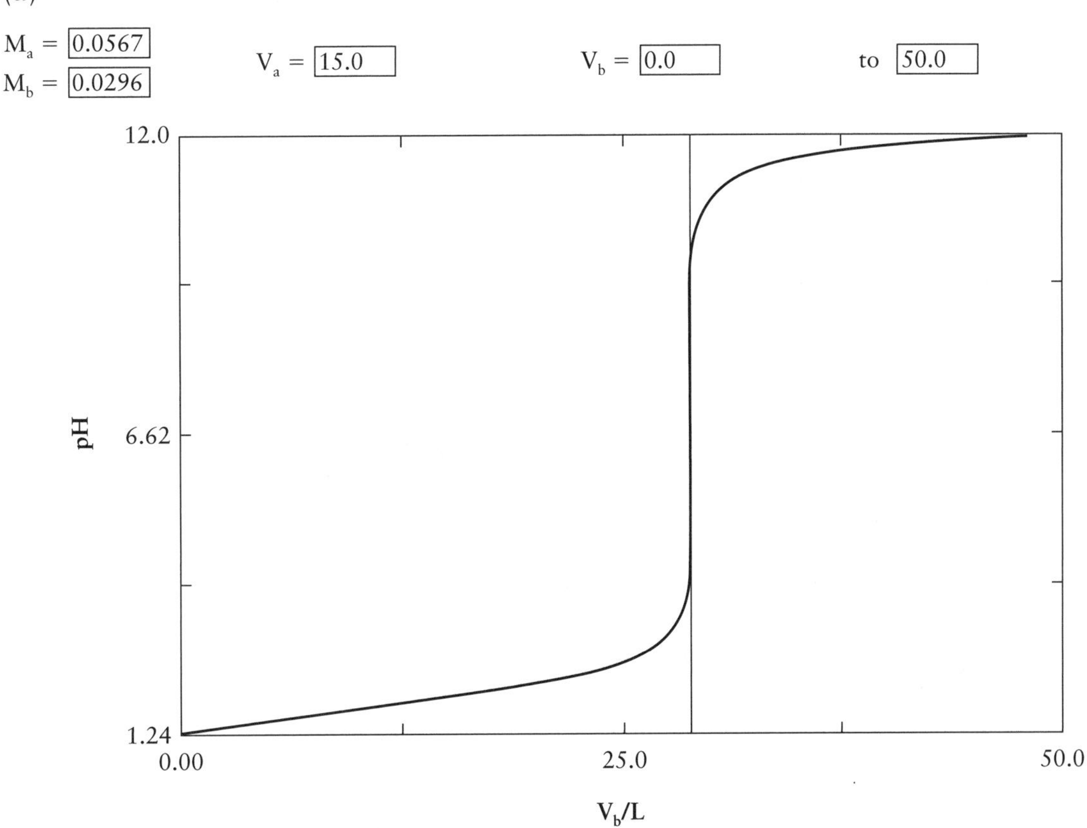

(b) 28.6 mL

(c) 1.24

(d) Because this is a titration of a strong acid with a strong base, the pH at the equivalence point will be 7.00.

11.85 The strong acid, HCl, will protonate the HCO_2^- ion.

moles of HCl = 0.0040 L × 0.070 mol·L^{-1} = 2.8×10^{-4} mol HCl (H^+)

moles of HCO_2^- = 0.0600 L × 0.10 mol·L^{-1} = 6.0×10^{-3} mol HCO_2^-

After protonation, there are $(6.0 - 0.28) \times 10^{-3}$ mol HCO_2^- = 5.7×10^{-3} mol HCO_2^- and 2.8×10^{-4} mol HCOOH.

$$[HCO_2^-] = \frac{5.7 \times 10^{-3}\text{ mol}}{0.0640\text{ L}} = 8.9 \times 10^{-2}\text{ mol·L}^{-1}$$

$$[\text{HCOOH}] = \frac{2.8 \times 10^{-4}\ \text{mol}}{0.0640\ \text{L}} = 4.4 \times 10^{-3}\ \text{mol}\cdot\text{L}^{-1}$$

Concentration ($\text{mol}\cdot\text{L}^{-1}$)	$HCOOH(aq)$	+ $H_2O(l)$ ⇌	$H_3O^+(aq)$	+ $HCO_2^-(aq)$
initial	4.4×10^{-3}	—	0	8.9×10^{-2}
change	$-x$	—	$+x$	$+x$
equilibrium	$4.4 \times 10^{-3} - x$	—	x	$8.9 \times 10^{-2} + x$

$$K_a = 1.8 \times 10^{-4} = \frac{[H_3O^+][HCO_2^-]}{[HCOOH]} = \frac{(x)(8.9 \times 10^{-2} + x)}{(4.4 \times 10^{-3} - x)}$$

Assume that $+x$ and $-x$ are negligible; so $1.8 \times 10^{-4} = \frac{(8.9 \times 10^{-2}) \times (x)}{4.4 \times 10^{-3}}$

$$= 20.2x$$

$[H_3O^+] = x = 8.9 \times 10^{-6}\ \text{mol}\cdot\text{L}^{-1}$

$[HCOOH] = (4.4 \times 10^{-3}) - (8.9 \times 10^{-6}) \approx 4.4 \times 10^{-3}\ \text{mol}\cdot\text{L}^{-1}$

$\text{pH} = -\log(8.9 \times 10^{-6}) = 5.05$

11.87 Let novocaine = N; $N(aq) + H_2O(l) \rightleftharpoons HN^+(aq) + OH^-(aq)$

$$K_b = \frac{[HN^+][OH^-]}{[N]}$$

$\text{p}K_a = \text{p}K_w - \text{p}K_b = 14.00 - 5.05 = 8.95$

$$\text{pH} = \text{p}K_a + \log\left(\frac{[N]}{[HN^+]}\right)$$

$$\log\left(\frac{[N]}{[HN^+]}\right) = \text{pH} - \text{p}K_a = 7.4 - 8.95 = -1.55$$

Therefore, the ratio of the concentrations of novocaine and its conjugate acid is $[N]/[HN^+] = 10^{-1.55} = 2.8 \times 10^{-2}$.

11.89 $CH_3COOH(aq) + H_2O(l) \rightleftharpoons H_3O^+(aq) + CH_3CO_2^-(aq)$

$$K_a = \frac{[H_3O^+][CH_3CO_2^-]}{[CH_3COOH]}$$

$$\text{pH} = \text{p}K_a + \log\frac{[CH_3CO_2^-]}{[CH_3COOH]} = \text{p}K_a + \log\left(\frac{0.50}{0.150}\right)$$

$= 4.75 + 0.52 = 5.27$ (initial pH)

(a) $(0.0100\ \text{L})(1.2\ \text{mol}\cdot\text{L}^{-1}) = 1.2 \times 10^{-2}$ mol HCl added (a strong acid)

Produces 1.2×10^{-2} mol CH_3COOH from $CH_3CO_2^-$ after adding HCl:

$$[CH_3COOH] = \frac{(0.100\ \text{L})(0.150\ \text{mol}\cdot\text{L}^{-1})}{0.110\ \text{L}} + \frac{1.2 \times 10^{-2}\ \text{mol}}{0.110\ \text{L}}$$

$$= 0.245\ \text{mol}\cdot\text{L}^{-1}$$

$$[CH_3CO_2^-] = \frac{(0.100\ L)(0.50\ mol\cdot L^{-1})}{0.110\ L} - \frac{1.2 \times 10^{-2}\ mol}{0.110\ L}$$
$$= 0.345\ mol\cdot L^{-1}$$

$$pH = 4.75 + \log\left(\frac{0.345}{0.245}\right) = 4.75 + 0.15 = 4.90 \text{ (after adding HCl)}$$

(b) $(0.0500\ L)(0.095\ mol\cdot L^{-1}) = 4.7 \times 10^{-3}$ mol NaOH added (a strong base)
Produces 4.7×10^{-3} mol $CH_3CO_2^-$ from CH_3COOH after adding NaOH:

$$[CH_3COOH] = \frac{(0.100\ L)(0.150\ mol\cdot L^{-1})}{0.1500\ L} - \frac{4.7 \times 10^{-3}\ mol}{0.150\ L}$$
$$= 6.9 \times 10^{-2}\ mol\cdot L^{-1}$$

$$[CH_3CO_2^-] = \frac{(0.100\ L)(0.50\ mol\cdot L^{-1})}{0.150\ L} + \frac{4.7 \times 10^{-3}\ mol}{0.150\ L}$$
$$= 3.6 \times 10^{-1}\ mol\cdot L^{-1}$$

$$pH = 4.75 + \log\left(\frac{3.6 \times 10^{-1}}{6.9 \times 10^{-2}}\right) = 4.75 + 0.72 = 5.47 \quad \text{(after adding base)}$$

11.91 $CaF_2(s) \rightleftharpoons Ca^{2+}(aq) + 2\ F^-(aq)$
$[Ca^{2+}][F^-]^2 = 4.0 \times 10^{-11}$
$[Ca^{2+}](5 \times 10^{-5})^2 = 4.0 \times 10^{-11}$ $\quad [Ca^{2+}] = 0.016\ mol\cdot L^{-1}$
The maximum concentration of $[Ca^{2+}]$ allowed will be $0.016\ mol\cdot L^{-1}$.

11.93 moles of $Ag^+ = (0.100\ L)(1.0 \times 10^{-4}\ mol\cdot L^{-1}) = 1.0 \times 10^{-5}$ mol Ag^+
moles of $CO_3^{2-} = (0.100\ L)(1.0 \times 10^{-4}\ mol\cdot L^{-1}) = 1.0 \times 10^{-5}$ mol CO_3^{2-}
$Ag_2CO_3(s) \rightleftharpoons 2\ Ag^+(aq) + CO_3^{2-}(aq)$
$[Ag^+]^2[CO_3^{2-}] = Q_{sp}$

$$\left(\frac{1.0 \times 10^{-5}}{0.200}\right)^2\left(\frac{1.0 \times 10^{-5}}{0.200}\right) = 1.3 \times 10^{-13} = Q_{sp}$$

Because Q_{sp} (1.3×10^{-13}) is less than K_{sp} (6.2×10^{-12}), no precipitate will form.

11.95 The K_{sp} values from Tables 11.5 and 11.7 are Cu^{2+}, 1.3×10^{-36}; Co^{2+}, 5×10^{-22}; Cd^{2+}, 4×10^{-29}. All the salts have the same expression for K_{sp}, $K_{sp} = [M^{2+}][S^{2-}]$, so the compound with the smallest K_{sp} will precipitate first, in this case CuS.

11.97 (a) First, calculate the stoichiometry of the reaction of the strong base with the weak acid, according to the equation
$H_2SO_3(aq) + OH^-(aq) \longrightarrow HSO_3^-(aq) + H_2O(l)$

mol OH^- = (0.0250 L)(0.150 mol·L^{-1}) = 0.003 75 mol

mol H_2SO_3 = (0.0250 L)(0.125 mol·L^{-1}) = 0.003 12 mol

Because there is more OH^- than H_2SO_3, all of the H_2SO_3 will react to form HSO_3^-, with excess OH^- leftover.

0.003 12 mol of H_2SO_3 will form 0.003 12 mol HSO_3^-.

mol OH^- remaining = 0.003 75 mol − 0.003 12 mol = 0.000 63 mol remaining

This excess OH^- will then react with HSO_3^- to form SO_3^{2-}:

$HSO_3^-(aq) + OH^-(aq) \longrightarrow SO_3^{2-}(aq) + H_2O(l)$

We then allow this reaction to go to completion. All the OH^- will be consumed. 0.003 12 mol HSO_3^- will react with 0.000 63 mol OH^- to produce 0.000 63 mol of SO_3^{2-} and leave 0.003 12 − 0.000 63 mol = 0.002 49 mol HSO_3^-. The concentrations of these species will be

$$[HSO_3^-] = \frac{0.002\ 49\ \text{mol}}{0.0500\ \text{L}} = 0.0498\ \text{mol}\cdot\text{L}^{-1}$$

$$[SO_3^{2-}] = \frac{0.000\ 63\ \text{mol}}{0.0500\ \text{L}} = 0.013\ \text{mol}\cdot\text{L}^{-1}$$

This places us in the second buffer region of H_2SO_3, so the following equation will dominate the equilibrium (for which $K_{a2} = 1.2 \times 10^{-7}$):

Concentration (mol·L^{-1})	$HSO_3^-(aq)$	+ $H_2O(l)$ $\rightleftharpoons$	$SO_3^{2-}(aq)$	+ $H_3O^+(aq)$
initial	0.0498	—	0.013	0
change	$-x$	—	$+x$	$+x$
final	$0.0498 - x$	—	$0.013 + x$	$+x$

$$K_a = \frac{[SO_3^{2-}][OH^-]}{[HSO_3^-]} = \frac{(0.013 + x)(x)}{(0.0498 - x)} = 1.2 \times 10^{-7}$$

If $x \ll 0.013$, this simplifies to

$$\frac{0.013x}{0.0498} = 1.2 \times 10^{-7}$$

$x = 4.6 \times 10^{-7}$

$[H_3O^+] = 4.6 \times 10^{-7}$, pH = 6.34

This number is in the region in which the autoprotolysis of water may influence the value, but a more rigorous calculation taking this into account gives the same value.

(b) The problem is worked as in (a) but with additional OH^-, more SO_3^{2-} will be formed:

total mol OH^- = (0.0450 L)(0.150 mol·L^{-1}) = 0.006 75 mol

mol H_2SO_3 = (0.0250 L)(0.125 mol·L^{-1}) = 0.003 12 mol

Again, all the H_2SO_3 will react to form HSO_3^-, with excess OH^- leftover.

0.003 12 mol H_2SO_3 will form 0.003 12 mol HSO_3^-

mol OH^- remaining = 0.006 75 mol − 0.003 12 mol = 0.003 63 mol remaining

This excess OH^- will then react with HSO_3^- to form SO_3^{2-}:

$HSO_3^-(aq) + OH^-(aq) \longrightarrow SO_3^{2-}(aq) + H_2O(l)$

We then allow this reaction to go to completion. All the HSO_3^- will be consumed. 0.003 12 mol HSO_3^- will react to form 0.003 12 mol SO_3^{2-} and 0.003 63 − 0.003 12 mol = 0.000 51 OH^- will remain. The concentrations of these species will be

$$[SO_3^{2-}] = \frac{0.003\ 12\ \text{mol}}{0.0700\ \text{L}} = 0.0446\ \text{mol}\cdot\text{L}^{-1}$$

$$[OH^-] = \frac{0.000\ 51\ \text{mol}}{0.0700\ \text{L}} = 0.0073\ \text{mol}\cdot\text{L}^{-1}$$

This places us in first base hydrolysis region of SO_3^{2-}:

Concentration (mol·L^{-1})	$SO_3^{2-}(aq)$	+ $H_2O(l)$ $\rightleftharpoons$	$HSO_3^-(aq)$	+ $OH^-(aq)$
initial	0.0446	—	0	0.0073
change	$-x$	—	$+x$	$+x$
final	$0.0446 - x$	—	$+x$	$0.0073 + x$

$$K_{b1} = \frac{K_w}{K_{a2}} = \frac{1.00 \times 10^{-14}}{1.2 \times 10^{-7}} = 8.3 \times 10^{-8}$$

$$K_a = \frac{[HSO_3^-][OH^-]}{[SO_3^{2-}]} = \frac{(x)(0.0073 + x)}{(0.0446 - x)} = 8.3 \times 10^{-8}$$

If $x \ll 0.0073$, this simplifies to

$$\frac{0.0073x}{0.0446} = 8.3 \times 10^{-8}$$

$x = 5.1 \times 10^{-7}$

$[OH^-] = 0.0073 + 5.1 \times 10^{-7} = 0.0073$, pOH = 2.14

pH = 14.00 − 2.14 = 11.86

x is small compared to 0.0073 and so the assumption was valid. What this means here is that the pH is controlled by the amount of excess base added to the solution.

11.99 (1) first stoichiometric point: $OH^-(aq) + H_2SO_4(aq) \longrightarrow H_2O(l) + HSO_4^-(aq)$

Then, because the volume of the solution has doubled, $[HSO_4^-]$ = 0.10 M,

Concentration ($mol \cdot L^{-1}$)	$HSO_4^-(aq)$ +	$H_2O(l)$ $\rightleftharpoons$	$SO_4^{2-}(aq)$ +	$H_3O^+(aq)$
initial	0.10	—	0	0
change	$-x$	—	$+x$	$+x$
equilibrium	$0.10 - x$	—	x	x

$$K_{a2} = 0.012 = \frac{x^2}{0.10 - x}$$

$$0.0012 - 0.012x = x^2$$

$$x^2 + 0.012x - 0.0012 = 0$$

$$x = \frac{-0.012 + \sqrt{(0.012)^2 + (4)(0.0012)}}{2}$$

$$x = [H_3O^+] = 0.029\ mol \cdot L^{-1}$$

$$pH = -\log(0.029) = 1.54$$

(2) second stoichiometric point:

$HSO_4^-(aq) + OH^-(aq) \longrightarrow SO_4^{2-}(aq) + H_2O(l)$. Because the volume of the solution has increased by an equal amount,

Concentration ($mol \cdot L^{-1}$)	$SO_4^{2-}(aq)$ +	$H_2O(l)$ $\rightleftharpoons$	$HSO_4^-(aq)$ +	$OH^-(aq)$
initial	0.067	—	0	1×10^{-7}
change	$-x$	—	$+x$	$+x$
equilibrium	$0.067 - x$	—	x	$1 \times 10^{-7} + x$

$$K_b = 8.3 \times 10^{-3} = \frac{(x)(1 \times 10^{-7} + x)}{0.067 - x}$$

$$(5.6 \times 10^{-14}) - (8.3 \times 10^{-13}\,x) = 1 \times 10^{-7}\,x + x^2$$

$$x^2 + (1 \times 10^{-7}x) - (5.6 \times 10^{-14}) = 0$$

$$x = \frac{-1 \times 10^{-7} + \sqrt{(1 \times 10^{-7})^2 + (4)(5.6 \times 10^{-14})}}{2}$$

$$x = [HSO_4^-] = 1.9 \times 10^{-7}\ mol \cdot L^{-1}$$

$$[OH^-] = (1.0 \times 10^{-7}) + (1.9 \times 10^{-7}) = 2.9 \times 10^{-7}\ mol \cdot L^{-1}$$

$$pOH = -\log(2.9 \times 10^{-7}) = 6.54, \quad pH = 14.00 - 6.54 = 7.46$$

11.101 (a) $PbCl_2(s) \rightleftharpoons Pb^{2+}(aq) + 2\ Cl^-(aq)$

$$[Pb^{2+}][Cl^-]^2 = K_{sp} = 1.6 \times 10^{-5}$$

$$[0.010][Cl^-]^2 = 1.6 \times 10^{-5}$$

$$[Cl^-]^2 = 1.6 \times 10^{-3}$$

$[Cl^-] = 4.0 \times 10^{-2}\ mol \cdot L^{-1}$; will precipitate lead (II) ion

$$Ag^+(aq) + Cl^-(aq) \rightleftharpoons AgCl(s)$$

$$[Ag^+][Cl^-] = K_{sp} = 1.6 \times 10^{-10}$$

$[0.010][Cl^-] = 1.6 \times 10^{-10}$

$[Cl^-] = 1.6 \times 10^{-8}\ mol \cdot L^{-1}$; will precipitate Ag^+

(b) From part (a), Ag^+ will precipitate first.

(c) $[Ag^+](4.0 \times 10^{-2}) = 1.6 \times 10^{-10}$

$$[Ag^+] = \frac{1.6 \times 10^{-10}}{4.0 \times 10^{-2}} = 4.0 \times 10^{-9}\ mol \cdot L^{-1}$$

(d) percentage Ag^+ remaining $= \dfrac{4.0 \times 10^{-9}}{0.010} \times 10^2 = 4.0 \times 10^{-5}$ % unprecipitated; virtually 100% of the first cation (Ag^+) is precipitated

11.103 The K_{sp} value for PbF_2 obtained from Table 11.5 is 3.7×10^{-8}. Using this value, the $\Delta G°$ of the dissolution reaction can be obtained from

$\Delta G° = -RT \ln K$.

$\Delta G° = -(8.314\ J \cdot K^{-1} \cdot mol^{-1})(298.2\ K)\ln(3.7 \times 10^{-8})$

$\Delta G° = +42.43\ kJ \cdot mol^{-1}$

From the Appendices we find that $\Delta G°_f(F^-, aq) = -278.79\ kJ \cdot mol^{-1}$ and $\Delta G°_f(Pb^{2+}, aq) = -24.43\ kJ \cdot mol^{-1}$.

$\Delta G° = +42.43\ kJ \cdot mol^{-1} = \Delta G°_f(Pb^{2+}, aq) + \Delta G°_f(F^-, aq) - \Delta G°_f(PbF_2, s)$

$+42.43\ kJ \cdot mol^{-1} = (-24.43\ kJ \cdot mol^{-1}) + (-278.79\ kJ \cdot mol^{-1}) - \Delta G°_f(PbF_2, s)$

$\Delta G°_f(PbF_2, s) = -345.65\ kJ \cdot mol^{-1}$

11.105 (a) The K_{sp} values are:

Cu_2S 2.0×10^{-47} $\qquad$ CuS 1.3×10^{-36}

Although the K_{sp} values seem to indicate that the Cu_2S is less soluble, we really need to calculate the molar solubilities because there are differing numbers of ions involved.

For Cu_2S: $\quad Cu_2S(s) \rightleftharpoons 2\ Cu^+(aq) + S^{2-}(aq)$

conc. change $\qquad +2x \qquad +x$

$K_{sp} = [Cu^+]^2[S^{2-}] = (2x)^2(x) = 4x^3 = 2.0 \times 10^{-47}$

$x = 1.7 \times 10^{-16}$

molar solubility of $Cu_2S = 1.7 \times 10^{-16}\ mol \cdot L^{-1}$

For CuS: $\quad CuS(s) \rightleftharpoons Cu^{2+}(aq) + S^{2-}(aq)$

$K_{sp} = [Cu^{2+}][S^{2-}] = (x)(x) = x^2 = 1.3 \times 10^{-36}$

$x = 1.1 \times 10^{-18}$

molar solubility of $CuS = 1.1 \times 10^{-18}\ mol \cdot L^{-1}$

The molar solubility is greater for Cu_2S than for CuS.

(b) One might be tempted to take the ratios of the concentrations found in (a) as the solution to this question ($3.4 \times 10^{-16} \div 1.1 \times 10^{-18} = 3.1 \times 10^{2}$), however, the common ion effect will suppress the dissolution of CuS. Because Cu_2S is more soluble, we will assume that equilibrium will establish the concentration of S^{2-} in solution. From (a) we find that $[S^{2-}] = 1.7 \times 10^{-16}\ mol \cdot L^{-1}$ and $[Cu^{+}] = 2 \times 1.7 \times 10^{-16}\ mol \cdot L^{-1} = 3.4 \times 10^{-16}\ mol \cdot L^{-1}$. If we use the $[S^{2-}]$ from the first equilibrium to calculate the Cu^{2+} concentration, we obtain

$$1.3 \times 10^{-36} = [Cu^{2+}][1.7 \times 10^{-16}]$$

$$[Cu^{2+}] = 7.6 \times 10^{-21}\ mol \cdot L^{-1}$$

From this value we can see that the concentration of S^{2-} will not be significantly changed by the dissolution of CuS: $1.7 \times 10^{-16} + 7.6 \times 10^{-18} \cong 1.7 \times 10^{-16}$

The ratio of Cu^{+} to Cu^{2+} will be $\dfrac{3.4 \times 10^{-16}}{7.6 \times 10^{-21}} = 4.5 \times 10^{4}$.

CHAPTER 12
ELECTROCHEMISTRY

12.1 In each case, first obtain the balanced half-reactions. Multiply the oxidation and reduction half-reactions by appropriate factors that will result in the same number of electrons being present in both half-reactions. Then add the half-reactions, canceling electrons in the process, to obtain the balanced equation for the whole reaction. Check to see that the final equation is balanced.

(a) $4[Cl_2(g) + 2\ e^- \longrightarrow 2\ Cl^-(aq)]$

$1[S_2O_3^{2-}(aq) + 5\ H_2O(l) \longrightarrow 2\ SO_4^{2-}(aq) + 10\ H^+(aq) + 8\ e^-]$

$4\ Cl_2(g) + S_2O_3^{2-}(aq) + 5\ H_2O(l) + 8\ e^- \longrightarrow$
$8\ Cl^-(aq) + 2\ SO_4^{2-}(aq) + 10\ H^+(aq) + 8\ e^-$

$4\ Cl_2(g) + S_2O_3^{2-}(aq) + 5\ H_2O(l) \longrightarrow 8\ Cl^-(aq) + 2\ SO_4^{2-}(aq) + 10\ H^+(aq)$

Cl_2 is the oxidizing agent and $S_2O_3^{2-}$ is the reducing agent.

(b) $2[MnO_4^-(aq) + 8\ H^+(aq) + 5\ e^- \longrightarrow Mn^{2+}(aq) + 4\ H_2O(l)]$

$5[H_2SO_3(aq) + H_2O(l) \longrightarrow HSO_4^-(aq) + 3\ H^+(aq) + 2\ e^-]$

$2\ MnO_4^-(aq) + 16\ H^+(aq) + 5\ H_2SO_3(aq) + 5\ H_2O(l) + 10\ e^- \longrightarrow$
$2\ Mn^{2+}(aq) + 8\ H_2O(l) + 5\ HSO_4^-(aq) + 15\ H^+(aq) + 10\ e^-$

$2\ MnO_4^-(aq) + H^+(aq) + 5\ H_2SO_3(aq) \longrightarrow$
$2\ Mn^{2+}(aq) + 3\ H_2O(l) + 5\ HSO_4^-(aq)$

MnO_4^- is the oxidizing agent and H_2SO_3 is the reducing agent.

(c) $Cl_2(g) + 2\ e^- \longrightarrow 2\ Cl^-(aq)$

$H_2S(aq) \longrightarrow S(s) + 2\ H^+(aq) + 2\ e^-$

$Cl_2(g) + H_2S(aq) + 2\ e^- \longrightarrow 2\ Cl^-(aq) + S(s) + 2\ H^+(aq) + 2\ e^-$

$Cl_2(g) + H_2S(aq) \longrightarrow 2\ Cl^-(aq) + S(s) + 2\ H^+(aq)$

Cl_2 is the oxidizing agent and H_2S is the reducing agent.

(d) $Cl_2(g) + 2\ e^- \longrightarrow 2\ Cl^-(aq)$

$2\ H_2O(l) + Cl_2(g) \longrightarrow 2\ HOCl(aq) + 2\ H^+(aq) + 2\ e^-$

$2\ H_2O(l) + Cl_2(g) + 2\ e^- \longrightarrow 2\ HOCl(aq) + 2\ H^+(aq) + 2\ Cl^-(aq) + 2\ e^-$

or $H_2O(l) + Cl_2(g) \longrightarrow HOCl(aq) + H^+(aq) + Cl^-(aq)$

Cl_2 is both the oxidizing and the reducing agent.

12.3 (a) $O_3(g) \longrightarrow O_2(g)$

$O_3(g) \longrightarrow O_2(g) + H_2O(l)$ (balances O's)

$2\ H_2O(l) + O_3(g) \longrightarrow O_2(g) + H_2O(l) + 2\ OH^-(aq)$ (balances H's)

$H_2O(l) + O_3(g) \longrightarrow O_2(g) + 2\ OH^-(aq)$ (cancels H_2O)

$H_2O(l) + O_3(g) + 2\ e^- \longrightarrow O_2(g) + 2\ OH^-(aq)$ (balances charge);

$Br^-(aq) \longrightarrow BrO_3^-(aq)$

$3\ H_2O(l) + Br^-(aq) \longrightarrow BrO_3^-(aq)$ (balances O's)

$6\ OH^-(aq) + 3\ H_2O(l) + Br^-(aq) \longrightarrow BrO_3^-(aq) + 6\ H_2O(l)$ (balances H's)

$6\ OH^-(aq) + 3\ H_2O(l) + Br^-(aq) \longrightarrow$

$BrO_3^-(aq) + 6\ H_2O(l) + 6\ e^-$ (balances charge)

Combining half-reactions yields

$3[H_2O(l) + O_3(g) + 2\ e^- \longrightarrow O_2(g) + 2\ OH^-(aq)]$

$6\ OH^-(aq) + 3\ H_2O(l) + Br^-(aq) \longrightarrow BrO_3^-(aq) + 6\ H_2O(l) + 6\ e^-$

$6\ H_2O(l) + 3\ O_3(g) + 6\ OH^-(aq) + Br^-(aq) + 6\ e^- \longrightarrow$

$3\ O_2(g) + 6\ OH^-(aq) + BrO_3^-(aq) + 6\ H_2O(l) + 6\ e^-$

and $3\ O_3(g) + Br^-(aq) \longrightarrow 3\ O_2(g) + BrO_3^-(aq)$

O_3 is the oxidizing agent and Br^- is the reducing agent.

(b) $Br_2(l) + 2\ e^- \longrightarrow 2\ Br^-(aq)$ (balanced reduction half-reaction)

$Br_2(l) + 6\ H_2O(l) \longrightarrow 2\ BrO_3^-(aq)$ (O's balanced); then

$Br_2(l) + 6\ H_2O(l) + 12\ OH^-(aq) \longrightarrow$

$2\ BrO_3^-(aq) + 12\ H_2O(l)$ (H's balanced); and

$Br_2(l) + 12\ OH^-(aq) \longrightarrow$

$2\ BrO_3^-(aq) + 6\ H_2O(l) + 10\ e^-$ (electrons balanced)

Combining half-reactions yields

$5[Br_2(l) + 2\ e^- \longrightarrow 2\ Br^-(aq)]$

$1[Br_2(l) + 12\ OH^-(aq) \longrightarrow 2\ BrO_3^-(aq) + 6\ H_2O(l) + 10\ e^-]$

$6\ Br_2(l) + 12\ OH^-(aq) + 10\ e^- \longrightarrow$

$10\ Br^-(aq) + 2\ BrO_3^-(aq) + 6\ H_2O(l) + 10\ e^-$

$6\ Br_2(l) + 12\ OH^-(aq) \longrightarrow 10\ Br^-(aq) + 2\ BrO_3^-(aq) + 6\ H_2O(l)$

Dividing by 2 gives

$3\ Br_2(l) + 6\ OH^-(aq) \longrightarrow 5\ Br^-(aq) + BrO_3^-(aq) + 3\ H_2O(l)$

Br_2 is both the oxidizing agent and the reducing agent.

(c) $Cr^{3+}(aq) + 4\ H_2O(l) \longrightarrow CrO_4^{2-}(aq)$ (O's balanced); then

$Cr^{3+}(aq) + 4\ H_2O(l) + 8\ OH^-(aq) \longrightarrow CrO_4^{2-}(aq) + 8\ H_2O(l)$ (H's balanced); and

$Cr^{3+}(aq) + 8\ OH^-(aq) \longrightarrow CrO_4^{2-}(aq) + 4\ H_2O(l) + 3\ e^-$ (charge balanced)

$MnO_2(s) \longrightarrow Mn^{2+}(aq) + 2\ H_2O(l)$; then

$MnO_2(s) + 4\ H_2O(l) \longrightarrow Mn^{2+}(aq) + 2\ H_2O(l) + 4\ OH^-(aq)$ (H's balanced); and

$MnO_2(s) + 2\ H_2O(l) + 2\ e^- \longrightarrow Mn^{2+}(aq) + 4\ OH^-(aq)$ (charge balanced)

Combining half-reactions yields

$$2[Cr^{3+}(aq) + 8\ OH^-(aq) \longrightarrow CrO_4^{2-}(aq) + 4\ H_2O(l) + 3\ e^-]$$
$$3[MnO_2(s) + 2\ H_2O(l) + 2\ e^- \longrightarrow Mn^{2+}(aq) + 4\ OH^-(aq)]$$
$$2\ Cr^{3+}(aq) + 16\ OH^-(aq) + 3\ MnO_2(s) + 6\ H_2O(l) + 6\ e^- \longrightarrow 2\ CrO_4^{2-}(aq) + 8\ H_2O(l) + 3\ Mn^{2+}(aq) + 12\ OH^-(aq) + 6\ e^-$$
$$2\ Cr^{3+}(aq) + 4\ OH^-(aq) + 3\ MnO_2(s) \longrightarrow 2\ CrO_4^{2-}(aq) + 2\ H_2O(l) + 3\ Mn^{2+}(aq)$$

Cr^{3+} is the reducing agent and MnO_2 is the oxidizing agent.

(d) $3[P_4(s) + 8\ OH^-(aq) \longrightarrow 4\ H_2PO_2^-(aq) + 4\ e^-]$

$$P_4(s) + 12\ H_2O(l) + 12\ e^- \longrightarrow 4\ PH_3(g) + 12\ OH^-(aq)$$
$$4\ P_4(s) + 12\ H_2O(l) + 24\ OH^-(aq) + 12\ e^- \longrightarrow 12\ H_2PO_2^-(aq) + 4\ PH_3(g) + 12\ OH^-(aq) + 12\ e^-$$
$$4\ P_4(s) + 12\ H_2O(l) + 12\ OH^-(aq) \longrightarrow 12\ H_2PO_2^-(aq) + 4\ PH_3(g)$$

or $P_4(s) + 3\ H_2O(l) + 3\ OH^-(aq) \longrightarrow 3\ H_2PO_2^-(aq) + PH_3(g)$

$P_4(s)$ is both the oxidizing and the reducing agent.

12.5 The half-reactions are

$MnO_4^-(aq) \longrightarrow Mn^{2+}(aq) + 4\ H_2O(l)$ (O's balanced)

$MnO_4^-(aq) + 8\ H^+(aq) \longrightarrow Mn^{2+}(aq) + 4\ H_2O(l)$ (H's balanced)

$MnO_4^-(aq) + 8\ H^+(aq) + 5\ e^- \longrightarrow Mn^{2+}(aq) + 4\ H_2O(l)$ (charge balanced)

$C_6H_{12}O_6(aq) + 6\ H_2O(l) \longrightarrow 6\ CO_2(g)$ (O's balanced)

$C_6H_{12}O_6(aq) + 6\ H_2O(l) \longrightarrow 6\ CO_2(g) + 24\ H^+(aq)$ (H's balanced)

$C_6H_{12}O_6(aq) + 6\ H_2O(l) \longrightarrow 6\ CO_2(g) + 24\ H^+(aq) + 24\ e^-$ (charge balanced)

Adding half-reactions yields

$$24[MnO_4^-(aq) + 8\ H^+(aq) + 5\ e^- \longrightarrow Mn^{2+}(aq) + 4\ H_2O(l)]$$
$$5[C_6H_{12}O_6(aq) + 6\ H_2O(l) \longrightarrow 6\ CO_2(g) + 24\ H^+(aq) + 24\ e^-]$$
$$24\ MnO_4^-(aq) + 192\ H^+(aq) + 5\ C_6H_{12}O_6(aq) + 30\ H_2O(l) + 120\ e^- \longrightarrow 24\ Mn^{2+}(aq) + 96\ H_2O(l) + 30\ CO_2(g) + 120\ H^+(aq) + 120\ e^-$$
$$24\ MnO_4^-(aq) + 72\ H^+(aq) + 5\ C_6H_{12}O_6(aq) \longrightarrow 24\ Mn^{2+}(aq) + 66\ H_2O(l) + 30\ CO_2(g)$$

12.7 $P_4S_3(aq) \longrightarrow H_3PO_4(aq) + SO_4^{2-}(aq)$

For the oxidation of P_4S_3, both the P and S atoms are oxidized. The assignment of oxidation states to the P and S atoms is complicated by the presence of P—P bonds in the molecule, which leads to non-integral values. As long as we are consistent in out assignments, the end result should be the same. We will assume that S in P_4S_3 is $^{2-}$ and, therefore, loses 8 electrons on going to S^{+6} in the sulfate ion.

Because P_4S_3 is a neutral molecule and, if S has an oxidation number of -2, then each phosphorus atom will have an oxidation number of $+1.5$. Phosphorus in phosphoric acid has an oxidation number of $+5$. so each P atom of P_4S_3 must lose 3.5 electrons. The total number of electrons lost is $(4 \times 3.5) + (3 \times 8) = 38$.

$P_4S_3(aq) \longrightarrow 4\ H_3PO_4(aq) + 3\ SO_4^{2-}(aq) + 38\ e^-$

We balance the charge by adding H^+ in an acidic solution:

$P_4S_3(aq) \longrightarrow 4\ H_3PO_4(aq) + 3\ SO_4^{2-}(aq) + 44\ H^+(aq) + 38\ e^-$

The final balance is achieved by adding water to provide the oxygen and hydrogen atoms:

$P_4S_3(aq) + 28\ H_2O(l) \longrightarrow 4\ H_3PO_4(aq) + 3\ SO_4^{2-}(aq) + 44\ H^+(aq) + 38\ e^-$

The other half-reaction is simpler.

$NO_3^-(aq) \longrightarrow NO(g)$

N has an oxidation number of $+5$ in the nitrate ion and $+2$ in nitric oxide. Each nitrogen atom gains three electrons in the course of the reaction.

$NO_3^-(aq) + 3\ e^- \longrightarrow NO(g)$

Charge balance is again achieved by adding H^+:

$NO_3^-(aq) + 4\ H^+(aq) + 3\ e^- \longrightarrow NO(g)$

The number of hydrogen and oxygen atoms is completed by the addition of water:

$NO_3^-(aq) + 4\ H^+(aq) + 3\ e^- \longrightarrow NO(aq) + 2\ H_2O(l)$

Combining the two half-reactions gives

$$\begin{array}{l} \quad 38[NO_3^-(aq) + 4\ H^+(aq) + 3\ e^- \longrightarrow NO(g) + 2\ H_2O(l)] \\ \underline{+3[P_4S_3(aq) + 28\ H_2O(l) \longrightarrow 4\ H_3PO_4(aq) + 3\ SO_4^{2-}(aq) + 44\ H^+(aq) + 38\ e^-]} \\ 3\ P_4S_3(aq) + 38\ NO_3^-(aq) + 20\ H^+(aq) + 8\ H_2O(l) \longrightarrow \\ \qquad\qquad 12\ H_3PO_4(aq) + 9\ SO_4^{2-}(aq) + 38\ NO(g) \end{array}$$

12.9 (a) $Ag^+(aq) + e^- \longrightarrow Ag(s) \quad E°(cathode) = +0.80\ V$

$Ni^{2+}(aq) + 2\ e^- \longrightarrow Ni(s) \quad E°(anode) = -0.23\ V$

Reversing the anode half-reaction yields

$Ni(s) \longrightarrow Ni^{2+}(aq) + 2\ e^-$

and the cell reaction is, upon addition of the half-reactions,

$$2\ Ag^+(aq) + Ni(s) \longrightarrow Ag(s) + Ni^{2+}(aq) \quad E°_{cell} = +0.80\ V - (-0.23)\ V = +1.03\ V$$

(b) $2\ H^+(aq) + 2\ e^- \longrightarrow H_2(g) \quad E°(anode) = 0.00\ V$

$Cl_2(g) + 2\ e^- \longrightarrow 2\ Cl^-(aq) \quad E°(cathode) = +1.36\ V$

Therefore, at the anode, after reversal,

$H_2(g) \longrightarrow 2\ H^+(aq) + 2\ e^-$

and, the cell reaction is, upon addition of the half-reactions,

$Cl_2(g) + H_2(g) \longrightarrow 2\,H^+(aq) + 2\,Cl^-(aq) \quad E°_{cell} = +1.36\text{ V} - 0.00\text{ V}$

$= +1.36\text{ V}$

(c) $Cu^{2+}(aq) + 2\,e^- \longrightarrow Cu(s) \quad E°(\text{anode}) = +0.34\text{ V}$

$Ce^{4+}(aq) + e^- \longrightarrow Ce^{3+}(aq) \quad E°(\text{cathode}) = +1.61\text{ V}$

Therefore, at the anode, after reversal,

$Cu(s) \longrightarrow Cu^{2+}(aq) + 2\,e^-$

and, the cell reaction is, upon addition of the half-reactions,

$2\,Ce^{4+}(aq) + Cu(s) \longrightarrow Cu^{2+}(aq) + 2\,Ce^{3+}(aq) \quad E°_{cell} = 1.61\text{ V} - (0.34\text{ V})$

$= +1.27\text{ V}$

(d) $O_2(g) + 2\,H_2O(l) + 4\,e^- \longrightarrow 4\,OH^-(aq) \quad E°(\text{cathode}) = 0.40\text{ V}$

$O_2(g) + 4\,H^+(aq) + 4\,e^- \longrightarrow 2\,H_2O(l) \quad E°(\text{anode}) = 1.23\text{ V}$

Reversing the anode half-reaction yields

$2\,H_2O(l) \longrightarrow O_2(g) + 4\,H^+(aq) + 4\,e^-$

and the cell reaction is, upon addition of the half-reactions,

$4\,H_2O(l) \longrightarrow 4\,H^+(aq) + 4\,OH^-(aq) \quad E°_{cell} = 0.40\text{ V} - 1.23\text{ V} = -0.83\text{ V}$

or, $H_2O(l) \longrightarrow H^+(aq) + OH^-(aq)$

Note: This balanced equation corresponds to the cell notation given. The spontaneous process is the reverse of this reaction.

(e) $Sn^{4+}(aq) + 2\,e^- \longrightarrow Sn^{2+}(aq) \quad E°(\text{anode}) = +0.15\text{ V}$

$Hg_2Cl_2(s) + 2\,e^- \longrightarrow 2\,Hg(l) + 2\,Cl^-(aq) \quad E°(\text{cathode}) = +0.27\text{ V}$

Therefore, at the anode, after reversal,

$Sn^{2+}(aq) \longrightarrow Sn^{4+}(aq) + 2\,e^-$

and the cell reaction is, upon addition of the half-reactions,

$Sn^{2+}(aq) + Hg_2Cl_2(s) \longrightarrow 2\,Hg(l) + 2\,Cl^-(aq) + Sn^{4+}(aq)$

$E°_{cell} = 0.27\text{ V} - 0.15\text{ V} = 0.12\text{ V}$

12.11 (a) $Ni^{2+}(aq) + 2\,e^- \longrightarrow Ni(s) \quad E°(\text{cathode}) = -0.23\text{ V}$

$Zn^{2+}(aq) + 2\,e^- \longrightarrow Zn(s) \quad E°(\text{anode}) = -0.76\text{ V}$

Reversing the anode reaction yields

$Zn(s) \longrightarrow Zn^{2+}(aq) + 2\,e^-$ (at anode); then, upon addition,

$Ni^{2+}(aq) + Zn(s) \longrightarrow Ni(s) + Zn^{2+}(aq)$ (overall cell)

$E°_{cell} = -0.23\text{ V} - (-0.76\text{ V}) = +0.53\text{ V}$

and $Zn(s)|Zn^{2+}(aq)||Ni^{2+}(aq)|Ni(s)$

(b) $2[Ce^{4+}(aq) + e^- \longrightarrow Ce^{3+}(aq)] \quad E°(\text{cathode}) = +1.61\text{ V}$

$I_2(s) + 2\,e^- \longrightarrow 2\,I^-(aq) \quad E°(\text{anode}) = +0.54\text{ V}$

Reversing the anode reaction yields

$2\ I^-(aq) \longrightarrow 2\ e^- + I_2(s)$ (at anode); then, upon addition,

$2\ I^-(aq) + 2\ Ce^{4+}(aq) \longrightarrow 2\ Ce^{3+}(aq) + I_2(s)$ (overall cell)

$E°_{cell} = +1.61\ V - 0.54\ V = +1.07\ V$

and $Pt(s) | I^-(aq) | I_2(s) || Ce^{4+}(aq), Ce^{3+}(aq) | Pt(s)$

An inert electrode such as Pt is necessary when both oxidized and reduced species are in the same solution.

(c) $Cl_2(g) + 2\ e^- \longrightarrow 2\ Cl^-(aq)$ $E°(\text{cathode}) = +1.36\ V$

$2\ H^+(aq) + 2\ e^- \longrightarrow H_2(g)$ $E°(\text{anode}) = 0.00\ V$

Reversing the anode reaction yields

$H_2(g) \longrightarrow 2\ H^+(aq) + 2\ e^-$ (at anode); then, upon addition,

$H_2(g) + Cl_2(g) \longrightarrow 2\ HCl(aq)$ (overall cell) $E°_{cell} = +1.36\ V - 0.00\ V$

$= +1.36\ V$

and $Pt(s) | H_2(g) | H^+(aq) || Cl^-(aq) | Cl_2(g) | Pt(s)$

An inert electrode such as Pt is necessary for gas/ion electrode reactions.

(d) $3[Au^+(aq) + e^- \longrightarrow Au(s)]$ $E°(\text{cathode}) = +1.69\ V$

$Au^{3+}(aq) + 3\ e^- \longrightarrow Au(s)$ $E°(\text{anode}) = +1.40\ V$

Reversing the anode reaction yields

$Au(s) \longrightarrow Au^{3+}(aq) + 3\ e^-$ then, upon addition (anode),

$3\ Au^+(aq) \longrightarrow 2\ Au(s) + Au^{3+}(aq)$ (overall cell) $E°_{cell} = +1.69\ V - 1.40\ V$

$= +0.29\ V$

and $Au(s) | Au^{3+}(aq) || Au^+(aq) | Au(s)$

12.13 (a) $Ag^+(aq) + e^- \longrightarrow Ag(s)$ $E°(\text{cathode}) = +0.80\ V$

$AgBr(s) + e^- \longrightarrow Ag(s) + Br^-(aq)$ $E°(\text{anode}) = +0.07\ V$

Reversing the anode reaction yields

$Ag(s) + Br^-(aq) \longrightarrow AgBr(s) + e^-$ then, upon addition,

$Ag^+(aq) + Br^-(aq) \longrightarrow AgBr(s)$ (overall cell) $E°_{cell} = +0.80\ V - 0.07\ V$

$= +0.73\ V$

This is the direction of the spontaneous cell reaction that could be used to study the given solubility equilibrium. The reverse of this cell reaction corresponds to the reaction as given. It is not spontaneous. Thus,

$AgBr(s) \longrightarrow Ag^+(aq) + Br^-(aq)$ $E°_{cell} = -0.73\ V$

For this, the cathode and anode reactions are reversed relative to those above. A cell diagram for the nonspontaneous process is

$Ag(s) | Ag^+(aq) || Br^-(aq) | AgBr(s) | Ag(s)$

(b) To conform to the notation of this chapter, the neutralization is rewritten as

$H^+(aq) + OH^- \longrightarrow H_2O(l)$

$O_2(g) + 4\ H^+(aq) + 4\ e^- \longrightarrow 2\ H_2O(l)$ $E°(\text{cathode}) = +1.23$ V

$O_2(g) + 2\ H_2O(l) + 4\ e^- \longrightarrow 4\ OH^-(aq)$ $E°(\text{anode}) = +0.40$ V

Reversing the anode reaction yields

$4\ OH^-(aq) \longrightarrow O_2(g) + 2\ H_2O(l) + 4\ e^-$; then, upon addition,

$4\ H^+(aq) + 4\ OH^-(aq) \longrightarrow 4\ H_2O(l)$

or $H^+(aq) + OH^-(aq) \longrightarrow H_2O(l)$ (overall cell) $E° = +1.23\ V - 0.40\ V$
$= +0.83$ V

and $Pt(s) | O_2(g) | OH^-(aq) || H^+(aq) | O_2(g) | Pt(s)$

(c) $Cd(OH)_2(s) + 2\ e^- \longrightarrow Cd(s) + 2\ OH^-(aq)$ $E°(\text{anode}) = -0.81$ V

$Ni(OH)_3(s) + e^- \longrightarrow Ni(OH)_2(s) + OH^-(aq)$ $E°(\text{cathode}) = +0.49$ V

Reversing the anode reaction and multiplying the cathode reaction by 2 yields

$Cd(s) + 2\ OH^-(aq) \longrightarrow Cd(OH)_2(s) + 2\ e^-$

$2\ Ni(OH)_3 + 2\ e^- \longrightarrow 2\ Ni(OH)_2(s) + 2\ OH^-(aq)$ then, upon addition,

$2\ Ni(OH)_2(s) + Cd(s) \longrightarrow Cd(OH)_2(s) + 2\ Ni(OH)_2(s)$

and $Cd(s) | Cd(OH)_2(s) | KOH(aq) || Ni(OH)_3(s) | Ni(OH)_2(s) | Ni(s)$

12.15 (a) $MnO_4^-(aq) + 8\ H^+(aq) + 5\ e^- \longrightarrow Mn^{2+}(aq) + 4\ H_2O(l)$ (cathode half-reaction)

$5[Fe^{3+}(aq) + e^- \longrightarrow Fe^{2+}(aq)]$ (anode half-reaction)

(b) Reversing the anode reaction and adding the two equations yields

$MnO_4^-(aq) + 5\ Fe^{2+}(aq) + 8\ H^+(aq) \longrightarrow Mn^{2+}(aq) + 5\ Fe^{3+}(aq) + 4\ H_2O(l)$

The cell diagram is

$Pt(s) | Fe^{2+}(aq), Fe^{3+}(aq) || MnO_4^-(aq), Mn^{2+}(aq), H^+(aq) | Pt(s)$

12.17 A galvanic cell has a positive potential difference; therefore, identify as cathode and anode the electrodes that make $E°(\text{cell})$ positive upon calculating

$$E°(\text{cell}) = E°(\text{cathode}) - E°(\text{anode})$$

There are only two possibilities: If your first guess gives a negative $E°(\text{cell})$, switch your identification.

(a) $Cu^{2+}(aq) + 2\ e^- \longrightarrow Cu(s)$ $E°(\text{cathode}) = +0.34$ V

$Cr^{3+}(aq) + e^- \longrightarrow Cr^{2+}(aq)$ $E°(\text{anode}) = -0.41$ V

$E°(\text{cell}) = +0.34\ V - (-0.41\ V) = +0.75$ V

(b) $AgCl(s) + e^- \longrightarrow Ag(s) + Cl^-(aq)$ $E°(\text{cathode}) = +0.22$ V

$AgI(s) + e^- \longrightarrow Ag(s) + I^-(aq)$ $E°(\text{anode}) = -0.15$ V

$E°(\text{cell}) = +0.22V - (-0.15V) = +0.37$ V

(c) $Hg_2^{2+}(aq) + 2\,e^- \longrightarrow 2\,Hg(l)$ $E°(\text{cathode}) = +0.79$ V

$Hg_2Cl_2(s) + 2\,e^- \longrightarrow 2\,Hg(l) + 2\,Cl^-(aq)$ $E°(\text{anode}) = +0.27$ V

$E°(\text{cell}) = +0.79\text{ V} - (+0.27\text{ V}) = +0.52\text{ V}$

(d) $Pb^{4+}(aq) + 2\,e^- \longrightarrow Pb^{2+}(aq)$ $E°(\text{cathode}) = +1.67$ V

$Sn^{4+}(aq) + 2\,e^- \longrightarrow Sn^{2+}(aq)$ $E°(\text{anode}) = +0.15$ V

$E°(\text{cell}) = +1.67\text{ V} - (+0.15\text{ V}) = +1.52\text{ V}$

12.19 See Exercise 12.17 solutions for $E°$ (cell) values. In each case, $\Delta G°_r = -nFE°$. $1\text{ V} = 1\text{ J}\cdot\text{C}^{-1}$. n is determined by balancing the equation for the cell reaction constructed from the half-reactions given in Exercise 12.17.

(a) $Cu^{2+}(aq) + 2\,Cr^{2+}(aq) \longrightarrow Cu(s) + 2\,Cr^{3+}(aq)$, $n = 2$

$E°_{cell} = +0.75\text{ V}$ and $\Delta G°_r = -nFE° = -(2)(9.6485 \times 10^4\text{ C}\cdot\text{mol}^{-1})(0.75\text{ J}\cdot\text{C}^{-1})$
$= -145\text{ kJ}\cdot\text{mol}^{-1}$

(b) $AgCl(s) + I^-(aq) \longrightarrow AgI(s) + Cl^-(aq)$, $n = 1$

$E°_{cell} = +0.37\text{ V}$ and $\Delta G°_r = -nFE°$
$= -1 \times 9.6485 \times 10^4\text{ C}\cdot\text{mol}^{-1} \times 0.37\text{ J}\cdot\text{C}^{-1} = -36\text{ kJ}\cdot\text{mol}^{-1}$

(c) $Hg_2^{2+}(aq) + 2\,Cl^-(aq) \longrightarrow Hg_2Cl_2(s)$, $n = 2$

$E°_{cell} = +0.52\text{ V}$ and $\Delta G°_r = -nFE° = -(2)(9.6485 \times 10^4\text{ C}\cdot\text{mol}^{-1})(0.52\text{ J}\cdot\text{C}^{-1})$
$= -100\text{ kJ}\cdot\text{mol}^{-1}$

(d) $Pb^{4+}(aq) + Sn^{2+}(aq) \longrightarrow Pb^{2+}(aq) + Sn^{4+}(aq)$, $n = 2$

$E°_{cell} = +1.52\text{ V}$ and $\Delta G°_r = -nFE° = -(2)(9.6485 \times 10^4\text{ C}\cdot\text{mol}^{-1})(1.52\text{ J}\cdot\text{C}^{-1})$
$= -293\text{ kJ}\cdot\text{mol}^{-1}$

12.21 The cell, as written $Cu(s)|Cu^{2+}(aq)||M^{2+}(aq)|M(s)$, makes the Cu/Cu^{2+} electrode the anode, because this is where oxidation is occurring; the M^{2+}/M electrode is the cathode. The calculation is

$E° = E°(\text{cathode}) - E°(\text{anode})$

$-1.25\text{ V} = E°(\text{cathode}) - (+0.34\text{ V})$

$E°(\text{cathode}) = -0.91\text{ V}$

12.23 Refer to Appendix 2B. The more negative (less positive) the standard reduction potential, the stronger is the metal as a reducing agent.

(a) Cu < Fe < Zn < Cr

(b) Mg < Na < K < Li

(c) V < Ti < Al < U

(d) Au < Ag < Sn < Ni

12.25 In each case, identify the couple with the more positive reduction potential. This will be the couple at which reduction occurs, and therefore which contains the oxidizing agent. The other couple contains the reducing agent.

(a) Co^{2+}/Co $E° = -0.28$ V, Co^{2+} is the oxidizing agent (cathode)

Ti^{3+}/Ti^{2+} $E° = -0.37$ V, Ti^{2+} is the reducing agent (anode)

$Pt(s)|Ti^{2+}(aq), Ti^{3+}(aq)||Co^{2+}(aq)|Co(s)$

$E°_{cell} = E°(cathode) - E°(anode) = -0.28\ V - (-0.37\ V) = +0.09\ V$

(b) U^{3+}/U $E° = -1.79$ V, U^{3+} is the oxidizing agent (cathode)

La^{3+}/La $E° = -2.52$ V, La is the reducing agent (anode)

$La(s)|La^{3+}(aq)||U^{3+}(aq)|U(s)$

$E°_{cell} = -1.79\ V - (-2.52\ V) = +0.73\ V$

(c) Fe^{3+}/Fe^{2+} $E° = +0.77$ V, Fe^{3+} is the oxidizing agent (cathode)

H^+/H_2 $E° = 0.00$ V, H_2 is the reducing agent (anode)

$Pt(s)|H_2(g)|H^+(aq)||Fe^{2+}(aq), Fe^{3+}(aq)|Pt(s)$

$E°_{cell} = +0.77\ V - 0.00\ V = +0.77\ V$

(d) $O_3/O_2,OH^-$ $E° = +1.24$ V, O_3 is the oxidizing agent (cathode)

Ag^+/Ag $E° = +0.80$ V, Ag is the reducing agent (anode)

$Ag(s)|Ag^+(aq)||OH^-(aq)|O_3(g), O_2(g)|Pt(s)$

$E°_{cell} = +1.24\ V - 0.80\ V = +0.44\ V$

12.27 (a) H_2/H^+ $E° = 0.00$ V; Ni^{2+}/Ni $E° = -0.23$ V

No, Ni is the better reducing agent; its couple has a more negative $E°$.

(b) Cr^{3+}/Cr $E° = -0.74$ V; Pb^{2+}/Pb $E° = -0.13$ V

Yes, Cr is the better reducing agent; its couple has a more negative $E°$.

$2[Cr(s) \longrightarrow Cr^{3+}(aq) + 3\ e^-]$ (anode)

$3[Pb^{2+}(aq) + 2\ e^- \longrightarrow Pb(s)]$ (cathode)

$3\ Pb^{2+}(aq) + 2\ Cr(s) \longrightarrow 3\ Pb(s) + 2\ Cr^{3+}(aq)$

$E°_{cell} = E°(cathode) - E°(anode) = -0.13\ V - (-0.74\ V) = +0.61\ V$

(c) $MnO_4^-, H^+/Mn^{2+}$ $E° = +1.51$ V; Cu^{2+}/Cu $E° = +0.34$ V

Yes, MnO_4^- is the better oxidizing agent; its couple has a more positive $E°$.

$5[Cu(s) \longrightarrow Cu^{2+}(aq) + 2\ e^-]$ (anode)

$2[MnO_4^-(aq) + 8\ H^+(aq) + 5\ e^- \longrightarrow Mn^{2+}(aq) + 4\ H_2O(l)]$ (cathode)

$5\ Cu(s) + 2\ MnO_4^-(aq) + 16\ H^+(aq) \longrightarrow 5\ Cu^{2+}(aq) + 2\ Mn^{2+}(aq) + 8\ H_2O(l)$

$E°_{cell} = 1.51\ V - 0.34\ V = 1.17\ V$

(d) Fe^{3+}/Fe^{2+} $E° = +0.77$ V; Hg_2^{2+}/Hg $E° = +0.79$ V

No, Hg_2^{2+} is the better oxidizing agent; its couple has the more positive $E°$.

12.29 (a) $E°(Cl_2, Cl^-) = +1.36$ V (cathode)

$E°(Br_2, Br^-) = +1.09$ V (anode)

Because $E°(Cl_2, Cl^-) > E°(Br_2, Br^-)$ the reaction is spontaneous as written.

$E°_{cell} = +1.36\text{ V} - 1.09\text{ V} = +0.27\text{ V}$

$Cl_2(g)$ is the oxidizing agent.

(b) $E°(Ce^{4+}/Ce^{3+}) = +1.61\text{ V}$ (anode)

$E°(MnO_4^-/Mn^{2+}) = +1.51\text{ V}$ (cathode)

Because $E°(Ce^{4+}/Ce^{3+}) > E°(MnO_4^-/Mn^{2+})$, the reaction is not spontaneous as written.

(c) $E°(Pb^{4+}/Pb^{2+}) = +1.67\text{ V}$ (anode)

$E°(Pb^{2+}/Pb) = -0.13\text{ V}$ (cathode)

Because $E°(Pb^{4+}/Pb^{2+}) > E°(Pb^{2+}/Pb)$, the reaction is not spontaneous as written.

(d) $E°(NO_3^-/NO_2/H^+) = +0.80\text{ V}$ (cathode)

$E°(Zn^{2+}/Zn) = -0.76\text{ V}$ (anode)

Because $E°(NO_3^-/NO_2/H^+) > E°(Zn^{2+}/Zn)$, the reaction is spontaneous as written.

$E°_{cell} = +0.80\text{ V} - (-0.76\text{ V}) = +1.56\text{ V}$

NO_3^- is the oxidizing agent.

12.31 $E°(Br_2, Br^-) = +1.09\text{ V}$ $E°(Cl_2, Cl^-) = +1.36\text{ V}$ $E°(O_2, H^+, H_2O) = +1.23\text{ V}$

O_2 could be used because $E°(O_2, H^+, H_2O) > E°(Br_2, Br^-)$. It is not used because that reaction is so much slower than the one with Cl_2.

12.33 (a) $3\,Au^+(aq) \rightleftharpoons 2\,Au(s) + Au^{3+}(aq)$

(b) $Au^+(aq) + e^- \longrightarrow Au(s)$ $E° = +1.69\text{ V}$

$Au^{3+}(aq) + 3\,e^- \longrightarrow Au(s)$ $E° = +1.40\text{ V}$

Multiplying the first equation by three and subtracting the second equation gives the net equation desired. The potential is given simply by subtracting the second from the first:

$E° = 1.69\text{ V} - 1.40\text{ V} = +0.29\text{ V}$

Because $E°$ is positive, the process should be spontaneous.

12.35 The appropriate half-reactions are:

$U^{4+} + e^- \longrightarrow U^{3+}$ $E° = -0.61$ (A)

$U^{3+} + 3\,e^- \longrightarrow U$ $E° = -1.79$ (B)

(A) and (B) add to give the desired half-reaction (C):

$U^{4+} + 4\,e^- \longrightarrow U$ $E° = ?$ (C)

In order to calculate the potential of a *half-reaction*, we need to convert the $E°$ values into $\Delta G°$ values:

$\Delta G°(A) = -nFE°(A) = -1F(-0.61\text{ V})$

$\Delta G°(B) = -nFE°(B) = -3F(-1.79\text{ V})$

$\Delta G°(C) = -nFE°(C) = -4FE°(C)$

$\Delta G°(C) = \Delta G°(A) + \Delta G°(B)$

$-4FE°(C) = -1F(-0.61\text{ V}) + [-3F(-1.79\text{ V})]$

The constant F will cancel from both sides, leaving:

$-4E°(C) = -1(-0.61\text{ V}) - 3(-1.79\text{ V})$

$E°(C) = -[0.61\text{ V} + 5.37\text{ V}]/4 = -1.50\text{ V}$

12.37 (a) $E°_{cell} = E°(\text{cathode}) - E°(\text{anode}) = 0.34\text{ V} - (-0.41\text{ V}) = 0.75\text{ V}$, and

$$\ln K = \frac{nFE°}{RT} \text{ at } 25°\text{C} = \frac{nE°}{0.025\ 69\text{ V}} \quad \ln K = \frac{(2)(0.75\text{ V})}{0.025\ 69\text{ V}} = 58 \quad \text{and}$$

$K = 10^{25}$

(b) $Ti^{2+}(aq) + 2\ e^- \longrightarrow Ti(s) \quad E°(\text{cathode}) = -1.63\text{ V}$

$Mn^{2+}(aq) + 2\ e^- \longrightarrow Mn(s) \quad E°(\text{anode}) = -1.18\text{ V}$

Note: These equations represent the cathode and anode half-reactions for the overall reaction as written. The spontaneous direction of this reaction under standard conditions is the opposite of that given.

$E°_{cell} = E°(\text{cathode}) - E°(\text{anode}) = -1.63\text{ V} - (-1.18\text{ V}) = -0.45\text{ V}$, and

$$\ln K = \frac{nFE°}{RT} \text{ at } 25°\text{C} = \frac{nE°}{0.025\ 69\text{ V}} \quad \ln K = \frac{(2)(-0.45\text{ V})}{0.025\ 69\text{ V}} = -35 \quad \text{and}$$

$K = 10^{-15}$

(c) $E°$ for $Hg_2^{2+}/Hg = +0.79\text{ V} > E°$ for $Pb^{2+}/Pb = -0.13\text{ V}$

Therefore, Hg_2^{2+} is reduced and Pb is oxidized.

$Hg_2^{2+}(aq) + Pb(s) \longrightarrow Pb^{2+}(aq) + 2\ Hg(l)$

$Hg_2^{2+} + 2\ e^- \longrightarrow 2\ Hg(l) \quad E°(\text{cathode}) = 0.79\text{ V}$

$Pb^{2+} + 2\ e^- \longrightarrow Pb(s) \quad E°(\text{anode}) = -0.13\text{ V}$

$E°_{cell} = E°(\text{cathode}) - E°(\text{anode}) = 0.79\text{ V} - (-0.13\text{ V}) = 0.92\text{ V}$, and

$$\ln K = \frac{nFE°}{RT} \text{ at } 25°\text{C} = \frac{nE°}{0.025\ 69\text{ V}} \quad \ln K = \frac{(2)(0.92\text{ V})}{0.025\ 69\text{ V}} = +72 \quad \text{and}$$

$K = 10^{31}$

(d) $In^{3+}(aq) + e^- \longrightarrow In^{2+}(aq) \quad E°(\text{cathode}) = -0.49\text{ V}$

$U^{4+}(aq) + e^- \longrightarrow U^{3+}(aq) \quad E°(\text{anode}) = -0.61\text{ V}$

$E°_{cell} = E°(\text{cathode}) - E°(\text{anode}) = -0.49\text{ V} - (-0.61\text{ V}) = +0.12\text{ V}$

$$\ln K = \frac{nFE°}{RT} = \frac{nE°}{0.025\ 69\text{ V}} \text{ at } 25°\text{C} \quad \ln K = \frac{1 \times (+0.12\text{ V})}{0.025\ 69\text{ V}} = 4.7 \quad K = 1 \times 10^2$$

12.39 Consider the half-reactions involved. Construct a cell reaction from them and calculate its standard potential. If positive, the oxidation will occur.

$S_2O_8^{2-}(aq) + 2\,e^- \longrightarrow 2\,SO_4^{2-}(aq) \quad E°(\text{cathode}) = +2.05\text{ V}$

$Ag^{2+}(aq) + e^- \longrightarrow Ag^+(aq) \quad E°(\text{anode}) = +1.98\text{ V}$

Reverse the anode reaction and multiply by 2.

$$S_2O_8^{2-} + 2\,e^- \longrightarrow 2\,SO_4^{2-}$$
$$2[Ag^+ \longrightarrow Ag^{2+} + e^-]$$
$$S_2O_8^{2-}(aq) + Ag^+(aq) \longrightarrow 2\,Ag^{2+}(aq) + 2\,SO_4^{2-}(aq) \quad E°_{cell} = 2.05\text{ V} - (+1.98\text{ V}) = +0.07\text{ V}$$

Yes, the oxidation will work.

$$\ln K = \frac{nFE°}{RT}, \text{ at } 25°\text{C} = \frac{nE°}{0.025\ 69\text{ V}} = \frac{(2)(0.07\text{ V})}{0.025\ 69\text{ V}} = 5 \quad \text{and } K = 10^2$$

12.41 (a) $Pb^{4+}(aq) + 2e^- \longrightarrow Pb^{2+}(aq) \quad E°(\text{cathode}) = +1.67\text{ V}$

$Sn^{2+}(aq) \longrightarrow Sn^{4+}(aq) + 2\,e^- \quad E°(\text{anode}) = +0.15\text{ V}$

$$Pb^{4+}(aq) + Sn^{2+}(aq) \longrightarrow Pb^{2+}(aq) + Sn^{4+}(aq) \quad E°_{cell} = 1.67\text{ V} - (0.15\text{ V}) = +1.52\text{ V}$$

$$\text{Then, } E = E° - \left(\frac{0.025\ 693\text{ V}}{n}\right)\ln Q; \quad 1.33\text{ V} = 1.52\text{ V} - \left(\frac{0.025\ 693\text{ V}}{2}\right)\ln Q$$

$$\ln Q = \frac{1.52\text{ V} - 1.33\text{ V}}{0.0129\text{ V}} = \frac{0.19\text{ V}}{0.0129\text{ V}} = 15 \quad Q = 10^6$$

(b) $2[Cr_2O_7^{2-}(aq) + 14\,H^+(aq) + 6\,e^- \longrightarrow 2\,Cr^{3+}(aq) + 7\,H_2O(l)] \quad E°(\text{cathode}) = 1.33\text{ V}$

$3[2\,H_2O(l) \longrightarrow O_2(g) + 4\,H^+(aq) + 4\,e^-] \quad E°(\text{anode}) = +1.23\text{ V}$

$$2\,Cr_2O_7^{2-}(aq) + 16\,H^+(aq) \longrightarrow 4\,Cr^{3+}(aq) + 8\,H_2O(l) + 3\,O_2(g) \quad E°_{cell} = 0.10\text{ V}$$

$$\text{Then, } E = E° - \left(\frac{0.0257\text{ V}}{n}\right)\ln Q; \quad 0.10\text{ V} = +0.10\text{ V} - \left(\frac{0.0257\text{ V}}{12}\right)\ln Q$$

$\ln Q = 0.00 \quad Q = 1.0$

12.43 (a) $Cu^{2+}(aq, 0.010\text{ M}) + 2\,e^- \longrightarrow Cu(s) \quad (\text{cathode})$

$Cu^{2+}(aq, 0.0010\text{ M}) + 2\,e^- \longrightarrow Cu(s) \quad (\text{anode})$

$$Cu^{2+}(aq, 0.010\text{ M}) \longrightarrow Cu^{2+}(aq, 0.0010\text{ M}),\ n = 2$$

$E°_{cell} = E°(\text{cathode}) - E°(\text{anode}) = 0\text{ V}$

$$E_{cell} = E°_{cell} - \left(\frac{RT}{nF}\right)\ln Q = -\left(\frac{0.025\ 693\text{ V}}{2}\right)\ln Q \text{ at } 25°\text{C}$$

$$E_{cell} = -\left(\frac{0.025\ 693\text{ V}}{2}\right)\ln\left(\frac{0.0010\text{ M}}{0.010\text{ M}}\right) = +0.030\text{ V}$$

(b) at pH = 3.0, $[H^+] = 1 \times 10^{-3}$ M

at pH = 4.0, $[H^+] = 1 \times 10^{-4}$ M

Cell reaction is $H^+(aq, 1 \times 10^{-3}\ M) \longrightarrow H^+(aq, 1 \times 10^{-4}\ M)$, $n = 1$

$$E°_{cell} = 0\ V \quad E_{cell} = E°_{cell} - \left(\frac{RT}{nF}\right) \ln Q = -\left(\frac{0.025\ 693\ V}{1}\right) \ln\left(\frac{1 \times 10^{-4}}{1 \times 10^{-3}}\right)$$

$$= +6 \times 10^{-2}\ V$$

12.45 In each case, $E°_{cell} = E°(\text{cathode}) - E°(\text{anode})$. Recall that the values for $E°$ at the electrodes refer to the electrode potential for the half-reaction written as a reduction reaction. In balancing the cell reaction, the half-reaction at the anode is reversed. However, this does not reverse the sign of electrode potential used at the anode, because the value always refers to the reduction potential.

(a) $2\ H^+(aq, 1.0\ M) + 2\ e^- \longrightarrow H_2(g, 1\ atm) \quad E°(\text{cathode}) = 0.00\ V$

$H_2(g, 1\ atm) \longrightarrow 2\ H^+(aq, 0.075\ M) + 2\ e^- \quad E°(\text{anode}) = 0.00\ V$

$2\ H^+(aq, 1.0\ M) + H_2(g, 1\ atm) \longrightarrow 2\ H^+(aq, 0.075\ M) + H_2(g, 1\ atm)$

$E°_{cell} = 0.00\ V$

$$\text{Then, } E = E° - \left(\frac{0.025\ 693\ V}{n}\right) \ln\left(\frac{[H^+, 0.075\ M]^2 P_{H_2}}{[H^+, 1.0\ M]^2 P_{H_2}}\right)$$

$$E = 0.00\ V - \left(\frac{0.025\ 693\ V}{2}\right) \ln\left(\frac{(0.075\ M)^2 \times 1\ atm}{(1.0\ M)^2 \times 1\ atm}\right)$$

$$E = -0.0129\ V \ln (0.075)^2 = +0.067\ V$$

(b) $Ni^{2+}(aq) + 2\ e^- \longrightarrow Ni(s) \quad E°(\text{cathode}) = -0.23\ V$

$Zn(s) \longrightarrow Zn^{2+}(aq) + 2\ e^- \quad E°(\text{anode}) = -0.76\ V$

$Ni^{2+}(aq) + Zn(s) \longrightarrow Ni(s) + Zn^{2+}(aq) \quad E°_{cell} = +0.53\ V$

$$\text{Then, } E = E° - \left(\frac{0.025\ 693\ V}{n}\right) \ln\left(\frac{[Zn^{2+}]}{[Ni^{2+}]}\right)$$

$$E = 0.53\ V - \left(\frac{0.025\ 693\ V}{2}\right) \ln\left(\frac{0.37}{0.059}\right) = 0.53\ V - 0.02\ V = 0.51\ V$$

(c) $2\ H^+(aq) + 2\ e^- \longrightarrow H_2(g) \quad E°(\text{cathode}) = 0.00\ V$

$2\ Cl^-(aq) \longrightarrow Cl_2(g) + 2\ e^- \quad E°(\text{anode}) = +1.36\ V$

$2\ H^+(aq) + 2\ Cl^-(aq) \longrightarrow H_2(g) + Cl_2(g) \quad E°_{cell} = -1.36\ V$

Then,

$$E = E° - \left(\frac{0.025\ 693\ V}{n}\right) \ln\left(\frac{P_{H_2}P_{Cl_2}}{[H^+]^2[Cl^-]^2}\right)$$

$$E = -1.36\ V - \left(\frac{0.025\ 693\ V}{2}\right) \ln\left(\frac{\left(\frac{125}{760}\right)\left(\frac{250}{760}\right)}{(0.85)^2(1.0)^2}\right)(1.01325)^2$$

$$E = -1.36\ V + 0.03\ V$$

$$= -1.33\ V$$

(d) $Sn^{4+}(aq, 0.867\ \text{M}) + 2\ e^- \longrightarrow Sn^{2+}(aq, 0.55\ \text{M})$ $E°(\text{cathode}) = +0.15$ V

$Sn(s) \longrightarrow Sn^{2+}(aq, 0.277\ \text{M}) + 2\ e^-$ $E°(\text{anode}) = -0.14$ V

$Sn^{4+}(aq, 0.867\ \text{M}) + Sn(s) \longrightarrow Sn^{2+}(aq, 0.55\ ) + Sn^{2+}(aq, 0.277\ \text{M})$

$E°_{\text{cell}} = 0.29$ V

$$E = E° - \left(\frac{0.025\ 693\ \text{V}}{2}\right) \ln\left(\frac{(0.55)(0.277)}{(0.867)}\right)$$

$$E = 0.29\ \text{V} + 0.02\ \text{V} = 0.31\ \text{V}$$

12.47 In each case, obtain the balanced equation for the cell reaction from the half-cell reactions at the electrodes, by reversing the reduction equation for the half-reaction at the anode, multiplying the half-reaction equations by an appropriate factor to balance the number of electrons, and then adding the half-reactions. Calculate $E°_{\text{cell}} = E°(\text{cathode}) - E°(\text{anode})$. Then write the Nernst equation for the cell reaction and solve for the unknown.

(a) $Hg_2Cl_2(s) + 2\ e^- \longrightarrow 2\ Hg(l) + 2\ Cl^-(aq)$ $E°(\text{cathode}) = +0.27$ V

$H_2(g) \longrightarrow 2\ H^+(aq) + 2\ e^-$ $E°(\text{anode}) = 0.00$ V

$H_2(g) + Hg_2Cl_2(s) \longrightarrow 2\ H^+(aq) + 2\ Hg(l) + 2\ Cl^-(aq)$ $E°_{\text{cell}} = +0.27$ V

$$E = E° - \left(\frac{0.025\ 693\ \text{V}}{n}\right) \ln\left(\frac{[H^+]^2[Cl^-]^2}{[H_2]}\right)$$

$$0.33\ \text{V} = 0.27\ \text{V} - \left(\frac{0.025\ 693\ \text{V}}{2}\right) \ln\left(\frac{[H^+]^2(1)^2}{(1)}\right)$$

$$= 0.27\ \text{V} - (0.0129\ \text{V})\ln[H^+]^2$$

$$0.06\ \text{V} = -0.0257\ \text{V}\ \ln[H^+] = -0.0257\ \text{V} \times (2.303\ \log\ [H^+])$$

$$\text{pH} = \frac{0.06\ \text{V}}{(2.303)(0.025\ 693\ \text{V})} = 1$$

(b) $2[MnO_4^-(aq) + 8\ H^+(aq) + 5\ e^- \longrightarrow Mn^{2+}(aq) + 4\ H_2O(l)]$

$E°(\text{cathode}) = +1.51$ V

$5[2\ Cl^-(aq) \longrightarrow Cl_2(g) + 2\ e^-]$ $E°(\text{anode}) = +1.36$ V

$2\ MnO_4^-(aq) + 16\ H^+(aq) + 10\ Cl^-(aq) \longrightarrow 5\ Cl_2(g) + 2\ Mn^{2+}(aq) + 8\ H_2O(l)$ $E°_{\text{cell}} = +0.15$ V

$$E = E° - \left(\frac{0.0257\ \text{V}}{n}\right) \ln\left(\frac{[Cl_2]^5[Mn^{2+}]^2}{[MnO_4]^2[H^+]^{16}[Cl^-]^{10}}\right)$$

$$-0.30\ \text{V} = +0.15\ \text{V} - \left(\frac{0.0257\ \text{V}}{10}\right) \ln\left(\frac{(1)^5(0.10)^2}{(0.010)^2(1 \times 10^{-4})^{16}(Cl^-)^{10}}\right)$$

$$-0.45\ \text{V} = -(0.002\ 5693\ \text{V})\ \log\left(\frac{1 \times 10^{-2}}{(1 \times 10^{-4})(1 \times 10^{-64})[Cl^-]^{10}}\right)$$

$$= -0.002\ 5693\ \text{V}\left[\ln\ (1 \times 10^{66}) + \ln\left(\frac{1}{[Cl^-]^{10}}\right)\right]$$

$$= -0.390\ \text{V} + (0.0025\ 693\ \text{V})\ \ln[Cl^-]^{10}$$

$$-0.0594 \text{ V} = 0.002\,5693 \text{ V} \ln[\text{Cl}^-]^{10}$$
$$= (0.025\,693 \text{ V}) \ln[\text{Cl}^-]$$
$$\ln[\text{Cl}^-] = \frac{-0.06 \text{ V}}{0.025\,693 \text{ V}} = -2$$
$$[\text{Cl}^-] = 10^{-1} \text{ mol}\cdot\text{L}^{-1}$$

12.49 To calculate this value, we need to determine the $E°$ value for the solubility reaction:

$Hg_2Cl_2(s) \rightleftharpoons Hg_2^{2+}(aq) + 2\ Cl^-(aq) \qquad E° = ?$

The relationship $\Delta G° = -nRT \ln K = -nFE°$ can be used to calculate the value of K_{sp}. The equations that will add to give the net equation we want are

$Hg_2Cl_2(s) + 2\ e^- \longrightarrow 2\ Hg(l) + 2\ Cl^-(aq) \qquad E° = +0.27 \text{ V}$

$2\ Hg(1) \longrightarrow Hg_2^{2+}(aq) + 2\ e^- \qquad E° = 0.79 \text{ V}$

Notice that the second equation is reversed from the reduction reaction given in the Appendix, and consequently the $E°$ value is changed in sign.

Adding these two equations together gives the desired net reaction, and summing the $E°$ values will give the $E°$ value for that process:

$E° = (+0.29 \text{ V}) + (-0.79 \text{ V}) = -0.50 \text{ V}$

$$\ln K_{sp} = \frac{nFE°}{RT} = \frac{(2)(9.65 \times 10^4 \text{ C}\cdot\text{mol}^{-1})(-0.50 \text{ V})}{(8.314 \text{ J}\cdot\text{K}^{-1}\cdot\text{mol}^{-1})(298 \text{ K})} = -38.95$$

$K_{sp} = 1.2 \times 10^{-17}$

(b) This value is similar to the value reported in Table 11.5 (1.3×10^{-18}).

12.51 For the standard calomel electrode, $E° = +0.27$ V. If this were set equal to 0, all other potentials would also be decreased by 0.27 V. (a) Therefore, the standard hydrogen electrode's standard reduction potential would be 0.00 V − 0.27 V or −0.27 V. (b) The standard reduction potential for Cu^{2+}/Cu would be 0.34 V − 0.27 V or +0.07 V.

12.53 $Fe^{3+}(aq) + 3\ e^- \longrightarrow Fe(s) \quad E° = -0.04 \text{ V}$

$Cr^{3+}(aq) + 3\ e^- \longrightarrow Cr(s) \quad E° = -0.74 \text{ V}$

$Fe^{2+}(aq) + 2\ e^- \longrightarrow Fe(s) \quad E° = -0.44 \text{ V}$

$Cr^{2+}(aq) + 2\ e^- \longrightarrow Cr(s) \quad E° = -0.91 \text{ V}$

Comparison of the reduction potentials shows that Cr is more easily oxidized than Fe, so the presence of Cr retards the rusting of Fe. At the position of the scratch, the gap is filled with oxidation products of Cr, thereby preventing contact of air and water with the iron.

12.55 (a) $Fe_2O_3 \cdot H_2O$ (b) H_2O and O_2 jointly oxidize iron. (c) Water is more highly conducting if it contains dissolved ions, so the rate of rusting is increased.

12.57 (a) aluminum or magnesium; both are below titanium in the electrochemical series.

(b) cost, availability, and toxicity of products in the environment

(c) $Cu^{2+} + 2\ e^- \longrightarrow Cu(s)$ $E° = +0.34$ V

$Cu^+ + e^- \longrightarrow Cu(s)$ $E° = +0.52$ V

$Fe^{3+} + 3\ e^- \longrightarrow Fe(s)$ $E° = -0.04$ V

$Fe^{2+} + 2\ e^- \longrightarrow Fe(s)$ $E° = -0.44$ V

Fe could act as the anode of an electrochemical cell if Cu^{2+} or Cu^+ were present; therefore, it could be oxidized at the point of contact. Water with dissolved ions would act as the electrolyte.

12.59 The strategy is to consider the possible competing cathode and anode reactions. At the cathode, choose the reduction reaction with the most positive (least negative) standard reduction potential ($E°$ value). At the anode, choose the oxidation reaction with the least positive (most negative) standard reduction potential ($E°$ value, as given in the table). Then calculate $E°_{cell} = E°(\text{cathode}) - E°(\text{anode})$. The negative of this value is the minimum potential that must be supplied.

(a) cathode: $Ni^{2+}(aq) + 2\ e^- \longrightarrow Ni(s)$ $E° = -0.23$ V

(rather than $2\ H_2O(l) + 2\ e^- \longrightarrow H_2(g) + 2\ OH^-(aq)$ $E° = -0.83$ V)

(b) anode: $2\ H_2O(l) \longrightarrow O_2(g) + 4\ H^+(aq) + 4\ e^-$ $E° = +1.23$ V

(the SO_4^{2-} ion will not oxidize)

(c) $E°_{cell} = E°(\text{cathode}) - E°(\text{anode}) = -0.23\ \text{V} - (+0.81\ \text{V}) = -1.04$ V

Therefore E (supplied) must be $> +1.04$ V (1.04 V is the minimum).

12.61 In each case, compare the reduction potential of the ion to the reduction potential of water ($E° = -0.42$ V) and choose the process with the least negative $E°$ value.

(a) $Mn^{2+}(aq) + 2\ e^- \longrightarrow Mn(s)$ $E° = -1.18$ V

(b) $Al^{3+}(aq) + 3\ e^- \longrightarrow Al(s)$ $E° = -1.66$ V

The reactions in (a) and (b) evolve hydrogen rather than yield a metallic deposit because water is reduced, according to $2\ H_2O(l) + 2\ e^- \longrightarrow H_2(g) + 2\ OH^-(aq)$ ($E° = -0.42$ V, at pH = 7)

(c) $Ni^{2+}(aq) + 2\ e^- \longrightarrow Ni(s)$ $E° = -0.23$ V

(d) $Au^{3+}(aq) + 3\ e^- \longrightarrow Au(s)$ $E° = +1.69$ V

In (c) and (d) the metal ion will be reduced.

12.63 $4500\ \text{C} \div 9.65 \times 10^4\ \text{C}\cdot\text{F}^{-1} = 0.047\ \text{F} = 0.047\ \text{mol e}^-$

(a) $(0.047 \div 3)\ \text{mol Bi}^{3+} + 0.047\ \text{mol e}^- \longrightarrow$

$(0.047 \div 3)\ \text{mol Bi, or } 0.016\ \text{mol Bi} = 3.3\ \text{g}$

(b) $0.047\ \text{mol H}^+ + 0.047\ \text{mol e}^- \longrightarrow (0.047 \div 2)\ \text{mol H}_2$

$0.024\ \text{mol H}_2 \times 24.45\ \text{L}\cdot\text{mol}^{-1}\ (\text{at } 298\ \text{K}) = 0.59\ \text{L}$

(c) $(0.047 \div 3)\ \text{mol Co}^{3+} + 0.047\ \text{mol e}^- \longrightarrow$

$(0.047 \div 3)\ \text{mol Co} = 0.016\ \text{mol Co or } 0.94\ \text{g}$

12.65 (a) $\text{Ag}^+(\text{aq}) + \text{e}^- \longrightarrow \text{Ag(s)}$

$$\text{time} = (1.00\ \text{Ag})\left(\frac{1\ \text{mol Ag}}{107.98\ \text{g Ag}}\right)\left(\frac{1\ \text{mol e}^-}{1\ \text{mol Ag}}\right)$$
$$\left(\frac{9.65 \times 10^4\ \text{C}}{1\ \text{mol e}^-}\right)\left(\frac{1\ \text{A}\cdot\text{s}}{1\ \text{C}}\right)\left(\frac{1}{0.0178}\right) = 5.0 \times 10^4\ \text{s or } 14\ \text{h}$$

(b) $\text{Cu}^{2+}(\text{aq}) + 2\ \text{e}^- \longrightarrow \text{Cu(s)}$

$$\text{mass Cu} = (5.0 \times 10^4\ \text{s})(0.0178\ \text{A})\left(\frac{1\ \text{C}}{1\ \text{A}\cdot\text{s}}\right)\left(\frac{1\ \text{mol e}^-}{9.65 \times 10^4\ \text{C}}\right)$$
$$\left(\frac{0.50\ \text{mol Cu}}{1\ \text{mol e}^-}\right)\left(\frac{63.5\ \text{g Cu}}{1\ \text{mol Cu}}\right) = 0.29\ \text{g Cu}$$

12.67 (a) $\text{Cr(VI)} + 6\ \text{e}^- \longrightarrow \text{Cr(s)}$

$$\text{current} = \frac{\text{charge}}{\text{time}}$$
$$= \frac{4.0\ \text{g Cr}\left(\frac{1\ \text{mol Cr}}{52.00\ \text{g Cr}}\right)\left(\frac{6\ \text{mol e}^-}{1\ \text{mol Cr}}\right)\left(\frac{9.65 \times 10^4\ \text{C}}{1\ \text{mol e}^-}\right)}{24\ \text{h} \times 3600\ \text{s}\cdot\text{h}^{-1}}$$
$$= 0.52\ \text{C}\cdot\text{s}^{-1} = 0.52\ \text{A}$$

(b) $\text{Na}^+ + \text{e}^- \longrightarrow \text{Na(s)}$

$$\text{current} = \frac{4.0\ \text{g Na}\left(\frac{1\ \text{mol Na}}{22.99\ \text{g Na}}\right)\left(\frac{1\ \text{mol e}^-}{1\ \text{mol Na}}\right)\left(\frac{9.65 \times 10^4\ \text{C}}{1\ \text{mol e}^-}\right)}{24\ \text{h} \times 3600\ \text{s}\cdot\text{h}^{-1}}$$
$$= 0.19\ \text{C}\cdot\text{s}^{-1} = 0.19\ \text{A}$$

12.69 $\text{Ru}^{n+}(\text{aq}) + n\ \text{e}^- \longrightarrow \text{Ru(s)}$; solve for n

$$\text{moles of Ru} = (0.0310\ \text{g Ru})\left(\frac{1\ \text{mol}}{101.07\ \text{g Ru}}\right) = 3.07 \times 10^{-4}\ \text{mol}$$

$$\text{total charge} = (500\ \text{s})(120\ \text{mA})\left(\frac{10^{-3}\ \text{A}}{1\ \text{mA}}\right)\left(\frac{1\ \text{C}\cdot\text{s}^{-1}}{1\ \text{A}}\right) = 60\ \text{C}$$

$$\text{moles of e}^- = (60\ \text{C})\left(\frac{1\ \text{mol e}^-}{96\ 500\ \text{C}}\right) = 6.2 \times 10^{-4}\ \text{mol e}^-$$

$$n = \frac{6.2 \times 10^{-4} \text{ mol e}^-}{3.07 \times 10^{-4} \text{ mol}} = \frac{2 \text{ mol charge}}{1 \text{ mol}}$$

Therefore, oxidation number of Ru^{2+} is +2.

12.71 $Hf^{n+} + n\,e^- \longrightarrow Hf(s)$; solve for n.

charge consumed = $15.0 \text{ C}\cdot\text{s}^{-1} \times 2.00 \text{ h} \times 3600 \text{ s}\cdot\text{h}^{-1} = 1.08 \times 10^5 \text{ C}$

$$\text{moles of charge consumed} = (1.08 \times 10^5 \text{ C})\left(\frac{1 \text{ mol e}^-}{9.65 \times 10^4 \text{ C}}\right) = 1.12 \text{ mol e}^-$$

$$\text{moles of Hf plated} = (50.0 \text{ g Hf})\left(\frac{1 \text{ mol Hf}}{178.49 \text{ g Hf}}\right) = 0.280 \text{ mol Hf}$$

$$\text{Then, } n = \frac{1.12 \text{ mol e}^-}{0.280 \text{ mol Hf}} = 4.00 \text{ mol e}^-/\text{mol Hf}$$

Therefore, the oxidation number is 4, that is, Hf^{4+}.

12.73 $MCl_3 \longrightarrow M^{3+} + 3\,Cl^-$ $\quad M^{3+} + 3\,e^- \longrightarrow M(s)$

First, determine the number of moles of electrons consumed; the number of moles of M^{3+} reduced is one-third of this number.

$$\text{charge used} = (6.63 \text{ h})\left(\frac{3600 \text{ s}}{1 \text{ h}}\right)\left(\frac{0.700 \text{ C}}{1 \text{ s}}\right) = 1.67 \times 10^4 \text{ C}$$

$$\text{number of moles of e}^- = (1.67 \times 10^4 \text{ C})\left(\frac{1 \text{ mol e}^-}{9.65 \times 10^4 \text{ C}}\right) = 0.173$$

$$\text{numbef of moles of M}^{3+} \text{ (and M)} = 0.173 \text{ mol e}^- \times \frac{1 \text{ mol M}^{3+}}{3 \text{ mol e}^-}$$

$$= 0.0577$$

$$\text{molar mass M} = \frac{3.00 \text{ g}}{0.0577 \text{ mol}} = 52.0 \text{ g}\cdot\text{mol}^{-1} \text{ (Cr)}$$

12.75 (a) The electrolyte is KOH(aq)/HgO(s), which will have the consistency of a moist paste.

(b) The oxidizing agent is HgO(s).

(c) $HgO(s) + Zn(s) \longrightarrow Hg(l) + ZnO(s)$

12.77 See Box 12.1.

The anode reaction is $Zn(s) \longrightarrow Zn^{2+}(aq) + 2\,e^-$; this reaction supplies the electrons to the external circuit. The cathode reaction is $MnO_2(s) + H_2O(l) + e^- \longrightarrow MnO(OH)_2(s) + OH^-(aq)$. The $OH^-(aq)$ produced reacts with $NH_4^+(aq)$ from the $NH_4Cl(aq)$ present: $NH_4^+(aq) + OH^-(aq) \longrightarrow H_2O(l) + NH_3(g)$. The

$NH_3(g)$ produced complexes with the $Zn^{2+}(aq)$ produced in the anode reaction $Zn^{2+}(aq) + 4\ NH_3(g) \longrightarrow [Zn(NH_3)_4]^{2+}(aq)$. The overall reaction is complicated.

12.79 See Table 12.2. (a) KOH(aq) (b) In the charging process, the cell reaction is the reverse of what occurs in discharge. Therefore, at the anode, $2\ Ni(OH)_2(s) + 2\ OH^-(aq) \longrightarrow 2\ Ni(OH)_3 + 2\ e^-$.

12.81

$$2[Zn^{2+}(aq) + 2\ e^- \longrightarrow Zn(s)] \qquad E°(\text{cathode}) = -0.76\ V$$

$$M(s) \longrightarrow M^{4+}(aq) + 4\ e^- \qquad E°(\text{anode}) = x$$

$$M(s) + 2\ Zn^{2+}(aq) \longrightarrow 2\ Zn(s) + M^{4+}(aq) \qquad E°_{cell} = 0.16\ V$$

$$E°_{cell} = E°(\text{cathode}) - E°(\text{anode})$$

$$+0.16\ V = -0.76\ V - (x)$$

$$x = -0.92\ V = E°(M^{4+}/M)$$

12.83 The strategy is to find the $E°$ value for the solubility reaction and then find appropriate half-reactions that add to give that solubility reaction. One of these half-reactions is our unknown, the other is obtained from Appendix 2B:

$$Cu(IO_3)_2(s) + 2\ e^- \longrightarrow Cu(s) + 2\ IO_3^-(aq) \qquad E° = ? \qquad (A)$$

$$Cu(s) \longrightarrow Cu^{2+}(aq) + 2\ e^- \qquad E° = -0.34\ V \qquad (B)$$

$$Cu(IO_3)_2(s) + \longrightarrow Cu^{2+}(aq) + 2\ IO_3^-(aq) \qquad E° = \frac{RT \ln K_{sp}}{nF} \qquad (C)$$

$$E° = \frac{RT \ln K_{sp}}{nF}$$

$$= \frac{(8.314\ J \cdot K^{-1} \cdot mol^{-1})(298.2\ K) \ln(1.4 \times 10^{-7})}{2(9.65 \times 10^4\ C \cdot mol^{-1})}$$

$$= -0.20\ V$$

$$-0.20\ V = E°(A) + (-0.34\ V)$$

$$E°(A) = +0.14\ V$$

12.85 (a) In acidic solution, the relevant reactions are

$$O_2 + 4\ H^+ + 4\ e^- \longrightarrow 2\ H_2O \qquad E° = +1.23\ V$$

$$Ag \longrightarrow Ag^+ + e^- \qquad E° = -0.80\ V$$

Overall reaction:

$$O_2 + 4\ H^+ + 4\ Ag \longrightarrow 4\ Ag^+ + 2\ H_2O \qquad E° = +0.43\ V$$

Because the potential is positive, the reaction should be spontaneous and would be expected to occur. We should also consider the conditions; because air is only 20.95% O_2, the potential may be different from that calculated for standard conditions. If air is the source of oxygen, then it will be present at 0.2095 × 1.013 25 bar = 0.2123 bar.

$$E = E° - \frac{0.0592}{4} \log \frac{[Ag^+]^4}{P_{O_2}[H^+]^4}$$

$$= +0.43\text{ V} - \frac{0.0592}{4} \log \frac{[1.0]^4}{(0.2123)[1.0]^4}$$

$$= +0.43\text{ V} - \frac{0.0592}{4} \log \frac{1}{0.2123}$$

$$= +0.43\text{ V} - 0.010\text{ V}$$

$$= +0.42\text{ V}$$

The potential is still positive and the reaction is expected to be spontaneous.

(b) In basic solution, the relevant reactions are

$O_2 + 2\,H_2O + 4\,e^- \longrightarrow 4\,OH^-$ $\qquad E° = +0.40$ V

$Ag \longrightarrow Ag^+ + e^-$ $\qquad E° = -0.80$ V

Overall reaction:

$O_2 + 2\,H_2O + 4\,Ag \longrightarrow 4\,Ag^+ + 4\,OH^-$ $\qquad E° = -0.40$ V

This process as written is nonspontaneous and is not predicted to occur. However, AgOH forms an insoluble precipitate, changing the nature of the reaction. The K_{sp} value for AgOH is 1.5×10^{-8}. We use the Nernst equation to calculate the potential under these conditions:

$$E = E° - \frac{0.0592}{4} \log \frac{[Ag^+]^4[OH^-]^4}{P_{O_2}}$$

$$= -0.40\text{ V} - \frac{0.0592}{4} \log \frac{K_{sp}{}^4}{P_{O_2}}$$

$$= -0.40\text{ V} - \frac{0.0592}{4} \log \frac{(1.5 \times 10^{-8})^4}{0.2132}$$

$$= -0.40\text{ V} + 0.45\text{ V}$$

$$= +0.05\text{ V}$$

Under these condition, the potential is slightly positive and the oxidation should be spontaneous.

12.87 In each case, determine the cathode and anode half-reactions corresponding to the reaction *as written*. Look up the standard reduction potentials for these half-reactions and then calculate $E°_{cell} = E°(\text{cathode}) - E°(\text{anode})$. If $E°_{cell}$ is positive, the reaction is spontaneous under standard conditions.

(a) $E°_{cell} = E°(\text{cathode}) - E°(\text{anode}) = +0.96\text{ V} - (+0.79\text{ V}) = +0.17\text{ V}$

Therefore, spontaneous galvanic cell:

$Hg(l)\,|\,Hg_2^{2+}(aq)\,||\,NO_3^-(aq), H^+(aq)\,|\,NO(g)\,|\,Pt(s)$

$\Delta G°_r = -nFE° = -(6)(9.65 \times 10^4\ C\cdot mol^{-1})(+0.17\ J\cdot C^{-1}) = -98\ kJ\cdot mol^{-1}$

(b) $E°_{cell} = E°(cathode) - E°(anode) = +0.92\ V - (+1.09\ V) = -0.17\ V$

Therefore, not spontaneous.

(c) $E°_{cell} = E°(cathode) - E°(anode) = +1.33\ V - (+0.97\ V) = +0.36\ V$

Therefore, spontaneous galvanic cell.

$Pt(s)\,|\,Pu^{3+}(aq), Pu^{4+}(aq)\,||\,Cr_2O_7^{2-}(aq), Cr^{3+}(aq), H^-(aq)\,|\,Pt(s)$

$\Delta G°_r = -nFE° = -(6)(9.65 \times 10^4\ C\cdot mol^{-1})(0.36\ J\cdot C^{-1}) = -208\ kJ\cdot mol^{-1}$

12.89 $F_2(g) + 2\ e^- \longrightarrow 2\ F^-(aq) \quad E°(cathode) = +2.87\ V$

$2\ HF(aq) \longrightarrow F_2(g) + 2\ H^+(aq) + 2\ e^- \quad E°(anode) = +3.03\ V$

$2\ HF(aq) \longrightarrow 2\ H^+(aq) + 2\ F^-(aq) \quad E°_{cell} = -0.16\ V$

For the above reaction, $K = \dfrac{[H^+]^2[F^-]^2}{[HF]^2}$ and $\ln K = \dfrac{nFE°}{RT}$

$$\text{at } 25°C = \frac{nE°}{0.025\ 69\ V} = \frac{(2)(-0.16\ V)}{0.025\ 69\ V} = -12$$

$K = 10^{-5}$

$K_a = \sqrt{K} = \sqrt{10^{-5}} = 10^{-3}$

12.91 The wording of this exercise suggests that K^+ ions participate in an electrolyte concentration cell reaction. Therefore, $E°_{cell} = 0.00\ V$, because the two half cells would be identical under standard conditions.

Then,

$$E = E° - \left(\frac{0.0257\ V}{n}\right) \ln\left(\frac{[K_{out}^+]}{[K_{in}^+]}\right) = 0.00\ V - \left(\frac{0.0257\ V}{1}\right) \ln\left(\frac{1}{30}\right)$$

$$= +0.09\ V$$

$$\text{and} \quad E = 0.00\ V - \left(\frac{0.0257\ V}{1}\right) \ln\left(\frac{1}{20}\right) = +0.08\ V$$

The range of potentials is 0.08 V to 0.09 V.

12.93 $Ag^+(aq) + e^- \longrightarrow Ag \quad E°(cathode) = +0.80\ V$

$Fe^{2+}(aq) \longrightarrow Fe^{3+}(aq) + e^- \quad E°(anode) = +0.77\ V$

$Ag^+(aq) + Fe^{2+}(aq) \longrightarrow Fe^{3+}(aq) + Ag(s) \quad E°_{cell} = +0.03\ V$

$$E_{cell} = E°_{cell} - \left(\frac{0.0257\ V}{n}\right) \ln\left(\frac{[Fe^{3+}]}{[Ag^+][Fe^{2+}]}\right)$$

$$= 0.03\ V - (0.0257\ V) \ln\left(\frac{1}{(0.010)(0.0010)}\right) = 0.03\ V - 0.30\ V$$

$$= -0.27\ V$$

Comment: The cell changes from spontaneous to nonspontaneous as a function of concentration.

12.95 (a) $O_2(g) + 4\,H^+(aq) + 4\,e^- \longrightarrow 2\,H_2O(l)$ $\quad E°(\text{cathode}) = +1.23\text{ V}$

$2[Sn^{2+}(aq) \longrightarrow Sn^{4+}(aq) + 2\,e^-]$ $\quad E°(\text{anode}) = +0.15\text{ V}$

$E°_{cell} = +1.08\text{ V}$

(b) $O_2(g) + 4\,H^+(aq) + 2\,Sn^{2+}(aq) \longrightarrow 2\,Sn^{4+}(aq) + 2\,H_2O(l)$

(c) $$E_{cell} = E°_{cell} - \frac{0.025\,693\text{ V}}{n} \ln \frac{[Sn^{4+}]^2}{[Sn^{2+}]^2[H^+]^4[O_2]}$$

$$E_{cell} = 1.08\text{ V} - \frac{0.025\,693\text{ V}}{4} \ln \frac{(0.010)^2}{(0.10)^2(1.0 \times 10^{-4})^4(1)}$$

$$E_{cell} = 1.08\text{ V} - 0.21\text{ V} = 0.87\text{ V}$$

(d) $$\ln K = \frac{nFE°}{RT} = \frac{nE°}{0.025\,693\text{ V}} = \frac{(4)(0.87)}{0.025\,693} = 1.4 \times 10^2 \text{ and } K = 6 \times 10^{60}$$

(e) $$\Delta G_r = -nFE = -(4)(9.6485) \times 10^4\text{ C}\cdot\text{mol}^{-1})(0.87\text{ J}\cdot\text{C}^{-1})$$
$$= -3.4 \times 10^2\text{ kJ}\cdot\text{mol}^{-1}$$

(f) $$E_{cell} = E°_{cell} - \frac{0.025\,693\text{ V}}{n} \ln \frac{[Sn^{4+}]^2}{[Sn^{2+}]^2[H^+]^4[O_2]}$$

$$0.89\text{ V} = 1.08\text{ V} - \frac{0.025\,693\text{ V}}{4} \ln \frac{(0.010)^2}{(0.10)^2(H^+)^4(1)}$$

Solving, $[H^+] = 1.9 \times 10^{-4}\text{ mol}\cdot\text{L}^{-1}$

and $\text{pH} = -\log[H^+] = -\log(1.9 \times 10^{-4}) = 3.72$

12.97 buffer system $= HA \rightleftharpoons H^+ + A^-$

$$Q = \frac{(H^+)(A^-)}{(HA)}$$

Note: (H^+), as opposed to $[H^+]$, indicates a nonequilibrium molarity.

Because in a buffer system $(A) \approx (HA)$, we can write

$$Q = (H^+)$$

$$E_{cell} = E°_{cell} - \frac{RT}{nF} \ln (H^+)$$

$$0.060\text{ V} = E°_{cell} - \left(\frac{0.025\,693}{1}\right)(2.303)(\log(H^+))$$

Because $\log(H^+) = -[-\log(H^+)] = -\text{pH}$, we have

$$0.060\text{ V} = E°_{cell} - 0.0592 \times (-\text{pH})$$

$$0.060\text{ V} = E°_{cell} + 0.0592 \times \text{pH}$$

$$0.060\text{ V} = E°_{cell} + 0.0592 \times 9.40$$

$$0.060\text{ V} = E°_{cell} + 0.556\text{ V}$$

$E° = 0.060 \text{ V} - 0.556 \text{ V} = -0.496 \text{ V}$

Similarly, $0.22 \text{ V} = -0.496 \text{ V} + 0.0592 \text{ V} \times \text{pH}$

$$\text{pH} = \frac{0.22 \text{ V} + 0.496 \text{ V}}{0.0592 \text{ V}} = 12$$

12.99 A simplified electrochemical series for the relevant couples can be written:

Reducing agent	Oxidizing agents	$E°$(V)
	Au^{3+}	+1.40
	Ag^{+}	+0.80
Cu		+0.34
	Ni^{2+}	−0.23
	Co^{2+}	−0.28
	Fe^{2+}	−0.44

It is evident that, at least under standard conditions (by no means guaranteed in an industrial process!), Cu cannot reduce Ni^{2+}, Co^{2+}, and Fe^{2+}, leaving Au and Ag as "anode mud." (Chemical engineers will adjust conditions, which will be non-standard, to secure this effect.)

12.101 (A) $ClO_4^- + 2\,H^+ + 2\,e^- \longrightarrow ClO_3^- + H_2O \qquad E° = +1.23 \text{ V}$
(B) $ClO_4^- + H_2O + 2\,e^- \longrightarrow ClO_3^- + 2\,OH^- \qquad E° = +0.36 \text{ V}$

(a) The Nernst equation can be used to derive the potential as a function of pH:

$$E' = E° - \frac{RT}{nF} \ln Q$$

For (A), $E'(\text{A}) = 1.23 \text{ V} - \frac{0.059\,16}{2} \log \frac{[ClO_3^-]}{[ClO_4^-][H^+]^2}$

We are only interested in varying $[H^+]$, so the $[ClO_3^-]$ and $[ClO_4^-]$ will be left at the standard values of 1 M.

$$E'(\text{A}) = +1.23 \text{ V} - \frac{0.059\,16}{2} \log \frac{1}{[H^+]^2}$$
$$= +1.23 \text{ V} - \frac{0.059\,16}{2} \times 2 \log \frac{1}{[H^+]}$$
$$= +1.23 \text{ V} - 0.059\,16\,(-\log [H^+])$$
$$= +1.23 \text{ V} - 0.059\,16 \text{ pH}$$

Similarly, for (B):

$$E'(\text{B}) = +0.36 \text{ V} - \frac{0.059\,16}{2} \log \frac{[ClO_3^-][OH^-]^2}{[ClO_4^-]}$$

As above, we are only interested in varying $[OH^-]$, so the $[ClO_3^-]$ and $[ClO_4^-]$ will be left at the standard values of 1 M.

$$E'(B) = +0.36\ V - \frac{0.059\ 16}{2} \log [OH^-]^2$$

$$= +0.36\ V - \frac{0.059\ 16}{2} \times 2 \log [OH^-]$$

$$= +0.36\ V - 0.059\ 16 \times \log [OH^-]$$

$$= +0.36\ V + 0.059\ 16\ pOH$$

Because $pOH + pH = pK_w = 14.00$, we can write:

$$pOH = 14.00 - pH$$

$$E'(B) = +0.36\ V + 0.059\ 16[14.00 - pH]$$

$$= +0.36\ V + 0.83\ V - 0.059\ 16\ pH$$

$$= +1.19\ V - 0.059\ 16\ pH$$

If we compare this to $E°(A)$, we find that the equations are essentially the same. They should be identical, the difference being due to the limitation of the number of significant figures available for the calculations.

(b) From the discussion above, we can see that the potential in neutral solution should be the same, regardless of which half-reaction we use to calculate the value.

Using $E°(A) = +1.23\ V - 0.059\ 16\ pH = +1.23\ V - 0.059\ 16\ (7.00) = 0.82$
Using $E'(B) = +0.36\ V + 0.059\ 16\ pOH = +0.36 + 0.059\ 16\ (7.00) = 0.77$

Although these numbers differ slightly, they should be identical; again the difference lies in the limitation of the number of significant figures.

12.103 (a) $Fe^{2+} + 2\ e^- \longrightarrow Fe \quad E° = -0.44\ V$
$Mn^{2+} + 2\ e^- \longrightarrow Mn \quad E° = -1.18$

Because these are reduction reactions, we need a corresponding oxidation. The nitrate ion contains N in its highest oxidation state so it cannot be oxidized further. The logical choice is the oxidation of water. The appropriate reduction potential is

$$O_2 + 4\ H^+ + 4\ e^- \longrightarrow 2\ H_2O \quad E° = +1.23\ V$$

The two overall reactions will be:

$$2\ Fe^{2+} + 2\ H_2O \longrightarrow 2\ Fe + O_2 + 4\ H^+ \quad E° = -0.44\ V - 1.23\ V = -1.67\ V$$

$$2\ Mn^{2+} + 2\ H_2O \longrightarrow 2\ Mn + O_2 + 4\ H^+ \quad E° = -1.18\ V - 1.23\ V = -2.41\ V$$

(b) The actual potentials, however, will differ from these standard potentials because the concentrations of the metal ions and hydrogen ions are not 1 M, and the pressure of O_2 is not 1 bar. To calculate the actual values, the Nernst equation is used.

For the Fe reaction: $E = -1.67\ \text{V} - \dfrac{0.059\ 16}{4} \log \dfrac{P_{O_2}[H^+]^4}{[Fe^{2+}]^2}$

In an open beaker, with the metal ions dissolved in water with pH = 5.00, the pressure of O_2 will be $0.2095 \times 1.00\ \text{atm} \times 0.987\ \text{bar}\cdot\text{atm}^{-1} = 0.207\ \text{bar}$. Substituting the specific values will give

$$E = -1.67\ \text{V} - \frac{0.059\ 16}{4} \log \frac{(0.207)(1.00 \times 10^{-5})^4}{(0.100)^2}$$

$$= -1.67\ \text{V} + 0.28\ \text{V} = -1.39\ \text{V}$$

For the Mn reaction:

$$E = -2.41\ \text{V} - \frac{0.059\ 16}{4} \log \frac{(0.207)(1.00 \times 10^{-5})^4}{(0.150)^2}$$

$$= -2.41\ \text{V} + 0.28\ \text{V} = -2.13\ \text{V}$$

In order to plate out iron from this mixture, 1.39 V must be applied, and 2.13 V must be applied to cause the reduction of Mn^{2+}.

(c) Because the potential for reducing iron(II) is more positive than the potential for reducing manganese(II), the iron will plate out first.

(d) The answer to this question is obtained from the Nernst equation by determining the concentration of Fe^{2+} when the applied potential reaches 2.13 V:

$$-2.13\ \text{V} = -1.67\ \text{V} - \frac{0.059\ 16}{4} \log \frac{(0.207)[H^+]^4}{[Fe^{2+}]^2}$$

$$-0.46\ \text{V} = -\frac{0.059\ 16}{4} \log \frac{(0.207)[H^+]}{[Fe^{2+}]^2}$$

$$31.10 = \log 0.207 + \log \frac{[H^+]^4}{[Fe^{2+}]^2}$$

$$31.78 = \log \frac{[H^+]^4}{[Fe^{2+}]^2}$$

$$\frac{[H^+]^4}{[Fe^{2+}]^2} = 6.0 \times 10^{31}$$

$$\frac{[H^+]^2}{[Fe^{2+}]} = 7.7 \times 10^{15}$$

For the last ratio to be 7.7×10^{15}, essentially all of the Fe^{2+} must be converted to Fe(s). This means that $[H^+]$ will essentially be $0.200\ \text{mol}\cdot\text{L}^{-1}$. Substituting this number gives $[Fe^{2+}] = 5 \times 10^{-18}\ \text{mol}\cdot\text{L}^{-1}$. We can say that the iron is quantitatively precipitated by this point.

We might note, however, that the potential of 2.13 V is now no longer the potential at which Mn^{2+} will begin to be reduced. Because the reduction of Fe^{2+}

has produced a considerable amount of acid, the original reduction of Mn^{2+} should be recalculated:

$$E = -2.41\text{ V} - \frac{0.059\ 16}{4}\log\frac{(0.207)(0.200)^4}{(0.150)^2} = -2.38\text{ V}$$

Thus, even less iron will remain in solution.

12.105 (a) Addition of an electron to any molecule should have the electron enter the molecule's lowest unoccupied molecular orbital (LUMO) first. (b) For CH_3X, one would predict that the LUMO would be antibonding between C and one of the atoms attached to it. Because the C—H bond strength ($412\text{ kJ}\cdot\text{mol}^{-1}$) is greater than all of the C—X bond strengths given (C—Cl, $338\text{ kJ}\cdot\text{mol}^{-1}$; C—Br, $276\text{ kJ}\cdot\text{mol}^{-1}$; C—I, $238\text{ kJ}\cdot\text{mol}^{-1}$) we would expect the LUMO to be the antibonding orbital for the C—X bond. Adding an electron to this orbital should then result in a weakening of the C—X bond. The result is the elimination of X^- and the formation of a CH_3 radical:

$$CH_3X + e^- \longrightarrow CH_3 + X^-$$

(c) We would expect this reduction process to follow the C—X bond strengths so that the formation of X^- and generation of CH_3 radicals would be easiest for X = I, followed by Br, and then Cl.

CHAPTER 13
CHEMICAL KINETICS

13.1 (a) $\text{rate}(N_2) = \text{rate}(H_2) \times \left(\dfrac{1 \text{ mol } N_2}{3 \text{ mol } H_2}\right) = \dfrac{1}{3} \times \text{rate}(H_2)$

(b) $\text{rate }(NH_3) = \text{rate}(H_2) \times \left(\dfrac{2 \text{ mol } NH_3}{3 \text{ mol } H_2}\right) = \dfrac{2}{3} \times \text{rate}(H_2)$

(c) $\text{rate}(NH_3) = \text{rate}(N_2) \times \left(\dfrac{2 \text{ mol } NH_3}{1 \text{ mol } N_2}\right) = 2 \times \text{rate}(N_2)$

13.3 (a) rate of formation of dichromate ions $= \left(\dfrac{0.14 \text{ mol } Cr_2O_7^{2-}}{\text{L}\cdot\text{s}}\right)\left(\dfrac{2 \text{ mol } CrO_4^{2-}}{1 \text{ mol } Cr_2O_7^{2-}}\right)$

$= 0.28 \text{ mol}\cdot\text{L}^{-1}\cdot\text{s}^{-1}$

(b) $0.14 \text{ mol}\cdot\text{L}^{-1}\cdot\text{s}^{-1} \div 1 = 0.14 \text{ mol}\cdot\text{L}^{-1}\cdot\text{s}^{-1}$

13.5 (a) rate of formation of $O_2 = \left(6.5 \times 10^{-3} \dfrac{\text{mol } NO_2}{\text{L}\cdot\text{s}}\right) \times \left(\dfrac{1 \text{ mol } O_2}{2 \text{ mol } NO_2}\right)$

$= 3.3 \times 10^{-3} \text{ (mol } O_2)\cdot\text{L}^{-1}\cdot\text{s}^{-1}$

(b) $6.5 \times 10^{-3} \text{ mol}\cdot\text{L}^{-1}\cdot\text{s}^{-1} \div 2 = 3.3 \times 10^{-3} \text{ mol}\cdot\text{L}^{-1}\cdot\text{s}^{-1}$

13.7 (a) and (c)

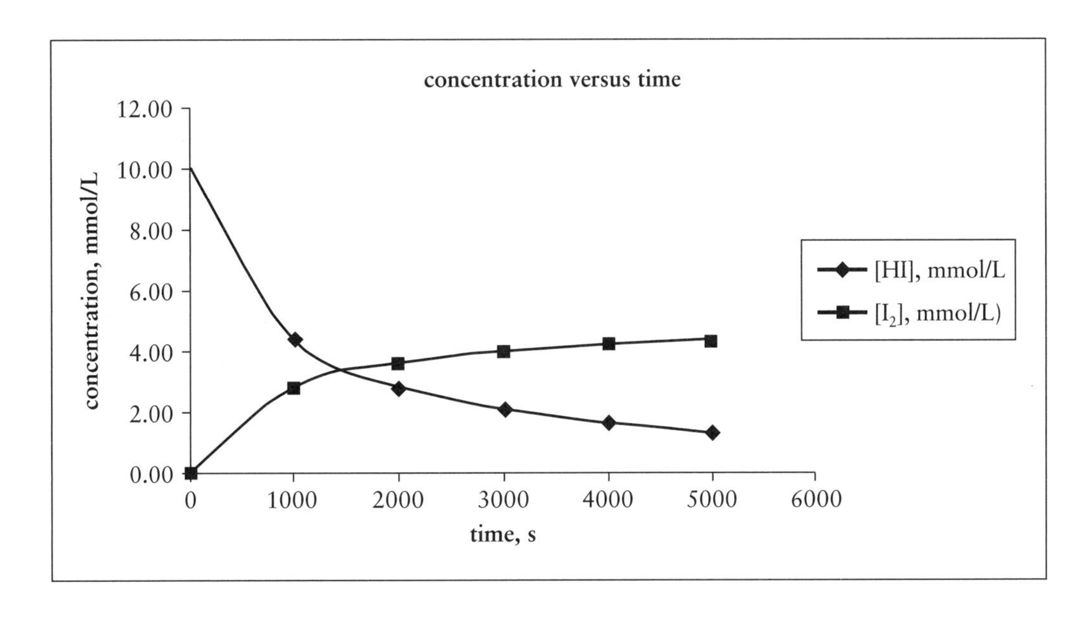

Note that the curves for the $[I_2]$ and $[H_2]$ are identical and only the $[I_2]$ curve is shown.

(b) The rates at individual points are given by the slopes of the lines tangent to the points in question. If these are determined graphically, there may be some variation from the numbers given below.

time, s	rate, $mmol \cdot L^{-1} \cdot s^{-1}$
0	−0.0060
1000	−0.003
2000	−0.000 98
3000	−0.000 61
4000	−0.000 40
5000	−0.000 31

13.9 For A $\longrightarrow$ products, rate = $(\text{mol A}) \cdot L^{-1} \cdot s^{-1}$

(a) rate $[(\text{mol A}) \cdot L^{-1} \cdot s^{-1}] = k_0[A]^0 = k_0$, so units of k_0 are $(\text{mol A}) \cdot L^{-1} \cdot s^{-1}$ (same as the units for the rate, in this case)

(b) rate $[(\text{mol A}) \cdot L^{-1} \cdot s^{-1}] = k_1[A]$, so units of k_1 are $\dfrac{(\text{mol A}) \cdot L^{-1} \cdot s^{-1}}{(\text{mol A}) \cdot L^{-1}} = s^{-1}$

(c) rate $[(\text{mol A}) \cdot L^{-1} \cdot s^{-1}] = k_2[A]^2$, so units of k_2 are $\dfrac{(\text{mol A}) \cdot L^{-1} \cdot s^{-1}}{[(\text{mol A}) \cdot L^{-1}]^2} =$

$L \cdot (\text{mol A})^{-1} \cdot s^{-1}$

13.11 From the units of the rate constant, k, it follows that the reaction is first order, thus rate $= k[N_2O_5]$.

$$[N_2O_5] = \left(\frac{3.45 \text{ g } N_2O_5}{0.750 \text{ L}}\right)\left(\frac{1 \text{ mol } N_2O_5}{108.02 \text{ g } N_2O_5}\right) = 0.0426 \text{ mol} \cdot L^{-1}$$

rate $= 5.2 \times 10^{-3} \text{ s}^{-1} \times 0.0426 \text{ mol} \cdot L^{-1} = 2.2 \times 10^{-4} \text{ (mol } N_2O_5) \cdot L^{-1} \cdot s^{-1}$

13.13 From the units of the rate constant, it follows that the reaction is second order; therefore,

rate $= k[H_2][I_2]$

$$= (0.063 \text{ L} \cdot \text{mol}^{-1} \cdot s^{-1})\left(\frac{0.45 \text{ g } H_2}{0.750 \text{ L}}\right)\left(\frac{1 \text{ mol } H_2}{2.016 \text{ g } H_2}\right)\left(\frac{0.16 \text{ g } I_2}{0.750 \text{ L}}\right)\left(\frac{1 \text{ mol } I_2}{253.8 \text{ g } I_2}\right)$$

$$= 1.6 \times 10^{-5} \text{ mol} \cdot L^{-1} \cdot s^{-1}$$

(b) rate(new) $= k \times 2 \times [H_2]_{initial}[I_2] = 2 \times$ rate(initial), so, by a factor of 2

13.15 Because the rate increased in direct proportion to the concentrations of both reactants, the rate is first order in both reactants. rate $= k[CH_3Br][OH^-]$

13.17 When the concentration of ICl is doubled, the rate doubles (experiments 1 and 2). Therefore, the reaction is first order in ICl. When the concentration of H_2 is tripled, the rate triples (experiments 2 and 3); thus, the reaction is first order in H_2.

(a) rate $= k[\text{ICl}][\text{H}_2]$

(b) $k = \left(\dfrac{22 \times 10^{-7}\ \text{mol}}{\text{L}\cdot\text{s}}\right)\left(\dfrac{\text{L}}{3.0 \times 10^{-3}\ \text{mol}}\right)\left(\dfrac{\text{L}}{4.5 \times 10^{-3}\ \text{mol}}\right)$

$= 0.16\ \text{L}\cdot\text{mol}^{-1}\cdot\text{s}^{-1}$

(c) rate $= \left(\dfrac{0.16\ \text{L}}{\text{mol}\cdot\text{s}}\right)\left(\dfrac{4.7 \times 10^{-3}\ \text{mol}}{\text{L}}\right)\left(\dfrac{2.7 \times 10^{-3}\ \text{mol}}{\text{L}}\right)$

$= 2.0 \times 10^{-6}\ \text{mol}\cdot\text{L}^{-1}\cdot\text{s}^{-1}$

13.19 (a) Doubling the concentration of A (experiments 1 and 2) doubles the rate; therefore, the reaction is first order in A. Increasing the concentration of B by the ratio 3.02/1.25 (experiments 2 and 3) increases the rate by $(3.02/1.25)^2$; hence, the reaction is second order in B. Tripling the concentration of C (experiment 3 and 4) increases the rate by $3^2 = 9$; thus, the reaction is second order in C. Therefore, rate $= k[\text{A}][\text{B}]^2[\text{C}]^2$.

(b) overall order = 5

(c) $k = \dfrac{\text{rate}}{[\text{A}][\text{B}]^2[\text{C}]^2}$

Using the data from experiment 4, we get

$k = \left(\dfrac{0.457\ \text{mol}}{\text{L}\cdot\text{s}}\right)\left(\dfrac{\text{L}}{1.25 \times 10^{-3}\ \text{mol}}\right)\left(\dfrac{\text{L}}{3.02 \times 10^{-3}\ \text{mol}}\right)^2\left(\dfrac{\text{L}}{3.75 \times 10^{-3}\ \text{mol}}\right)^2$

$= 2.85 \times 10^{12}\ \text{L}^4\cdot\text{mol}^{-4}\cdot\text{s}^{-1}$

From experiment 3, we get

$k = \left(\dfrac{5.08 \times 10^{-2}\ \text{mol}}{\text{L}\cdot\text{s}}\right)\left(\dfrac{\text{L}}{1.25 \times 10^{-3}\ \text{mol}}\right)\left(\dfrac{\text{L}}{3.02 \times 10^{-3}\ \text{mol}}\right)^2\left(\dfrac{\text{L}}{1.25 \times 10^{-3}\ \text{mol}}\right)^2$

$= 2.85 \times 10^{12}\ \text{L}^4\cdot\text{mol}^{-4}\cdot\text{s}^{-1}$ (Checks!)

(d) rate $= \left(\dfrac{2.85 \times 10^{12}\ \text{L}^4}{\text{mol}^4\cdot\text{s}}\right)\left(\dfrac{3.01 \times 10^{-3}\ \text{mol}}{\text{L}}\right)\left(\dfrac{1.00 \times 10^{-3}\ \text{mol}}{\text{L}}\right)^2\left(\dfrac{1.15 \times 10^{-3}\ \text{mol}}{\text{L}}\right)^2$

$= 1.13 \times 10^{-2}\ \text{mol}\cdot\text{L}^{-1}\cdot\text{s}^{-1}$

13.21 (a) $k = \dfrac{0.693}{t_{1/2}} = \dfrac{0.693}{1000\ \text{s}} = 6.93 \times 10^{-4}\ \text{s}^{-1}$

(b) We use $\ln\left(\dfrac{[\text{A}]_0}{[\text{A}]_t}\right) = kt$ and solve for k.

$$k = \frac{\ln([\text{A}]_0/[\text{A}]_t)}{t} = \frac{\ln\left(\dfrac{0.67\ \text{mol}\cdot\text{L}^{-1}}{0.53\ \text{mol}\cdot\text{L}^{-1}}\right)}{25\ \text{s}} = 9.4 \times 10^{-3}\ \text{s}^{-1}$$

(c) $[A]_t = \left(\frac{0.153 \text{ mol A}}{\text{L}}\right) - \left[\left(\frac{2 \text{ mol A}}{1 \text{ mol B}}\right)\left(\frac{0.034 \text{ mol B}}{\text{L}}\right)\right]$

$= 0.085 \text{ (mol A)}\cdot\text{L}^{-1}$

$$k = \frac{\ln\left(\frac{0.153 \text{ mol}\cdot\text{L}^{-1}}{0.085 \text{ mol}\cdot\text{L}^{-1}}\right)}{115 \text{ s}} = 5.1 \times 10^{-3} \text{ s}^{-1}$$

13.23 (a) $t_{1/2} = \frac{0.693}{k} = \left(\frac{0.693 \text{ s}}{3.7 \times 10^{-5}}\right)\left(\frac{1 \text{ min}}{60 \text{ s}}\right)\left(\frac{1 \text{ h}}{60 \text{ min}}\right) = 5.2 \text{ h}$

(b) $[A]_t = [A]_0 \, e^{-kt}$

$t = 3.5 \text{ h} \times 3600 \text{ s}\cdot\text{h}^{-1} = 1.3 \times 10^4 \text{ s}$

$[N_2O_5] = 0.0567 \text{ mol}\cdot\text{L}^{-1} \times e^{-(3.7\times10^{-5} \text{ s}^{-1})(1.3\times10^4 \text{ s})} = 3.5 \times 10^{-2} \text{ mol}\cdot\text{L}^{-1}$

(c) Solve for t from $\ln\left(\frac{[A]_0}{[A]_t}\right) = kt$, which gives

$$t = \frac{\ln\left(\frac{[A]_0}{[A]_t}\right)}{k} = \frac{\ln\left(\frac{[N_2O_5]_0}{[N_2O_5]_t}\right)}{k} = \frac{\ln\left(\frac{0.0567}{0.0135}\right)}{3.7 \times 10^{-5} \text{ s}^{-1}} = 3.9 \times 10^4 \text{ s}$$

$$= (3.9 \times 10^4 \text{ s})\left(\frac{1 \text{ min}}{60 \text{ s}}\right) = 6.5 \times 10^2 \text{ min}$$

13.25 (a) $\frac{[A]}{[A]_0} = \frac{1}{4} = \left(\frac{1}{2}\right)^2$; so the time elapsed is 2 half-lives.

$t = 2 \times 355 \text{ s} = 710 \text{ s}$

(b) Because 15% is not a multiple of $\frac{1}{2}$, we cannot work directly from the half-life. But $k = 0.693/t_{1/2}$

so $k = \frac{0.693}{355 \text{ s}} = 1.95 \times 10^{-3} \text{ s}^{-1}$

Then [see the solution to Exercise 13.23(c)],

$$t = \frac{\ln\left(\frac{[A]_0}{[A]_t}\right)}{k} = \frac{\ln\left(\frac{1}{0.15}\right)}{1.95 \times 10^{-3} \text{ s}^{-1}} = 9.7 \times 10^2$$

(c) $t = \frac{\ln \frac{[A]_0}{\frac{1}{9}[A]_0}}{k} = \frac{\ln 9}{1.95 \times 10^{-3} \text{ s}^{-1}} = 1.1 \times 10^3 \text{ s}$

13.27 (a) $t_{1/2} = \frac{0.693}{k} = \frac{0.693}{2.81 \times 10^{-3} \text{ min}^{-1}} = 247 \text{ min}$

(b) See the solutions to Exercises 13.31(c) and 13.33(c).

$$t = \frac{\ln\left(\frac{[SO_2Cl_2]_0}{[SO_2Cl_2]_t}\right)}{k} = \frac{\ln 10}{2.81 \times 10^{-3}\ \text{min}^{-1}} = 819\ \text{min}$$

(c) $[A]_t = [A]_0\, e^{-kt}$

Because the vessel is sealed, masses and concentrations are proportional, and we write

$$\begin{aligned}(\text{mass left})_t &= (\text{mass})_0\, e^{-kt} \\ &= 14.0\ \text{g} \times e^{-(2.81\times10^{-3}\ \text{min}^{-1}\times 60\ \text{min}\cdot\text{h}^{-1}\times 1.5\ \text{h})} \\ &= 10.9\ \text{g}\end{aligned}$$

Note: Knowledge of the volume of the vessel is not required. However, we could have converted mass to concentration, solved for the new concentration at 1.5 h, and finally converted back to the new (remaining) mass. But this is not necessary.

13.29 (a) We first calculate the concentration of A at 3.0 min.

$$\begin{aligned}[A]_t &= [A]_0 - \left(\frac{1\ \text{mol A}}{3\ \text{mol B}}\right) \times [B]_t \\ &= 0.015\ \text{mol}\cdot\text{L}^{-1} - \left(\frac{1\ \text{mol A}}{3\ \text{mol B}}\right) \times 0.018\ (\text{mol B})\cdot\text{L}^{-1} \\ &= 0.009\ \text{mol}\cdot\text{L}^{-1}\end{aligned}$$

The rate constant is then determined from the first-order integrated rate law.

$$k = \frac{\ln\left(\frac{[A]_0}{[A]_t}\right)}{t} = \frac{\ln\left(\frac{0.015}{0.009}\right)}{3.0\ \text{min}} = 0.17\ \text{min}^{-1}$$

(b) $[A]_t = 0.015\ \text{mol}\cdot\text{L}^{-1} - \left(\frac{1\ \text{mol A}}{3\ \text{mol B}}\right) \times 0.030\ (\text{mol B})\cdot\text{L}^{-1}$

$= 0.005\ \text{mol}\cdot\text{L}^{-1}$

$$t = \frac{\ln\left(\frac{[A]_0}{[A]_t}\right)}{k} = \frac{\ln\left(\frac{0.015}{0.005}\right)}{0.17\ \text{min}^{-1}} = 6.5\ \text{min}$$

additional time = 6.5 min − 3.0 min = 3.5 min

13.31 (a) Draw up the following table and plot 1/[HI] against time.

time, s	[HI], $\text{mol}\cdot\text{L}^{-1}$	1/[HI] $\text{L}\cdot\text{mol}^{-1}$
0	1.00	1.00
1000	0.112	8.93
2000	0.061	16
3000	0.041	24
4000	0.031	32

Equation 17.b in the text can be rearranged as

$$\frac{1}{[\mathrm{A}]_t} = \frac{1 + [\mathrm{A}]_0\, kt}{[\mathrm{A}]_0} = \frac{1}{[\mathrm{A}]_0} + kt$$

Thus, if the reaction is second order, a plot of 1/[HI] against time should give a straight line of slope k.

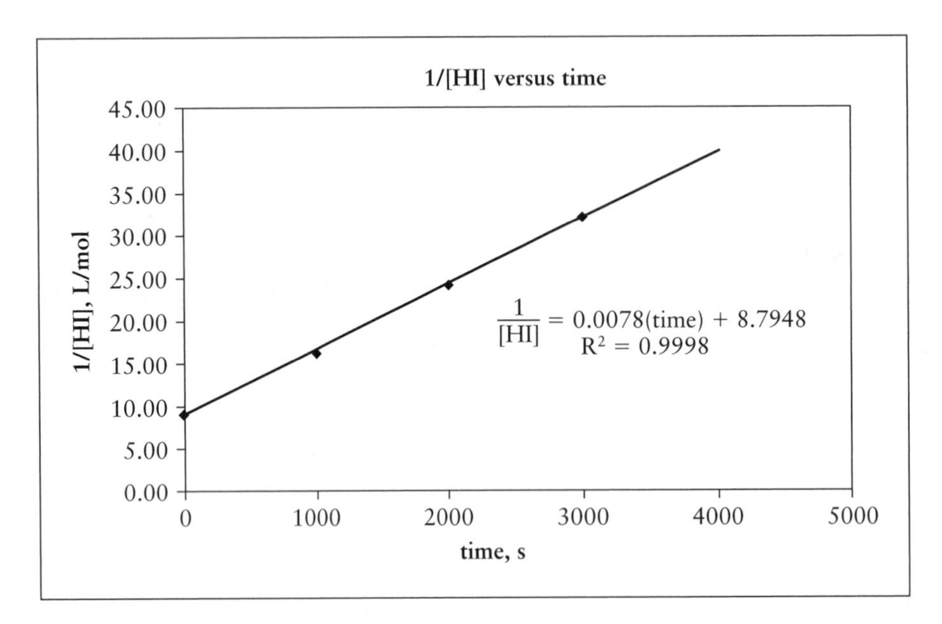

As can be seen from the graph, the data fit the equation for a second-order reaction quite well. The slope is determined by a least squares fit of the data by the graphing program.

(b) slope $= k = 7.8 \times 10^{-3}\ \mathrm{L{\cdot}mol^{-1}{\cdot}s^{-1}}$

13.33 It is convenient to obtain an expression for the half-life of a second-order reaction. We work with Eq. 17.b.

$$[\mathrm{A}]_t = \frac{[\mathrm{A}]_0}{1 + [\mathrm{A}]_0 kt} \qquad (17.\mathrm{b})$$

$$\frac{[\mathrm{A}]_{t_{1/2}}}{[\mathrm{A}]_0} = \frac{1}{2} = \frac{1}{1 + [\mathrm{A}]_0 k t_{1/2}}$$

Therefore, $1 + [\mathrm{A}]_0 k t_{1/2} = 2$, or $[\mathrm{A}]_0 k t_{1/2} = 1$, or

$$t_{1/2} = \frac{1}{k[\mathrm{A}]_0} \quad \text{and} \quad k = \frac{1}{t_{1/2}[\mathrm{A}]_0}$$

It is also convenient to rewrite Eq. 17.b to solve for t. We take reciprocals:

$$\frac{1}{[\mathrm{A}]_t} = \frac{1}{[\mathrm{A}]_0} + kt$$

giving

$$t = \frac{\dfrac{1}{[\mathrm{A}]_t} - \dfrac{1}{[\mathrm{A}]_0}}{k}$$

$$\text{(a) } k = \frac{1}{t_{1/2}[\text{A}]_0} = \frac{1}{(50.5\ \text{s})(0.84\ \text{mol}\cdot\text{L}^{-1})} = 0.024\ \text{L}\cdot\text{mol}^{-1}\cdot\text{s}^{-1}$$

$$t = \frac{\dfrac{1}{[\text{A}]} - \dfrac{1}{[\text{A}]_0}}{k} = \frac{\dfrac{16}{[\text{A}]_0} - \dfrac{1}{[\text{A}]_0}}{k} = \frac{15}{k[\text{A}]_0}$$

$$= \frac{15}{(0.024\ \text{L}\cdot\text{mol}^{-1}\cdot\text{s}^{-1})(0.84\ \text{mol}\cdot\text{L}^{-1})} = 7.4 \times 10^2\ \text{s}$$

$$\text{(b) } t = \frac{\dfrac{4}{[\text{A}]_0} - \dfrac{1}{[\text{A}]_0}}{k} = \frac{3}{k[\text{A}]_0}$$

$$= \frac{3}{(0.024\ \text{L}\cdot\text{mol}^{-1}\cdot\text{s}^{-1})(0.84\ \text{mol}\cdot\text{L}^{-1})} = 1.5 \times 10^2\ \text{s}$$

$$\text{(c) } t = \frac{\dfrac{5}{[\text{A}]_0} - \dfrac{1}{[\text{A}]_0}}{k} = \frac{4}{k[\text{A}]_0}$$

$$= \frac{4}{(0.024\ \text{L}\cdot\text{mol}^{-1}\cdot\text{s}^{-1})(0.84\ \text{mol}\cdot\text{L}^{-1})} = 2.0 \times 10^2\ \text{s}$$

13.35 See the solution to Exercise 13.33 for the derivation of the formulas needed here.

$$\text{(a) } t = \frac{\dfrac{1}{[\text{A}]} - \dfrac{1}{[\text{A}]_0}}{k} = \frac{\dfrac{1\ \text{L}}{0.080\ \text{mol}} - \dfrac{1\ \text{L}}{0.10\ \text{mol}}}{0.015\ \text{L}\cdot\text{mol}^{-1}\cdot\text{min}^{-1}} = 1.7 \times 10^2\ \text{min}$$

$$\text{(b) } [\text{A}] = \frac{0.15\ \text{mol A}}{\text{L}} - \left[\left(\frac{0.19\ \text{mol B}}{\text{L}}\right)\left(\frac{1\ \text{mol A}}{2\ \text{mol B}}\right)\right]$$

$$= 0.055(\text{mol A})\cdot\text{L}^{-1} = 0.37[\text{A}]_0$$

$$t = \frac{\dfrac{1}{[\text{A}]_t} - \dfrac{1}{[\text{A}]_0}}{k}$$

$$= \frac{\dfrac{1}{0.055\ \text{mol}\cdot\text{L}^{-1}} - \dfrac{1}{0.15\ \text{mol}\cdot\text{L}^{-1}}}{0.0035\ \text{L}\cdot\text{mol}^{-1}\cdot\text{min}^{-1}}$$

$$= 3.3 \times 10^3\ \text{min}$$

13.37 (a) Make the following table and graph:

T, K	$1/T$, K^{-1}	k, s^{-1}	$\ln k$
750	0.001 33	1.8×10^{-4}	−8.62
800	0.001 25	2.7×10^{-3}	−5.91
850	0.001 18	3.0×10^{-2}	−3.51
900	0.001 11	2.6×10^{-1}	−1.35

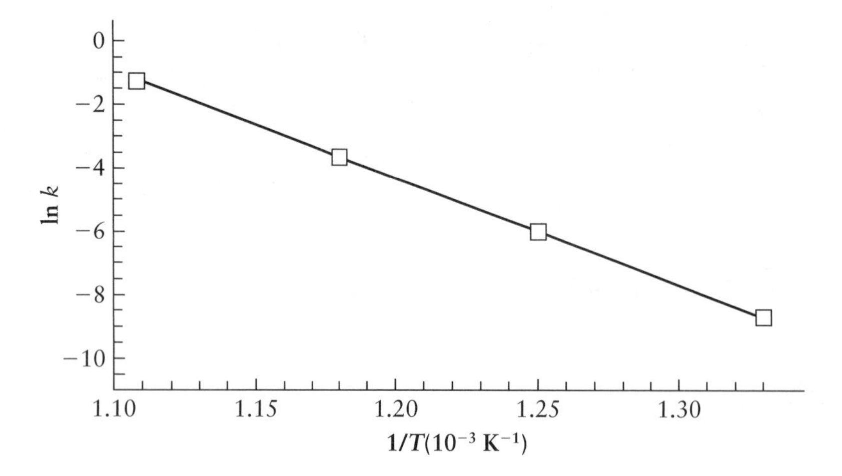

$\ln k = \ln A - E_a/RT$

The slope of the plot can be determined from a graphing program using a least squares fitting routine or from two points on the graph. The slope from the graphing program used here gave a value for the slope of -3.26×10^4. From two points:

$$\text{slope of plot} = -E_a/R = \frac{[-1.35 - (-8.62)]\ \text{K}}{0.001\ 11 - 0.001\ 33} = -3.3 \times 10^4\ \text{K}$$

$$E_a = (3.3 \times 10^4\ \text{K})(8.31 \times 10^{-3}\ \text{kJ}\cdot\text{mol}^{-1}\cdot\text{K}^{-1}) = 2.7 \times 10^2\ \text{kJ}\cdot\text{mol}^{-1}$$

(b) $$\ln\left(\frac{k'}{k}\right) = \frac{E_a}{R}\left(\frac{1}{T} - \frac{1}{T'}\right)$$

$$= \left(\frac{2.7 \times 10^2\ \text{kJ}\cdot\text{mol}^{-1}}{8.31 \times 10^{-3}\ \text{kJ}\cdot\text{K}^{-1}\cdot\text{mol}^{-1}}\right)\left(\frac{1}{750\ \text{K}} - \frac{1}{873\ \text{K}}\right)$$

$$= 6.2$$

$$\frac{k'}{k} = 4.9 \times 10^2$$

$$k' = (4.9 \times 10^2)(1.8 \times 10^{-4}\ \text{s}^{-1}) = 8.8 \times 10^{-2}\ \text{s}^{-1}$$

An approximate value of k' can also be obtained from the plot itself and yields the same value.

13.39 We use $\ln\left(\frac{k'}{k}\right) = \frac{E_a}{R}\left(\frac{1}{T} - \frac{1}{T'}\right) = \frac{E_a}{R}\left(\frac{T' - T}{T'T}\right)$

$$\ln\left(\frac{k'}{k}\right) = \ln\left(\frac{0.87\ \text{s}^{-1}}{0.76\ \text{s}^{-1}}\right)$$

$$= \left(\frac{E_a}{8.31 \times 10^{-3}\ \text{kJ}\cdot\text{K}^{-1}\cdot\text{mol}^{-1}}\right)\left(\frac{1030\ \text{K} - 1000\ \text{K}}{1030\ \text{K} \times 1000\ \text{K}}\right)$$

$$E_a = \frac{(8.31 \times 10^{-3}\ \text{kJ}\cdot\text{K}^{-1}\cdot\text{mol}^{-1})(1000\ \text{K})(1030\ \text{K})(0.14)}{30\ \text{K}} = 4.0 \times 10^2\ \text{kJ}\cdot\text{mol}^{-1}$$

13.41 We use $\ln\left(\frac{k'}{k}\right) = \frac{E_a}{R}\left(\frac{1}{T} - \frac{1}{T'}\right) = \frac{E_a}{R}\left(\frac{T' - T}{TT'}\right)$

k' = rate constant at 700°C, $T' = (700 + 273)\text{ K} = 973\text{ K}$

$$\ln\left(\frac{k'}{k}\right) = \left(\frac{315\text{ kJ}\cdot\text{mol}^{-1}}{8.314 \times 10^{-3}\text{ kJ}\cdot\text{K}^{-1}\cdot\text{mol}^{-1}}\right)\left(\frac{973\text{ K} - 1073\text{ K}}{973\text{ K} \times 1073\text{ K}}\right)$$

$$= -3.63; \quad \frac{k'}{k} = 0.026$$

$k' = 0.026 \times 9.7 \times 10^{10}\text{ L}\cdot\text{mol}^{-1}\cdot\text{s}^{-1} = 2.5 \times 10^{9}\text{ L}\cdot\text{mol}^{-1}\cdot\text{s}^{-1}$

13.43 $\ln\left(\frac{k'}{k}\right) = \frac{E_a}{R}\left(\frac{1}{T} - \frac{1}{T'}\right) = \frac{E_a}{R}\left(\frac{T' - T}{TT'}\right)$

$$= \left(\frac{103\text{ kJ}\cdot\text{mol}^{-1}}{8.314 \times 10^{-3}\text{ kJ}\cdot\text{K}^{-1}\cdot\text{mol}^{-1}}\right)\left(\frac{323\text{ K} - 318\text{ K}}{318\text{ K} \times 323\text{ K}}\right) = 0.60$$

$\frac{k'}{k} = 1.8$

$k' = 1.8 \times 5.1 \times 10^{-4}\text{ s}^{-1} = 9.2 \times 10^{-4}\text{ s}^{-1}$

13.45 The overall reaction is $CH_2{=}CHCOOH + HCl \longrightarrow ClCH_2CH_2COOH$. The intermediates include chloride ion, $CH_2{=}CHC(OH)_2^{+}$ and $ClCH_2CHC(OH)_2$.

13.47 The first elementary reaction is the rate-controlling step, because it is the slow step. The second elementary reaction is fast and does not affect the overall reaction order, which is second order as a result of the fact that the rate-controlling step is bimolecular.

rate $= k[NO][Br_2]$

13.49 The overall rate is determined by the slow step. rate $= k_3[COCl][Cl_2]$. But COCl is an intermediate and its concentration has to be eliminated.

$k_2[Cl][CO] = k'_2[COCl]$

$$[COCl] = \frac{k_2}{k'_2}[Cl][CO]$$

but we cannot leave [Cl] in the expression either:

$k_1[Cl_2] = k'_1[Cl]^2$, giving $[Cl] = \sqrt{\frac{k_1}{k'_1}}\,[Cl_2]^{1/2}$

$[COCl] = \left(\frac{k_2}{k'_2}\right)\sqrt{\frac{k_1}{k'_1}}\,[Cl_2]^{1/2}[CO]$, therefore

rate $= k_3(k_2/k'_2)(k_1/k'_1)^{1/2}[CO][Cl_2]^{3/2} = k[CO][Cl_2]^{3/2}$

13.51 If mechanism (a) were correct, the rate law would be rate $= k_2[NO_2][CO]$. But this expression does not agree with the experimental result and can be eliminated as a possibility. Mechanism (b) has rate $= k_2[NO_2]^2$ from the slow step. Step 2 does not influence the overall rate, but it is necessary to achieve the correct overall reaction; thus this mechanism agrees with the experimental data $k = k_2$. Mechanism (c) is not correct, which can be seen from the rate expression for the slow step, rate $= k_2[NO_3][CO]$. [CO] cannot be eliminated from this expression to yield the experimental result, which does not contain [CO].

13.53 (a) True; (b) False. At equilibrium, the *rates* of the forward and reverse reactions are equal, *not the rate constants.* (c) True. (d) False. Increasing the concentration of a reactant causes the rate to increase by providing more reacting molecules. It does not affect the rate constant of the reaction.

13.55 (a) The equilibrium constant will be given by the ratio of the rate constant of the forward reaction to the rate constant of the reverse reaction:

$$K = \frac{k}{k'} = \frac{265\ \text{L}\cdot\text{mol}^{-1}\cdot\text{min}^{-1}}{392\ \text{L}\cdot\text{mol}^{-1}\cdot\text{min}^{-1}} = 0.676$$

(b) The reaction profile corresponds to a plot similar to that shown in Fig. 13.31a. The reaction is endothermic—the reverse reaction has a lower activation barrier than the forward reaction.

(c) Raising the temperature will increase the rate constant of the reaction with the higher activation barrier more than it will the rate constant of the reaction with the lower energy barrier. We expect the rate of the forward reaction to go up substantially more than for the reverse reaction in this case. k will increase more than k' and consequently the equilibrium constant K will increase. This is consistent with Le Chatelier's principle.

13.57 (a) cat = catalyzed, uncat = uncatalyzed $E_{a,cat} = \frac{75}{125} E_{a,uncat} = 0.60\ E_{a,uncat}$

$$\frac{\text{rate(cat)}}{\text{rate(uncat)}} = \frac{k_{cat}}{k_{uncat}} = \frac{Ae^{-E_{a,cat}/RT}}{Ae^{-E_a/RT}} = \frac{e^{-(0.40)E_a/RT}}{e^{-E_a/RT}} = e^{(0.40)E_a/RT}$$

$$e^{[(0.40)(125\ \text{kJ}\cdot\text{mol}^{-1})/(8.314\times10^{-3}\ \text{kJ}\cdot\text{K}^{-1}\cdot\text{mol}^{-1}\times298\ \text{K})]} = 6 \times 10^8$$

(b) The last step of the calculation in (a) is repeated with $T = 350$ K.

$$e^{[(0.40)(125\ \text{kJ}\cdot\text{mol}^{-1})/(8.314\times10^{-3}\ \text{kJ}\cdot\text{K}^{-1}\cdot\text{mol}^{-1}\times350\ \text{K})]} = 3 \times 10^7$$

The rate enhancement is lower at higher temperatures.

13.59 cat = catalyzed, uncat = uncatalyzed

$$\frac{\text{rate(cat)}}{\text{rate(uncat)}} = \frac{k_{cat}}{k_{uncat}} = 1000 = \frac{Ae^{-E_{a,cat}/RT}}{Ae^{-E_a/RT}} = \frac{e^{-E_{a,cat}/RT}}{e^{-E_a/RT}}$$

$$\ln 1000 = \frac{-E_{a,cat}}{RT} + \frac{E_a}{RT}$$

$$\begin{aligned} E_{a,cat} &= E_a - RT \ln 1000 \\ &= 98\ \text{kJ}\cdot\text{mol}^{-1} - (8.31 \times 10^{-3}\ \text{kJ}\cdot\text{K}^{-1}\cdot\text{mol}^{-1})(298\ \text{K})(\ln 1000) \\ &= 81\ \text{kJ}\cdot\text{mol}^{-1} \end{aligned}$$

13.61 Balanced equation: $CH_3OH(aq) + HBr(aq) \rightleftharpoons CH_3Br(aq) + H_2O(l)$
Intermediates: $CH_3OH_2^+$, CH_3^+

13.63 (a) False. A catalyst increases the rate of both the forward and reverse reactions by providing a completely different pathway. (b) True, although a catalyst may be poisoned and lose activity. (c) False. There is a completely different pathway provided for the reaction in the presence of a catalyst. (d) False. The position of the equilibrium is unaffected by the presence of a catalyst.

13.65 rate $= k[C_4H_9Br]^a[OH^-]^b$
Because the rate increases in direct proportion to $[C_4H_9Br]$, a must be 1. Because $[OH^-]$ has no effect on the rate, b must be 0. Therefore,
rate $= k[C_4H_9Br]^1[OH^-]^0 = k[C_4H_9Br]$
The reaction is first order in $[C_4H_9Br]$, zero order in $[OH^-]$, and first order overall.

13.67 (a) The easiest way to solve this problem is to set up a system of simultaneous equations.

$[H_2SeO_3]$	$[I^-]$	$[H^+]$	Rate, $\text{mol}\cdot\text{L}^{-1}\cdot\text{s}^{-1}$
0.020	0.020	0.010	8.0×10^{-6}
0.020	0.010	0.020	4.0×10^{-6}
0.020	0.030	0.030	2.4×10^{-4}
0.010	0.020	0.020	1.6×10^{-5}

We have the following general relationship:
rate $= k[H_2SeO_3]^x[I^-]^y[H^+]^z$, which can be rewritten for ease of computation as
$\ln(\text{rate}) = \ln k + x \ln [H_2SeO_3] + y \ln [I^-] + z \ln [H^+]$
Using the data above, we can create four equations, which should be enough to solve the system of four unknown variables:
$\ln(8.0 \times 10^{-6}) = \ln k + x \ln 0.020 + y \ln 0.020 + z \ln 0.010$
$\ln(4.0 \times 10^{-6}) = \ln k + x \ln 0.020 + y \ln 0.010 + z \ln 0.020$

$\ln(2.4 \times 10^{-4}) = \ln k + x \ln 0.020 + y \ln 0.030 + z \ln 0.030$

$\ln(1.6 \times 10^{-5}) = \ln k + x \ln 0.010 + y \ln 0.020 + z \ln 0.020$

which give, upon calculating the numerical logarithms:

$-11.74 = \ln k - 3.91\,x - 3.91\,y - 4.60\,z \qquad (1)$

$-12.40 = \ln k - 3.91\,x - 4.60\,y - 3.91\,z \qquad (2)$

$-8.33 = \ln k - 3.91\,x - 3.51\,y - 3.51\,z \qquad (3)$

$-11.04 = \ln k - 4.60\,x - 3.91\,y - 3.91\,z \qquad (4)$

Solving this set of simultaneous equations and rounding the x, y, and z answers to the nearest whole number gives $x = 1$, $y = 3$, and $z = 2$, with $k = 5.0 \times 10^5$ $\text{L}^5\cdot\text{mol}^{-5}\cdot\text{s}^{-1}$.

(b) With $[H_2SeO_3] = 0.035\ \text{mol}\cdot\text{L}^{-1}$, $[I^-] = 0.020\ \text{mol}\cdot\text{L}^{-1}$ and $[H^+] =$ $0.015\ \text{mol}\cdot\text{L}^{-1}$,

$$\text{rate} = (5.0 \times 10^5\ \text{L}^5\cdot\text{mol}^{-5}\cdot\text{s}^{-1})(0.035\ \text{mol}\cdot\text{L}^{-1})^1(0.020\ \text{mol}\cdot\text{L}^{-1})^3 (0.015\ \text{mol}\cdot\text{L}^{-1})^2$$

$$\text{rate} = 3.2 \times 10^{-5}\ \text{mol}^{-1}\cdot\text{L}^{-1}\cdot\text{s}^{-1}$$

13.69 (a) $$k = \frac{\dfrac{1}{[\text{A}]_t} - \dfrac{1}{[\text{A}]_0}}{t} = \frac{\dfrac{1}{0.0050\ \text{mol}\cdot\text{L}^{-1}} - \dfrac{1}{0.040\ \text{mol}\cdot\text{L}^{-1}}}{12\ \text{h}} = 15\ \text{L}\cdot\text{mol}^{-1}\cdot\text{h}^{-1}$$

(b) $$[EX_2] = 0.040\,\frac{\text{mol EX}_2}{\text{L}} - \left(0.070\,\frac{\text{mol X}}{\text{L}}\right)\left(\frac{1\ \text{mol EX}_2}{2\ \text{mol X}}\right) = 0.005\ \text{mol}\cdot\text{L}^{-1}$$

$$k = \frac{\dfrac{1}{[EX_2]_t} - \dfrac{1}{[EX_2]_0}}{t} = \frac{\dfrac{1}{0.005\ \text{mol}\cdot\text{L}^{-1}} - \dfrac{1}{0.040\ \text{mol}\cdot\text{L}^{-1}}}{15\ \text{h}}$$

$$= 10\ \text{L}\cdot\text{mol}^{-1}\cdot\text{h}^{-1}(1\ \text{sf})$$

13.71

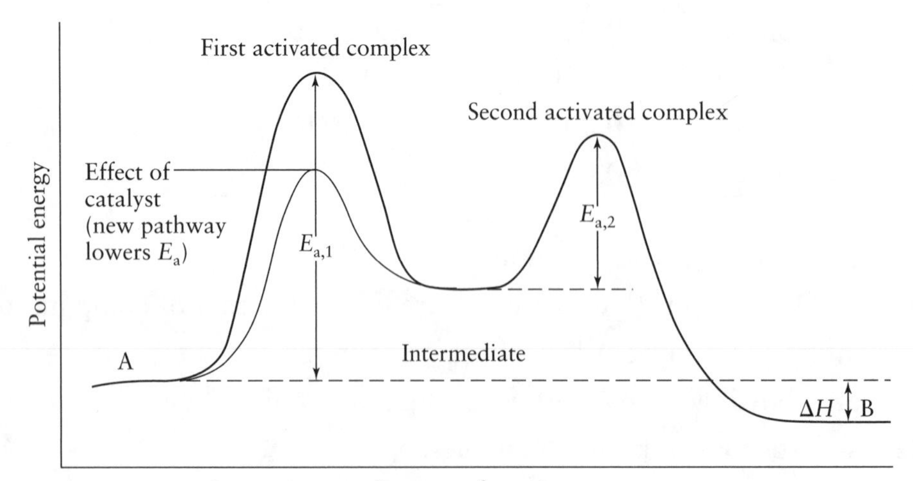

13.73 The overall reaction is $RCN + H_2O \longrightarrow RC(=O)NH_2$. The intermediates include

$$\underset{\displaystyle \text{OH}}{\text{R—}\overset{}{\text{C}}\text{=N}^-} \quad \underset{\displaystyle \text{OH}}{\text{R—}\text{C}\text{=N—H}}$$

The hydroxide ion serves as a catalyst for the reaction.

13.75 x = amount of original sample = 10.0 mg

n = number of half-lives

$\left(\frac{1}{2}\right)^n \times x$ = amount remaining

$\frac{10.9}{12.3} = 0.89$ half-lives

$\left(\frac{1}{2}\right)^{0.89} \times 10.0 \text{ mg} = 5.4 \text{ mg}$

13.77 (a) The objective is to reproduce the observed rate law. If step 2 is the slow step, if step 1 is a rapid equilibrium, and if step 3 is fast also, then our proposed rate law will be rate = $k_2[N_2O_2][H_2]$. Consider the equilibrium of Step 1: $k_1[NO]^2 = k_1'[N_2O_2]$

$[N_2O_2] = \frac{k}{k_1'}[NO]^2$ Substituting in our proposed rate law, we have

$$\text{rate} = k_2\left(\frac{k_1}{k_1'}\right)[NO]^2[H_2] = k[NO]^2[H_2] \text{ where } k = k_2\left(\frac{k_1}{k_1'}\right)$$

The assumptions made above reproduce the observed rate law; therefore, step 2 is the slow step.

(b)

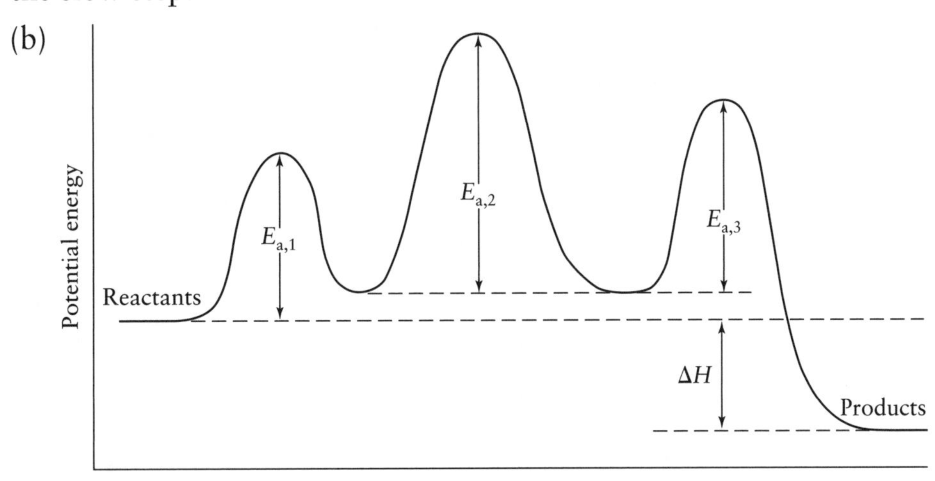

Note: The dips that represent the formation of the intermediate N_2O_2 and N_2O will not be at the same energy, but we have no information to determine which should be lower.

13.79 $$\frac{\text{rate at } 28°\text{C}}{\text{rate at } 5°\text{C}} = \frac{k'}{k} = \frac{t}{t'} = \frac{48 \text{ h}}{4 \text{ h}}$$

We use $\ln\left(\frac{k'}{k}\right) = \frac{E_a}{R}\left(\frac{1}{T} - \frac{1}{T'}\right)$ and solve for E_a.

$$E_a = \frac{R \ln\left(\frac{k'}{k}\right)}{\left(\frac{1}{T} - \frac{1}{T'}\right)} = \frac{(8.314 \times 10^{-3} \text{ kJ·K}^{-1}\text{·mol}^{-1}) \ln\left(\frac{48}{4}\right)}{\left(\frac{1}{278 \text{ K}} - \frac{1}{301 \text{ K}}\right)} = 75 \text{ kJ·mol}^{-1}$$

13.81 (a) ClO is the reaction intermediate; Cl is the catalyst.

(b) Cl, ClO, O, O_2

(c) Step 1 is initiating; step 2 is propagating.

(d) $Cl + Cl \longrightarrow Cl_2$

13.83

Concentration ($mol·L^{-1}$)	$2\ N_2O_5$ $\rightleftharpoons$	$4\ NO_2$ +	O_2
initial	P_0	0	0
change	$-x$	$+2x$	$+0.5x$
at time t	$P_0 - x$	$2x$	$0.5x$

Therefore, P_{total} at time $t = P_0 + 1.5x$. This allows calculation of x at each time, which in turn allows calculation of $P_{N_2O_5}(= P_0 - x)$ at these times. Converting the units to atmospheres by dividing by 101.325 $kPa·atm^{-1}$ and to $[N_2O_5]$ by dividing by RT allows us to make the following table:

t, min	x, kPa	$P_{N_2O_5}$, kPa	$P_{N_2O_5}$, atm	$[N_2O_5]$, $mol·L^{-1}$	$\ln [N_2O_5]$
0	0	27.3	0.269	0.0100	-4.605
5	10.9	16.4	0.162	6.01×10^{-3}	-5.114
10	17.5	9.85	0.0972	3.61×10^{-3}	-5.624
15	21.4	5.9	0.058	2.2×10^{-3}	-6.12
20	23.8	3.5	0.035	1.3×10^{-3}	-6.65
30	26.0	1.3	0.013	4.8×10^{-4}	-7.64

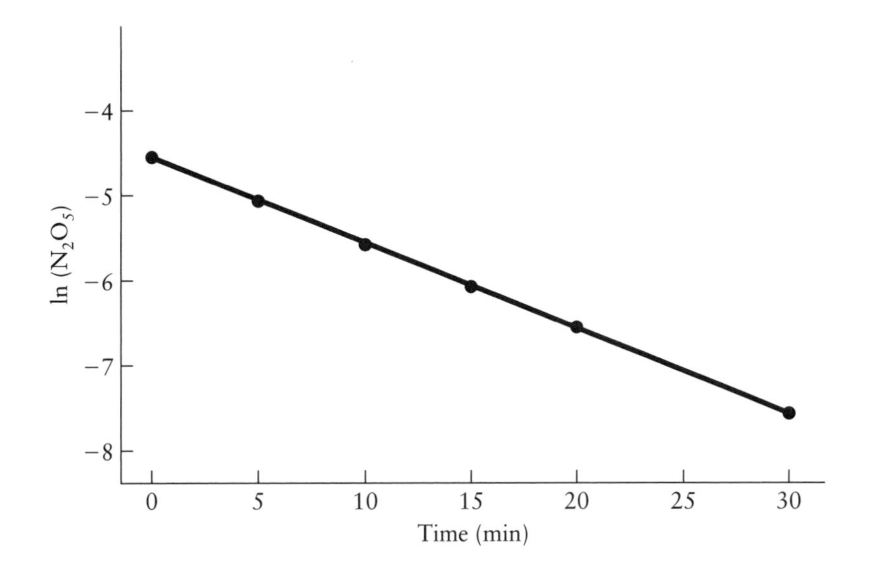

The data fit closely to a straight line; therefore, this is a first-order reaction. The rate constant can be obtained from the slope, which is

$$\frac{-4.605 - (-7.64)}{30\text{ min}} = 0.101\text{ min}^{-1} = k$$

rate $= k[N_2O_5] = 0.101\text{ min}^{-1}[N_2O_5]$, which gives the results in the table below.

t, min	rate, $\text{mol}\cdot\text{L}^{-1}\cdot\text{min}^{-1}$
0	1.01×10^{-3}
5	6.07×10^{-4}
10	3.65×10^{-4}
15	2.2×10^{-4}
20	1.3×10^{-4}
30	4.8×10^{-5}

13.85 For a third-order reaction,

$$t_{1/2} \propto \frac{1}{[A_0]^2} \quad \text{or} \quad t_{1/2} = \frac{\text{constant}}{[A_0]^2}$$

(a) The time necessary for the concentration to fall to one-half of the initial concentration is one half-life:

$$\text{first half-life} = t_1 = t_{1/2} = \frac{\text{constant}}{[A_0]^2}$$

(b) This time, $t_{1/4}$, is two half-lives, but because of different starting concentrations, the half-lives are not the same:

$$\text{second half-life} = t_2 = \frac{\text{constant}}{(\frac{1}{2}[A_0])^2} = \frac{4(\text{constant})}{[A_0]^2} = 4t_1$$

total time $= t_1 + t_2 = t_1 + 4t_1 = 5t_1 = t_{1/4}$

(c) This time, $t_{1/16}$, is four half-lives; again, the half-lives are not the same:

third half-life $= t_3 = \dfrac{\text{constant}}{(\frac{1}{4}[A_0])^2} = \dfrac{16(\text{constant})}{[A_0]^2} = 16t_1$

fourth half-life $= t_4 = \dfrac{\text{constant}}{(\frac{1}{8}[A_0])^2} = \dfrac{64(\text{constant})}{[A_0]^2} = 64t_1$

total time $= t_1 + t_2 + t_3 + t_4 = t_1 + 4t_1 + 16t_1 + 64t_1 = 85t_1 = t_{1/16}$

If t_1 is known, the times $t_{1/4}$ and $t_{1/16}$ can be calculated easily.

13.87 The reaction reaches equilibrium. Consider, for example, $A \rightleftharpoons B$. Assume that the reaction is first order in both directions:

rate(forward) $= k[A]$

rate(reverse) $= k'[B]$

At equilibrium, rate(forward) = rate(reverse), so

rate = 0 = rate(forward) − rate(reverse)

or $k[A] - k'[B] = 0$

or $k[A] - k'([A]_0 - [A]) = 0$

or $(k + k')[A] = k'[A]_0$

or $[A] = \dfrac{k'}{(k + k')}[A]_0$, which is not zero.

13.89 By analogy with the reaction in Exercise 13.82, the overall reaction here is

$CH_4(g) + Cl_2(g) \longrightarrow CH_3Cl + HCl$

(a) Initiation: $Cl_2 \longrightarrow 2\ Cl$

Propagation: $Cl + CH_4 \longrightarrow CH_3Cl + H$

$H + Cl_2 \longrightarrow HCl + Cl$

Termination: $Cl + Cl \longrightarrow Cl_2$

$H + H \longrightarrow H_2$

$H + Cl \longrightarrow HCl$

(b) CH_3Cl and HCl

13.91 The strategy for working this problem is to obtain the equilibrium constants for the reaction at two or more temperatures and then use those values to calculate $\Delta H°_r$ and $\Delta S°_r$. From Table 13.1 we can obtain K values at 4 temperatures:

$$K = \frac{k}{k'}$$

$$K_{500} = \frac{6.4 \times 10^{-9}\ L{\cdot}mol^{-1}{\cdot}s^{-1}}{4.3 \times 10^{-7}\ L{\cdot}mol^{-1}{\cdot}s^{-1}} = 0.015$$

$$K_{600} = \frac{9.7 \times 10^{-6}\ \mathrm{L \cdot mol^{-1} \cdot s^{-1}}}{4.4 \times 10^{-4}\ \mathrm{L \cdot mol^{-1} \cdot s^{-1}}} = 0.022$$

$$K_{700} = \frac{1.8 \times 10^{-3}\ \mathrm{L \cdot mol^{-1} \cdot s^{-1}}}{6.3 \times 10^{-2}\ \mathrm{L \cdot mol^{-1} \cdot s^{-1}}} = 0.028$$

$$K_{700} = \frac{9.7 \times 10^{-2}\ \mathrm{L \cdot mol^{-1} \cdot s^{-1}}}{2.6\ \mathrm{L \cdot mol^{-1} \cdot s^{-1}}} = 0.037$$

We can choose to calculate the desired quantities from any two of these points, or we can plot the data and determine the values from the slope and intercept of the graph:

$$\ln K = -\frac{\Delta H°_{\mathrm{r}}}{R}\left(\frac{1}{T_1}\right) + \frac{\Delta S°_{\mathrm{r}}}{R}$$

The plot should be ln K versus $\frac{1}{T}$. The slope will be $-\frac{\Delta H°_{\mathrm{r}}}{R}$ and the intercept will be $\frac{\Delta S°_{\mathrm{r}}}{R}$.

T(K)	$\frac{1}{T}$(K^{-1})	K	ln K
500	0.0200	0.015	−4.20
600	0.001 67	0.022	−3.82
700	0.001 43	0.028	−3.58
800	0.001 25	0.037	−3.30

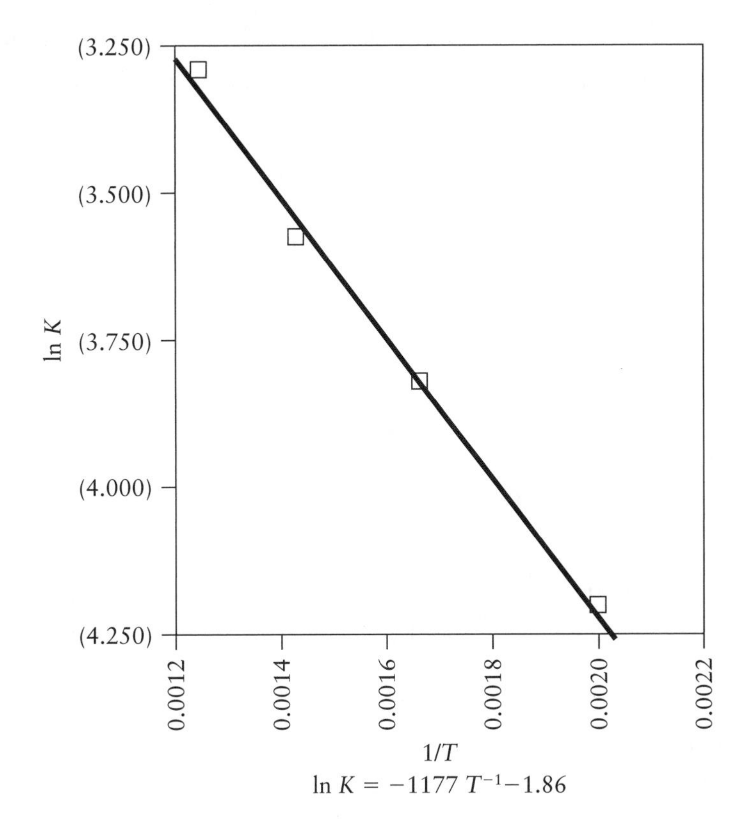

ln $K = -1177\ T^{-1} - 1.86$

$$-\frac{\Delta H^\circ_r}{R} = -1177, \Delta H^\circ_r = 9.8 \text{ kJ}\cdot\text{mol}^{-1}$$

$$\frac{\Delta S^\circ_r}{R} = -1.86, \Delta S^\circ_r = -15 \text{ J}\cdot\text{K}^{-1}\cdot\text{mol}^{-1}$$

13.93 In order for the reaction to be catalyzed heterogeneously, the reacting species must attach themselves to the surface of the catalyst. The concentration of the reactants is usually much greater than the number of active sites available on the catalyst so that the rate is determined by the surface area of the catalyst and not by the concentrations or pressures of reactants.

13.95 (a) $OCl^- + H_2O \underset{k_1'}{\overset{k_1}{\rightleftharpoons}} HOCl + OH^-$ fast equilibrium

$HOCl + I^- \xrightarrow{k_2} HOI + Cl^-$ very slow

$HOI + OH^- \underset{k_3'}{\overset{k_3}{\rightleftharpoons}} OI^- + H_2O$ fast equilibrium

The overall reaction is $OCl^- + I^- \longrightarrow OI^- + Cl^-$

(b) The rate law will be based upon the slow step of the reaction:

rate $= k_2[HOCl][I^-]$

Even though HOCl is a stable species because it is an intermediate in the reaction as written, technically we should not leave the rate law in this form. The concentration of HOCl can be expressed in terms of the reactants and products using the fast equilibrium approach:

$$K = \frac{k_1}{k_1'} = \frac{[HOCl][OH^-]}{[OCl^-]}$$

$$[HOCl] = \frac{k_1}{k_1'}\frac{[OCl^-]}{[OH^-]}$$

$$\text{rate} = \frac{k_2 k_1}{k_1'}\frac{[OCl^-][I^-]}{[OH^-]}$$

(c) An examination of the rate law shows that the rate is dependent upon the concentration of OH^-, which means that the rate will be dependent upon the pH of the solution.

(d) If the reaction is carried out in an organic solvent, then H_2O is no longer the solvent and its concentration must be included in calculating the equilibrium concentration of HOCl:

$$K = \frac{k_1}{k_1'} = \frac{[HOCl][OH^-]}{[OCl^-][H_2O]}$$

$$[HOCl] = \frac{k_1}{k_1'} \frac{[OCl^-][H_2O]}{[OH^-]}$$

$$\text{rate} = \frac{k_2 k_1}{k_1'} \frac{[OCl^-][I^-][H_2O]}{[OH^-]}$$

The rate of reaction will then show a dependence upon the concentration of water, which will be obscured when the reaction is carried out with water as the solvent.

CHAPTER 14
THE ELEMENTS: THE FIRST FOUR MAIN GROUPS

14.1 (a) carbon (b) lithium (c) indium (d) iodine

14.3 (a) sulfur (b) selenium (c) sodium (d) oxygen

14.5 iodine < bromine < chlorine

14.7 chlorine

14.9 antimony

14.11 (a) KCl because the ionic radius of K^+ is larger than that of Na^+
(b) Na—O. The higher charge on Mg^{2+} makes its ionic radius much smaller than that of Na^+.
(c) Tl(III). The ionic radius is shorter for a metal ion with a higher oxidation state.

14.13 $2\ K(s) + H_2(g) \longrightarrow 2\ KH(s)$

14.15 (a) saline (b) molecular (c) molecular (d) metallic

14.17 (a) acidic (b) amphoteric (c) acidic (d) basic

14.19 (a) CO_2 (b) SO_3 (c) B_2O_3

14.21 (a) $C_2H_2(g) + H_2(g) \longrightarrow H_2C{=}CH_2(g)$
Oxidation number of C in $C_2H_2 = -1$; of C in $H_2C{=}CH_2 = -2$, carbon has been reduced.
(b) $CO(g) + H_2O(g) \rightleftharpoons CO_2(g) + H_2(g)$
(c) $BaH_2(s) + 2\ H_2O(l) \longrightarrow Ba(OH)_2 + 2\ H_2(g)$

14.23 (a) $CH_4(g) + H_2O(g) \longrightarrow CO(g) + 3\,H_2(g)$

$$\Delta H°_r = \Delta H°_f(CO, g) - [\Delta H°_f(CH_4, g) + \Delta H°_f(H_2O, g)]$$
$$= (-110.53\ kJ\cdot mol^{-1}) - [(-74.81\ kJ\cdot mol^{-1}) + (-241.82\ kJ\cdot mol^{-1})]$$
$$= +206.10\ kJ\cdot mol^{-1}$$

(b) $\Delta S°_r = S°(CO, g) + 3S°(H_2, g) - [S°(CH_4, g) - S°(H_2O, g)]$

$$= 197.67\ J\cdot K^{-1}\cdot mol^{-1} + 3(130.68\ J\cdot K^{-1}\cdot mol^{-1}) - [186.26\ J\cdot K^{-1}\cdot mol^{-1} + 188.83\ J\cdot K^{-1}\cdot mol^{-1}]$$
$$= +214.62\ J\cdot K^{-1}\cdot mol^{-1}$$

(c) $\Delta G°_r = \Delta G°_f(CO, g) - [\Delta G°_f(CH_4, g) + \Delta G°_f(H_2O, g)]$

$$= (-137.17\ kJ\cdot mol^{-1}) - [(-50.72\ kJ\cdot mol^{-1}) + (-228.57\ kJ\cdot mol^{-1})]$$
$$= +142.12\ kJ\cdot mol^{-1}$$

$\Delta G°_r$ can also be calculated from $\Delta H°_r$ and $\Delta S°_r$:

$$\Delta G°_r = \Delta H°_r - T\Delta S°_r$$
$$= +206.10\ kJ\cdot mol^{-1} - (298\ K)(+214.62\ J\cdot K^{-1}\cdot mol^{-1})/(1000\ J\cdot kJ^{-1})$$
$$= +142.14\ kJ\cdot mol^{-1}$$

14.25 (a) At STP (273 K and 1.00 atm), 1.00 mol of $H_2(g)$, assumed to be ideal, occupies 22.4 L. Then for

$$CaH_2(s) + 2\,H_2O(l) \longrightarrow Ca(OH)_2(s) + H_2(g)$$

$$\text{volume of } H_2(g) = (10.0\ g\ CaH_2)\left(\frac{1\ mol\ CaH_2}{42.10\ g\ CaH_2}\right)\left(\frac{2\ mol\ H_2}{1\ mol\ CaH_2}\right)\left(\frac{22.4\ L\ H_2}{1\ mol\ H_2}\right)$$
$$= 10.6\ L\ H_2$$

(b) $$\text{volume of } H_2O(l) = (10.0\ g\ CaH_2)\left(\frac{1\ mol\ CaH_2}{42.10\ g\ CaH_2}\right)\left(\frac{2\ mol\ H_2O}{1\ mol\ CaH_2}\right)$$
$$\left(\frac{18.02\ g\ H_2O}{1\ mol\ H_2O}\right)\left(\frac{1.0\ mL\ H_2O}{1.0\ g\ H_2O}\right) = 8.56\ mL\ H_2O$$

14.27 (a) $H_2(g) + Cl_2(g) \xrightarrow{\text{light}} 2\,HCl(g)$

(b) $H_2(g) + 2\,Na(l) \xrightarrow{\Delta} 2\,NaH(s)$

(c) $P_4(s) + 6\,H_2(g) \longrightarrow 4\,PH_3(g)$

(d) $Cu(s) + H_2(g) \longrightarrow$ N.R.

14.29 (a)

$H_2(g) \longrightarrow 2\,H^+(aq) + 2\,e^-$	$E° = 0.00\ V$
$O_2(g) + 4\,H^+(aq) + 4\,e^- \longrightarrow 2\,H_2O(l)$	$E° = +1.23\ V$
$2\,H_2(g) + O_2(g) \longrightarrow 2\,H_2O(l)$	$E° = +1.23\ V$

The maximum potential possible is 1.23 V.

(b) The difficulty is isolating the two half cells but still maintaining electrical contact. Ions need to flow through the system to maintain charge balance in the reaction. In this case, a material that allows hydrogen ions but not hydrogen gas or oxygen gas to pass through would be necessary.

14.31 Lithium is the only Group 1 element that reacts directly with nitrogen to form lithium nitride:

$$6\ Li(s) + N_2(g) \xrightarrow{\Delta} 2\ Li_3N(s)$$

Lithium reacts with oxygen to form mainly the oxide:

$$4\ Li(s) + O_2(g) \longrightarrow 2\ Li_2O(s)$$

The other members of the group form mainly the peroxide or superoxide. Lithium exhibits the diagonal relationship that is common to many first members of a group. Li is similar in many of its compounds to the compounds of Mg. This is related to the small ionic radius of Li^+, 58 pm, which is closer to the ionic radius of Mg^{2+}, 72 pm, but substantially less than that of Na^+, 102 pm.

14.33 (a) $4\ Li(s) + O_2(g) \longrightarrow 2\ Li_2O(s)$

(b) $6\ Li(s) + N_2(g) \xrightarrow{\Delta} 2\ Li_3N(s)$

(c) $2\ Na(s) + 2\ H_2O(l) \longrightarrow 2\ NaOH(aq) + H_2(g)$

(d) $4\ KO_2(s) + 2\ H_2O(g) \longrightarrow 4\ KOH(s) + 3\ O_2(g)$

14.35 1 mol $Na_2CO_3 \cdot 10\ H_2O$ yields 1 mol Na_2CO_3 in water.

$$\begin{aligned}\text{mass of } Na_2CO_3 \cdot 10\ H_2O &= 0.500\ L \times 0.135\ mol \cdot L^{-1} \\ &\quad \times 286.15\ g\ Na_2CO_3 \cdot 10\ H_2O \cdot mol^{-1} \\ &= 19.3\ g\ Na_2CO_3 \cdot 10\ H_2O\end{aligned}$$

14.37 $Mg(s) + 2\ H_2O(l) \longrightarrow Mg(OH)_2 + H_2(g)$

14.39 (a) $CaO(s) + H_2O(l) \longrightarrow Ca(OH)_2(s)$

(b)
$$\begin{aligned}\Delta G^\circ_r &= \Delta G^\circ_f(Ca(OH)_2, s) - [\Delta G^\circ_f(CaO, s) + \Delta G^\circ_f(H_2O, l)] \\ &= -898.49\ kJ \cdot mol^{-1} - [(-604.03\ kJ \cdot mol^{-1}) + (-237.13\ kJ \cdot mol^{-1})] \\ &= -57.33\ kJ \cdot mol^{-1}\end{aligned}$$

14.41 Be is the weakest reducing agent; Mg is stronger, but weaker than the remaining members of the group, all of which have approximately the same reducing strength. This effect is related to the very small radius of the Be^{2+} ion, 27 pm; its strong polarizing power introduces much covalent character into its compounds.

Thus, Be attracts electrons more strongly and does not release them as readily as other members of the group. Mg^{2+} is also a small ion, 58 pm, so the same reasoning applies to it also, but to a lesser extent. The remaining ions of the group are considerably larger, release electrons more readily, and are better reducing agents.

14.43 (a) $2\,Al(s) + 2\,OH^-(aq) + 6\,H_2O(l) \longrightarrow 2[Al(OH)_4]^-(aq) + 3\,H_2(g)$

(b) $Be(s) + 2\,OH^-(aq) + 2\,H_2O(l) \longrightarrow [Be(OH)_4]^{2-}(aq) + H_2(g)$

The similarity of Be and Al in chemical reactions is an example of the diagonal relationship in the periodic table; namely, the similar chemical behavior of elements that are diagonal neighbors, such as Be and Al.

14.45 (a) $Mg(OH)_2(s) + 2\,HCl(aq) \longrightarrow MgCl_2(aq) + 2\,H_2O(l)$

(b) $Ca(s) + 2\,H_2O(l) \longrightarrow Ca(OH)_2(aq) + H_2(g)$

(c) $BaCO_3(s) \xrightarrow{\Delta} BaO(s) + CO_2(g)$

14.47 (a) $:\ddot{\underset{..}{Cl}}{-}Be{-}\ddot{\underset{..}{Cl}}:$

$Mg^{2+}[:\ddot{\underset{..}{Cl}}:]^-[:\ddot{\underset{..}{Cl}}:]^-$

$MgCl_2$ is ionic; $BeCl_2$ is a molecular compound.

(b) 180° (c) *sp*

14.49 (a) $CaCO_3(s) \xrightarrow{\Delta} CaO(s) + CO_2(g)$

$$\Delta H°_r = (-635.09\ kJ\cdot mol^{-1}) + (-393.51\ kJ\cdot mol^{-1}) - (-1206.9\ kJ\cdot mol^{-1})$$
$$= +178.3\ kJ\cdot mol^{-1}$$

$$\Delta S°_r = (39.75\ J\cdot K^{-1}\cdot mol^{-1}) + (213.74\ J\cdot K^{-1}\cdot mol^{-1}) - (92.9\ J\cdot K^{-1}\cdot mol^{-1})$$
$$= +160.6\ J\cdot K^{-1}\cdot mol^{-1}$$

(b) The temperature at which the equilibrium constant becomes greater than 1 is the temperature at which $\Delta G°_r$ crosses over between positive and negative values. Therefore, it is the temperature at which $\Delta G°_r = 0$.

$$\Delta G°_r = 0 = \Delta H°_r - T\Delta S°_r = -RT \ln K \text{ (for } K = 1, \ln K = 0)$$

$$T = \frac{\Delta H°_r}{\Delta S°_r} = \frac{178\ 300\ J\cdot mol^{-1}}{160.6\ J\cdot K^{-1}\cdot mol^{-1}} = 1110\ K = 837°C$$

Consequently, at temperatures above 1110 K, $\ln K > 0$ and K becomes greater than 1.

14.51 The overall equation for the electrolytic reduction in the Hall process is

$4\,Al^{3+}(melt) + 6\,O^{2-}(melt) + 3\,C(s, gr) \longrightarrow 4\,Al(s) + 3\,CO_2(g)$

14.53 (a) $B_2O_3(s) + 3\ Mg(l) \xrightarrow{\Delta} 2\ B(s) + 3\ MgO(s)$

(b) $2\ Al(s) + 3\ Cl_2(g) \longrightarrow 2\ AlCl_3(s)$

(c) $4\ Al(s) + 3\ O_2(g) \longrightarrow 2\ Al_2O_3(s)$

14.55 (a) The hydrate of $AlCl_3$, that is, $AlCl_3 \cdot 6H_2O$, functions as a deodorant and antiperspirant.

(b) α-Alumina is corundum. It is used as an abrasive in sandpaper.

(c) $B(OH)_3$ is an antiseptic and insecticide.

14.57 The cathode reaction is $Al^{3+}(melt) + 3\ e^- \longrightarrow Al(l)$

$$\text{charge consumed} = (24.0\ \text{h})\left(\frac{3600\ \text{s}}{1\ \text{h}}\right)(3.5 \times 10^6\ \text{C}\cdot\text{s}^{-1}) = 3.0 \times 10^{11}\ \text{C}$$

$$\text{mass of Al produced} = (3.0 \times 10^{11}\ \text{C})\left(\frac{1\ \text{mol e}^-}{9.65 \times 10^4\ \text{C}}\right)\left(\frac{1\ \text{mol Al}}{3\ \text{mol e}^-}\right)\left(\frac{26.98\ \text{g Al}}{1\ \text{mol Al}}\right)$$

$$= 2.8 \times 10^7\ \text{g Al}$$

14.59 (a) We want $E°$ for $Tl^{3+}(aq) + 3\ e^- \longrightarrow Tl(s)$, $n = 3$.

This reaction is the reverse of the formation reaction:

$Tl(s) \longrightarrow Tl^{3+}(aq) + 3\ e^-$

Therefore, for the Tl^{3+}/Tl couple, $\Delta G°_r = -215\ \text{kJ}\cdot\text{mol}^{-1}$

$$\Delta G°_r = -nFE° = -215\ \text{kJ}\cdot\text{mol}^{-1}$$

$$E° = \frac{\Delta G°_r}{-nF} = \frac{-2.15 \times 10^5\ \text{J}\cdot\text{mol}^{-1}}{-3 \times 9.65 \times 10^4\ \text{C}\cdot\text{mol}^{-1}} = +0.743\ \text{V}$$

(b) Using the potential from part (a) and the potential from Appendix 2B for the reduction of Tl^+, we can determine if Tl^+ will be spontaneously disproportionate in solution. The equation of interest is

$3\ Tl^+(aq) \longrightarrow 2\ Tl(s) + Tl^{3+}(aq)$

The half-reactions to combine are

$Tl^+ \longrightarrow Tl^{3+} + 2\ e^-$ (1)

$Tl^+ + e^- \longrightarrow Tl$ (2)

The potential for reaction (1) must be obtained using the $\Delta G°$ values of the two known half-reactions:

$Tl \longrightarrow Tl^{3+} + 3\ e^-$ $\quad \Delta G° = +215\ \text{kJ}\cdot\text{mol}^{-1}$

$Tl^+ + e^- \longrightarrow Tl$ $\quad \Delta G° = -nFE°$

$$= -1(9.65 \times 10^4\ \text{J}\cdot\text{V}^{-1}\cdot\text{mol}^{-1})(-0.34\ \text{V})/(1000\ \text{J}\cdot\text{kJ}^{-1})$$

$$= +33\ \text{kJ}\cdot\text{mol}^{-1}$$

$\Delta G°$ for the combined half-reaction $Tl^+ \longrightarrow Tl^{3+} + 2\ e^-$ is the sum of these two numbers:

$+215\ kJ \cdot mol^{-1} + 33\ kJ \cdot mol^{-1} = +248\ kJ \cdot mol^{-1}$

We can now combine this with the reduction process for Tl^+ to get the desired equation:

$$
\begin{array}{ll}
Tl^+ \longrightarrow Tl^{3+} + 2\ e^- & \Delta G° = +248\ kJ \cdot mol^{-1} \\
+2(Tl^+ + e^- \longrightarrow Tl) & \Delta G° = 2(+33\ kJ \cdot mol^{-1}) \\
\hline
3\ Tl^+ \longrightarrow 2\ Tl + Tl^{3+} & \Delta G° = +248\ kJ \cdot mol^{-1} + 2(+33\ kJ \cdot mol^{-1}) \\
 & = +314\ kJ \cdot mol^{-1}
\end{array}
$$

This is not spontaneous.

14.61 Silicon occurs widely in the Earth's crust in the form of silicates in rocks and as silicon dioxide in sand. It is obtained from quartzite, a form of quartz (SiO_2), by the following processes:

(1) reduction in an electric arc furnace

$$SiO_2(s) + 2\ C(s) \longrightarrow Si(s,\ crude) + 2\ CO(g)$$

(2) purification of the crude product in two steps

$$Si(s,\ crude) + 2\ Cl_2(g) \longrightarrow SiCl_4(l)$$

followed by reduction with hydrogen to the pure element

$$SiCl_4(l) + 2\ H_2(g) \longrightarrow Si(s,\ pure) + 4\ HCl(g)$$

14.63 In diamond, carbon is sp^3 hybridized and forms a tetrahedral, three-dimensional network structure, which is extremely rigid. Graphite carbon is sp^2 hybridized and planar, and its application as a lubricant results from the fact that the two-dimensional sheets can "slide" across one another, thereby reducing friction. In graphite, the unhybridized *p*-electrons are free to move from one carbon atom to another, which results in its high electrical conductivity. In diamond, all electrons are localized in sp^3 hybridized C—C σ-bonds, so diamond is a poor conductor of electricity.

14.65 (a) $SiCl_4(l) + 2\ H_2(g) \longrightarrow Si(s) + 4\ HCl(g)$

(b) $SiO_2(s) + 3\ C(s) \xrightarrow{2000°C} SiC(s) + 2\ CO(g)$

(c) $Ge(s) + 2\ F_2(g) \longrightarrow GeF_4(s)$

(d) $CaC_2(s) + 2\ H_2O(l) \longrightarrow Ca(OH)_2(s) + C_2H_2(g)$

14.67

$$\left[\;:\ddot{O}:\;\;\;:\ddot{O}-\underset{|}{\overset{|}{Si}}-\ddot{O}:\;\;\;:\ddot{O}:\;\right]^{4-}$$

Formal charges: Si = 0, O = −1

oxidation numbers: Si = +4, O = −2

This is an AX_4 VSEPR structure; therefore, the shape is tetrahedral.

14.69 $SiO_2(s) + 2\ C(s) \longrightarrow Si(s) + 2\ CO(g)$

$$\begin{aligned}\Delta H^\circ_r &= \Delta H^\circ_f(\text{products}) - \Delta H^\circ_f(\text{reactants})\\ &= [(2)(-110.53\ \text{kJ}\cdot\text{mol}^{-1})] - [-910.94\ \text{kJ}\cdot\text{mol}^{-1}]\\ &= +689.88\ \text{kJ}\cdot\text{mol}^{-1}\end{aligned}$$

$$\begin{aligned}\Delta S^\circ_r &= S^\circ(\text{products}) - S^\circ(\text{reactants})\\ &= [18.83\ \text{J}\cdot\text{K}^{-1}\cdot\text{mol}^{-1} + (2)(197.67\ \text{J}\cdot\text{K}^{-1}\cdot\text{mol}^{-1})]\\ &\quad - [41.84\ \text{J}\cdot\text{K}^{-1}\cdot\text{mol}^{-1} + (2)(5.740\ \text{J}\cdot\text{K}^{-1}\cdot\text{mol}^{-1})]\\ &= +360.85\ \text{J}\cdot\text{K}^{-1}\cdot\text{mol}^{-1}\end{aligned}$$

$$\begin{aligned}\Delta G^\circ_r &= \Delta H^\circ_r - T\Delta S^\circ_r\\ &= 689.88\ \text{kJ}\cdot\text{mol}^{-1} - (298.15\ \text{K})(360.85\ \text{J}\cdot\text{K}^{-1}\cdot\text{mol}^{-1})/1000\ \text{J}\cdot\text{kJ}^{-1}\\ &= +5.8229 \times 10^2\ \text{kJ}\cdot\text{mol}^{-1}\end{aligned}$$

The temperature at which the equilibrium constant becomes greater than 1 is the temperature at which $\Delta G^\circ_r = -RT \ln K = 0$, because $\ln 1 = 0$. Above this temperature, the equilibrium constant is greater than 1. $\Delta G^\circ_r = 0$ when $T\Delta S^\circ_r = \Delta H^\circ_r$, or

$$T = \frac{\Delta H^\circ_r}{\Delta S^\circ_r} = \frac{+689.88 \times 10^3\ \text{J}\cdot\text{mol}^{-1}}{+360.85\ \text{J}\cdot\text{K}^{-1}\cdot\text{mol}^{-1}} = 1912\ \text{K}$$

14.71

$$\begin{aligned}\text{mass of HF} &= (2.00 \times 10^{-3}\ \text{g})\left(\frac{1\ \text{mol SiO}_2}{60.09\ \text{g SiO}_2}\right)\left(\frac{6\ \text{mol HF}}{1\ \text{mol SiO}_2}\right)\left(\frac{20.01\ \text{g HF}}{1\ \text{mol HF}}\right)\\ &= 4.00 \times 10^{-3}\ \text{g HF} = 4.00\ \text{mg HF}\end{aligned}$$

14.73 (a) The $Si_2O_7^{6-}$ ion is built from two SiO_4^{4-} tetrahedral ions in which the silicate tetrahedra share one O atom. This is the only case in which one O is shared.

(b) The pyroxenes, for example, jade, $NaAl(SiO_3)_2$, consist of chains of SiO_4 units in which two O atoms are shared by neighboring units. The repeating unit has the formula SiO_3^{2-}. See Fig. 14.44.

14.75 SiF_6^{2-}

14.77 Ionic fluorides react with water to liberate HF, which then reacts with the glass. Glass bottles used to store metal fluorides become brittle and may disintegrate upon standing on the shelf.

14.79 The iron ions impart a deep red color to the clay, which is not desirable for the manufacture of fine china; a white base is aesthetically more pleasing.

14.81 In the majority of its reactions, hydrogen acts as a reducing agent. Examples are $2\ H_2(g) + O_2(g) \longrightarrow 2\ H_2O(l)$ and various ore reduction processes, such as $NiO(s) + H_2(g) \xrightarrow{\Delta} Ni(s) + H_2O(g)$. With highly electropositive elements, such as the alkali metals, $H_2(g)$ acts as an oxidizing agent and forms metal hydrides, for example, $2\ K(s) + H_2(g) \longrightarrow 2\ KH(s)$.

14.83 (a)

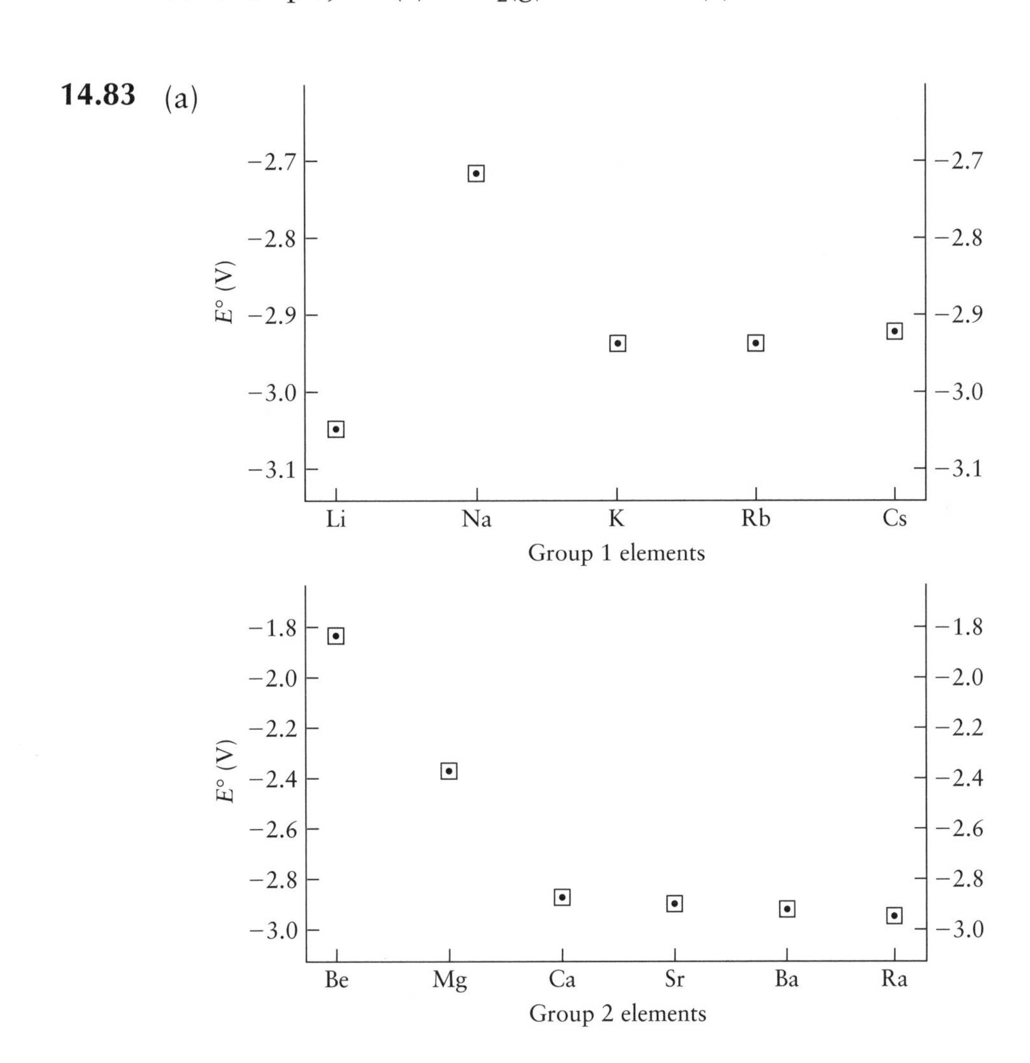

(b) For both groups, the trend in standard potentials with increasing atomic number is overall downward (they become more negative), but lithium is anomalous. This overall downward trend makes sense, because we expect that it is easier to remove electrons that are farther away from the nuclei. However, because there are several factors that influence ease of removal, the trend is not smooth. The potentials are a net composite of the free energies of sublimation of solids, dissociation of gaseous molecules, ionization enthalpies, and enthalpies of hydration of gaseous ions. The origin of the anomalously strong reducing power of Li is the strongly exothermic energy of hydration of the very small Li^+ ion, which favors the ionization of the element in aqueous solution.

14.85 (a, b)

$H_2(g) + F_2(g) \longrightarrow 2\,HF(g)$, explosive $\quad \Delta G^\circ_f = -273.2\ kJ\cdot mol^{-1}$

$H_2(g) + Cl_2(g) \longrightarrow 2\,HCl(g)$, explosive $\quad \Delta G^\circ_f = -95.30\ kJ\cdot mol^{-1}$

$H_2(g) + Br_2(g) \longrightarrow 2\,HBr(g)$, vigorous $\quad \Delta G^\circ_f = -53.45\ kJ\cdot mol^{-1}$

$H_2(g) + I_2(g) \longrightarrow 2\,HI(g)$, less vigorous $\quad \Delta G^\circ_f = +1.70\ kJ\cdot mol^{-1}$

The word *vigor* as used here has two components, a thermodynamic one and a kinetic one. ΔG°_f is most negative for HF and becomes slightly positive for HI. So, the formation of HF, HCl, and HBr are all thermodynamically spontaneous. The kinetics parallel, in this case, the thermodynamic spontaneity, although this parallel behavior is not necessarily true in other systems.

(c) hydrofluoric acid
hydrochloric acid
hydrobromic acid
hydroiodic acid

14.87 (a) The oxide ion, O^{2-}, in CaO acts as a Lewis base and reacts with the Lewis acid, SiO_2, in a Lewis acid-base reaction:

$$CaO(s) + SiO_2(s) \longrightarrow CaSiO_3(l)$$

SiO_2, which is an impurity in iron ore, is removed by this reaction. The calcium oxide in this reaction can be obtained from limestone [$CaCO_3(s) \xrightarrow{\Delta} CaO(s) + CO_2(g)$]. This is the reason that limestone is important in the iron industry.

(b) $CaO(s) + CO_2(g) \longrightarrow CaCO_3(s)$ (not an efficient preparation of $CaCO_3$, because of the weak Lewis acidity of CO_2)

14.89 (a) H_3BO_3, acid; $B(OH)_4^-$, conjugate base

(b) $B(OH)_3(aq) + 2\,H_2O(l) \rightleftharpoons H_3O^+(aq) + B(OH)_4^-$

14.91 (a)

Element		Ionization energy, $kJ \cdot mol^{-1}$	Atomic radius, pm
Group 13	B	799	88
	Al	577	143
	Ga	577	153
	In	556	167
	Tl	590 (Gen. decreasing ↓)	171 (Increasing ↓)
Group 14	C	1090	77
	Si	786	118
	Ge	784	122
	Sn	707	158
	Pb	716 (Gen. decreasing ↓)	175 (Increasing ↓)

(b) The ionization energies generally decrease down a group. As the atomic number of an element increases, atomic shells and subshells that are farther from the nucleus are filled. The outermost valence electrons are consequently easier to remove. The radii increase down a group for the same reason. The radii are primarily determined by the outer shell electrons, which are farther from the nucleus in the heavy elements.

(c) The trends correlate well with elemental properties; for example, the greater ease of outermost electron removal correlates with increased metallic character, that is, the ability to form positive ions by losing one or more electrons.

14.93 (a) $2\,H_2O(l) + 2\,e^- \longrightarrow H_2(g) + 2\,OH^-(aq)$

(b) cathode, because this process is reduction

(c) At STP (273 K and 1 atm), 1 mol H_2 occupies 22.4 L.

$$V_{H_2(g)} = (10.0\text{ A})(30\text{ min})\left(\frac{60\text{ s}}{1\text{ min}}\right)\left(\frac{1\text{ mol e}^-}{9.65 \times 10^4\text{ A}\cdot\text{s}}\right)\left(\frac{1\text{ mol H}_2}{2\text{ mol e}^-}\right)\left(\frac{22.4\text{ L H}_2}{1\text{ mol H}_2}\right)$$

$$= 2.1\text{ L H}_2$$

14.95 In the majority of its reactions, hydrogen acts as a reducing agent, that is, $H_2(g) \longrightarrow 2\,H^+(aq) + 2\,e^-$, $E° = 0$ V. In these reactions, hydrogen resembles Group 1 elements, such as Na and K. However, as described in the text and in the answer to Exercise 14.79, it may also act as an oxidizing agent; that is, $H_2(g) + 2\,e^- \longrightarrow 2\,H^-(aq)$, $E° = -2.25$ V. In these reactions, hydrogen resembles Group 17 elements, such as Cl and Br. Consequently, H_2 will oxidize elements with standard reduction potentials more negative than -2.25 V, such as the alkali and alkaline earth metals (except Be). The compounds formed are hydrides and contain the H^- ion; the singly charged negative ion is reminiscent of the halide ions. Hydrogen also forms diatomic molecules and covalent bonds like the halogens.

The atomic radius of H is 78 pm, which compares rather well to that of F (64 pm) but not as well to that of Li (157 pm). The ionization energy of H is 1310 $kJ\cdot mol^{-1}$, which is similar to that of F (1680 $kJ\cdot mol^{-1}$) but not similar to that of Li (519 $kJ\cdot mol^{-1}$). The electron affinity of H is +73 $kJ\cdot mol^{-1}$, that of F is +328 $kJ\cdot mol^{-1}$, and that of Li is 60 $kJ\cdot mol^{-1}$. So in its atomic radius and ionization energy, H more closely resembles the Period 2 halogen, fluorine, in Group 17, than the Period 2 alkali metal, lithium, in Group 1; whereas in electron affinity, it more closely resembles lithium, Group 1. In electronegativity, H does not resemble elements in either Group 1 or Group 17, although its electronegativity is somewhat closer to those of Group 1. Consequently, hydrogen could be placed in either Group 1 or Group 17. But it is best to think of hydrogen as a unique element that has properties in common with both metals and nonmetals; therefore, it should probably be centered in the periodic table, as it is shown in the table in the text.

14.97 (a)

Ion	Radius, pm	Polarizing ability (×1000)	Ion	Radius, pm	Polarizing ability (×1000)
Li^+	58	17	Be^{2+}	27	74
Na^+	102	9.80	Mg^{2+}	72	28
K^+	138	7.25	Ca^{2+}	100	20.0
Rb^+	149	6.71	Sr^{2+}	116	17.2
Cs^+	170	5.88	Ba^{2+}	136	14.7

(b) These data roughly support the diagonal relationship. Li^+ is more like Mg^{2+} than Be^{2+}, and Na^+ is more like Ca^{2+} than Mg^{2+}; but further down the group, the correlation fails. Charge divided by r^3 would be a better measure of polarizing ability.

14.99 $H_2(g) + Br_2(l) \longrightarrow 2\ HBr(g)$

$$\text{number of moles of HBr} = (0.120\ \text{L H}_2)\left(\frac{1\ \text{mol H}_2}{22.4\ \text{L H}_2}\right)\left(\frac{2\ \text{mol HBr}}{1\ \text{mol H}_2}\right)$$

$$= 0.0107\ \text{mol}$$

$$\text{molar concentration of HBr} = \frac{0.0107\ \text{mol}}{0.150\ \text{L}} = 0.0713\ \text{mol}\cdot\text{L}^{-1}$$

14.101 The smaller the cation, the greater is the ability of the cation to polarize and weaken the carbonate ion, CO_3^{2-}. On that basis, we would predict that within a group the carbonates of the first members of the group are less stable than those of the later members. Thus, $Li_2CO_3 < Na_2CO_3 < K_2CO_3 < Rb_2CO_3 < Cs_2CO_3$

and $BeCO_3 < MgCO_3 < CaCO_3 < SrCO_3 < BaCO_3$. Between groups, we would expect the stability of the carbonates in one period to decrease from Group 1 to Group 13 because of the smaller size of Group 13 ions. Thus, $Al_2(CO_3)_2 < MgCO_3 < Na_2CO_3$. Carbonates of Group 13 M^{3+} ions are, in fact, so unstable that they do not exist. Tl_2CO_3 where Tl has an oxidation number of +1 is known, however.

14.103 (a) NaH, like other saline hydrides, reacts with O—H bonds to liberate H_2. For example, reaction of NaH with H_2O gives NaOH(aq) and H_2(g). The corresponding reaction with methanol is

$$NaH(s) + CH_3OH(l) \longrightarrow NaOCH_3(alc) + H_2(g)$$

Sodium methoxide (also known as sodium methylate), $NaOCH_3$, is a strong base that is used in organic chemistry.

(b) $Na^+ \; {}^-\!:\!\ddot{\underset{..}{O}}\!-\!\underset{H}{\overset{H}{\underset{|}{\overset{|}{C}}}}\!-\!H$

(c) $NaOCH_3(alc) + H_2O(l) \longrightarrow NaOH(aq) + HOCH_3(aq)$

14.105 (a) The unit cell described will contain a total of 4 B atoms and 4 N atoms. The volume of the cell is $(361.5\ pm)^3 = (3.615 \times 10^{-8}\ cm)^3$ = or $4.724 \times 10^{-23}\ cm^{-3}$. The mass in the unit cell will be $(4 \times 10.81\ g \cdot mol^{-1} + 4 \times 14.01\ g \cdot mol^{-1}) \div 6.022 \times 10^{23}\ mol^{-1} = 1.649 \times 10^{-22}\ g$.

$$d = \frac{1.649 \times 10^{-22}\ g}{4.724 \times 10^{-23}\ cm^3} = 3.491\ g \cdot cm^{-3}$$

(b) Because the density of cubic boron nitride is greater than that of hexagonal BN, we would expect the cubic form to be favored at high pressures, exactly as found for the cubic (diamond) and hexagonal (graphite) forms of carbon.

14.107 The metal hydride compounds have a molecular orbital structure that is very asymmetric. Because the hydrogen atom is much more electronegative than the metal atom, its orbital lies much lower in energy. Consequently, when a bond is formed, it is a strongly ionic bond with the electrons heavily localized on the H atom.

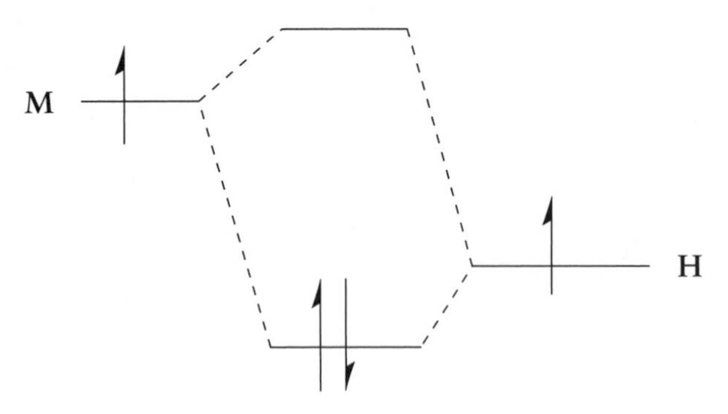

However, in the case of the lighter *p*-block elements, the electronegativity difference is not so great and the bonding is much more covalent. If anything, the *p*-block element is more electronegative than hydrogen; the bond polarity would lie in the other direction but would be much less pronounced than in the saline hydrides.

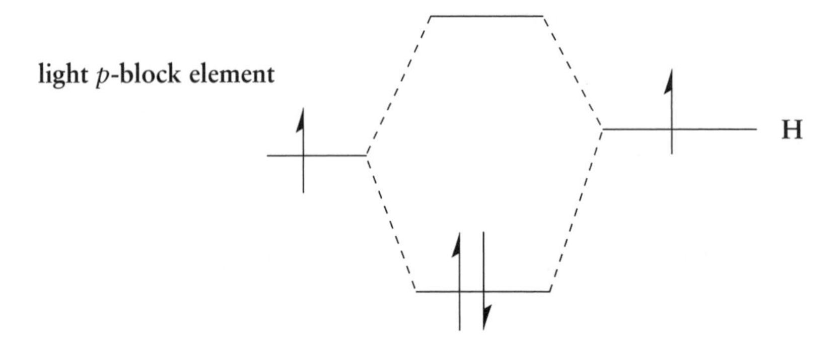

14.109 (a) Diborane, B_2H_6, and $Al_2Cl_6(g)$ have the same basic structure in the way in which the atoms are arranged in space.

(b) The bonding between the boron atoms and the bridging hydrogen atoms is electron deficient. There are three atoms and only two electrons to hold them together in a 3-center-2-electron bond. The B—H—B bonds may be better represented as curved lines connecting all three atoms:

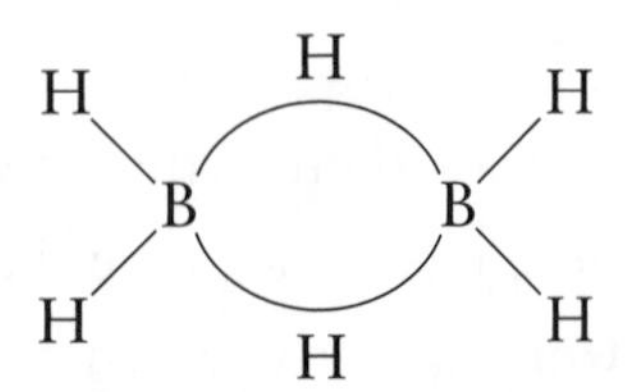

The bonding in Al_2Cl_6 is conventional in that all the bonds involve two atoms and two electrons. Here, the lone pair of a Cl atom is donated to an adjacent Al.

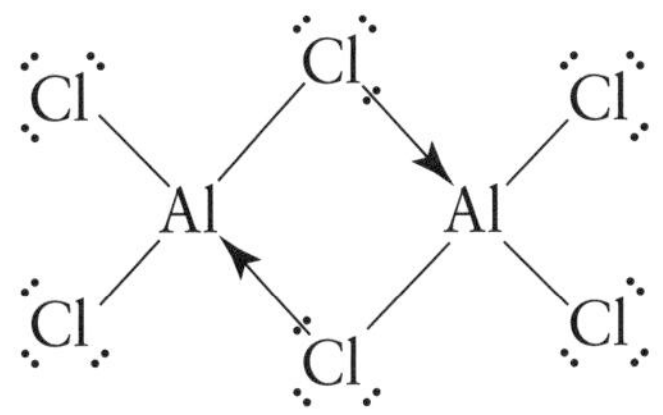

(c) The hybridization is sp^3 at the B and Al atoms.

(d) The molecules are not planar. The Group 13 element and the terminal atoms to which it is bound lie in a plane that is perpendicular to the plane that contains the main group element and the bridging atoms.

14.111 The reaction of interest is

$$SnO(s) + \tfrac{1}{2}\,O_2(g) \longrightarrow SnO_2(s)$$

To find the standard reduction potential, we can use the values for the free energy of the reaction.

$$\Delta G^\circ_r = \Delta G^\circ_f(SnO_2) - \Delta G^\circ_f(SnO)$$
$$= -519.6\ kJ\cdot mol^{-1} - (-256.9\ kJ\cdot mol^{-1})$$
$$= -262.7\ kJ\cdot mol^{-1}$$

Then using the relationship $\Delta G^\circ_r = -nFE^\circ$:

$$-262.7 \times 10^3\ J\cdot mol^{-1} = -(2)(96485\ J\cdot V^{-1}\cdot mol^{-1})(E^\circ)$$
$$E^\circ = +1.36\ V$$

14.113 (a) By viewing the unit cell from different directions, it is clear that it belongs to a hexagonal crystal system. (b) There are eight carbonate ions on edges, giving $\frac{1}{4} \times 8 = 2$ carbonate ions, plus there are four carbonate ions completely within the unit cell. The total number of carbonate ions in the unit cell is six. Calcium ions lie at the corners of the unit cell ($\frac{1}{8} \times 8 = 1$) as well as on the edges ($4 \times \frac{1}{4}$) and within the cell (4). The total number of calcium ions in the cell is six, agreeing with the overall stoichiometry of calcite, $CaCO_3$.

14.115 There are two Ca^{2+} ions located completely within the cell plus four located on faces ($4 \times \frac{1}{2}$) for a total of four Ca^{2+} ions in the unit cell. Similarly, there are two sulfate ions located completely within the unit cell and four on faces, also giving two sulfate ions in the unit cell. There are eight water molecules completely inside the unit cell. The overall formula is $Ca_4(SO_4)_4(H_2O)_8$ or $CaSO_4\cdot 2H_2O$.

CHAPTER 15
THE ELEMENTS: THE LAST FOUR MAIN GROUPS

15.1

−3	NH_3, Li_3N, $LiNH_2$, NH_2^-
−2	H_2NNH_2
−1	N_2H_2, NH_2OH
0	N_2
+1	N_2O, N_2F_2
+2	NO
+3	NF_3, NO_2^-, NO^+
+4	NO_2, N_2O_4
+5	HNO_3, NO_3^-, NO_2F

15.3 $CO(NH_2)_2 + 2\ H_2O \longrightarrow (NH_4)_2CO_3$

$$\text{mass of }(NH_4)_2CO_3 = (5.0\text{ kg urea})\left(\frac{10^3\text{ g}}{1\text{ kg}}\right)\left(\frac{1\text{ mol}}{60.06\text{ g}}\right)\left(\frac{1\text{ mol }(NH_4)_2CO_3}{1\text{ mol urea}}\right)\left(\frac{96.09\text{ g }(NH_4)_2CO_3}{1\text{ mol }(NH_4)_2CO_3}\right)$$

$$= 8.0 \times 10^3\text{ g (or 8.0 kg)}(NH_4)_2CO_3$$

15.5 (a) 1 mole of $N_2(g)$ occupies 22.4 L at STP. For the reaction $Pb(N_3)_2 \longrightarrow Pb + 3\ N_2$, the volume of $N_2(g)$ produced is

$$(1.0\text{ g }Pb(N_3)_2)\left(\frac{1\text{ mol }Pb(N_3)_2}{291.25\text{ g }Pb(N_3)_2}\right)\left(\frac{3\text{ mol }N_2}{1\text{ mol }Pb(N_3)_2}\right)\left(\frac{22.4\text{ L }N_2}{1\text{ mol }N_2}\right)$$

$$= 0.23\text{ L }N_2(g)$$

(b) $Hg(N_3)_2$ would produce a larger volume, because its molar mass is less. Note that molar mass occurs in the denominator in this calculation.

(c) Metal azides are good explosives because the azid ion is thermodynamically unstable with respect to the production of $N_2(g)$. This is because the N—N triple bond is so strong and also because the production of a gas is favored entropically.

15.7 N_2O: $H_2N_2O_2$; $N_2O(g) + H_2O(l) \longrightarrow H_2N_2O_2(aq)$

N_2O_3: HNO_2; $N_2O_3(g) + H_2O(l) \longrightarrow 2\ HNO_2(aq)$

N_2O_5: HNO_3; $N_2O_5(g) + H_2O(l) \longrightarrow 2\ HNO_3(aq)$

15.9 (a)

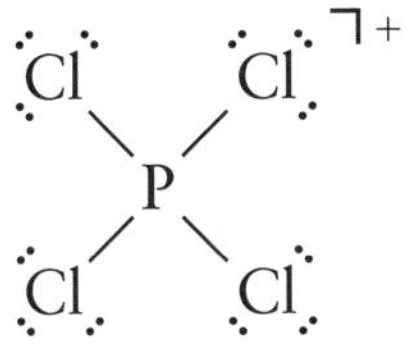

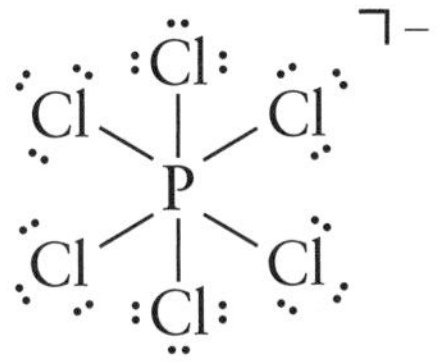

PCl_4^+, AX_4
tetrahedral

PCl_6^-, AX_6
octahedral

(b) and (c) The PX_6^- ions (where X = halogen) are octahedral. The calculation of the X—X separation can easily be done using the Pythagorean theorem.

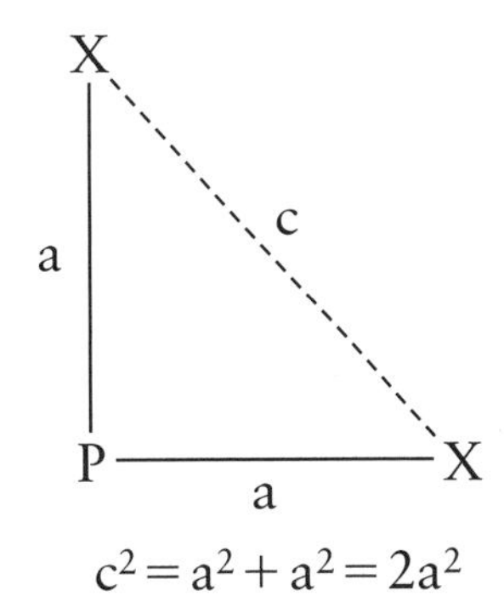

$$c^2 = a^2 + a^2 = 2a^2$$

For Br, a = 220 pm, giving c = 311 pm. For Cl, a = 204 pm, giving c = 288 pm. Both these distances are much shorter than twice the van der Waals radius (Cl, 362 pm; Br, 390 pm) but are substantially longer than twice the atomic radius (Cl, 198; Br, 228 pm). It is not clear from these numbers that the steric interaction is the main consideration. Weaker P—Br bond energies may also play a role in the nonobservation of the PBr_6^- ion.

15.11 (a) The reaction is $4\ NH_3(g) + 5\ O_2(g) \rightarrow 4\ NO(g) + 6\ H_2O(g)$

$\Delta H^\circ_r = 4(+90.25\ kJ\cdot mol^{-1}) + 6(-241.82\ kJ\cdot mol^{-1}) - [4(-46.11\ kJ\cdot mol^{-1})]$

$= -905.48\ kJ\cdot mol^{-1}$

$\Delta S^\circ_r = 4(210.76\ J\cdot K^{-1}\cdot mol^{-1}) + 6(188.83\ J\cdot K^{-1}\cdot mol^{-1})$

$- [4(192.45\ J\cdot K^{-1}\cdot mol^{-1}) + 5(205.14\ J\cdot K^{-1}\cdot mol^{-1})]$

$\Delta S^\circ_r = +180.52\ J\cdot K^{-1}\cdot mol^{-1}$

$\Delta G^\circ_r = 4(+86.55\ kJ\cdot mol^{-1}) + 6(-228.57\ kJ\cdot mol^{-1}) - [4(-16.45\ kJ\cdot mol^{-1})]$

$\Delta G^\circ_r = -959.42\ kJ\cdot mol^{-1}$ at 298 K

At 1000°C (1273 K),

$\Delta G^\circ_r = -905.48\ kJ\cdot mol^{-1} - (1273\ K)(+180.52\ J\cdot K^{-1}\cdot mol^{-1})/(1000\ J\cdot kJ^{-1})$

$= -1135.28\ kJ\cdot mol^{-1}$

(b) The reaction is spontaneous over all temperatures.

15.13 (a) The reaction is $3\ NO_2(g) + H_2O(l) \rightarrow 2\ HNO_3(aq) + NO(g)$

$$\Delta H°_r = 2(-207.36\ kJ\cdot mol^{-1}) + 90.25\ kJ\cdot mol^{-1} - [3(33.18\ kJ\cdot mol^{-1}) + (285.83\ kJ\cdot mol^{-1})]$$
$$= -138.18\ kJ\cdot mol^{-1}$$

(b) $HNO_3(l) \rightarrow HNO_3(aq)$

$$\Delta H°_r = -207.36\ kJ\cdot mol^{-1} - (-174.10\ kJ\cdot mol^{-1})$$
$$= -33.26\ kJ\cdot mol^{-1}$$

15.15 (a) $4\ Li(s) + O_2(g) \xrightarrow{\Delta} 2\ Li_2O(s)$

(b) $2\ Na(s) + 2\ H_2O(l) \longrightarrow 2\ NaOH(aq) + H_2(g)$

(c) $2\ F_2(g) + 2\ H_2O(l) \longrightarrow 4\ HF(aq) + O_2(g)$

(d) $2\ H_2O(l) \longrightarrow O_2(g) + 4\ H^+(aq) + 4\ e^-$

15.17 (a) $2\ H_2S(g) + 3\ O_2(g) \xrightarrow{\Delta} 2\ SO_2(g) + 2\ H_2O(g)$

(b) $CaO(s) + H_2O(l) \longrightarrow Ca(OH)_2(aq)$

(c) $2\ H_2S(g) + SO_2(g) \xrightarrow{300°C,\ Al_2O_3} 3\ S(s) + 2\ H_2O(l)$

15.19 (a) H

Ö—Ö

H

Each O in H_2O_2 is an AX_2E_2 structure; therefore, the bond angle is predicted to be $<109.5°$. In actuality, it is 97°.

(b)–(e), The reduction potential of H_2O_2 is $+1.78$ V in acidic solution. It should, therefore, be able to oxidize any ion that has a reduction potential that is less than $+1.78$ V. For the ions listed, Cu^+ and Mn^{2+} will be oxidized. It would require an input of 1.98 V to oxidize Ag^+ to Ag^{2+} and 2.87 V to oxidize F^-.

15.21 $2\ H_2O_2(l) \longrightarrow 2\ H_2O(l) + O_2(g)$

Assume 3% by mass. At 273 K and 1.00 atm, 1 mol O_2 has a volume of 22.4 L.

$$\text{volume of } O_2 = (500\ mL)\left(\frac{1.0\ g}{1.0\ mL}\right)\left(\frac{0.03\ g\ H_2O_2}{1.0\ g\ soln}\right)$$
$$\left(\frac{1\ mol\ H_2O_2}{34.02\ g\ H_2O_2}\right)\left(\frac{1\ mol\ O_2}{2\ mol\ H_2O_2}\right)\left(\frac{22.4\ L\ O_2}{1\ mol\ O_2}\right) = 5\ L\ O_2$$

15.23 $O_2^{2-} + H_2O \rightleftharpoons HO_2^- + OH^-$ essentially complete

$$HO_2^- + H_2O \rightleftharpoons H_2O_2 + OH^- \quad K_b = \frac{K_w}{K_{a1}}$$

$K_{a1} = 1.8 \times 10^{-12} \quad K_b = \dfrac{1.00 \times 10^{-14}}{1.8 \times 10^{-12}} = 5.6 \times 10^{-3}$

Because this K_b is relatively small, we can assume that essentially all the OH^- is formed in the first ionization; therefore,

$$[OH^-] = \left(\frac{2.00 \text{ g } Na_2O_2}{0.200 \text{ L}}\right)\left(\frac{1 \text{ mol } Na_2O_2}{77.98 \text{ g } Na_2O_2}\right)\left(\frac{1 \text{ mol } OH^-}{1 \text{ mol } Na_2O_2}\right)$$
$$= 0.128 \text{ mol}\cdot\text{L}^{-1}$$

$pOH = -\log(0.128) = 0.893 \quad pH = 14.00 - 0.893 = 13.11$

If we do not ignore the second ionization, then the additional contribution to $[OH^-]$ can be approximately calculated as follows:

$$K_b = \frac{[H_2O_2][OH^-]}{[HO_2^-]} = \frac{x(0.128 + x)}{(0.128 - x)} = 5.6 \times 10^{-3}$$

To a first approximation, $x = 5.6 \times 10^{-3} \text{ mol}\cdot\text{L}^{-1}$

To a second approximation, $x = \dfrac{K_b(0.128 - 0.0056)}{(0.128 + 0.0056)} = 0.005$

Then $[OH^-] = 0.128 + 0.005 = 0.133$; $pOH = -\log(0.133) = 0.876$; and $pH = 13.12$. The difference between calculations is slight.

15.25 The weaker the H—X bond, the stronger the acid. H_2Te has the weakest bond; H_2O, the strongest. Therefore, the acid strengths are

$H_2Te > H_2Se > H_2S > H_2O$

15.27 $\frac{3}{2} O_2(g) \longrightarrow O_3(g)$

(a) $\Delta H^\circ_f(O_3, g) = 142.7 \text{ kJ}\cdot\text{mol}^{-1}$

$\Delta S^\circ_f(O_3, g) = 238.93 \text{ J}\cdot\text{K}^{-1}\cdot\text{mol}^{-1} - \frac{3}{2} \times 205.14 \text{ J}\cdot\text{K}^{-1}\cdot\text{mol}^{-1}$
$= -68.78 \text{ J}\cdot\text{K}^{-1}\cdot\text{mol}^{-1}$

(b) $\Delta G^\circ_f(O_3, g) \approx \Delta H^\circ_f(O_3, g, 25°C) - T\Delta S^\circ_f(O_3, g, 25°C)$

Because $\Delta G^\circ_f(O_3, g)$ is positive at all temperatures, the reaction is not spontaneous at any temperature. It is less favored at high temperatures.

(c) Because the reaction entropy is negative, the $-T\Delta S^\circ_f$ term is always positive; so the entropy contribution to ΔG°_f is always positive, and the entropy does not favor the spontaneous formation of ozone.

15.29 (a) The reaction is $H_2SO_4(l) \longrightarrow H_2SO_4(aq)$

where $H_2SO_4(aq)$ is $H^+(aq) + HSO_4^-(aq)$ because H_2SO_4 is a strong acid. (Note: $\Delta H^\circ_f(H^+)$ is defined as 0).

$\Delta H^\circ_r = -887.34 \text{ kJ}\cdot\text{mol}^{-1} - (-813.99 \text{ kJ}\cdot\text{mol}^{-1}) = -73.35 \text{ kJ}\cdot\text{mol}^{-1}$

(b) The number of moles of H_2SO_4 is $10.00 \div 98.07 \text{ g}\cdot\text{mol}^{-1} = 0.1020 \text{ mol}$

The amount of heat generated should be

$0.1020 \text{ mol} \times -73.35 \text{ kJ}\cdot\text{mol}^{-1} = -7.482 \text{ kJ}.$

The heat capacity of water is $4.18 \text{ J}\cdot(°\text{C})^{-1}\cdot\text{g}^{-1}$. Adding 7.482 kJ of heat to the water should raise the temperature by

$$\Delta t = \frac{7.482 \text{ kJ} \times 1000 \text{ J}\cdot\text{kJ}^{-1}}{4.18 \text{ J}\cdot(°\text{C})^{-1}\cdot\text{g}^{-1} \times 500.0 \text{ g}} = 3.56°$$

The final temperature should be 25.0°C + 3.56°C = 28.6°C.

15.31 Fluorine comes from the minerals fluorspar, CaF_2; cryolite, Na_3AlF_6; and the fluorapatites, $Ca_5F(PO_4)_3$. The free element is prepared from HF and KF by electrolysis, but the HF and the KF needed for the electrolysis are prepared in the laboratory. Chlorine primarily comes from the mineral rock salt, NaCl. The pure element is obtained by electrolysis of liquid NaCl.

Bromine is found in seawater and brine wells as the Br^- ion; it is also found as a component of saline deposits; the pure element is obtained by oxidation of $Br^-(aq)$ by $Cl_2(g)$.

Iodine is found in seawater, seaweed, and brine wells as the I^- ion; the pure element is obtained by oxidation of $I^-(aq)$ by $Cl_2(g)$.

15.33 (a) HIO(aq) H = +1, O = −2; therefore, I = +1

(b) ClO_2 O = −2; therefore, Cl = +4

(c) Cl_2O_7 O = −2; therefore, Cl = +14/2 = +7

(d) $NaIO_3$ Na = +1, O = −2; therefore, I = +5

15.35 (a) $4\ KClO_3(l) \xrightarrow{\Delta} 3\ KClO_4(s) + KCl(s)$

(b) $Br_2(l) + H_2O(l) \longrightarrow HBrO(aq) + HBr(aq)$

(c) $NaCl(s) + H_2SO_4(aq) \longrightarrow NaHSO_4(aq) + HCl(g)$

(d) (a) and (b) are redox reactions. In (a), Cl is both oxidized and reduced. In (b), Br is both oxidized and reduced. (c) is a Brønsted acid-base reaction; H_2SO_4 is the acid, and Cl^- the base.

15.37 (a) $HClO < HClO_2 < HClO_3 < HClO_4$ ($HClO_4$ is strongest; HClO, weakest)

(b) The oxidation number of Cl increases from HClO to $HClO_4$. In $HClO_4$, chlorine has its highest oxidation number of +7, so $HClO_4$ will be the strongest oxidizing agent.

15.39 [Lewis structure: O bonded to two Cl atoms, with lone pairs]

AX_2E_2, angular, slightly less than 109°

15.41

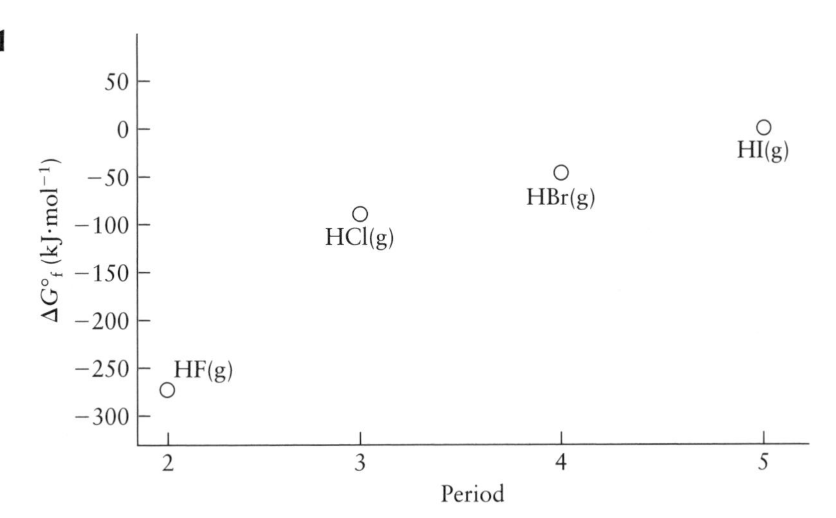

The thermodynamic stability of the hydrogen halides decreases down the group. The $\Delta G°_f$ values of HCl, HBr, and HI fit nicely on a straight line; whereas HF is anomalous. In other properties, HF is also the anomalous member of the group, in particular, its acidity. Also see Exercise 15.42.

15.43 $Cl_2(g) + 2\ e^- \longrightarrow 2\ Cl^-(aq) \quad E° = +1.36\ V$

$MnO_4^-(aq) + 8\ H^+(aq) + 5\ e^- \longrightarrow Mn^{2+}(aq) + 4\ H_2O(l) \quad E° = +1.51\ V$

$E°_{cell} = (1.36 - 1.51)\ V = -0.15\ V$

Because $E°_{cell}$ is negative, $Cl_2(g)$ will not oxidize Mn^{2+} to form the permanganate ion in an acidic solution.

15.45 $\frac{1}{2}\ H_2(g) + \frac{1}{2}\ I_2(s) \rightleftharpoons HI(g)$

$\Delta G°_r = \Delta G°_f(HI, g) = 1.70\ kJ\cdot mol^{-1}$

$\Delta G°_r = -RT \ln K_p$

$$\ln K_p = \frac{-1.70 \times 10^3\ J\cdot mol^{-1}}{8.314\ J\cdot mol^{-1}\cdot K^{-1} \times 298\ K} = -0.686$$

$K_p = 0.504$

15.47 $F^-(aq) + PbCl_2(s) \longrightarrow Cl^-(aq) + PbClF(s)$

$$\text{molarity of } F^- \text{ ions} = \left(\frac{0.765\ g\ PbClF}{0.0250\ L}\right)\left(\frac{1\ mol\ PbClF}{261.64\ g\ PbClF}\right)\left(\frac{1\ mol\ F^-}{1\ mol\ PbClF}\right)$$

$$= 0.117\ mol\cdot L^{-1}$$

15.49 $3\ NH_4ClO_4(s) + 3\ Al(s) \longrightarrow Al_2O_3(s) + AlCl_3(s) + 6\ H_2O(g) + 3\ NO(g)$

The number of moles of NH_4ClO_4 is

$1.00\ kg \times 1000\ g\cdot kg^{-1} \div 117.49\ g\cdot mol^{-1} = 8.51\ mol$

The number of moles of Al is

$1.00\ kg \times 1000\ g\cdot kg^{-1} \div 26.98\ g\cdot mol^{-1} = 37.06\ mol$

The limiting reagent is the NH_4ClO_4.

The standard enthalpy for the reaction is given by

$$\Delta H^\circ_r = (-1675.7\ \text{kJ}\cdot\text{mol}^{-1}) + (-704.2\ \text{kJ}\cdot\text{mol}^{-1}) + 6(-241.82\ \text{kJ}\cdot\text{mol}^{-1}) + 3(90.25\ \text{kJ}\cdot\text{mol}^{-1}) - [3(-295.31\ \text{kJ}\cdot\text{mol}^{-1})]$$
$$= -2674.1\ \text{kJ}\cdot\text{mol}^{-1}$$

This value is the amount of heat released for 3 mol NH_4ClO_4. The amount released for 8.51 mol will be

$$8.51\ \text{mol NH}_4\text{ClO}_4 \times \frac{-2674.1\ \text{kJ}\cdot\text{mol}^{-1}}{3\ \text{mol NH}_4\text{ClO}_4} = -7.59 \times 10^3\ \text{kJ}$$

There will be 7.59×10^3 kJ of heat released.

15.51 Helium occurs as a component of natural gases found under rock formations in certain locations, especially some in Texas. Argon is obtained by distillation of liquid air.

15.53 (a) KrF_2: F = −1; therefore, Kr = +2
(b) XeF_6: F = −1; therefore, Xe = +6
(c) KrF_4: F = −1; therefore, Kr = +4
(d) XeO_4^{2-}: O = −2, $N_{ox}(Xe) - 8 = -2$; therefore, $N_{ox}(Xe) = +6$

15.55 $XeF_4 + 4\ H^+ + 4\ e^- \longrightarrow Xe + 4\ HF$

15.57 Because H_4XeO_6 has more highly electronegative O atoms bonded to Xe, we predict that H_4XeO_6 is more acidic than H_2XeO_4.

15.59 A sol is colloid comprised of solid particles suspended in a liquid. Muddy water is a type of sol. A foam is a suspension of a gas in a solid or liquid. Styrofoam, foam rubber, soapsuds, and aerogels are all types of foams.

15.61 (a) both a sol and an emulsion (b) a foam (c) a sol

15.63 In fluorescence, the light that is emitted is of lower energy than the light that is absorbed, and the fluorescence stops as soon as the exciting radiation is stopped. In phosphorescence, the phosphorescing molecules remain excited for a period of time after the stimulus has stopped.

15.65 The mercury atoms are an energy transfer agent. They absorb the energy from a high-voltage discharge and emit ultraviolet light in the region of 254 and 185 nm.

This emitted light then excites a fluorescent material that emits radiation in the visible region of the spectrum, which is the light observed when the lamp is turned on.

15.67 (a) and (b). The formal charges are given under each atom.

$$[\ddot{\underset{..}{N}}{=}N{=}\ddot{\underset{..}{N}}]^-$$

$$-1 \quad +1 \quad -1$$

(c) The value of $-\frac{1}{3}$ is an average oxidation number based solely upon the number of nitrogen atoms and the overall charge. From the Lewis structure, we can see that the molecule is asymmetric and it is possible that the different nitrogen atoms may have different oxidation numbers. It would be extreme to state that the terminal nitrogens have oxidation numbers of -1 and the central nitrogen atom has an oxidation number of $+1$, but assigning the same oxidation number of $-\frac{1}{3}$ to all three atoms is also not strictly accurate.

(d) This situation most often arises when an element has bonds to other atoms of the same type.

15.69 The larger the value of $E°$ for the reduction $X_2 + 2\,e^- \longrightarrow 2\,X^-$, the greater the oxidizing strength of the halogen X_2. From Appendix 2B,

$F_2 + 2\,e^- \longrightarrow 2\,F^- \quad E° = +2.87$ V

$Cl_2 + 2\,e^- \longrightarrow 2\,Cl^- \quad E° = +1.36$ V

$Br_2 + 2\,e^- \longrightarrow 2\,Br^- \quad E° = +1.09$ V

$I_2 + 2\,e^- \longrightarrow 2\,I^- \quad E° = +0.54$ V

Thus, $I_2 < Br_2 < Cl_2 < F_2$.

15.71 oxidation:

$As_2S_3(s) + 8\,H_2O(l) \longrightarrow 2\,AsO_4^{3-}(aq) + 3\,S^{2-}(aq) + 16\,H^+(aq) + 4\,e^-$

reduction: $H_2O_2(aq) + 2\,H^+(aq) + 2\,e^- \longrightarrow 2\,H_2O(l)$

Multiply the reduction reaction by 2, cancel electrons, and add.

overall:

$As_2S_3(s) + 2\,H_2O_2(aq) + 4\,H_2O(l) \longrightarrow 2\,AsO_4^{3-}(aq) + 3\,S^{2-}(aq) + 12\,H^+(aq)$

or $As_2S_3(s) + 2\,H_2O_2(aq) + 4\,H_2O(l) \longrightarrow 2\,H_3AsO_4(aq) + 3\,H_2S(aq)$

15.73 This ratio, $\Delta H_{vap}/T_b$, is the entropy of vaporization. Hydrogen bonding is much stronger in $H_2O(l)$ than in $H_2S(l)$. Thus $H_2O(l)$ has a more ordered arrangement than $H_2S(l)$. Consequently, the change in entropy upon transformation to the gaseous state is greater for H_2O than for H_2S.

15.75 $Ca_3(PO_4)_2(s) + 3\ H_2SO_4(l) \longrightarrow 2\ H_3PO_4(l) + 3\ CaSO_4(s)$

$$V = (1000 \text{ kg } H_3PO_4)\left(\frac{10^3 \text{ g}}{1 \text{ kg}}\right)\left(\frac{1 \text{ mol } H_3PO_4}{97.99 \text{ g } H_3PO_4}\right)$$

$$\left(\frac{3 \text{ mol } H_2SO_4}{2 \text{ mol } H_3PO_4}\right)\left(\frac{98.08 \text{ g } H_2SO_4}{1 \text{ mol } H_2SO_4}\right)\left(\frac{1 \text{ mL}}{1.84 \text{ g}}\right)\left(\frac{10^{-3} \text{ L}}{1 \text{ mL}}\right)$$

$$= 8.16 \times 10^2 \text{ L conc. } H_2SO_4$$

15.77 The two molecules are shown below:

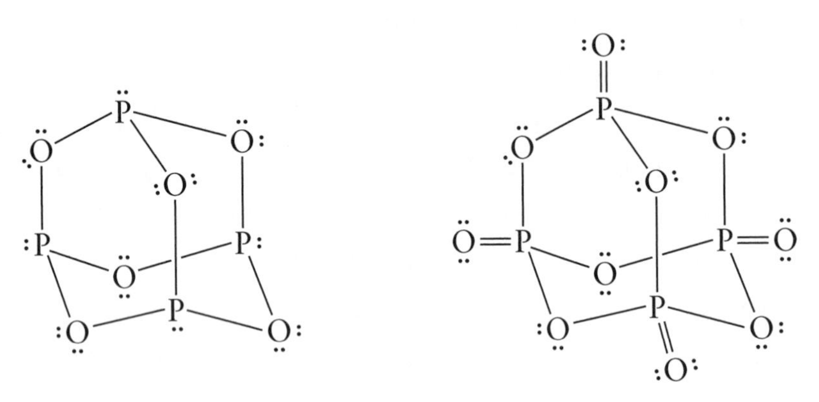

Phosphorus(III) oxide   Phosphorus(V) oxide

The basic structure of the two molecules is the same. The phosphorus atoms lie in a tetrahedral arrangement in which there are bridging oxygen atoms to the other phosphorus atoms. In phosphorus(V) oxide, there is an additional terminal oxygen atom bonded to each phosphorus atom. In phorphorus(III) oxide, each oxygen atom has a formal charge of 0 as does each phosphorus atom. In phosphorus(V) oxide, this is also true. According to the Lewis structures, all P—O bonds in phosphorus(III) oxide have a bond order of 1, while in phosphorus(V) oxide, the terminal oxygen atoms have a bond order of 2 between them and the phosphorus atoms to which they are attached. If one examines the molecular parameters, one sees that all of the $P—O_{bridging}$ distances for phosphorus(III) oxide are slightly longer than those of phosphorus(V) oxide (163.8 pm versus 160.4 pm). This is expected because the radius of phosphorus(V) should be smaller than that of phosphorus(III). The terminal P=O distances in phosphorus(V) oxide are considerably shorter (142.9 pm); this agrees with the higher bond order between phosphorus and these atoms.

15.79 (a) $SO_2(g) + H_2O(l) \longrightarrow H_2SO_3(l)$ This is a Lewis acid-base reaction. SO_2 is the acid and H_2O is the base.

(b) $2\ F_2(g) + 2\ NaOH(aq) \longrightarrow OF_2(g) + 2\ NaF(aq) + H_2O(l)$ This is a redox reaction illustrating the oxidizing ability of F_2 in basic solution and is used for the preparation of $OF_2(g)$. O is oxidized and F is reduced.

(c) $S_2O_3^{2-}(aq) + 4\ Cl_2(g) + 13\ H_2O(l) \longrightarrow 2\ HSO_4^-(aq) + 8\ H_3O^+(aq) + 8\ Cl^-(aq)$ This is a redox reaction illustrating the oxidizing power of $Cl_2(g)$ in acidic solution. S is oxidized and Cl is reduced.

(d) $2\ XeF_6(s) + 16\ OH^-(aq) \longrightarrow XeO_6^{4-}(aq) + Xe(g) + 12\ F^-(aq) + 8\ H_2O(l) + O_2(g)$ This is a redox reaction that is also a disproportionation reaction in that Xe goes from oxidation number +6 to +8 and to 0. Xe is both oxidized and reduced.

15.81 (a) $I_2(s) + 3\ F_2(g) \longrightarrow 2\ IF_3(s)$; $I_2(s) + 5\ F_2(g) \longrightarrow 2\ IF_5(s)$, etc.

(b) $I_2(aq) + I^-(aq) \longrightarrow I_3^-(aq)$

(c) $Cl_2(g) + H_2O(l) \longrightarrow HCl(aq) + HOCl(aq)$

But there are competing reactions, such as $Cl_2(g) + H_2O(l) \longrightarrow 2\ HCl(aq) + \frac{1}{2}\ O_2(g)$. The predominant reaction is determined by the temperature and pH.

(d) $2\ F_2(g) + 2\ H_2O(l) \longrightarrow 4\ HF(aq) + O_2(g)$

15.83 The heads of matches consist of a paste of potassium chlorate ($KClO_3$), antimony sulfide (Sb_2S_3), sulfur, and powdered glass. The striking strip contains red phosphorus. When the match is struck against the red phosphorus surface, a reaction of the red phosphorus and potassium chlorate causes the match to ignite. The Sb_2S_3 and sulfur are the fuels that are consumed by combustion after the ignition. The powdered glass helps to produce the friction required for ignition.

15.85 Orpiment is As_2S_3 and realgar is As_4S_4. Orpiment is yellow and realgar is orange-red. They are both used as pigments.

15.87 (a) $[\ddot{\underset{\cdot\cdot}{N}}{=}N{=}\ddot{\underset{\cdot\cdot}{N}}]^-$

AX_2, linear, 180°

(b) F^-, 133 pm; N_3^-, 148 pm; Cl^-, 181 pm; therefore, between fluorine and chlorine

(c) HCl, HBr, and HI are all strong acids. For HF, $K_a = 3.5 \times 10^{-4}$, so HF is slightly more acidic than HN_3. The small size of the azide ion suggests that the H—N bond in HN_3 is similar in strength to that of the H—F bond, so it is expected to be a weak acid.

(d) ionic: NaN_3, $Pb(N_3)_2$, AgN_3, etc.
covalent: HN_3, $B(N_3)_3$, FN_3, etc.

15.89 $$[ClO^-] = (0.028\ 34\ L)\left(\frac{0.110\ mol\ S_2O_3^{2-}}{1\ L\ Na_2S_2O_3}\right)\left(\frac{1\ mol\ I_2}{2\ mol\ S_2O_3^{2-}}\right)\left(\frac{1\ mol\ ClO^-}{1\ mol\ I_2}\right)\left(\frac{1}{0.010\ 00\ L\ ClO^-}\right)$$
$$= 0.156\ mol \cdot L^{-1}$$

15.91 The solubility of the ionic halides is determined by a variety of factors, especially the lattice enthalpy and enthalpy of hydration. There is a delicate balance between the two factors, with the lattice enthalpy usually being the determining one. Lattice enthalpies decrease from chloride to iodide, so water molecules can more readily separate the ions in the latter. Less ionic halides, such as the silver halides, generally have a much lower solubility, and the trend in solubility is the reverse of the more ionic halides. For the less ionic halides, the covalent character of the bond allows the ion pairs to persist in water. The ions are not easily hydrated, making them less soluble. The polarizability of the halide ions, and thus, the covalency of their bonding, increases down the group.

15.93 To answer this question, we can compare equilibrium "vapor pressures" of water over each of these reagents. The one with the lowest equilibrium vapor pressure will be the better drying agent. The two reactions of interest are

$$CaO(s) + H_2O(g) \longrightarrow Ca(OH)_2(s)$$
$$P_4O_{10}(s) + 6\ H_2O(g) \longrightarrow 4\ H_3PO_4(s)$$

First, the free energies of the reactions are calculated and from these the equilibrium pressure of water can be obtained.

For CaO:

$$\Delta G^\circ_r = -898.49\ kJ \cdot mol^{-1} - [(-604.03\ kJ \cdot mol^{-1}) + (-228.57\ kJ \cdot mol^{-1})]$$
$$= -65.89\ kJ \cdot mol^{-1}$$

$$\Delta G^\circ_r = -RT \ln K$$

$$K = e^{-\frac{\Delta G^\circ}{RT}} = e^{-\frac{-65\ 890\ J \cdot mol^{-1}}{(8.314\ J \cdot K^{-1} \cdot mol^{-1})(298\ K)}} = 3.5 \times 10^{11}$$

$$K = \frac{1}{P_{H_2O}} = 3.5 \times 10^{11}$$

$P_{H_2O} = 2.8 \times 10^{-12}$ bar

For P_4O_{10}:

$$\Delta G°_r = 4(-1119.2 \text{ kJ·mol}^{-1}) - [(-2697.0 \text{ kJ·mol}^{-1}) + 6(-228.57 \text{ kJ·mol}^{-1})]$$
$$= -408.4 \text{ kJ·mol}^{-1}$$

$$K = e^{-\frac{\Delta G°}{RT}} = e^{-\frac{-408\,400 \text{ J·mol}^{-1}}{(8.314 \text{ J·K}^{-1}\text{·mol}^{-1})(298 \text{ K})}} = 3.9 \times 10^{71}$$

$$K = \frac{1}{(P_{H_2O})^6} = 3.9 \times 10^{71}$$

$P_{H_2O} = 1.2 \times 10^{-12}$ bar

Because the pressure of water possible above CaO is greater than that above P_4O_{10}, CaO will be a poorer drying agent.

15.95 (a) $Ag^+ + e^- \longrightarrow Ag \qquad E° = +0.80$ V

$I_2 + 2\,e^- \longrightarrow 2\,I^- \qquad E° = +0.54$ V

For $2\,Ag^+ + 2\,I^- \longrightarrow Ag + I_2$, $E° = +0.80 \text{ V} - 0.54 \text{ V} = +0.26$ V

The process should be spontaneous.

(b) The formation of AgI precipitate means that the concentration of Ag^+ ions is never high enough to achieve the conditions necessary for the redox reaction to take place. The solubility product K_{sp} limits the concentrations in solution, so that the actual redox potential is not the value calculated, which represents the values when $[Ag^+] = 1$ M and $[I^-] = 1$ M. If we use the concentrations established by the solubility equilibrium and the Nernst equation, we can calculate the actual redox potential:

$$E = E° - \frac{RT}{nF} \ln Q$$

where, in this case, $Q = \frac{1}{K_{sp}^{\;2}}$ for the reaction as written

$K_{sp} = 1.5 \times 10^{-16}$ for AgI

$$E = +0.26 \text{ V} - \frac{(8.314 \text{ J·K}^{-1}\text{·mol}^{-1})(298 \text{ K})}{(2)(96\,485 \text{ J·V}^{-1}\text{·mol}^{-1})} \ln \frac{1}{(1.5 \times 10^{-16})^2}$$
$$= +0.26 \text{ V} - \frac{(8.314 \text{ J·K}^{-1}\text{·mol}^{-1})(298 \text{ K})}{(96\,485 \text{ J·V}^{-1}\text{·mol}^{-1})} \ln \frac{1}{(1.5 \times 10^{-16})}$$
$$= +0.26 \text{ V} - 0.94 \text{ V}$$
$$= -0.68 \text{ V}$$

The fact that the concentrations of Ag^+ and I^- are limited in solution means that the redox potential for a spontaneous reaction is never achieved.

15.97 The molecular orbital diagram for NO^+ should have the oxygen orbitals slightly lower in energy than the nitrogen orbitals, because oxygen is more electronegative. This will cause the bonding to be more ionic than in either N_2 or O_2. There is an ambiguity, however, in that the MO diagram could be similar to either that of N_2 or that of O_2. Refer to Figures 3.34 and 3.35 where you will see that the σ_{2p} or the π_{2p} have different relative energies. There are consequently two possibilities for the orbital energy diagram:

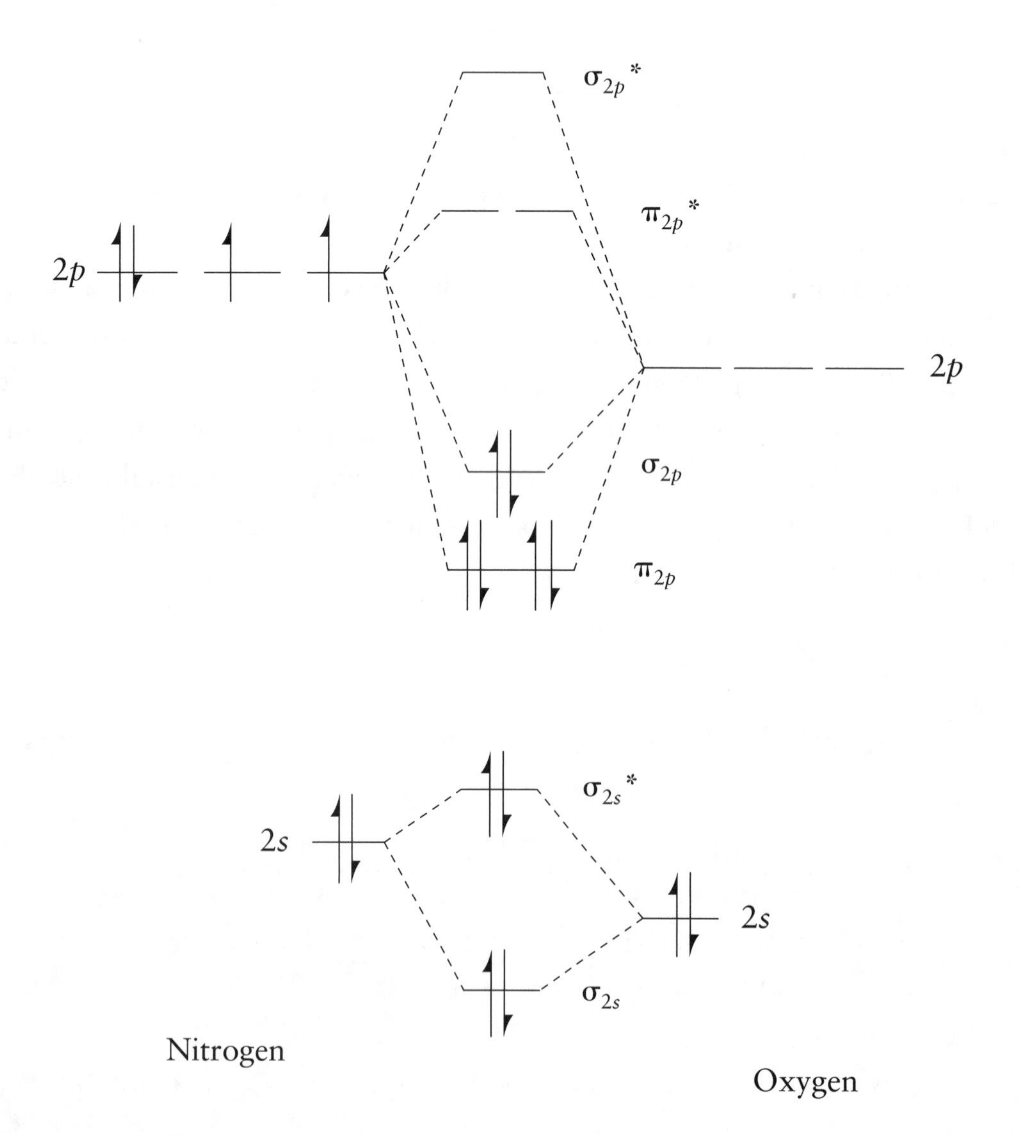

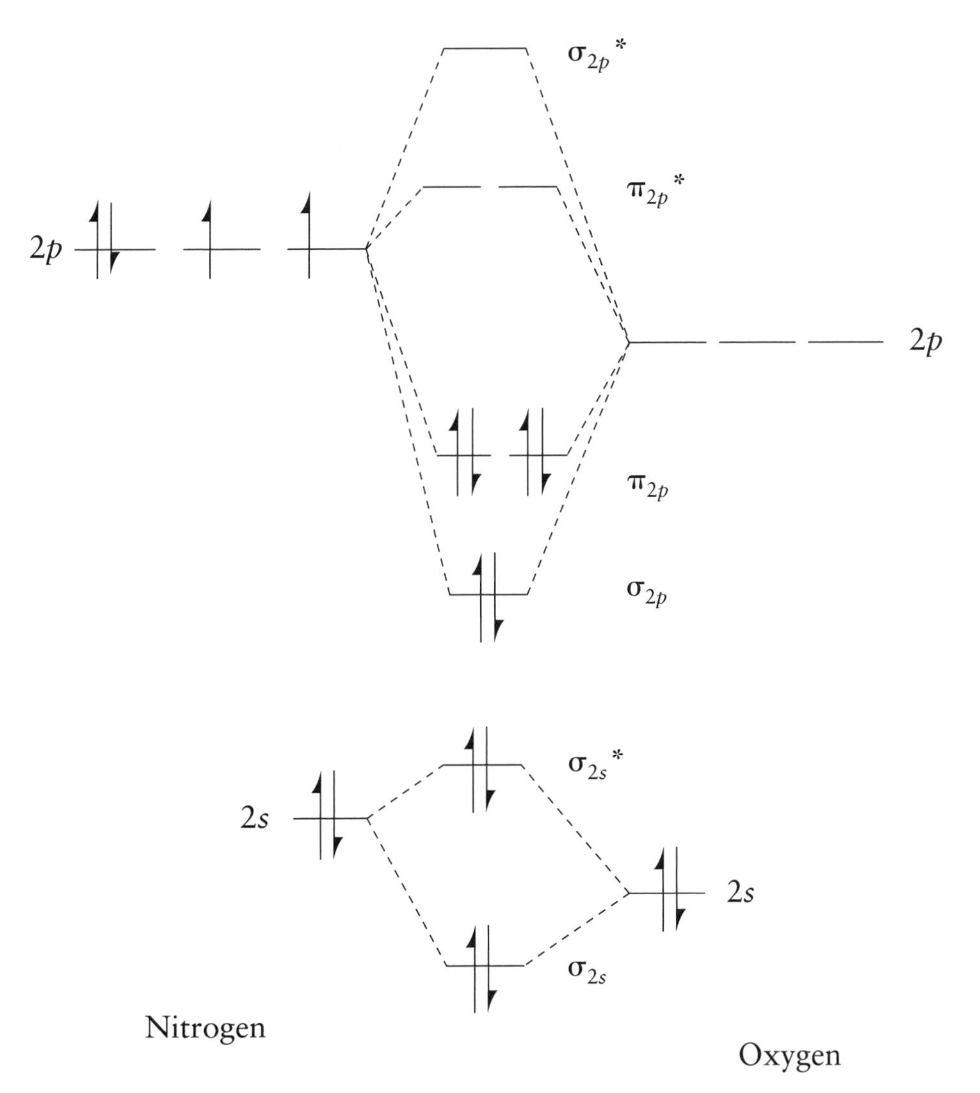

(b) The two orbital diagrams predict the same bond order (3) and same magnetic properties (diamagnetic), so these properties cannot be used to determine which diagram is the correct one. That must be determined by more complex spectroscopic measurements.

15.99 (a) Pyrite adopts a face-centered cubic unit cell.

(b) The iron atoms lie at the corners and at the face centers of the unit cell. Eight sulfur atoms lie completely within the unit cell.

(c) The coordination number of iron is six (octahedral).

(d) Each sulfur atom is bonded to one other sulfur atom and three iron atoms.

(e) It is best to consider Fe to have an oxidation number of +2. The sulfur atoms are bonded to each other in S_2^{2-}. From the formula alone, it is tempting to conclude that the oxidation number on Fe is +4, with two S^{2-} ions. However, Fe^{4+} is very rare because it is extremely reactive. It would be a very strong oxidant and S^{2-} is very easily oxidized.

CHAPTER 16
THE *d* BLOCK: METALS IN TRANSITION

16.1 Elements at the left of the *d* block tend to have strongly negative standard potentials.

16.3 (a) Sc (b) Au (c) Nb

(d) One might expect osmium to be larger than ruthenium because it is a third row transition metal and ruthenium is in row two; however, because of the lanthanide contraction, they are about the same size (Os, 135 pm; Ru, 134 pm).

16.5 (a) Ti (b) Cu (c) Zn

(d) Because ruthenium is larger, one expects it to have a lower first ionization potential, which is observed ($711\ kJ \cdot mol^{-1}$ vs $759\ kJ \cdot mol^{-1}$).

(e) One might expect the third row transition metal to have a lower first ionization energy; however, due to the lanthanide contraction, the ionization potential for ruthenium is less ($711\ kJ \cdot mol^{-1}$ vs $840\ kJ \cdot mol^{-1}$).

16.7 Hg is much more dense than Cd, because the shrinkage in atomic radius that occurs between $Z = 58$ and $Z = 71$ (the lanthanide contraction) causes the atoms following the rare earths to be smaller than might have been expected for their atomic masses and atomic numbers. Zn and Cd have densities that are not too dissimilar, because the radius of Cd is subject only to a smaller *d*-block contraction.

16.9 (a) Proceeding down a group in the *d* block (for example, from Cr to Mo to W), there is an increasing probability of finding the elements in higher oxidation states. That is, higher oxidation states become more stable on going down a group.

(b) The trend for the *p*-block elements is reversed. Because of the inert pair effect, the higher oxidation states tend to be less stable as one descends a group.

16.11 In MO_3, M has an oxidation number of $+6$. Of these three elements, the $+6$ oxidation state is most stable for Cr. See Fig. 16.7.

16.13 (a) Ti(s), $MgCl_2$(s)

$TiCl_4(g) + 2\ Mg(l) \longrightarrow Ti(s) + 2\ MgCl_2(s)$

(b) Co^{2+}(aq), HCO_3^-(aq), NO_3^-(aq)

$CoCO_3(s) + HNO_3\ (aq) \longrightarrow Co^{2+}(aq) + HCO_3^-(aq) + NO_3^-(aq)$

(c) V(s), CaO(s)

$V_2O_5(s) + 5\ Ca(l) \xrightarrow{\Delta} 2\ V(s) + 5\ CaO(s)$

16.15 (a) titanium(IV) oxide, TiO_2

(b) iron(III) oxide, Fe_2O_3

(c) manganese(IV) oxide, MnO_2

16.17 (a) $V^{2+} + 2\ e^- \longrightarrow V(s) \quad E° = -1.19\ V$

$V^{3+} + e^- \longrightarrow V^{2+} \quad E° = -0.26\ V$

$2\ H^+ + 2\ e^- \longrightarrow H_2(g) \quad E° = 0.00\ V$

Therefore, V(s) will be oxidized to V^{3+}. The products are V^{3+}, H_2, and Cl^-.

(b) $Hg_2^{2+} + 2\ e^- \longrightarrow 2\ Hg \quad E° = +0.79\ V$

$Hg^{2+} + 2\ e^- \longrightarrow Hg \quad E° = +1.62\ V$

$2\ H^+ + 2\ e^- \longrightarrow H_2(g) \quad E° = 0.00\ V$

Therefore, no reaction.

(c) $Co^{2+} + 2\ e^- \longrightarrow Co(s) \quad E° = -0.28\ V$

$Co^{3+} + e^- \longrightarrow Co^{2+} \quad E° = +1.81\ V$

$2\ H^+ + 2\ e^- \longrightarrow H_2(g) \quad E° = 0.00\ V$

Therefore, Co(s) will be oxidized to Co^{2+}. The products are Co^{2+}, H_2, and Cl^-. The further oxidation to Co^{3+} is not favorable by reaction of H^+ with Co^{2+}. However, Co^{2+} is oxidized in air to Co^{3+}.

16.19 (a) More than one kind of reduction occurs.

In Zone C,

$Fe_2O_3(s) + 3\ CO(g) \longrightarrow 2\ Fe(s) + 3\ CO_2(g)$

In Zone D,

$3\ Fe_2O_3(s) + CO(g) \longrightarrow 2\ Fe_3O_4(s) + CO_2(g)$

$Fe_3O_4(s) + CO(g) \longrightarrow 3\ FeO(s) + CO_2(g)$

These reactions combine to give

$Fe_2O_3(s) + CO(g) \longrightarrow 2\ FeO(s) + CO_2(g)$

In Zone C,

$FeO(s) + CO(g) \longrightarrow Fe(s) + CO_2(g)$

(b) $TiCl_4(g) + 2\ Mg(l) \xrightarrow{\Delta} Ti(s) + 2\ MgCl_2(s)$

(c) $CaO(s) + SiO_2(s) \xrightarrow{\Delta} CaSiO_3(l)$

16.21 (a) Cr^{3+} ions in water form the complex $[Cr(H_2O)_6]^{3+}(aq)$, which behaves as a Brønsted acid:

$$[Cr(H_2O)_6]^{3+}(aq) + H_2O(l) \rightleftharpoons [Cr(H_2O)_5OH]^{2+}(aq) + H_3O^+(aq)$$

(b) The gelatinous precipitate is the hydroxide $Cr(OH)_3$. The precipitate dissolves as the $Cr(OH)_4^-$ complex ion is formed:

$$Cr^{3+}(aq) + 3\ OH^-(aq) \longrightarrow Cr(OH)_3(s)$$

$$Cr(OH)_3(s) + OH^-(aq) \longrightarrow Cr(OH)_4^-(aq)$$

16.23 (a) $2\ ZnS(s) + 3\ O_2(g) \xrightarrow{\Delta} 2\ ZnO(s) + 2\ SO_2(g)$

followed by $ZnO(s) + C(s) \xrightarrow{\Delta} Zn(l) + CO(g)$

(b) $HgS(s) + O_2(g) \xrightarrow{\Delta} Hg(g) + SO_2(g)$

16.25

$$Cu^+ + e^- \longrightarrow Cu \qquad E° = +0.52\ V$$

$$Cu^+ \longrightarrow Cu^{2+} + e^- \qquad E° = +0.15\ V$$

$$2\ Cu^+ \longrightarrow Cu^{2+} + Cu \quad E° = +0.37\ V$$

$$\Delta G°_r = -nFE° = -RT \ln K \quad (n = 1)$$

$$nFE° = RT \ln K$$

$$\ln K = \frac{(1)(9.65 \times 10^4\ C \cdot mol^{-1})(0.37\ V)}{(8.314\ J \cdot mol^{-1} \cdot K^{-1})(298\ K)} = 14$$

$$K = 10^6$$

16.27 (a) hexacyanoferrate(II) ion

Let x = the oxidation number to be determined

$$x(Fe) + 6 \times (-1) = -4$$

$$x(Fe) = -4 - (-6) = +2$$

(b) hexaamminecobalt(III) ion

$$x(Co) + 6 \times (0) = +3$$

$$x(Co) = +3$$

(c) aquapentacyanocobaltate(III) ion

$$x(Co) + 5 \times (-1) + 1 \times (0) = -2$$

$$x(Co) = -2 - (-5) = +3$$

(d) pentaamminesulfatocobalt(III) ion

$$x(Co) + 1 \times (-2) + 5 \times (0) = +1$$

$$x(Co) = +1 - (-2) = +3$$

16.29 (a) $K_3[Cr(CN)_6]$

(b) $[Co(NH_3)_5(SO_4)]Cl$

(c) $[Co(NH_3)_4(H_2O)_2]Br_3$

(d) $Na[Fe(H_2O)_2(C_2O_4)_2]$

16.31 The molecule $HN(CH_2CH_2NH_2)_2$ has three nitrogen atoms, each with a lone pair of electrons that may be used for bonding to a metal center. The molecule can thus function as a tridentate ligand. The CO_3^{2-} ion can bind to a metal ion through either one or two oxygen atoms. It may, therefore, serve as a mono- or bidentate ligand. H_2O is always a monodenate ligand. The oxalate ion can bind through two oxygen atoms and is usually a bidentate ligand.

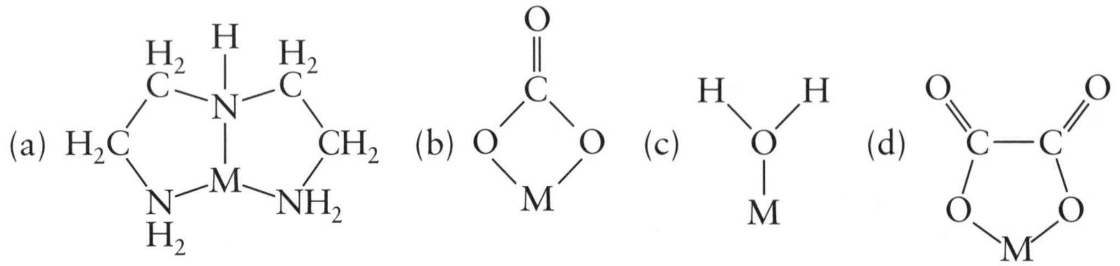

(a) (b) (c) (d)

16.33 As shown below, only the molecule (b) can function as a chelating ligand. The two amine groups in (a) and (c) are arranged so that they would not be able to coordinate simultaneously to the same metal center. It is possible for each of the amine groups in (a) and (c) to coordinate to two different metal centers, however. This is not classified as chelating. When a single ligand binds to two different metal centers, it is known as a *bridging* ligand.

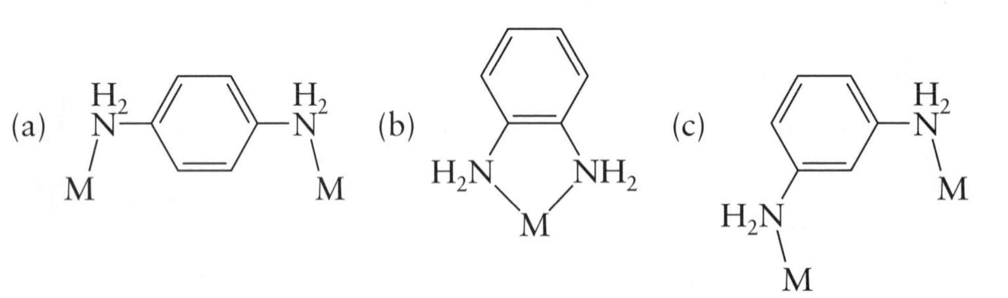

(a) (b) (c)

16.35 (a) 4 (b) 2 (c) 6 (en is bidentate) (d) 6 (EDTA is hexadentate)

16.37 (a) structural isomers, linkage isomers
(b) structural isomers, ionization isomers
(c) structural isomers, linkage isomers
(d) structural isomers, ionization isomers

16.39 $[Co(H_2O)_6]Cl_3$, $[CoCl(H_2O)_5]Cl_2 \cdot H_2O$, $[CoCl_2(H_2O)_4]Cl \cdot 2H_2O$, and $[CoCl_3(H_2O)_3] \cdot 3H_2O$

16.41 (a) yes

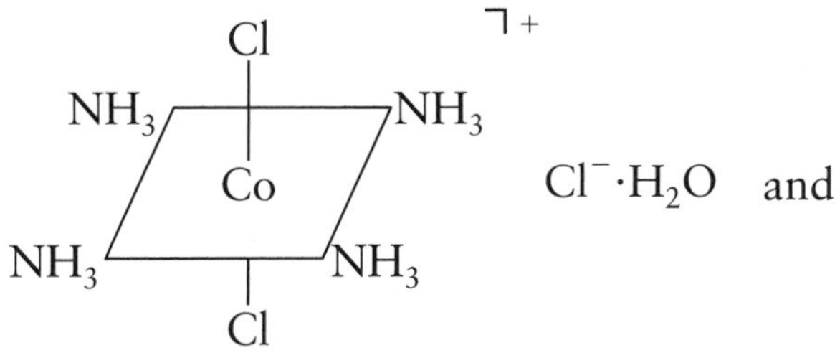

and

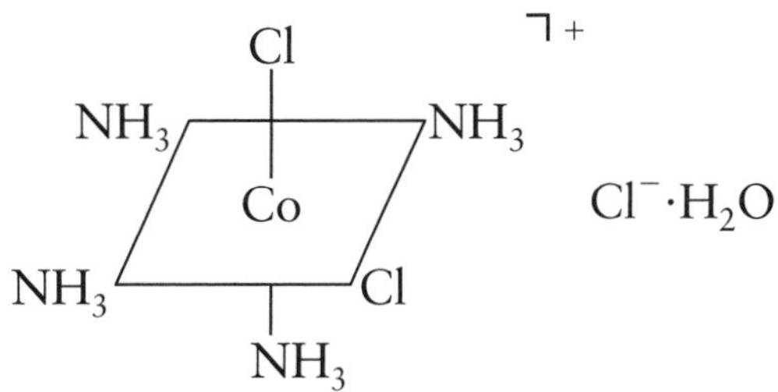

trans-tetraamminedichlorocobalt(III) chloride monohydrate

cis-tetraamminedichlorocobalt(III) chloride monohydrate

(b) no

(c) yes

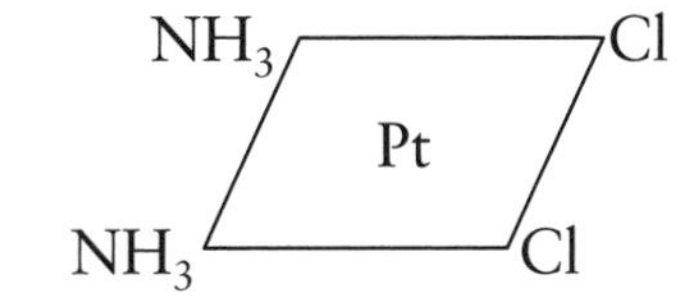

cis-diamminedichloroplatinum(II)

and

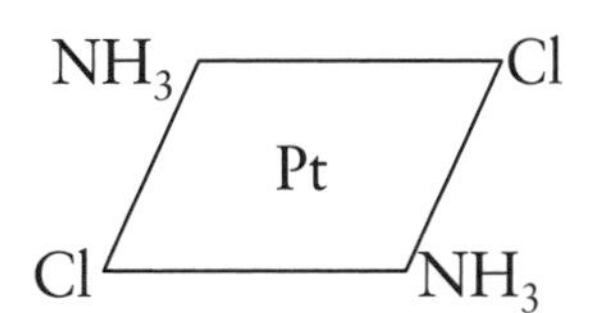

trans-diamminedichloroplatinum(II)

16.43

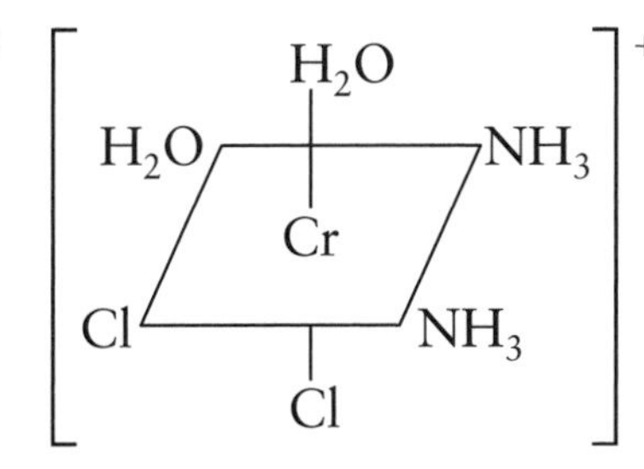

The species is optically active as it has a nonsuperimposable mirror image.

16.45 First complex:

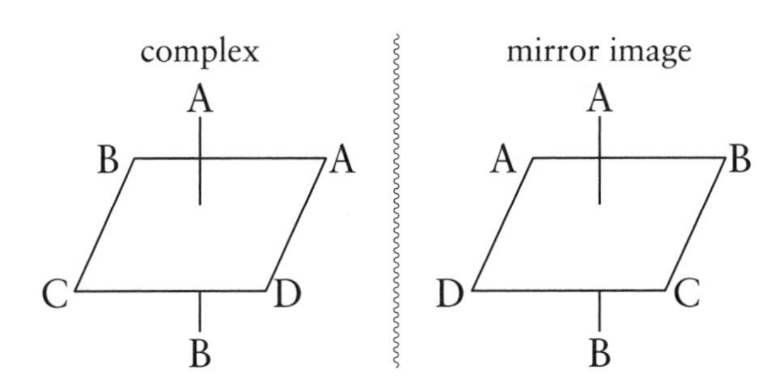

No rotation will make the complex and its mirror image match; therefore, it is chiral.

Second complex:

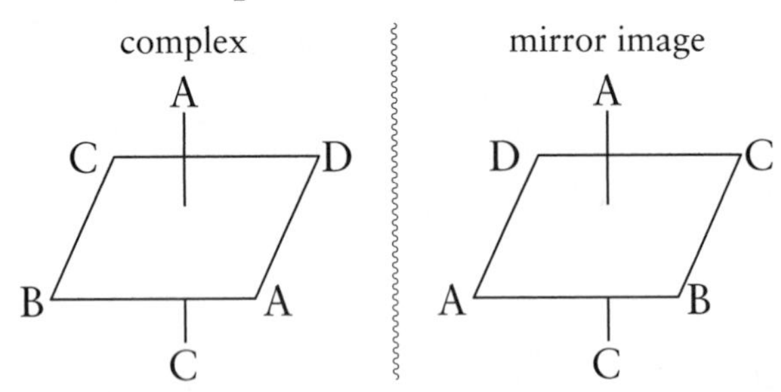

A double rotation shows that the complex and its mirror image are superimposable; therefore, it is not chiral.

The two complexes are not enantiomers; they are not even isomers.

16.47 (a) 1; (b) 6; (c) 5; (d) 3; (e) 6; (f) 6

16.49 (a) 2; (b) 5; (c) 8; (d) 10; (e) 0; (f) 0

16.51 (a) octahedral: strong-field ligand, 6 e^-

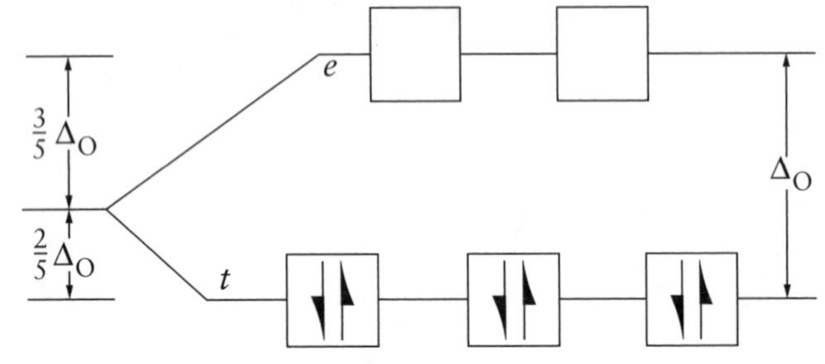

diamagnetic
no unpaired electrons

(b) tetrahedral: weak-field ligand, 8 e^-

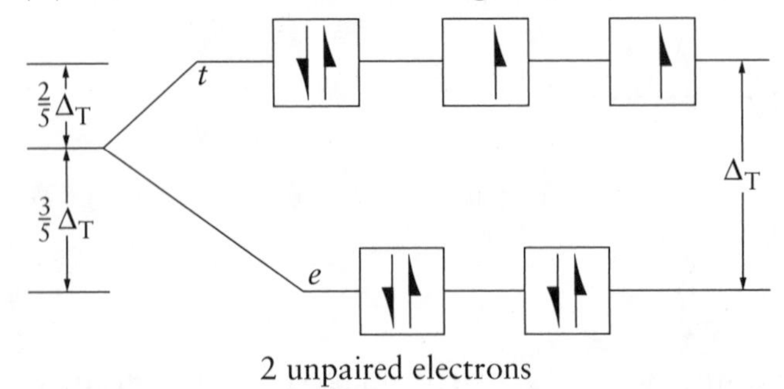

2 unpaired electrons

(c) octahedral: weak-field ligand, 5 e^-

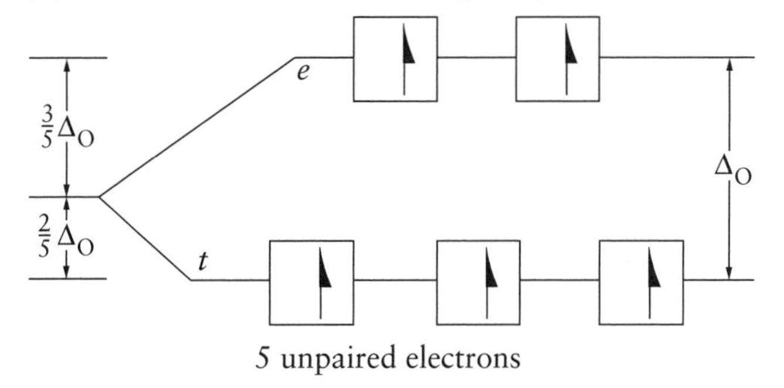

(d) octahedral: strong-field ligand, 5 e^-

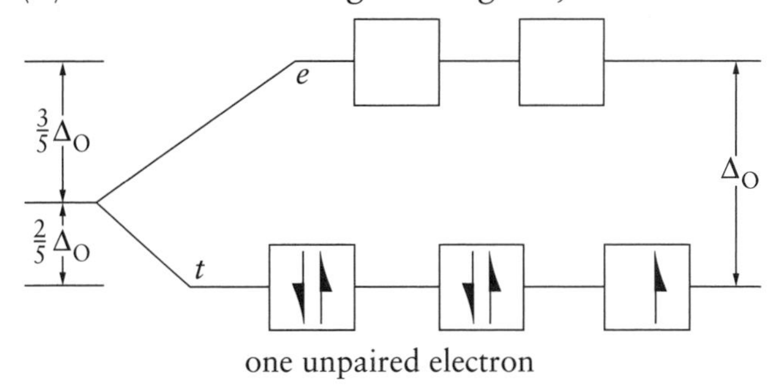

16.53 (a) $[Co(en)_3]^{3+}$ 6 e^- ↑↓ ↑↓ ↑↓ 0 unpaired electrons

(b) $[Mn(CN)_6]^{3-}$ 4 e^- ↑↓ ↑ ↑ 2 unpaired electrons

16.55 Weak-field ligands do not interact strongly with the *d*-electrons in the metal ion, so they produce only a small crystal field splitting of the *d*-electron energy states. The opposite is true of strong-field ligands. With weak-field ligands, unpaired electrons remain unpaired if there are unfilled orbitals; hence, a weak-field ligand is likely to lead to a high-spin complex. Strong-field ligands cause electrons to pair up with electrons in lower energy orbitals. A strong-field ligand is likely to lead to a low-spin complex. Ligands arranged in the spectrochemical series help to distinguish strong-field and weak-field ligands. Measurement of magnetic susceptibility (paramagnetism) can be used to determine the number of unpaired electrons, which, in turn, establishes whether the associated ligand is weak-field or strong-field in nature.

16.57 (a) $[CoF_6]^{3-}$ 6 e$^-$ F$^-$ is a weak-field ligand; therefore,

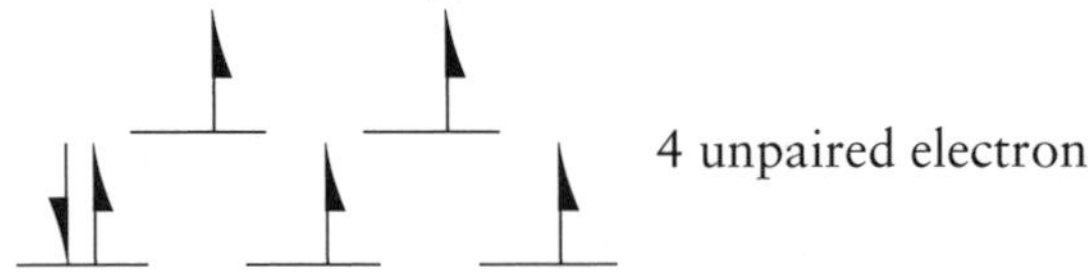

4 unpaired electrons

(b) $[Co(en)_3]^{3+}$ 6 e$^-$ en is a strong-field ligand; therefore,

0 unpaired electrons

Because F$^-$ is a weak-field ligand and en a strong-field ligand, the splitting between levels is less in (a) than in (b). Therefore, (a) will absorb light of longer wavelength than will (b) and consequently will display a shorter wavelength color. Blue light is shorter in wavelength than yellow light, so (a) $[CoF_6]^{3-}$ is blue and (b) $[Co(en)_3]^{3+}$ is yellow.

16.59 In Zn^{2+}, the 3d-orbitals are filled (d^{10}). Therefore, there can be no electronic transitions between the t and e levels; hence, no visible light is absorbed and the aqueous ion is colorless. The d^{10} configuration has no unpaired electrons, so Zn compounds would not be paramagnetic.

16.61 (a) $\Delta_O = \dfrac{hc}{\lambda} = \dfrac{(6.63 \times 10^{-34}\ \text{J}\cdot\text{s}^{-1})(3.00 \times 10^{8}\ \text{m}\cdot\text{s}^{-1})}{740 \times 10^{-9}\ \text{m}} = 2.69 \times 10^{-19}\ \text{J}$

(b) $\Delta_O = \dfrac{hc}{\lambda} = \dfrac{(6.63 \times 10^{-34}\ \text{J}\cdot\text{s}^{-1})(3.00 \times 10^{8}\ \text{m}\cdot\text{s}^{-1})}{460 \times 10^{-9}\ \text{m}} = 4.32 \times 10^{-19}\ \text{J}$

(c) $\Delta_O = \dfrac{hc}{\lambda} = \dfrac{(6.63 \times 10^{-34}\ \text{J}\cdot\text{s}^{-1})(3.00 \times 10^{8}\ \text{m}\cdot\text{s}^{-1})}{575 \times 10^{-9}\ \text{m}} = 3.46 \times 10^{-19}\ \text{J}$

These numbers can be multiplied by 6.02×10^{23} to obtain $\text{kJ}\cdot\text{mol}^{-1}$.

(a) $2.69 \times 10^{-19}\ \text{J} \times 6.02 \times 10^{23}\ \text{mol}^{-1} = 162\ \text{kJ}\cdot\text{mol}^{-1}$

(b) $4.32 \times 10^{-19}\ \text{J} \times 6.02 \times 10^{23}\ \text{mol}^{-1} = 260\ \text{kJ}\cdot\text{mol}^{-1}$

(c) $3.46 \times 10^{-19}\ \text{J} \times 6.02 \times 10^{23}\ \text{mol}^{-1} = 208\ \text{kJ}\cdot\text{mol}^{-1}$

$Cl < H_2O < NH_3$ (spectrochemical series)

16.63 The e_g set, which is comprised of the $d_{x^2-y^2}$ and d_{z^2} orbitals

16.65 (a) The CN$^-$ ion is a π-acid ligand accepting electrons into the empty π^* orbital created by the C—N multiple bond. (b) The Cl$^-$ ion has extra lone pairs in addition to the one that is used to form the σ-bond to the metal, and so it can act as a π-base, donating electrons in a p-orbital to an empty d-orbital on the metal.

(c) H_2O, like Cl^-, also has an "extra" lone pair of electrons that can be donated to a metal center, making it a weak π-base; (d) en is neither a π-acid nor a π-base, because it does not have any empty π-type antibonding orbitals nor does it have any extra lone pairs of electrons to donate. $Cl^- < H_2O < en < CN^-$. Note that the spectrochemical series orders the ligands as π-bases < σ-bond only ligands < π-acceptors.

16.67 Nonbinding or slightly antibonding. In a complex that forms only σ-bonds, the t_{2g} set of orbitals is nonbonding. If the ligands can function as weak π-donors (those close to the middle of the spectrochemical series, such as H_2O), the t_{2g} set becomes slightly antibonding by interacting with the filled p-orbitals on the ligands.

16.69 Antibonding. The e_g set of orbitals on an octahedral metal ion are always antibonding because of interactions with ligand orbitals that form the σ-bonds. This is true regardless of whether the ligands are π-acceptors, π-donors, or neither.

16.71 Water has two lone pairs of electrons. Once one of these is used to form the σ-bond to the metal ion, the second may be used to form a π-bond. This causes the t_{2g} set of orbitals to move up in energy, making Δ smaller; therefore, water is a weak field ligand. Ammonia does not have this extra lone pair of electrons and consequently cannot function as a π-donor ligand.

16.73 (a) CO

(b) $Fe_3O_4(s) + CO(g) \longrightarrow 3\ FeO(s) + CO_2(g)$ (Zone D)

$Fe_2O_3(s) + CO(g) \longrightarrow 2\ FeO(s) + CO_2(g)$ (Zone C)

followed by $FeO(s) + CO(g) \longrightarrow Fe(s) + CO_2(g)$ (Zone C)

$Fe_2O_3(s) + 3\ CO(g) \longrightarrow 2\ Fe(s) + 3\ CO_2(g)$ (Zone D)

(c) carbon

16.75 The major impurity is carbon; it is removed by oxidation of the carbon to CO_2, followed by capture of the CO_2 by base to form a slag.

16.77 Copper and zinc

16.79 Alloys are usually (1) harder and more brittle, and (2) poorer conductors of electricity than the metals from which they are made.

16.81 (a) A ferrofluid is a mixture containing a viscous, nonpolar liquid (oil), a detergent that can form micelles, and a finely powdered magnetic material, such as Fe_3O_4. When placed in a magnetic field, the magnetic particles in the fluid tend to align with the field, which means that the position and motion of the liquid can be controlled by the applied magnetic field. (b) At first glance, the oxidation state in Fe_3O_4 appears to be the nonintegral value of $+8/3$. Another way to view this is that the compound is composed of both Fe^{2+} and Fe^{3+} in a 1:2 ratio. The compound could also be written as $FeO \cdot Fe_2O_3$.

16.83 The compound is ferromagnetic below T_C because the magnetism is higher. Above the Curie temperature, the compound is a simple paramagnet with randomly oriented spins, but below that temperature, the spins align and the magnetism increases.

16.85 (a) $[PtBrCl(NH_3)_2]$

cis-Diamminebromochloroplatinum(II) *trans*-Diamminebromochloroplatinum(II)

(b) If the compound were tetrahedral, there would be only one compound, not two.

16.87 (a) The first, $[Ni(SO_4)(en)_2]Cl_2$, will give a precipitate of AgCl when $AgNO_3$ is added; the second will not.
(b) The second, $[NiCl_2(en)_2]I_2$, will show free I_2 when mildly oxidized with, for example, Br_2, but the first will not.

16.89 (a), (b)

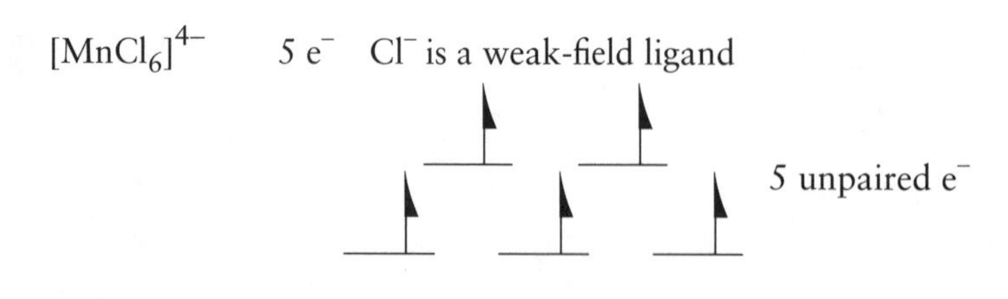

$[Mn(CN)_6]^{4-}$ 5 e^- CN^- is a strong-field ligand

1 unpaired e^-

(c) A weak-field ligand absorbs long-wavelength light; therefore, $[MnCl_6]^{4-}$ transmits shorter wavelengths and $[Mn(CN)_6]^{4-}$ transmits longer wavelengths.

16.91 4, 5, 6, and 7

16.93 High spin Mn^{2+} ions have a d^5 configuration with 5 unpaired electrons, as shown below.

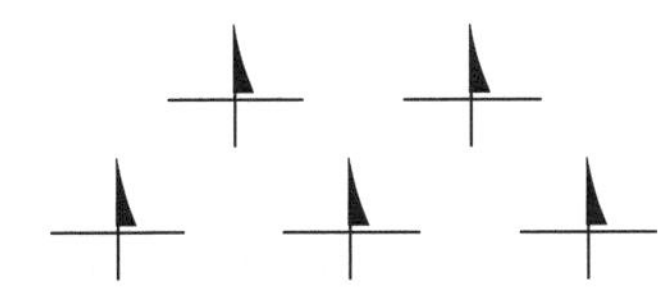

In order for light to be absorbed in the visible region of the spectrum, an electron from the t_{2g} set has to be moved into the e_g set of orbitals. Because these orbitals each already have an electron in them, it would be necessary to pair the electrons, which requires additional energy. This extra energy requirement makes the absorption of the visible light less probable, and so the complexes usually are only faintly colored.

16.95 Consider the formation reaction of a metal complex from the metal ion and its ligands:

$M^{x+} + nL \longrightarrow ML_n^{\ x+}$

It can be seen that there is a considerable loss in entropy upon creating the complex. Polydentate ligands that have two or more places on a ligand that can bind to a metal, do not have nearly so unfavorable an entropy change—the coordination of the second site on the ligand occurs intramolecularly, once the first part of the ligand is attached. Thus, in general, metal complexes with chelating ligands have much higher formation constants than complexes with all monodentate ligands. This is known as the *chelation effect;* it is so strong that once one end of a polydentate ligand binds to a metal center, the other end usually binds extremely rapidly.

16.97 The correct structure for $[Co(NH_3)_6]Cl_3$ consists of four ions, $Co(NH_3)_6^{\ 3+}$, and 3 Cl^- in aqueous solution. The chloride ions can be easily precipitated as AgCl. This would not be possible if they were bonded to the other (NH_3) ligands. If the structure were $Co(NH_3—NH_3—Cl)_3$, VSEPR theory would predict that the Co^{3+} ion would have a trigonal planar ligand arrangement. The splitting of the *d*-orbital energies would not be the same as the octahedral arrangement and would lead to different spectroscopic and magnetic properties inconsistent with the experimental evidence. In addition, neither optical nor geometrical isomers would be observed.

16.99 The ligand field strength of the halide ions shows them to lie in the order $I^- <$ $Br^- < Cl^- < F^-$, which shows that the less electronegative the ligand is, the smaller the value of Δ for the complex. The value of Δ correlates with the ability of the ligand's extra lone pairs of electrons to interact with the t_{2g} set of the octahedral metal ion. This means that the less electronegative the ligand is, the easier it is for the ligand to donate electrons to the metal ion.

16.101 In order to determine these relationships, we need to consider the types of interactions that the ligands on the ends of the spectrochemical series will have with the metal ions. Those that are weak-field (form high spin complexes, π-bases) have extra lone pairs of electrons that can be donated to a metal ion in a π fashion. The strong-field ligands (form low spin complexes, π-acids) accept electrons from the metals. The complexes that will be more stable will be produced in general by the match between ligand and metal. Thus, the early transition metals in high oxidation states will have few or no electrons in the *d*-orbitals. These metal ions will become stabilized by ligands that can donate more electrons to the metal—the *d*-orbitals that are empty can readily accept electrons. The more stable complexes will be formed with the weak-field ligands. The opposite is true for metals with many electrons, which are the ones at the right side of the periodic table, in low oxidation states. These metals generally have most of the *d*-orbitals filled, so they would, in fact, be destabilized by π donation. They form instead more stable complexes with the π-acceptor ligands that can remove some of the electron density from the metal ions.

16.103 (a) The calculation is as follows:

Cu atoms:	1/8 × 8 atoms at the corners of the unit cell	1 atom
	¼ × 8 atoms on the edges of the unit cell	2 atoms
	Total	3 atoms
Y atoms	1 atom in the cell center	1 atom
Ba atoms	2 atoms completely within the unit cell	2 atoms
O atoms	¼ × 12 atoms on the edges of the unit cell	3 atoms
	½ × 8 atoms on the faces of the unit cell	4 atoms
	Total	7 atoms

(b) 382 pm × 388 pm × 1351 pm

(c) Orthorhombic ($a \neq b \neq c$; $\alpha = \beta = \delta = 90°$). The cell is close to tetragonal, but for a tetragonal cell, two of the edge lengths must be exactly equal and here they are slightly different.

16.105 (a) $PtCl_2(NH_3)_2$, *cis*-diamminedichloroplatinum(II)

(b) The only other isomer is the trans form. Neither the cis nor the trans form is optically active.

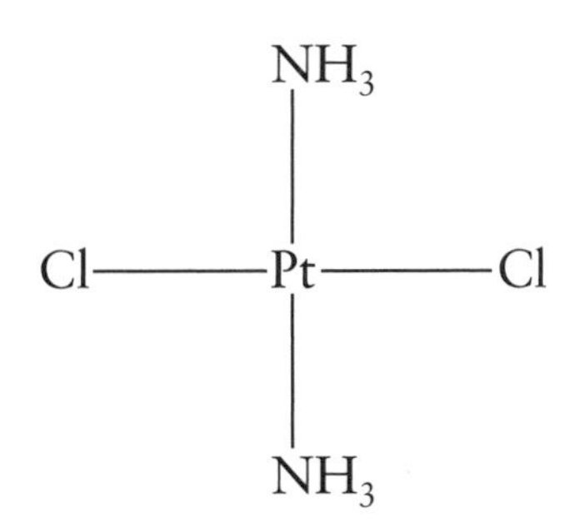

(c) square planar

16.107 (a) ZrO_2: there are 4 zirconium atoms in the unit cell. (1/8 × 8 corner atoms + ½ × six face-centered atoms) and 8 oxygen atoms (All O atoms lie completely inside the unit cell.)

(b) face-centered cubic

(c) eight

(d) coordination number = 4, tetrahedral

(e) An obvious difference is that zirconia is composed of two different types of atoms, whereas diamond is made up solely of carbon; however, both structures are based upon a face-centered cubic unit cell. In zirconia, all of the tetrahedral holes created in the cubic close-packed arrangement of zirconium atoms are occupied by oxide ligands. In the case of diamond, only half of the tetrahedral holes are filled. The consequence is that the zirconium atoms are connected to twice as many oxygen atoms as carbon is connected to other carbon atoms.

CHAPTER 17
NUCLEAR CHEMISTRY

17.1 $\lambda = \frac{c}{\nu}$, $E = N_A h\nu$, 1 Hz = 1 s^{-1}

(a) $\lambda = \frac{3.00 \times 10^8\ m\cdot s^{-1}}{6.7 \times 10^{20}\ s^{-1}} = 4.5 \times 10^{-13}\ m$

$E = 6.02 \times 10^{23}\ mol^{-1} \times 6.63 \times 10^{-34}\ J\cdot s \times 6.7 \times 10^{20}\ s^{-1}$
$= 2.7 \times 10^{11}\ J\cdot mol^{-1}$

(b) $\lambda = \frac{3.00 \times 10^8\ m\cdot s^{-1}}{3.2 \times 10^{22}\ s^{-1}} = 9.4 \times 10^{-15}\ m$

$E = 6.02 \times 10^{23}\ mol^{-1} \times 6.63 \times 10^{-34}\ J\cdot s \times 3.2 \times 10^{22}\ s^{-1}$
$= 1.3 \times 10^{13}\ J\cdot mol^{-1}$

(c) $\lambda = \frac{3.00 \times 10^8\ m\cdot s^{-1}}{4.1 \times 10^{21}\ s^{-1}} = 7.3 \times 10^{-14}\ m$

$E = 6.02 \times 10^{23}\ mol^{-1} \times 6.63 \times 10^{-34}\ J\cdot s \times 4.1 \times 10^{21}\ s^{-1}$
$= 1.6 \times 10^{12}\ J\cdot mol^{-1}$

(d) $\lambda = \frac{3.00 \times 10^8\ m\cdot s^{-1}}{8.6 \times 10^{19}\ s^{-1}} = 3.5 \times 10^{-12}\ m$

$E = 6.02 \times 10^{23}\ mol^{-1} \times 6.63 \times 10^{-34}\ J\cdot s \times 8.6 \times 10^{19}\ s^{-1}$
$= 3.4 \times 10^{10}\ J\cdot mol^{-1}$

17.3 We assume that all the change in energy goes into the energy of the γ ray emitted. Then, in each case,

$$\nu = \frac{\Delta E}{h}, \qquad \lambda = \frac{c}{\nu}$$

$$\text{energy of 1 MeV} = \left(\frac{10^6\ eV}{1\ MeV}\right)\left(\frac{1.602 \times 10^{-19}\ J}{1\ eV}\right) = 1.602 \times 10^{-13}\ J\cdot MeV^{-1}$$

(a) $\Delta E = (1.33\ MeV)\left(\frac{1.602 \times 10^{-13}\ J}{1\ MeV}\right) = 2.13 \times 10^{-13}\ J$

$$\nu = \frac{\Delta E}{h} = \frac{2.13 \times 10^{-13}\ J}{6.63 \times 10^{-34}\ J\cdot s} = 3.21 \times 10^{20}\ s^{-1} = 3.21 \times 10^{20}\ Hz$$

$$\lambda = \frac{c}{\nu} = \frac{3.00 \times 10^8\ m\cdot s^{-1}}{3.21 \times 10^{20}\ s^{-1}} = 9.35 \times 10^{-13}\ m$$

(b) $\Delta E = (1.64 \text{ MeV})\left(\dfrac{1.602 \times 10^{-13}\text{ J}}{1 \text{ MeV}}\right) = 2.63 \times 10^{-13}\text{ J}$

$$\nu = \frac{\Delta E}{h} = \frac{2.63 \times 10^{-13}\text{ J}}{6.63 \times 10^{-34}\text{ J}\cdot\text{s}} = 3.97 \times 10^{20}\text{ s}^{-1} = 3.97 \times 10^{20}\text{ Hz}$$

$$\lambda = \frac{3.00 \times 10^{8}\text{ m}\cdot\text{s}^{-1}}{3.95 \times 10^{20}\text{ s}^{-1}} = 7.59 \times 10^{-13}\text{ m}$$

(c) $\Delta E = (1.10 \text{ MeV})\left(\dfrac{1.602 \times 10^{-13}\text{ J}}{1 \text{ MeV}}\right) = 1.76 \times 10^{-13}\text{ J}$

$$\nu = \frac{\Delta E}{h} = \frac{1.76 \times 10^{-13}\text{ J}}{6.63 \times 10^{-34}\text{ J}\cdot\text{s}} = 2.65 \times 10^{20}\text{ s}^{-1} = 2.65 \times 10^{20}\text{ Hz}$$

$$\lambda = \frac{c}{\nu} = \frac{3.00 \times 10^{8}\text{ m}\cdot\text{s}^{-1}}{2.65 \times 10^{20}\text{ s}^{-1}} = 1.13 \times 10^{-12}\text{ m}$$

17.5 (a) ${}^{3}_{1}\text{T} \longrightarrow {}^{0}_{-1}\text{e} + {}^{3}_{2}\text{He}$

(b) ${}^{83}_{39}\text{Y} \longrightarrow {}^{0}_{1}\text{e} + {}^{83}_{38}\text{Sr}$

(c) ${}^{87}_{36}\text{Kr} \longrightarrow {}^{0}_{-1}\text{e} + {}^{87}_{37}\text{Rb}$

(d) ${}^{225}_{91}\text{Pa} \longrightarrow {}^{4}_{2}\alpha + {}^{221}_{89}\text{Ac}$

17.7 (a) ${}^{8}_{5}\text{B} \longrightarrow {}^{0}_{1}\text{e} + {}^{A}_{Z}\text{E}$ $\quad A = 8 - 0 = 8, Z = 5 - 1 = 4$, E = Be

so, ${}^{8}_{5}\text{B} \longrightarrow {}^{0}_{1}\text{e} + {}^{8}_{4}\text{Be}$

(b) ${}^{63}_{28}\text{Ni} \longrightarrow {}^{0}_{-1}\text{e} + {}^{A}_{Z}\text{E}$ $\quad A = 63 - 0 = 63, Z = 28 - (-1) = 29$, E = Cu

so, ${}^{63}_{28}\text{Ni} \longrightarrow {}^{0}_{-1}\text{e} + {}^{63}_{29}\text{Cu}$

(c) ${}^{185}_{79}\text{Au} \longrightarrow {}^{4}_{2}\alpha + {}^{A}_{Z}\text{E}$ $\quad A = 185 - 4 = 181, Z = 79 - 2 = 77$, E = Ir

so, ${}^{185}_{79}\text{Au} \longrightarrow {}^{4}_{2}\alpha + {}^{181}_{77}\text{Ir}$

(d) ${}^{7}_{4}\text{Be} + {}^{0}_{-1}\text{e} \longrightarrow {}^{A}_{Z}\text{E}$ $\quad A = 7 + 0 = 7, Z = 4 - 1 = 3$, E = Li

so, ${}^{7}_{4}\text{Be} + {}^{0}_{-1}\text{e} \longrightarrow {}^{7}_{3}\text{Li}$

17.9 (a) ${}^{24}_{11}\text{Na} \longrightarrow {}^{24}_{12}\text{Mg} + {}^{0}_{-1}\text{e}$; a β particle is emitted.

(b) ${}^{128}_{50}\text{Sn} \longrightarrow {}^{128}_{51}\text{Sb} + {}^{0}_{-1}\text{e}$; a β particle is emitted.

(c) ${}^{140}_{57}\text{La} \longrightarrow {}^{140}_{56}\text{Ba} + {}^{0}_{1}\text{e}$; a positron ($\beta^{+}$) is emitted.

(d) ${}^{228}_{90}\text{Th} \longrightarrow {}^{224}_{88}\text{Ra} + {}^{4}_{2}\alpha$; an α particle is emitted.

17.11 (a) ${}^{11}_{5}\text{B} + {}^{4}_{2}\alpha \longrightarrow 2\,{}^{1}_{0}\text{n} + {}^{13}_{7}\text{N}$

(b) ${}^{35}_{17}\text{Cl} + {}^{2}_{1}\text{D} \longrightarrow {}^{1}_{0}\text{n} + {}^{36}_{18}\text{Ar}$

(c) ${}^{96}_{42}\text{Mo} + {}^{2}_{1}\text{D} \longrightarrow {}^{1}_{0}\text{n} + {}^{97}_{43}\text{Tc}$

(d) ${}^{45}_{21}\text{Sc} + {}^{1}_{0}\text{n} \longrightarrow {}^{4}_{2}\alpha + {}^{42}_{19}\text{K}$

17.13 (a) $A/Z = 68/29 = 2.34 > (A/Z)_{\text{based}}$; hence, ${}^{68}_{29}\text{Cu}$ is neutron rich, and β decay is most likely.

${}^{68}_{29}\text{Cu} \longrightarrow {}^{0}_{-1}\text{e} + {}^{68}_{30}\text{Zn}$

(b) $A/Z = 103/48 = 2.15 < (A/Z)_{\text{based}}$; therefore, ${}^{103}_{48}\text{Cd}$ is proton rich, and β^+ decay is most likely.

$${}^{103}_{48}\text{Cd} \longrightarrow {}^{0}_{1}\text{e} + {}^{103}_{47}\text{Ag}$$

(c) ${}^{243}_{97}\text{Bk}$ has $Z > 83$ and is proton rich; therefore, α decay is most likely.

$${}^{243}_{97}\text{Bk} \longrightarrow {}^{4}_{2}\alpha + {}^{239}_{95}\text{Am}$$

(d) ${}^{260}_{105}\text{Db}$ has $Z > 83$; therefore, α decay is most likely.

$${}^{260}_{105}\text{Db} \longrightarrow {}^{4}_{2}\alpha + {}^{256}_{103}\text{Lr}$$

17.15

α	${}^{235}_{92}\text{U} \longrightarrow {}^{4}_{2}\alpha + {}^{231}_{90}\text{Th}$
β	${}^{231}_{90}\text{Th} \longrightarrow {}^{0}_{-1}\text{e} + {}^{231}_{91}\text{Pa}$
α	${}^{231}_{91}\text{Pa} \longrightarrow {}^{4}_{2}\alpha + {}^{227}_{89}\text{Ac}$
β	${}^{227}_{89}\text{Ac} \longrightarrow {}^{0}_{-1}\text{e} + {}^{227}_{90}\text{Th}$
α	${}^{227}_{90}\text{Th} \longrightarrow {}^{4}_{2}\alpha + {}^{223}_{88}\text{Ra}$
α	${}^{223}_{88}\text{Ra} \longrightarrow {}^{4}_{2}\alpha + {}^{219}_{86}\text{Rn}$
α	${}^{219}_{86}\text{Rn} \longrightarrow {}^{4}_{2}\alpha + {}^{215}_{84}\text{Po}$
β	${}^{215}_{84}\text{Po} \longrightarrow {}^{0}_{-1}\text{e} + {}^{215}_{85}\text{At}$
α	${}^{215}_{85}\text{At} \longrightarrow {}^{4}_{2}\alpha + {}^{211}_{83}\text{Bi}$
β	${}^{211}_{83}\text{Bi} \longrightarrow {}^{0}_{-1}\text{e} + {}^{211}_{84}\text{Po}$
α	${}^{211}_{84}\text{Po} \longrightarrow {}^{4}_{2}\alpha + {}^{207}_{82}\text{Pb}$

17.17

$${}^{238}_{92}\text{U} \longrightarrow {}^{234}_{90}\text{Th} + {}^{4}_{2}\alpha$$
$${}^{234}_{90}\text{Th} \longrightarrow {}^{234}_{91}\text{Pa} + {}^{0}_{-1}\beta$$
$${}^{234}_{91}\text{Pa} \longrightarrow {}^{234}_{92}\text{U} + {}^{0}_{-1}\beta$$
$${}^{234}_{92}\text{U} \longrightarrow {}^{230}_{90}\text{Th} + {}^{4}_{2}\alpha$$
$${}^{230}_{90}\text{Th} \longrightarrow {}^{226}_{88}\text{Ra} + {}^{4}_{2}\alpha$$
$${}^{226}_{88}\text{Ra} \longrightarrow {}^{222}_{86}\text{Rn} + {}^{4}_{2}\alpha$$
$${}^{222}_{86}\text{Rn} \longrightarrow {}^{218}_{84}\text{Po} + {}^{4}_{2}\alpha$$
$${}^{218}_{84}\text{Po} \longrightarrow {}^{214}_{82}\text{Pb} + {}^{4}_{2}\alpha$$
$${}^{214}_{82}\text{Pb} \longrightarrow {}^{214}_{83}\text{Bi} + {}^{0}_{-1}\beta$$
$${}^{214}_{83}\text{Bi} \longrightarrow {}^{214}_{84}\text{Po} + {}^{0}_{-1}\beta$$
$${}^{214}_{84}\text{Po} \longrightarrow {}^{210}_{82}\text{Pb} + {}^{4}_{2}\alpha$$
$${}^{210}_{82}\text{Pb} \longrightarrow {}^{210}_{83}\text{Bi} + {}^{0}_{-1}\beta$$
$${}^{210}_{83}\text{Bi} \longrightarrow {}^{210}_{84}\text{Po} + {}^{0}_{-1}\beta$$
$${}^{210}_{84}\text{Po} \longrightarrow {}^{206}_{82}\text{Pb} + {}^{4}_{2}\alpha$$

17.19 To determine the charge and mass of the unknown particle, it helps to write ${}^{1}_{1}\text{p}$ and ${}^{1}_{0}\text{n}$ for the proton and neutron, respectively; and ${}^{0}_{-1}\text{e}$ and ${}^{0}_{1}\text{e}$ for the β particle and positron, respectively.

(a) $^{14}_{7}N + ^{4}_{2}\alpha \longrightarrow ^{17}_{8}O + ^{1}_{1}p$
(b) $^{248}_{96}Cm + ^{1}_{0}n \longrightarrow ^{249}_{97}Bk + ^{0}_{-1}e$
(c) $^{243}_{95}Am + ^{1}_{0}n \longrightarrow ^{244}_{96}Cm + ^{0}_{-1}e + \gamma$
(d) $^{13}_{6}C + ^{1}_{0}n \longrightarrow ^{14}_{6}C + \gamma$

17.21 (a) $^{20}_{10}Ne + ^{4}_{2}\alpha \longrightarrow ^{8}_{4}Be + ^{16}_{8}O$
(b) $^{20}_{10}Ne + ^{20}_{10}Ne \longrightarrow ^{16}_{8}O + ^{24}_{12}Mg$
(c) $^{44}_{20}Ca + ^{4}_{2}\alpha \longrightarrow \gamma + ^{48}_{22}Ti$
(d) $^{27}_{13}Al + ^{2}_{1}H \longrightarrow ^{1}_{1}p + ^{28}_{13}Al$

17.23 In each case, identify the unknown particle by performing a mass and charge balance as you did in the solutions to Exercises 17.5 and 17.7. Then write the complete nuclear equation.
(a) $^{14}_{7}N + ^{4}_{2}\alpha \longrightarrow ^{17}_{8}O + ^{1}_{1}p$
(b) $^{239}_{94}Pu + ^{1}_{0}n \longrightarrow ^{240}_{95}Am + ^{0}_{-1}e$

17.25 (a) untriquadium, Utq (b) unquadpentium, Uqp (c) binilunium, Bnu

17.27 $\text{activity} = (5.8 \times 10^5\ \text{Bq})\left(\dfrac{1\ \text{Ci}}{3.7 \times 10^{10}\ \text{Bq}}\right) = 1.6 \times 10^{-5}\ \text{Ci}$

17.29 1 Bq = 1 disintegration per second (dps)

(a) $(2.5\ \mu\text{Ci})\left(\dfrac{10^{-6}\ \text{Ci}}{1\ \mu\text{Ci}}\right)\left(\dfrac{3.7 \times 10^{10}\ \text{dps}}{1\ \text{Ci}}\right) = 9.2 \times 10^4\ \text{dps}$

$= 9.2 \times 10^4\ \text{Bq}$

(b) $117\ \text{Ci} = (117)(3.7 \times 10^{10}\ \text{dps}) = 4.3 \times 10^{12}\ \text{Bq}$

(c) $(7.2\ \text{mCi})\left(\dfrac{10^{-3}\ \text{Ci}}{1\ \text{mCi}}\right)\left(\dfrac{3.7 \times 10^{10}\ \text{dps}}{1\ \text{Ci}}\right) = 2.7 \times 10^8\ \text{dps}$

$= 2.7 \times 10^8\ \text{Bq}$

17.31 $\text{dose in rads} = 1.0\ \text{J}\cdot\text{kg}^{-1} \times \left(\dfrac{1\ \text{rad}}{10^{-2}\ \text{J}\cdot\text{kg}^{-1}}\right) = 1.0 \times 10^2\ \text{rad}$

$\text{dose equivalent in rems} = Q \times \text{dose in rads}$

$= \left(\dfrac{1\ \text{rem}}{1\ \text{rad}}\right)(1.0 \times 10^2\ \text{rad}) = 1.0 \times 10^2\ \text{rem}$

$1.0 \times 10^2\ \text{rem} \div 100\ \text{rem/Sv} = 1.0\ \text{Sv}$

17.33 $1.0\ \text{rad}\cdot\text{day}^{-1} = (1.0\ \text{rad}\cdot\text{day}^{-1})\left(\dfrac{1\ \text{rem}}{1\ \text{rad}}\right) = 1\ \text{rem}\cdot\text{day}^{-1}$

$$100 \text{ rem} = 1 \text{ rem}\cdot\text{day}^{-1} \times \text{time}$$
$$\text{time} = 100 \text{ day}$$

17.35 $k = \dfrac{0.693}{t_{1/2}}$

(a) $k = \dfrac{0.693}{12.3 \text{ y}} = 5.63 \times 10^{-2} \text{ y}^{-1}$

(b) $k = \dfrac{0.693}{0.84 \text{ s}} = 0.83 \text{ s}^{-1}$

(c) $k = \dfrac{0.693}{10.0 \text{ min}} = 0.0693 \text{ min}^{-1}$

17.37 In each case, $k = \dfrac{0.693}{t_{1/2}}$, initial activity $\propto N_0$, final activity $\propto N$, and $N = N_0 e^{-kt}$

Therefore, $\dfrac{\text{initial activity}}{\text{final activity}} = \dfrac{N_0}{N} = e^{kt}$ and $\ln\left(\dfrac{N_0}{N}\right) = kt$

Solving for t, $t = \left(\dfrac{1}{k}\right) \ln\left(\dfrac{N_0}{N}\right) = \left(\dfrac{1}{k}\right) \ln\left(\dfrac{\text{initial activity}}{\text{final activity}}\right)$

(a) $k = \dfrac{0.693}{1.60 \times 10^3 \text{ y}} = 4.33 \times 10^{-4} \text{ y}^{-1}$

$$t = \left(\frac{1}{4.33 \times 10^{-4} \text{ y}^{-1}}\right) \ln\left(\frac{1.0 \text{ Ci}}{0.10 \text{ Ci}}\right) = 5.3 \times 10^3 \text{ y}$$

(b) $k = \dfrac{0.693}{1.26 \times 10^9 \text{ y}} = 5.50 \times 10^{-10} \text{ y}^{-1}$

$$t = \left(\frac{1}{5.50 \times 10^{-10} \text{ y}^{-1}}\right) \ln\left(\frac{1.0 \times 10^{-6} \text{ Ci}}{10 \times 10^{-9} \text{ Ci}}\right) = 8.4 \times 10^9 \text{ y}$$

(c) $k = \dfrac{0.693}{5.26 \text{ y}} = 0.132 \text{ y}^{-1}$

$$t = \left(\frac{1}{0.132 \text{ y}^{-1}}\right) \ln\left(\frac{10 \text{ Ci}}{8 \text{ Ci}}\right) = 2\text{y}$$

17.39 We know that initial activity $\propto N_0$, and final activity $\propto N$. Therefore,

$$\frac{\text{final activity}}{\text{initial activity}} = \frac{N}{N_0} = e^{-kt}$$

$$k = \frac{0.693}{t_{1/2}} = \frac{0.693}{5.26 \text{ y}} = 0.132 \text{ y}^{-1}$$

$$\begin{aligned} \text{final activity} &= \text{initial activity} \times e^{-kt} \\ &= 4.4 \text{ Ci} \times e^{-(0.132 \text{ y}^{-1} \times 50 \text{ y})} \\ &= 6.0 \times 10^{-3} \text{ Ci} \end{aligned}$$

17.41 In each case, $k = \dfrac{0.693}{t_{1/2}}$, $N = N_0 e^{-kt}$, $\dfrac{N}{N_0} = e^{-kt}$, and the percentage remaining $= 100\% \times (N/N_0)$

(a) $k = \dfrac{0.693}{5.73 \times 10^3\ \text{y}} = 1.21 \times 10^{-4}\ \text{y}^{-1}$

percentage remaining $= 100\% \times e^{-(1.21 \times 10^{-4}\ \text{y}^{-1} \times 2000\ \text{y})} = 78.5\%$

(b) $k = \dfrac{0.693}{12.3\ \text{y}} = 0.0563\ \text{y}^{-1}$

percentage remaining $= 100\% \times e^{-(0.0563\ \text{y}^{-1} \times 11.0\ \text{y})} = 53.8\%$

17.43 (a) $t_{1/2} = 4.5 \times 10^9$ y, $k = \dfrac{0.693}{t_{1/2}} = \dfrac{0.693}{4.5 \times 10^9\ \text{y}} = 1.54 \times 10^{-10}\ \text{y}^{-1}$

$$\text{fraction remaining} = \frac{N}{N_0} = e^{-kt}$$
$$= e^{-(1.54 \times 10^{-10}\ \text{y}^{-1} \times 4.5 \times 10^9\ \text{y})}$$
$$= e^{-1.4} = 0.50$$

After 1 half-life, 50% remains.

(b) fraction remaining $= \dfrac{N}{N_0} = \dfrac{3}{5}$;

$t_{1/2} = 1.26 \times 10^9$ y, $k = \dfrac{0.693}{1.26 \times 10^9\ \text{y}} = 5.50 \times 10^{-10}\ \text{y}^{-1}$

$$\frac{3}{5} = e^{-kt}$$

$$\frac{3}{5} = e^{-(5.50 \times 10^{-10}\ \text{y}^{-1} \times x)}$$

$x = 9.3 \times 10^8$ y

17.45 Let dis = disintegrations

activity from "old" sample $= \dfrac{1500\ \text{dis}/0.250\ \text{g}}{10.0\ \text{h}} = 600\ \text{dis}\cdot\text{g}^{-1}\cdot\text{h}^{-1}$

activity from current sample $= 920\ \text{dis}\cdot\text{g}^{-1}\cdot\text{h}^{-1}$

$k = \dfrac{0.693}{t_{1/2}} = \dfrac{0.693}{5.73 \times 10^3\ \text{y}} = 1.21 \times 10^{-4}\ \text{y}^{-1}$

"old" activity $\propto N$, current activity $\propto N_0$

$$\frac{\text{"old" activity}}{\text{current activity}} = \frac{N}{N_0} = e^{-kt},\ \frac{N_0}{N} = e^{kt},\ \ln\left(\frac{N_0}{N}\right) = kt$$

Solve for t (= age):

$$t = \frac{\ln\left(\frac{N_0}{N}\right)}{k} = \frac{\ln\left(\frac{920}{600}\right)}{1.21 \times 10^{-4}\ y^{-1}} = 3.53 \times 10^3\ y$$

17.47 In each case, $k = \frac{0.693}{t_{1/2}\ (\text{in s})}$, activity in $Bq = k \times N$

$$\text{activity in Ci} = \frac{\text{activity in } Bq}{3.7 \times 10^{10}\ Bq \cdot Ci^{-1}}$$

Note: Bq(= disintegrating nuclei per second) has the units of nuclei$\cdot s^{-1}$

(a) $k = \left(\frac{0.693}{1.60 \times 10^3\ y}\right)\left(\frac{1\ y}{3.17 \times 10^7\ s}\right) = 1.37 \times 10^{-11}\ s^{-1}$

$$N = (1.0 \times 10^{-3}\ g)\left(\frac{1\ mol}{226\ g}\right)\left(\frac{6.02 \times 10^{23}\ \text{nuclei}}{1\ mol}\right) = 2.7 \times 10^{18}\ \text{nuclei}$$

$$\text{activity} = 1.37 \times 10^{-11}\ s^{-1} \times 2.7 \times 10^{18}\ \text{nuclei} \times \left(\frac{1\ Ci}{3.7 \times 10^{10}\ Bq}\right)$$

$$= 1.0 \times 10^{-3}\ Ci$$

(b) $k = \left(\frac{0.693}{28.1\ y}\right)\left(\frac{1\ y}{3.17 \times 10^7\ s}\right) = 7.80 \times 10^{-10}\ s^{-1}$

$$N = (2.0 \times 10^{-6}\ g)\left(\frac{1\ mol}{90\ g}\right)\left(\frac{6.02 \times 10^{23}\ \text{nuclei}}{1\ mol}\right) = 1.3 \times 10^{16}\ \text{nuclei}$$

$$\text{activity} = (7.80 \times 10^{-10}\ s^{-1})(1.3 \times 10^{16}\ \text{nuclei})\left(\frac{1\ Ci}{3.7 \times 10^{10}\ Bq}\right)$$

$$= 2.7 \times 10^{-4}\ Ci$$

(c) $k = \left(\frac{0.693}{2.6\ y}\right)\left(\frac{1\ y}{3.17 \times 10^7\ s}\right) = 8.4 \times 10^{-9}\ s^{-1}$

$$N = (0.43 \times 10^{-3}\ g)\left(\frac{1\ mol}{147\ g}\right)\left(\frac{6.02 \times 10^{23}\ \text{nuclei}}{1\ mol}\right) = 1.8 \times 10^{18}\ \text{nuclei}$$

$$\text{activity} = (8.4 \times 10^{-9}\ s^{-1})(1.8 \times 10^{18}\ \text{nuclei})\left(\frac{1\ Ci}{3.7 \times 10^{10}\ Bq}\right) = 0.41\ Ci$$

17.49 $k = \frac{0.693}{t_{1/2}} = \frac{0.693}{8.05\ d} = 0.0861\ d^{-1}$

$N = N_0 e^{-kt}$ and $\frac{N}{N_0} = e^{-kt}$

Taking the natural log of both sides gives

$$\ln\left(\frac{N}{N_0}\right) = -kt$$

Because activity is proportional to N (Eq. 2), we can write

$$\ln\left(\frac{\text{final activity}}{\text{initial activity}}\right) = -kt$$

Solving for t gives

$$t = -\left(\frac{1}{k}\right)\ln\left(\frac{\text{final activity}}{\text{initial activity}}\right) = -\left(\frac{1}{0.0861\ \text{d}^{-1}}\right)\ln\left(\frac{10}{500}\right) = 45\ \text{d}$$

17.51 (a) activity $\propto N$; and, because $\ln\left(\frac{N}{N_0}\right) = -kt$

$$\ln\left(\frac{\text{final activity}}{\text{initial activity}}\right) = -kt$$

$$\ln\left(\frac{32}{58}\right) = -k \times 12.3\ \text{d}$$

$$k = 0.048\ \text{d}^{-1}$$

$$t_{1/2} = \frac{0.693}{k} = \frac{0.693}{0.048\ \text{d}^{-1}} = 14\ \text{d}$$

(b) $\ln\left(\frac{N}{N_0}\right) = -0.048\ \text{d}^{-1} \times 30\ \text{d} = -1.4$

$$\frac{N}{N_0} = \text{fraction remaining} = 0.25$$

17.53 $10.0\ \text{Ci} = 10.0 \times 3.7 \times 10^{10}\ \text{nuclei}\cdot\text{s}^{-1} \times \frac{8.64 \times 10^{4}\ \text{s}}{1\ \text{d}}$

$$= 3.2 \times 10^{16}\ \text{nuclei}\cdot\text{d}^{-1}$$

$$k = \frac{0.693}{t_{1/2}} = \frac{0.693}{88\ \text{d}} = 7.9 \times 10^{-3}\ \text{d}^{-1}$$

$$\text{activity} = \text{rate} = 3.2 \times 10^{16}\ \text{nuclei}\cdot\text{d}^{-1} = 7.9 \times 10^{-3}\ \text{d}^{-1} \times \text{N}$$

$$N = \frac{3.2 \times 10^{16}\ \text{nuclei}\cdot\text{d}^{-1}}{7.9 \times 10^{-3}\ \text{d}^{-1}} = 4.1 \times 10^{18}\ \text{nuclei}$$

$$\text{mass of } ^{35}\text{S} = (35\ \text{u})\left(\frac{1.661 \times 10^{-24}\ \text{g}}{1\ \text{u}}\right)(4.1 \times 10^{18}\ \text{nuclei})$$

$$= 2.4 \times 10^{-4}\ \text{g}$$

17.55 If isotopically enriched water, such as $H_2{}^{18}O$, is used in the reaction, the label can be followed. Once the products are separated, a suitable technique, such as vibrational spectroscopy or mass spectrometry, can be used to determine whether the product has incorporated the ^{18}O. For example, if the methanol ends up with the

O atom from the water molecules, then its molar mass would be 34 $g \cdot mol^{-1}$, rather than 32 $g \cdot mol^{-1}$ found for methanol with elements present at their natural isotopic abundance.

17.57 The vibrational frequency is proportional to the reduced mass of the two atoms that form the bond according the equation:

$$\nu = \frac{1}{2\pi}\sqrt{\frac{k}{\mu}}$$

where $\mu = \dfrac{m_A m_B}{m_A + m_B}$

Because we are not given ν, it is easiest to make a relative comparison by taking the ratio of ν for the C—D molecule versus ν for the C—H molecule:

$$\frac{\nu_{C-D}}{\nu_{C-H}} = \frac{\frac{1}{2\pi}\sqrt{\frac{k}{\mu_{C-D}}}}{\frac{1}{2\pi}\sqrt{\frac{k}{\mu_{C-H}}}} = \sqrt{\frac{\mu_{C-H}}{\mu_{C-D}}} = \sqrt{\frac{\frac{m_C m_H}{m_C + m_H}}{\frac{m_C m_D}{m_C + m_D}}} = \sqrt{\frac{\frac{(12.01)(1.0079)}{12.01 + 1.0079}}{\frac{(12.01)(2.00)}{12.01 + 2.00}}}$$

$$= \sqrt{\frac{\left(\frac{12.02}{13.02}\right)}{\left(\frac{24.02}{14.01}\right)}} = 0.7338$$

We would thus expect the vibrational frequency for the C—D bond to be approximately 0.73 times the value for the C—H bond (lower in energy).

17.59 In each case, first calculate ΔE for the process described. Note whether ΔE is positive or negative, corresponding to energy added or removed from the system. Then calculate the change in mass from the change in energy using $\Delta E = (\Delta m)c^2$ or

$$\Delta m = \frac{\Delta E}{c^2}$$

(a) $\Delta E = 250\ g \times 0.39\ J \cdot (°C)^{-1} \cdot g^{-1} \times (250°C - 35°C) = 2.1 \times 10^4\ J$

$$\Delta m = \frac{2.1 \times 10^4\ J}{(3.00 \times 10^8\ m \cdot s^{-1})^2} = 2.3 \times 10^{-13}\ kg = 2.3 \times 10^{-10}\ g = \text{mass gained}$$

(b) $\Delta E = -n\Delta H°_{melt}$, where n = number of moles

$$\Delta E = -(6.01\ kJ \cdot mol^{-1})\left(\frac{50.0\ g}{18.02\ g \cdot mol^{-1}}\right) = -16.7\ kJ = -1.67 \times 10^4\ J$$

$$\Delta m = \frac{-1.67 \times 10^4\ J}{(3.00 \times 10^8\ m \cdot s^{-1})^2} = -1.86 \times 10^{-13}\ kg = -1.86 \times 10^{-10}\ g$$

= mass lost

(c) $\Delta E = 2\text{ mol} \times \Delta H°_f(\text{PCl}_5, \text{g}) = 2\text{ mol} \times (-374.9\text{ kJ}\cdot\text{mol}^{-1}) = -749.8\text{ kJ}$
$= -7.498 \times 10^5\text{ J}$

$$\Delta m = \frac{-7.498 \times 10^5\text{ J}}{(3.00 \times 10^8\text{ m}\cdot\text{s}^{-1})^2} = -8.33 \times 10^{-12}\text{ kg} = -8.33 \times 10^{-9}\text{ g} = \text{mass lost}$$

17.61 Remember to convert g to kg.

(a) $E = mc^2 = (1.0 \times 10^{-3}\text{ kg})(3.00 \times 10^8\text{ m}\cdot\text{s}^{-1})^2$
$= 9.0 \times 10^{13}\text{ kg}\cdot\text{m}^2\cdot\text{s}^{-2} = 9.0 \times 10^{13}\text{ J}$

(b) $E = mc^2 = (9.109 \times 10^{-31}\text{ kg})(3.00 \times 10^8\text{ m}\cdot\text{s}^{-1})^2$
$= 8.20 \times 10^{-14}\text{ kg}\cdot\text{m}^2\cdot\text{s}^{-2} = 8.20 \times 10^{-14}\text{ J}$

(c) $E = mc^2 = (1.0 \times 10^{-15}\text{ kg})(3.00 \times 10^8\text{ m}\cdot\text{s}^{-1})^2 = 90\text{ kg}\cdot\text{m}^2\cdot\text{s}^{-2} = 90\text{ J}$

(d) $E = mc^2$

$E = (1.673 \times 10^{-27}\text{ kg})(3.00 \times 10^8\text{ m}\cdot\text{s}^{-1})^2 = 1.51 \times 10^{-10}\text{ J}$

17.63 $$\Delta m = \frac{\Delta E}{c^2} = \frac{-3.9 \times 10^{26}\text{ J}\cdot\text{s}^{-1}}{(3.00 \times 10^8\text{ m}\cdot\text{s}^{-1})^2} = -4.3 \times 10^9\text{ kg}\cdot\text{s}^{-1}$$

17.65 $1\text{ u} = 1.6605 \times 10^{-27}\text{ kg}$

In each case, calculate the difference in mass between the nucleus and the free particles from which it may be considered to have been formed. Then obtain the binding energy from the relation $E_{\text{bind}} = \Delta mc^2$.

(a) ${}^{62}_{28}\text{Ni}$: $28\,{}^1\text{H} + 34\text{ n} \longrightarrow {}^{62}_{28}\text{Ni}$

$\Delta m = 61.928\text{ u} - (28 \times 1.0078\text{ u} + 34 \times 1.0087\text{ u}) = -0.585\,85\text{ u}$

$$\Delta m = (-0.585\,85\text{ u})\left(\frac{1.6605 \times 10^{-27}\text{ kg}}{1\text{ u}}\right) = -9.7280 \times 10^{-28}\text{ kg}$$

$E_{\text{bind}} = -(9.7280 \times 10^{-28}\text{ kg})(3.00 \times 10^8\text{ m}\cdot\text{s}^{-1})^2$
$= -8.7552 \times 10^{-11}\text{ kg}\cdot\text{m}^2\cdot\text{s}^{-2} = -8.7552 \times 10^{-11}\text{ J}$

$$E_{\text{bind}}/\text{nucleon} = \frac{-8.7552 \times 10^{-11}\text{ J}}{62\text{ nucleons}} = -1.4121 \times 10^{-12}\text{ J}\cdot\text{nucleon}^{-1}$$

(b) ${}^{239}_{94}\text{Pu}$: $94\,{}^1\text{H} + 145\text{ n} \longrightarrow {}^{239}_{94}\text{Pu}$

$\Delta m = 239.0522\text{ u} - (94 \times 1.0078\text{ u} + 145 \times 1.0087\text{ u}) = -1.9425\text{ u}$

$$\Delta m = -1.9425\text{ u} \times \left(\frac{1.6605 \times 10^{-27}\text{ kg}}{1\text{ u}}\right) = -3.2255 \times 10^{-27}\text{ kg}$$

$E_{\text{bind}} = -3.2255 \times 10^{-27}\text{ kg} \times (2.997 \times 10^8\text{ m}\cdot\text{s}^{-1})^2 = -2.897 \times 10^{-10}\text{ J}$

$$E_{\text{bind}}/\text{nucleon} = \frac{-2.897 \times 10^{-10}\text{ J}}{239\text{ nucleons}} = -1.212 \times 10^{-12}\text{ J}\cdot\text{nucleon}^{-1}$$

(c) ${}^2_1\text{H}$: ${}^1\text{H} + \text{n} \longrightarrow {}^2_1\text{H}$

$\Delta m = 2.0141\text{ u} - (1.0078\text{ u} + 1.0087\text{ u}) = -0.0024\text{ u}$

$$\Delta m = -0.0024\ \text{u} \times \left(\frac{1.6605 \times 10^{-27}\ \text{kg}}{1\ \text{u}}\right) = -4.0 \times 10^{-30}\ \text{kg}$$

$$E_{\text{bind}} = -4.0 \times 10^{-30}\ \text{kg} \times (3.00 \times 10^{8}\ \text{m}\cdot\text{s}^{-1})^2 = -3.6 \times 10^{-13}\ \text{J}$$

$$E_{\text{bind}}/\text{nucleon} = \frac{-3.6 \times 10^{-13}\ \text{J}}{2\ \text{nucleons}} = -1.8 \times 10^{-13}\ \text{J}\cdot\text{nucleon}^{-1}$$

(d) ${}^{3}_{1}\text{H}$: ${}^{1}\text{H} + 2\ \text{n} \longrightarrow {}^{3}_{1}\text{H}$

$$\Delta m = 3.016\ 05\ \text{u} - (1.0078\ \text{u} + 2 \times 1.0087\ \text{u}) = -0.009\ 15\ \text{u}$$

$$\Delta m = -0.009\ 15\ \text{u} \times \left(\frac{1.6605 \times 10^{-27}}{1\ \text{u}}\ \text{kg}\right) = -1.52 \times 10^{-29}\ \text{kg}$$

$$E_{\text{bind}} = -1.52 \times 10^{-29}\ \text{kg} \times (3.00 \times 10^{8}\ \text{m}\cdot\text{s}^{-1})^2 = -1.37 \times 10^{-12}\ \text{J}$$

$$E_{\text{bind}}/\text{nucleon} = \frac{-1.37 \times 10^{-12}\ \text{J}}{3\ \text{nucleons}} = -4.57 \times 10^{-13}\ \text{J}\cdot\text{nucleon}^{-1}$$

(e) ${}^{62}\text{Ni}$ is the most stable, because it has the largest binding energy per nucleon.

17.67 In each case, we first determine the change in mass, Δm = (mass of products) − (mass of reactants). We then calculate the energy released from $\Delta E = (\Delta m)c^2$.

(a) $\text{D} + \text{D} \longrightarrow {}^{3}\text{He} + \text{n}$

$$2.0141\ \text{u} + 2.0141\ \text{u} \longrightarrow 3.0160\ \text{u} + 1.0087\ \text{u}$$

$$4.0282\ \text{u} \longrightarrow 4.0247\ \text{u}$$

$$\Delta m = -0.0035\ \text{u}$$

$$\Delta m = (-0.0035\ \text{u})\left(\frac{1.661 \times 10^{-27}\ \text{kg}}{1\ \text{u}}\right) = -5.8 \times 10^{-30}\ \text{kg}$$

$$\Delta E = \Delta mc^2 = (-5.8 \times 10^{-30}\ \text{kg})(3.00 \times 10^{8}\ \text{m}\cdot\text{s}^{-1})^2 = -5.2 \times 10^{-13}\ \text{J}$$

$$\left(\frac{-5.2 \times 10^{-13}\ \text{J}}{4.0282\ \text{u}}\right)\left(\frac{1\ \text{u}}{1.661 \times 10^{-24}\ \text{g}}\right) = -7.8 \times 10^{10}\ \text{J}\cdot\text{g}^{-1}$$

(b) ${}^{3}\text{He} + \text{D} \longrightarrow {}^{4}\text{He} + {}^{1}_{1}\text{H}$

$$3.0160\ \text{u} + 2.0141\ \text{u} \longrightarrow 4.0026\ \text{u} + 1.0078\ \text{u}$$

$$5.0301\ \text{u} \longrightarrow 5.0104\ \text{u}$$

$$\Delta m = -0.0197\ \text{u}$$

$$\Delta m = -0.0197\ \text{u} \times \left(\frac{1.661 \times 10^{-27}\ \text{kg}}{1\ \text{u}}\right) = -3.27 \times 10^{-29}\ \text{kg}$$

$$\Delta E = \Delta mc^2 = -(3.27 \times 10^{-29}\ \text{kg})(3.00 \times 10^{8}\ \text{m}\cdot\text{s}^{-1})^2 = -2.94 \times 10^{-12}\ \text{J}$$

$$\left(\frac{-2.94 \times 10^{-12}\ \text{J}}{5.0301\ \text{u}}\right)\left(\frac{1\ \text{u}}{1.661 \times 10^{-24}\ \text{g}}\right) = -3.52 \times 10^{11}\ \text{J}\cdot\text{g}^{-1}$$

(c) ${}^{7}\text{Li} + {}^{1}_{1}\text{H} \longrightarrow 2\ {}^{4}\text{He}$

$$7.0160\ \text{u} + 1.0078\ \text{u} \longrightarrow 2(4.0026\ \text{u})$$

$$8.0238\ \text{u} \longrightarrow 8.0052\ \text{u}$$

$$\Delta m = -0.0186\ \text{u}$$

$$\Delta m = (-0.0186\ \text{u})\left(\frac{1.661 \times 10^{-27}\ \text{kg}}{1\ \text{u}}\right) = -3.09 \times 10^{-29}\ \text{kg}$$

$$\Delta E = \Delta mc^2 = (-3.09 \times 10^{-29}\ \text{kg})(3.00 \times 10^8\ \text{m}\cdot\text{s}^{-1})^2 = -2.78 \times 10^{-12}\ \text{J}$$

$$\left(\frac{-2.78 \times 10^{-12}\ \text{J}}{8.0238\ \text{u}}\right)\left(\frac{1\ \text{u}}{1.661 \times 10^{-24}\ \text{g}}\right) = -2.09 \times 10^{11}\ \text{J}\cdot\text{g}^{-1}$$

(d) $\text{D} + \text{T} \longrightarrow {}^{4}\text{He} + {}^{1}_{1}\text{H}$

$2.0141\ \text{u} + 3.0160\ \text{u} \longrightarrow 4.0026\ \text{u} + 1.0078\ \text{u}$

$5.0301\ \text{u} \longrightarrow 5.00104\ \text{u}$

$\Delta m = -0.0197\ \text{u}$

$$\Delta m = (-0.0197\ \text{u})\left(\frac{1.661 \times 10^{-27}\ \text{kg}}{1\ \text{u}}\right) = -3.27 \times 10^{-29}\ \text{kg}$$

$$\Delta E = \Delta mc^2 = (-3.27 \times 10^{-29}\ \text{kg})(3.00 \times 10^8\ \text{m}\cdot\text{s}^{-1})^2 = -2.94 \times 10^{-12}\ \text{J}$$

$$\left(\frac{-2.94 \times 10^{-12}\ \text{J}}{5.0301\ \text{u}}\right)\left(\frac{1\ \text{u}}{1.661 \times 10^{-24}\ \text{g}}\right) = -3.52 \times 10^{11}\ \text{J}\cdot\text{g}^{-1}$$

17.69 ${}^{24}_{11}\text{Na} \longrightarrow {}^{24}_{12}\text{Mg} + {}^{0}_{-1}\text{e}$

mass $({}^{24}_{11}\text{Na}) = 23.990\ 96\ \text{u}$

mass $({}^{24}_{12}\text{Mg}) = 23.985\ 04\ \text{u}$

The mass of the electron does not need to be explicitly included in the calculation because it is already included in the mass of Mg.

$$\Delta m = \text{mass}\ ({}^{24}_{12}\text{Mg}) - \text{mass}\ ({}^{24}_{11}\text{Na}) = 23.985\ 04\ \text{u} - 23.990\ 96\ \text{u} = -5.92 \times 10^{-3}\ \text{u}$$

$$\Delta m\ (\text{in kg}) = -5.92 \times 10^{-3}\ \text{u} \times 1.661 \times 10^{-27}\ \text{kg u}^{-1} = -9.83 \times 10^{-30}\ \text{kg}$$

(a) $\Delta E = \Delta mc^2 = -(9.83 \times 10^{-30}\ \text{kg})(3.00 \times 10^8\ \text{m}\cdot\text{s}^{-1}) = -8.85 \times 10^{-13}\ \text{J}$

(b) $\Delta E\ (\text{per nucleon}) = \dfrac{-8.85 \times 10^{-13}\ \text{J}}{24\ \text{nucleons}} = -3.69 \times 10^{-14}\ \text{J}\cdot\text{nucleon}^{-1}$

This simple calculation works because the number of nucleons is the same on both sides of the equation.

17.71 (a) ${}^{244}_{95}\text{Am} \longrightarrow {}^{134}_{53}\text{I} + {}^{107}_{42}\text{Mo} + 3\ {}^{1}_{0}\text{n}$

(b) ${}^{235}_{92}\text{U} + {}^{1}_{0}\text{n} \longrightarrow {}^{96}_{40}\text{Zr} + {}^{138}_{52}\text{Te} + 2\ {}^{1}_{0}\text{n}$

(c) ${}^{235}_{92}\text{U} + {}^{1}_{0}\text{n} \longrightarrow {}^{101}_{42}\text{Mo} + {}^{132}_{50}\text{Sn} + 3\ {}^{1}_{0}\text{n}$

17.73 (a) $1\ \text{Ci} = 3.7 \times 10^{10}$ decays per second (dps)

$$\text{decays per minute (dpm) for 4 pCi} = 4 \times 10^{-12}\ \text{Ci} \times 3.7 \times 10^{10}\ \text{dps} \times \left(\frac{60\ \text{s}}{1\ \text{min}}\right) = 9\ \text{dpm}$$

(b) $\text{volume(L)} = (2.0 \times 3.0 \times 2.5)\ \text{m}^3 \times \left(\frac{10^3\ \text{L}}{1\ \text{m}^3}\right) = 1.5 \times 10^4\ \text{L}$

$$\text{number of decays} = (1.5 \times 10^4\ \text{L})\left(\frac{4\ \text{pCi}}{1\ \text{L}}\right)\left(\frac{9\ \text{decays}\cdot\text{min}^{-1}}{4\ \text{pCi}}\right)(5.0\ \text{min})$$
$$= 7 \times 10^5\ \text{decays}$$

17.75 $N_0 = \text{number of } {}^{222}\text{Rn atoms} = 1.0 \times 10^{-5}\ \text{mol} \times 6.0 \times 10^{23}\ \text{atoms}\cdot\text{mol}^{-1}$
$= 6.0 \times 10^{18}\ \text{atoms}$

$$k = \frac{\ln 2}{t_{1/2}} = \frac{0.693}{3.82\ \text{d}} = 0.181\ \text{d}^{-1}$$

(a) $\text{rate of decay} = k \times N = \left(\frac{0.181}{\text{d}}\right)\left(\frac{1\ \text{d}}{8.64 \times 10^4\ \text{s}}\right)(6.0 \times 10^{18}\ \text{atoms})$
$= 1.26 \times 10^{13}\ \text{atoms}\cdot\text{s}^{-1}$ (dps or Bq)

$$\text{initial activity} = (1.26 \times 10^{13}\ \text{Bq})\left(\frac{1\ \text{Ci}}{3.7 \times 10^{10}\ \text{Bq}}\right)\left(\frac{1\ \text{pCi}}{10^{-12}\ \text{Ci}}\right)\left(\frac{1}{2000\ \text{m}^3}\right)\left(\frac{1\ \text{m}^3}{10^3\ \text{L}}\right)$$
$$= 1.7 \times 10^8\ \text{pCi}\cdot\text{L}^{-1}$$

(b) $N = N_0 e^{-kt} = 6.0 \times 10^{18}\ \text{atoms} \times e^{-0.181\ \text{d}^{-1} \times 1\ \text{d}} = 5.0 \times 10^{18}\ \text{atoms}$

(c) $\ln\left(\frac{\text{activity}}{\text{initial activity}}\right) = -kt$

$$t = -\left(\frac{1}{k}\right)\ln\left(\frac{\text{activity}}{\text{initial activity}}\right) = -\left(\frac{1}{0.181\ \text{d}^{-1}}\right)\ln\left(\frac{4}{1.70 \times 10^8}\right)$$
$$= 1 \times 10^2\ \text{days}$$

17.77 (a) At first thought, it might seem that a fusion bomb would be more suitable for excavation work, because the fusion process itself does not generate harmful radioactive waste products. However, in practice, fusion cannot be initiated in a bomb in the absence of the high temperatures that can only be generated by a fission bomb. So, there is no environmental advantage to the use of a fusion bomb. The fission bomb has the advantage that its destructive power can be more carefully controlled. It is possible to make small fission bombs whose destructive effect can be contained within a small area.

(b) The principal argument for the use of bombs in excavation is speed, and therefore cost-effectiveness, of the process. The principal argument against their use is environmental damage.

17.79 $k = \dfrac{0.693}{4.5 \times 10^9\ \text{y}} = 1.5 \times 10^{-10}\ \text{y}^{-1}$

$$t(=\text{age}) = -\left(\frac{1}{k}\right)\ln\left(\frac{N}{N_0}\right)$$

$$\frac{N}{N_0} = \frac{\text{mass of }^{238}\text{U}}{\text{initial mass of }^{238}\text{U}} = \frac{1}{1 + \dfrac{\text{mass of }^{206}\text{Pb}}{\text{mass of }^{238}\text{U}}}$$

(a) $\dfrac{N}{N_0} = \dfrac{1}{1 + 1.00} = \dfrac{1}{2.00}$, therefore age $= t_{1/2} = 4.5 \times 10^9$ y

(b) $\dfrac{N}{N_0} = \dfrac{1}{1 + \dfrac{1}{1.25}} = 0.556$

$$t(=\text{age}) = -\left(\frac{1}{1.5 \times 10^{-10}\ \text{y}^{-1}}\right)\ln(0.556) = 3.9 \times 10^9\ \text{y}$$

17.81 (a) activity $= (17.3\ \text{Ci})\left(\dfrac{3.7 \times 10^{10}\ \text{Bq}}{1\ \text{Ci}}\right)$

$= 6.4 \times 10^{11}\ \text{Bq} = 6.4 \times 10^{11}\ \text{nuclei}\cdot\text{s}^{-1}$

$$N = (2.0 \times 10^{-6}\ \text{g})\left(\frac{1\ \text{u}}{1.661 \times 10^{-24}\ \text{g}}\right)\left(\frac{1\ \text{nucleus}}{24\ \text{u}}\right) = 5.0 \times 10^{16}\ \text{nuclei}$$

$$k = \frac{\text{activity}}{N} = \frac{6.4 \times 10^{11}\ \text{nuclei}\cdot\text{s}^{-1}}{5.0 \times 10^{16}\ \text{nuclei}} = 1.3 \times 10^{-5}\ \text{s}^{-1} = 1.1\ \text{d}^{-1}$$

$$t_{1/2} = \frac{0.693}{k} = \frac{0.693}{1.3 \times 10^{-5}\ \text{s}^{-1}} = 5.3 \times 10^4\ \text{s} = 15\ \text{h} = 0.63\ \text{d}$$

(b) $m = m_0 e^{-kt} = 2.0\ \text{mg} \times e^{-1.11\ \text{d}^{-1} \times 2.0\ \text{d}} = 0.22\ \text{mg}$

17.83 (a) Radioactive substances which emit γ radiation are most effective for diagnosis because they are the least destructive of the types of radiation listed. Additionally, γ rays pass easily through body tissues and can be counted, whereas α and β particles are stopped by the body tissues. (b) α particles tend to be best for this application because they cause the most destruction. (c) and (d) ^{131}I, 8d (used to image the thyroid); ^{67}Ga, 78 h (used most often as the citrate complex); ^{99m}Tc, 6 h (used for various body tissues by varying the ligands attached to the Tc atom).

17.85 (a)

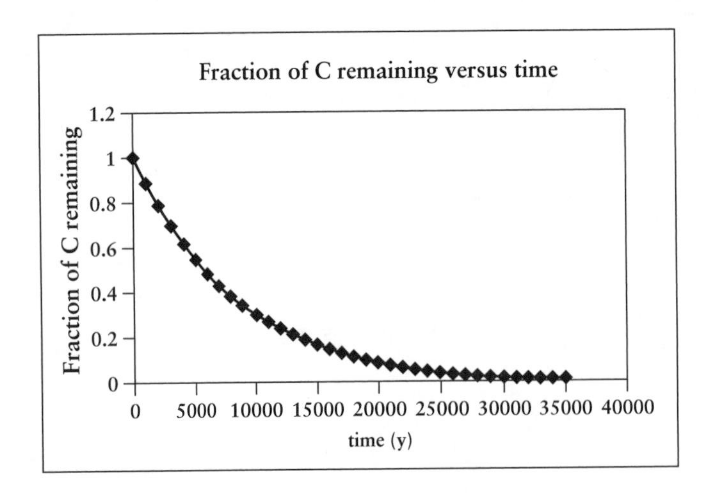

(b)

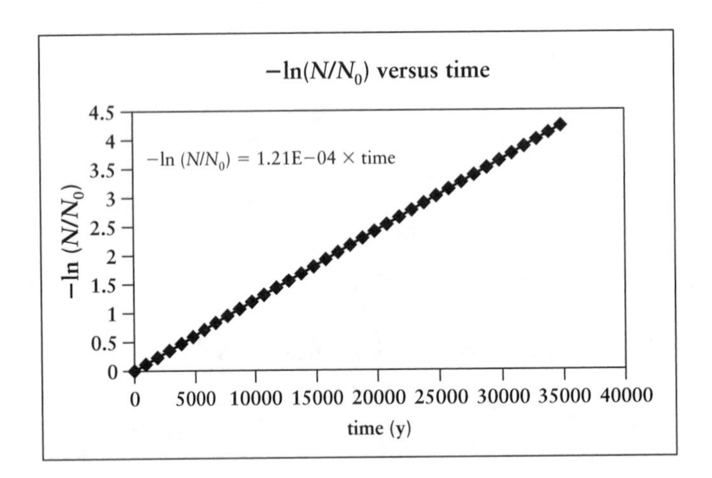

(c) The information can be obtained from the graphs or from the equation

$$-\ln \frac{N}{N_0} = kt$$

If we want to have less than 1% of the original amount of ^{14}C present, then we will want the value for which N/N_0 is 0.01 or less.

$-\ln 0.01 = (1.21 \times 10^{-4}\ y^{-1})(t)$

$t = 3.8 \times 10^{4}$ y

CHAPTER 18
ORGANIC CHEMISTRY I: THE HYDROCARBONS

18.1 (a)

$$\begin{array}{ccccccccccccc} & & H & & H & & & & & & H & & H \\ & & | & & | & & & & & & | & & | \\ H & - & C & - & C & - & C & \equiv & C & - & C & - & C & - & H \\ & & | & & | & & & & & & | & & | \\ & & H & & H & & & & & & H & & H \end{array}$$

alkyne

(b)

$$\begin{array}{ccccccccccccccc} & & H & & H & & H & & H & & H & & H \\ & & | & & | & & | & & | & & | & & | \\ H & - & C & - & C & - & C & - & C & - & C & - & C & - & H \\ & & | & & | & & | & & | & & | & & | \\ & & H & & H & & H & & H & & H & & H \end{array}$$

alkane

(c)

$$\begin{array}{ccccccccccc} & & & & & & H & & H \\ & & & & & & | & & | \\ H & - & C & = & C & - & C & - & C & - & H \\ & & | & & | & & | & & | \\ & & H & & H & & H & & H \end{array}$$

alkene

(d)

$$\begin{array}{ccccccccccccccccc} & & H & & H & & H & & H & & & & & & H \\ & & | & & | & & | & & | & & & & & & | \\ H & - & C & - & C & = & C & - & C & - & C & \equiv & C & - & C & - & H \\ & & | & & & & & & | & & & & & & | \\ & & H & & & & & & H & & & & & & H \end{array}$$

alkene and alkyne

(e)

$$\begin{array}{ccccccccccccc} & & H & & H & & H & & H & & H \\ & & | & & | & & | & & | & & | \\ H & - & C & - & C & - & C & - & C & = & C & - & H \\ & & | & & | & & | \\ & & H & & H & & H \end{array}$$

alkene

18.3 (a) $(CH_3)_3CH$ or C_4H_{10}, alkane; (b) $C_6H_7CH_3$ or C_7H_{10}, alkene; (c) C_6H_{12}, alkane; (d) C_6H_{12}, alkane

18.5 (a) $C_{12}H_{26}$, alkane; (b) $C_{13}H_{20}$, alkene; (c) C_7H_{14}, alkane; (d) $C_{14}H_8$, aromatic hydrocarbon

18.7 (a) propane; (b) butane; (c) heptane; (d) decane

18.9 (a) methyl; (b) pentyl; (c) propyl; (d) hexyl

18.11 (a) propane; (b) ethane; (c) pentane; (d) 2,3-dimethylbutane

18.13 (a) 4-methyl-2-pentene; (b) 2,3-dimethyl-2-phenylpentane

18.15 (a) $CH_2{=}CHCH(CH_3)CH_2CH_3$;
(b) $CH_3CH_2C(CH_3)_2CH(CH_2CH_3)(CH_2)_2CH_3$; (c) $CH{\equiv}C(CH_2)_2C(CH_3)_3$;
(d) $CH_3CH(CH_3)CH(CH_2CH_3)CH(CH_3)_2$

18.17 (a)

H H H CH_3 H H H H H
H—C—C—C—C——C—C—C—C—C—H
H H H CH_3 H H H H H

(b)

CH_3
H H CH_2 H H H H H
H—C≡C—C—C——C——C—C—C—C—C—H
H CH_2 CH_2 H H H H H
CH_2 CH_3
CH_3

(c)

H CH_3 H CH_3 H
H—C—C——C—C——C—H
H CH_3 H H H

(d)

CH_3CH_2 H
C=C
H CH_2CH_3

18.19 (a) (b) (c)

18.21 (a) hexenes:

H_2 H_2
H_3C C C C CH
H_2 H
1-Hexene

CH=CH
H_2C—CH_2 CH_3
H_3C
cis-2-Hexene

CH_3
H_3C CH=CH
CH_2—CH_2
trans-2-Hexene

CH_2 CH CH_3
CH_3 CH CH_2
cis-3-Hexene

CH_2 CH CH_3
CH_3 CH CH_2
trans-3-Hexene

pentenes:

4-Methyl-1-pentene   3-Methyl-1-pentene   2-Methyl-1-pentene

2-Methyl-2-pentene   3-Methyl-2-pentene   4-Methyl-2-pentene

butenes:

3,3-Dimethyl-1-butene   2,3-Dimethyl-1-butene   2,3-Dimethyl-2-butene

(b) cyclic molecules:

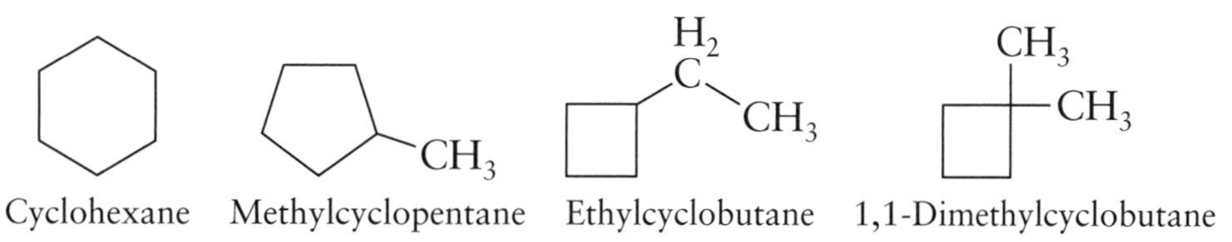

Cyclohexane   Methylcyclopentane   Ethylcyclobutane   1,1-Dimethylcyclobutane

The following structures are drawn to emphasize the stereochemistry

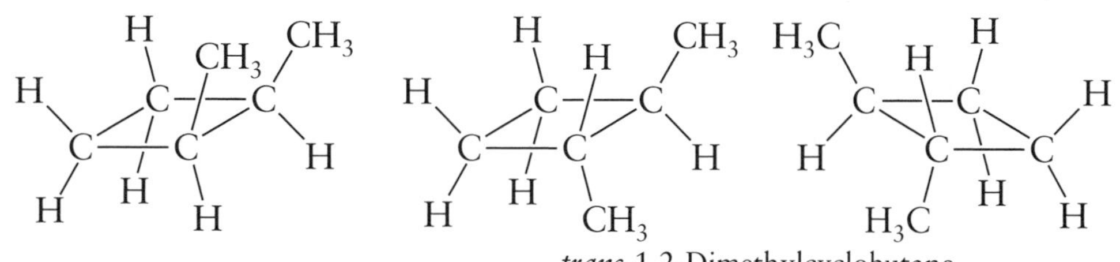

cis-1,2-Dimethylcyclobutane   *trans*-1,2-Dimethylcyclobutane (nonsuperimposable mirror images)

trans-1,3-Dimethylcyclobutane   *cis*-1,3-Dimethylcyclobutane

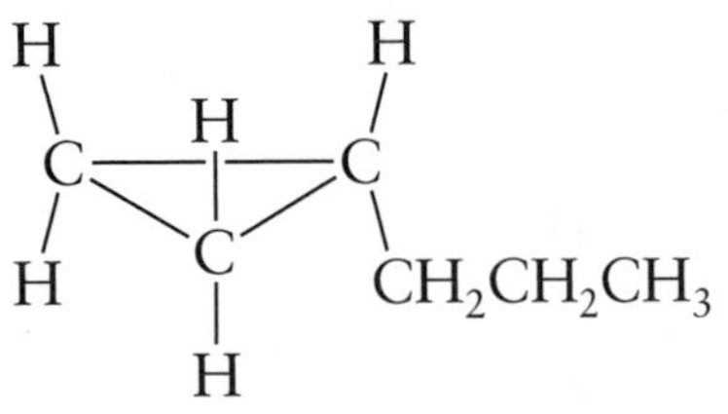

Propylcyclopropane

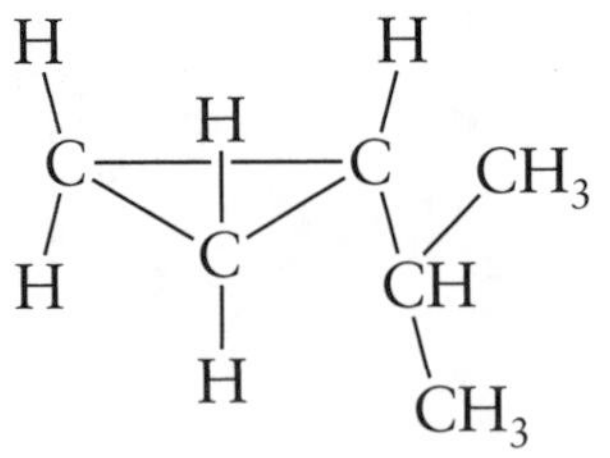

Isopropylcyclopropane or 2-Cyclopropylpropane

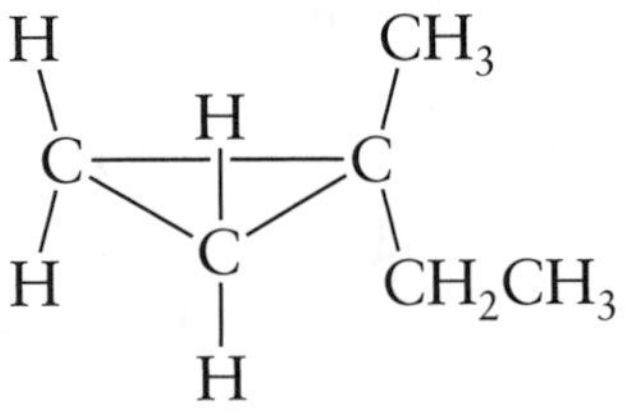

1-Ethyl-1-methylcyclopropane

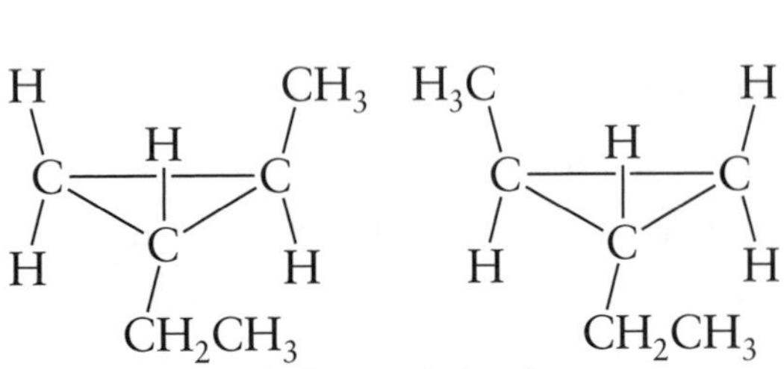

trans-1-Ethyl-2-methylcyclopropane (nonsuperimposable mirror images)

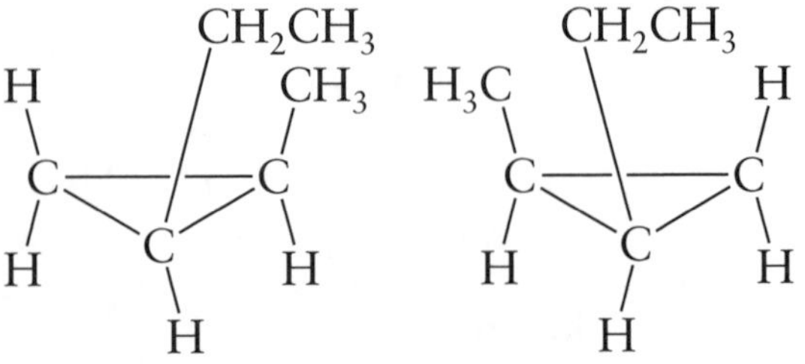

cis-1-Ethyl-2-methylcyclopropane (nonsuperimposable mirror images)

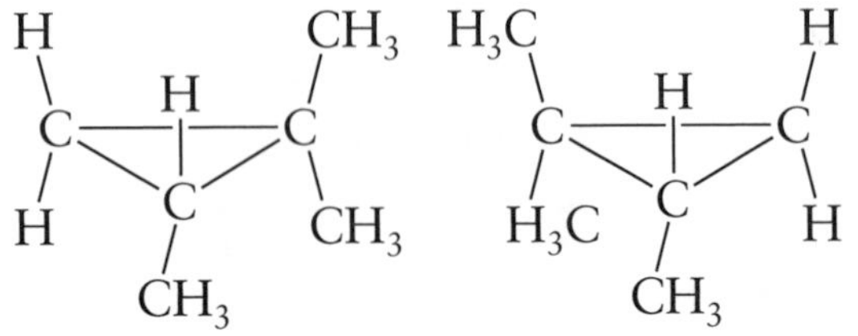

1,1,2-Trimethylcyclopropane (nonsuperimposable mirror images)

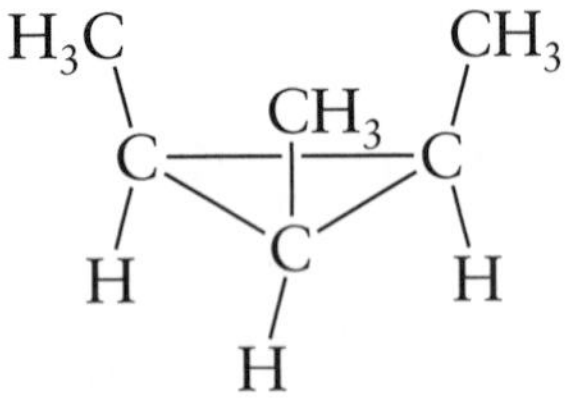

1,2,3-Trimethylcyclopropane (all cis isomer)

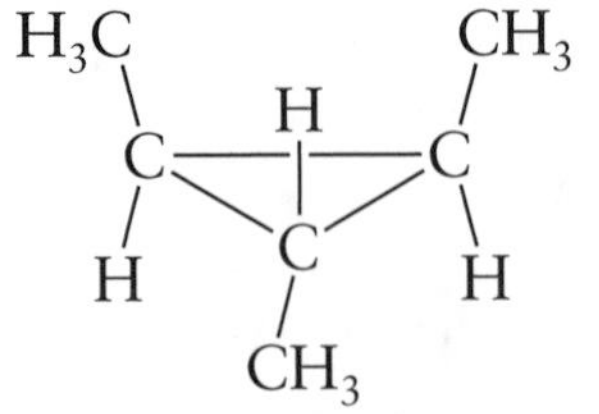

1,2,3-Trimethylcyclopropane (all cis,trans isomer)

18.23 (a) Butane is C_4H_{10}, cyclobutane is C_4H_8. Because they have different formulas, they are not isomers.

(b)

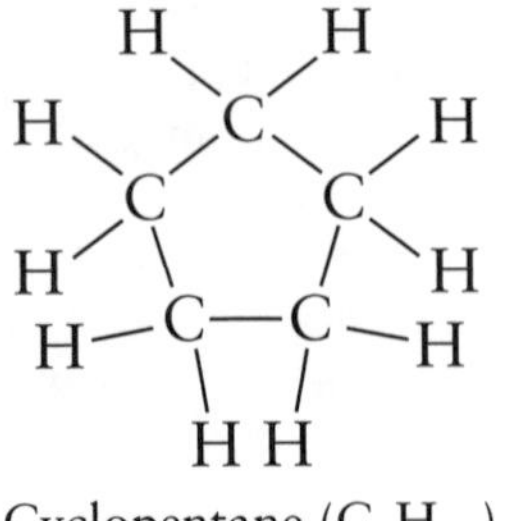

Cyclopentane (C_5H_{10})

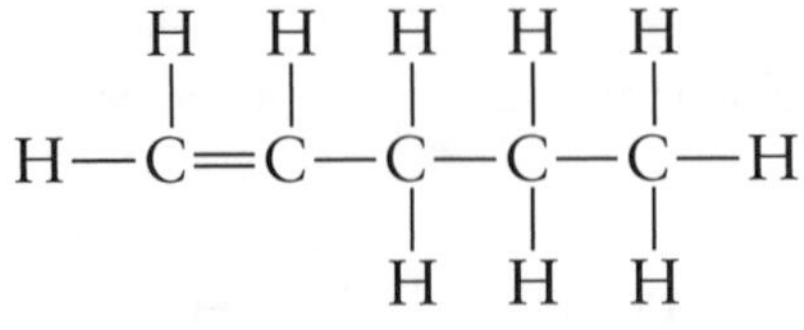

Pentene (C_5H_{10})

Same formula, but different structures; therefore, they are structural isomers.

(c) Same formula (C_5H_{10}), same structure (bonding arrangement is the same), but different geometry; therefore, they are geometrical isomers.

(d) Not isomers, because only their positions in space are different and these positions can be interchanged. Same molecule.

18.25 If only two isomeric products are formed and they are both branched, then the only possibilities are

$$\text{(a)}\ \begin{array}{c} \text{H} \\ | \\ \text{CH}_3\text{—C—CH}_3 \\ | \\ \text{CH}_3 \end{array} \quad \text{(b)}\ \begin{array}{c} \text{Cl} \\ | \\ \text{CH}_3\text{—C—CH}_3 \\ | \\ \text{CH}_3 \end{array} \quad \begin{array}{c} \text{H}\ \ \text{H} \\ |\ \ \ | \\ \text{Cl—C—C—CH}_3 \\ |\ \ \ | \\ \text{H}\ \ \text{CH}_3 \end{array}$$

Note: All methyl groups are equivalent.

18.27 (a), (c), and (d) are optically active.

$$\text{(a)}\ \begin{array}{c} \text{H} \\ | \\ \text{CH}_3\text{—C*—CH}_2\text{CH}_3 \\ | \\ \text{Br} \end{array} \quad \text{(c)}\ \begin{array}{c} \text{Br}\ \ \text{Cl} \\ |\ \ \ | \\ \text{H—C—C*—CH}_3 \\ |\ \ \ | \\ \text{H}\ \ \text{H} \end{array}$$

$$\text{(d)}\ \begin{array}{c} \text{Cl}\ \ \text{Cl}\ \ \text{H}\ \ \text{H}\ \ \text{H} \\ |\ \ \ |\ \ \ |\ \ \ |\ \ \ | \\ \text{H—C—C*—C—C—C—H} \\ |\ \ \ |\ \ \ |\ \ \ |\ \ \ | \\ \text{H}\ \ \text{H}\ \ \text{H}\ \ \text{H}\ \ \text{H} \end{array}$$

18.29 The difference can be traced to the weaker London forces that exist between branched molecules. Atoms in neighboring branched molecules cannot lie as close together as they do in the unbranched isomers. As a result of the molecules' irregular shape, the atoms in neighboring branched molecules are more effectively shielded from one another than they are in neighboring unbranched molecules.

18.31 The balanced equations are

$C_3H_8(g) + 5\ O_2(g) \longrightarrow 3\ CO_2(g) + 4\ H_2O(l)$

$C_4H_{10}(g) + 13/2\ O_2(g) \longrightarrow 4\ CO_2(g) + 5\ H_2O(l)$

$C_5H_{12}(g) + 8\ O_2(g) \longrightarrow 5\ CO_2(g) + 6\ H_2O(l)$

The enthalpies of combustion that correspond to these reactions are listed in the Appendix:

Compound	Enthalpy of combustion $kJ \cdot mol^{-1}$	Heat released per g $kJ \cdot g^{-1}$
Propane	−2220	50.3
Butane	−2878	49.5
Pentane	−3537	49.0

* Indicates the chiral carbon atoms.

The molar enthalpy of combustion increases with molar mass as might be expected, because the number of moles of CO_2 and H_2O formed will increase as the number of carbon and hydrogen atoms in the compounds increases. The heat released per gram of these hydrocarbons is essentially the same because the H to C ratio is similar in the three hydrocarbons.

18.33 There are nine possible products:

```
    H  H              H  H           H  Cl
    |  |              |  |           |  |
 H—C—C—Cl         Cl—C—C—Cl      H—C—C—Cl
    |  |              |  |           |  |
    H  H              H  H           H  H
```

one monochloro compound — two dichloro compounds

```
    H  Cl          Cl Cl          Cl Cl           Cl Cl
    |  |           |  |           |  |            |  |
 H—C—C—Cl      H—C—C—Cl      H—C—C—Cl      Cl—C—C—Cl
    |  |           |  |           |  |            |  |
    H  Cl          H  H           H  Cl           H  H
```

two trichloro compounds — two tetrachloro compounds

```
    Cl Cl             Cl Cl
    |  |              |  |
 Cl—C—C—Cl        Cl—C—C—Cl
    |  |              |  |
    H  Cl             Cl Cl
```

one pentachloro compound — one hexachloro compound

None of these form optical isomers.

18.35

```
   H  H  H  H  H                     H  H  H  H  H
   |  |  |  |  |                     |  |  |  |  |
H—C—C=C—C—C—H + HBr ⟶ H—C—C—C—C—C—H
   |        |  |                     |  |  |  |  |
   H        H  H                     H  H  Br H  H

   H  H  H  H  H                     H  H  H  H  H
   |  |  |  |  |                     |  |  |  |  |
H—C—C=C—C—C—H + HBr ⟶ H—C—C—C—C—C—H
   |        |  |                     |  |  |  |  |
   H        H  H                     H  Br H  H  H
```

The first reaction produces 3-bromopentane. The second reaction produces 2-bromopentane. These are addition reactions.

18.37 $C_6H_{11}Br + NaOCH_2CH_3 \longrightarrow C_6H_{10} + NaBr + HOCH_2CH_3$

```
   H   H    H   H                                   H    H
  H \ /      \ / H                               H  \   /
     C         C                                     C      H
    /  \     /  \                                  /   \   /
 H—C          C—H     +NaOCH2CH3              H—C        C
    |          |     ⟶                            |        ‖
 H—C          C—H     −NaBr                   H—C        C
    \  /     \  /      −HOCH2CH3                  \   /   \
      C        Br                                    C       H
   H / \ H  H                                    H  /  \
     H                                              H    H
```

Elimination reaction

18.39 $C_2H_4 + X_2 \longrightarrow C_2H_4X_2$

We will break one X—X bond and form two C—X bonds.

Using bond enthalpies:

Halogen	Cl	Br	I
X—X bond breakage ($kJ \cdot mol^{-1}$)	+242	+193	+151
C—X bond formation ($kJ \cdot mol^{-1}$)	−2(338)	−2(276)	−2(238)
Total ($kJ \cdot mol^{-1}$)	−434	−359	−325

The reaction is less exothermic as the halogen becomes lighter. In general, the reactivity, and also the danger associated with use of the halogens in reactions, decreases as one descends the periodic table.

18.41 (a) 1-ethyl-3-methylbenzene (b) 1,2,3,4,5-pentamethylbenzene (or, because there is only one possible structure, pentamethylbenzene)

18.43 (a) CH_3 (b) CH_3 (c) CH_3 (d) CH_3

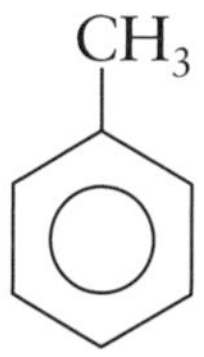

CH_3

Cl Cl

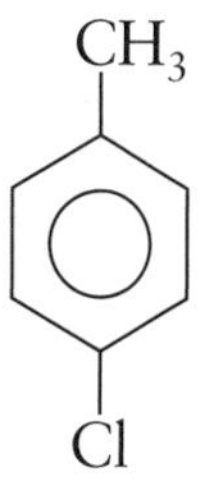

18.45 (a)–(b)

CH_3 Cl Cl

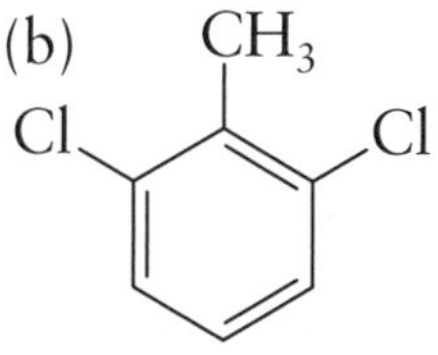

1,3-Dichloro-2-methylbenzene

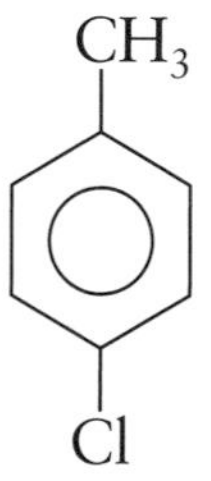

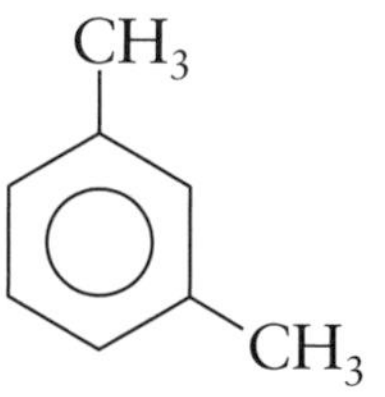

CH_3 Cl Cl

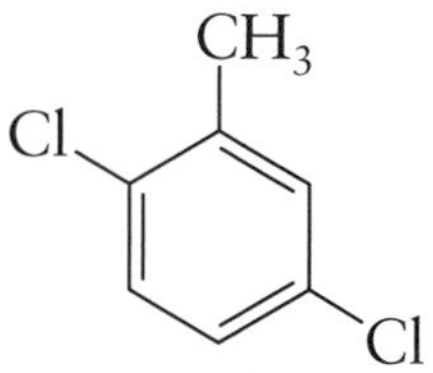

1,4-Dichloro-2-methylbenzene

CH_3 Cl Cl

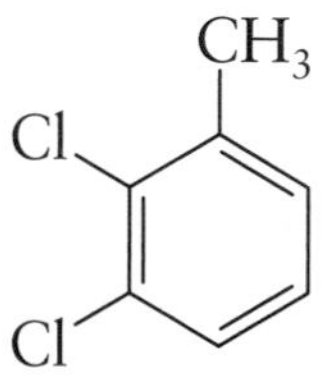

1,2-Dichloro-3-methylbenzene

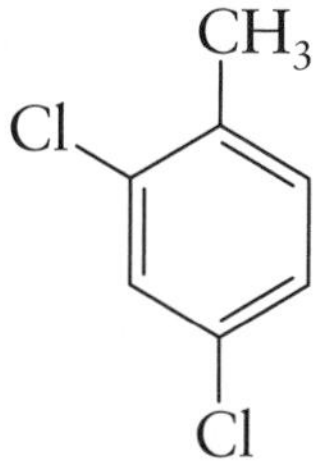

CH_3 Cl Cl

1,5-Dichloro-2-methylbenzene

CH_3 Cl Cl

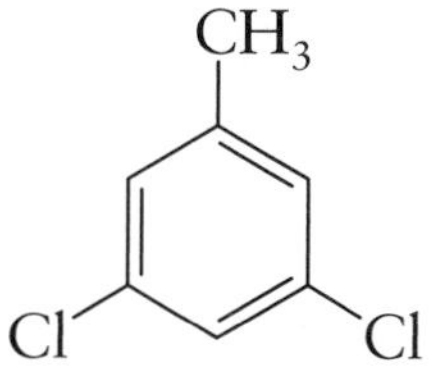

1,3-Dichloro-5-methylbenzene

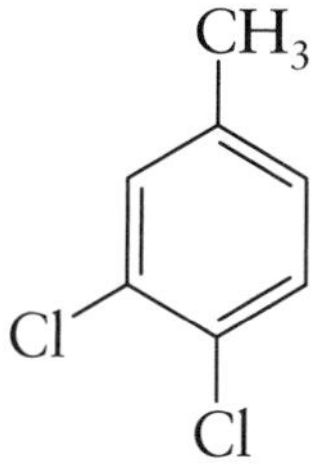

CH_3 Cl Cl

1,2-Dichloro-4-methylbenzene

All of these molecules will be at least slightly polar.

18.47

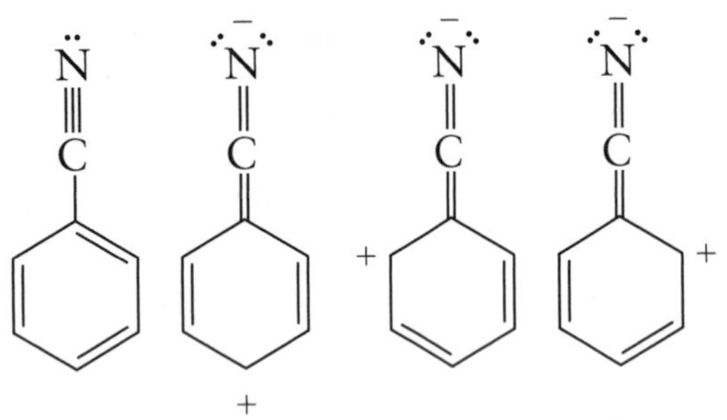

Electrophiles tend to avoid the ortho and para positions that develop slight + charges in the resonance forms.

18.49 Two:

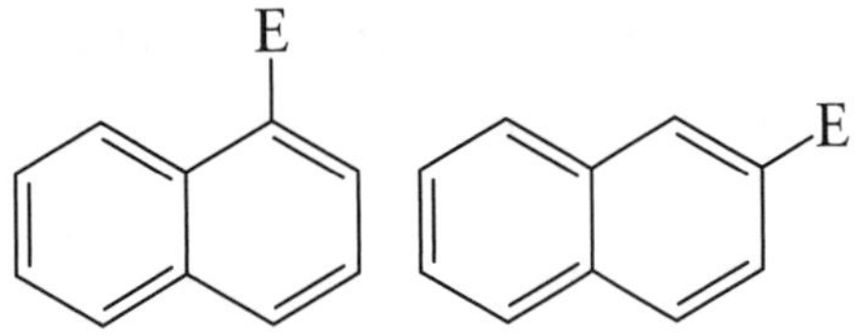

E = electrophile

18.51 (a) four σ-type single bonds; (b) two σ-type single bonds and one double bond, consisting of one σ-bond and one π-bond; (c) one σ-type single bond and one triple bond, consisting of one σ-bond and two π-bonds.

18.53 (a) $CH_4 + Cl_2 \xrightarrow{\text{light}} CH_3Cl + HCl$, substitution
(b) $CH_2{=}CH_2 + Br_2 \longrightarrow CH_2Br{-}CH_2Br$, addition

18.55 (a) $CH_3CH_3 + 2\ Cl_2 \xrightarrow{\text{light}} CH_2ClCH_2Cl + 2\ HCl$, substitution ($CH_2ClCH_3$ may also be produced)
(b) $CH_2CH_2 + Cl_2 \longrightarrow CH_2ClCH_2Cl$, addition
(c) $HC{\equiv}CH + 2\ Cl_2 \longrightarrow CHCl_2CHCl_2$, addition

18.57 The double bond in alkenes makes them more rigid than alkanes. The atoms of alkene molecules tend to be locked into a planar arrangement by the π-bond and therefore they cannot roll up into a ball as compactly as alkanes. Thus, they do not pack together as compactly as alkanes and so have lower boiling and melting points.

18.59 (a) 2-methyl-1-propene, no geometrical isomers; (b) *cis*-3-methyl-2-pentene, *trans*-3-methyl-2-pentene; (c) 1-hexyne, no geometrical isomers; (d) 3-hexyne, no geometrical isomers; (e) 2-hexyne, no geometrical isomers.

18.61

1-Heptene

trans-2-Heptene

cis-2-Heptene

trans-3-Heptene

cis-3-Heptene

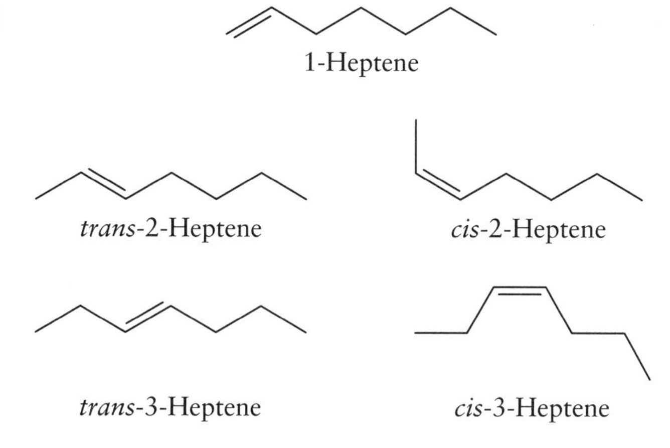

18.63 (a) $C_{10}H_{18}$; (b) naphthalene, , $C_{10}H_8$; (c) Yes. *Cis* and *trans* forms (relative to the C—C bond connecting the two six-membered rings) are possible.

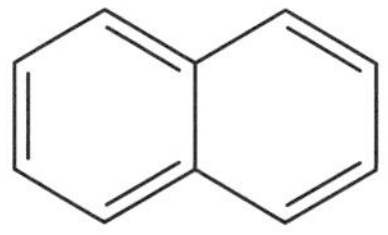

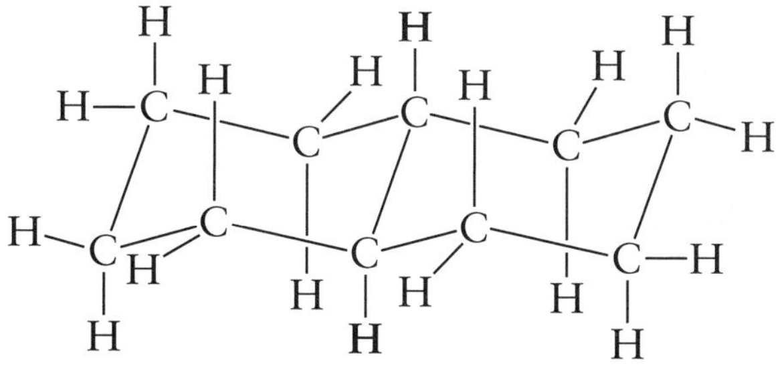

trans-Decalin

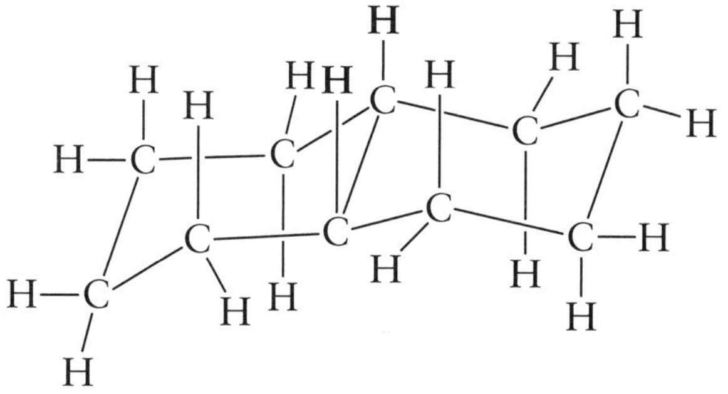

cis-Decalin

18.65 number of moles of H $= \left(\dfrac{3.32 \text{ g } H_2O}{18.02 \text{ g } H_2O/\text{mol } H_2O}\right)\left(\dfrac{2 \text{ mol H}}{1 \text{ mol } H_2O}\right)$

$= 0.368$ mol H

number of moles of C $= \left(\dfrac{6.48 \text{ g } CO_2}{44.01 \text{ g } CO_2/\text{mol } CO_2}\right)\left(\dfrac{1 \text{ mol C}}{1 \text{ mol } CO_2}\right)$

$= 0.147$ mol C

$$\frac{0.368 \text{ mol H}}{0.147 \text{ mol C}} = 2.50\left(\frac{\text{mol H}}{\text{mol C}}\right) = \frac{5 \text{ mol H}}{2 \text{ mol C}}$$

Therefore, the empirical formula is C_2H_5. The molecular formula might be C_4H_{10}, which matches the general formula for alkanes, C_nH_{2n+2}. The compound cannot be an alkene or alkyne, because they all have mol H/mol C ratios less than 2.5.

18.67 (a) 4-methyl-3-propylheptane

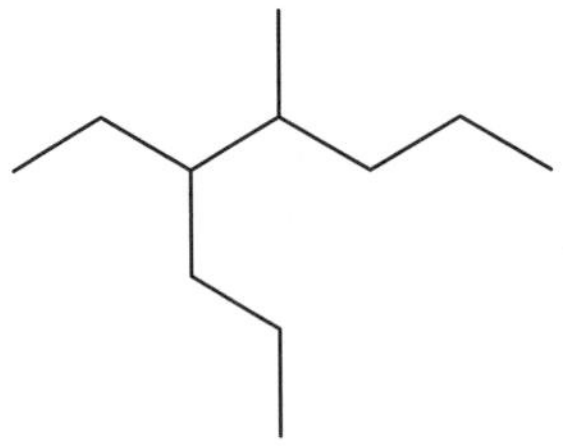

The longest chain has eight carbon atoms in it. The systematic name of the compound is 4-ethyl-5-methyloctane.

(b) 4,6-dimethyloctane

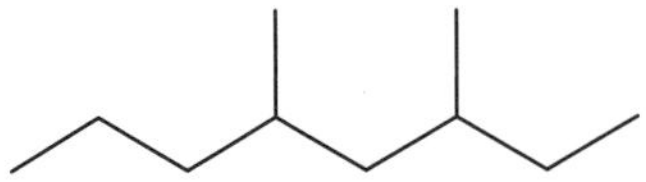

The compound name is almost correct, but the numbering scheme with the lowest numbers would be 3,5-dimethyloctane.

(c) 2,2-dimethyl-4-propylhexane

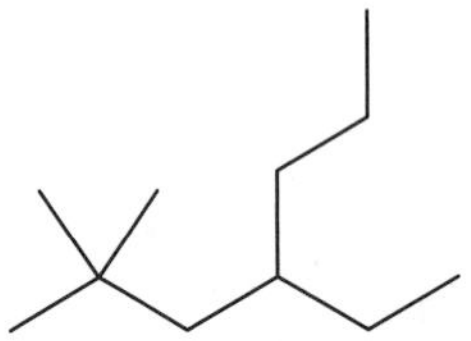

The longest carbon chain in the molecule is seven carbon atoms long. The systematic name is 2,2-dimethyl-4-propylheptane.

(d) 2,2-dimethyl-3-ethylhexane.

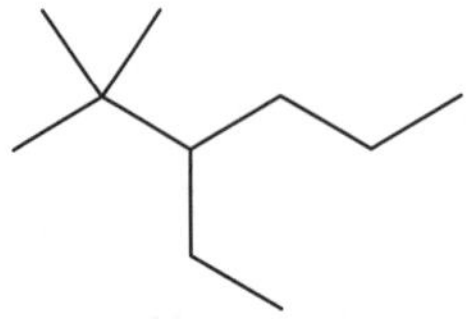

The name is essentially correct except that ethyl should be listed first. The systematic name is 3-ethyl-2,2-dimethylhexane.

18.69 These hydrocarbons are too volatile (they are all gases at room temperature) and would not remain in the liquid state.

18.71 Cracking is the process of breaking down hydrocarbons with many carbon atoms into smaller units, whereas alkylation is the process of combining smaller hydrocarbons into larger units. Both processes are carried out catalytically and both are used to convert hydrocarbons into units having 6 to 10 carbon atoms, suitable for use in gasoline.

18.73 $CH_3(CH_2)_2CH(OH)CH_3 \xrightarrow[120°C]{H_2SO_4} H_2O(g) + CH_3(CH_2)_2CH{=}CH_2$

18.75 The NO_2 group is a meta-directing group and the Br atom is an ortho-, para-directing group. Because the position para to Br is already substituted with the NO_2 function, further bromination will not occur there. The resonance forms show that the bromine atom will activate the position ortho to it, as expected. The NO_2 group will *deactivate* the group ortho to itself, which in essence enhances the reactivity of the position meta to the NO_2 group. This position is also the position that is ortho to the Br atom, so the effects of the Br and NO_2 reinforce each other. Bromination is thus expected to occur as shown:

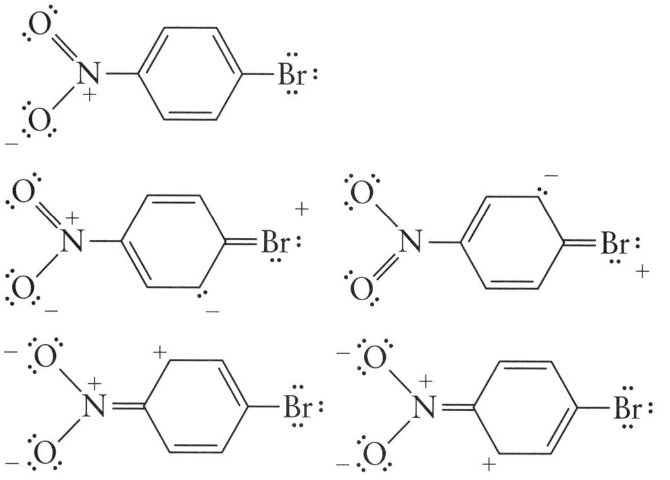

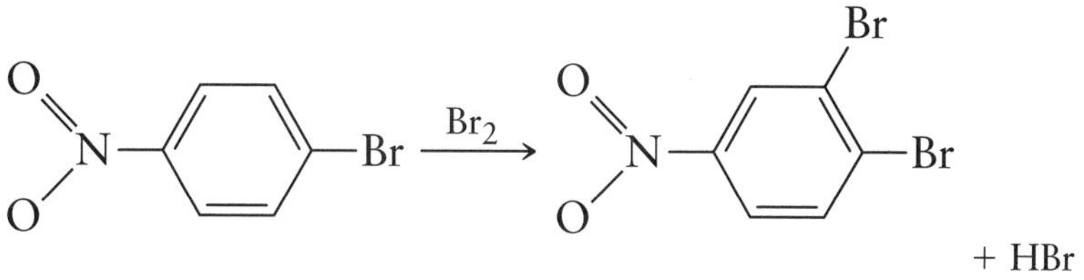

+ HBr

18.77 A mixture of ethane (0.500 bar) and hydrogen (0.300 bar) was placed over a metal catalyst and the equilibrium was established. The dominant reaction is the hydrogenation of ethane to methane. What will be the ratio of methane to ethane at equilibrium? (To simplify the problem use values for 298 K.)

The balanced equation is

$C_2H_6(g) + H_2(g) \longrightarrow 2\ CH_4(g)$

The equilibrium constant for this reaction can be obtained at 500°C from the enthalpy and entropy changes of the reaction:

$$\begin{aligned}\Delta H°_r &= 2\Delta H°_f(CH_4, g) - \Delta H°_f(C_2H_6, g)\\ &= 2(-74.81\ kJ\cdot mol^{-1}) - (-32.82\ kJ\cdot mol^{-1})\\ &= -116.80\ kJ\cdot mol^{-1}\end{aligned}$$

$$\Delta S°_r = 2S°_m(CH_4, g) - [S°_m(C_2H_6, g) + S_m(H_2, g)]$$
$$= 2(186.26\ J \cdot K^{-1} \cdot mol^{-1})$$
$$- [229.60\ J \cdot K^{-1} \cdot mol^{-1} + 130.68\ J \cdot K^{-1} \cdot mol^{-1}]$$
$$= +12.24\ J \cdot K^{-1} \cdot mol^{-1}$$
$$\Delta G°_r = \Delta H°_r - T\Delta S°_r = -116.80\ kJ \cdot mol^{-1}$$
$$- (778\ K)(12.24\ J \cdot K^{-1} \cdot mol^{-1})/1000\ J \cdot kJ^{-1}$$
$$= -126.32\ kJ \cdot mol^{-1}$$

The equilibrium constant is obtained from

$$\Delta G°_r = -RT \ln K$$
$$(-126.32\ kJ \cdot mol^{-1})(1000\ J \cdot kJ^{-1}) = -(8.314\ J \cdot K^{-1} \cdot mol^{-1})(778\ K) \ln K$$
$$\ln K = 190$$
$$K = 3 \times 10^{82}$$

Because the equilibrium constant is so large, the reaction essentially goes to completion. At equilibrium, then, there will be 0.300 bar of CH_4, 0.200 bar of C_2H_6, and essentially no H_2. The amount of H_2 can be calculated from the equilibrium expression:

$$\frac{P_{CH_4}{}^2}{P_{C_2H_6}P_{H_2}} = 3 \times 10^{82}$$
$$\frac{(0.300)^2}{(0.200)P_{H_2}} = 3 \times 10^{82}$$
$$P_{H_2} = \frac{(0.300)^2}{(0.200)(3 \times 10^{82})}$$
$$= 2 \times 10^{-83}\ \text{bar}$$

18.79 If the molecule contains two carbon centers that have four different substituents attached but are arranged such that they are mirror images of one another *within* the molecule, the molecule will not be optically active. Such an example is shown below in general, for a 1,2—X_2—1,2—Y_2—1,2—Z_2 substituted ethane. Many other examples are possible—the only criterion being that the carbon atoms that have four substituents must have a mirror image carbon center within the molecule.

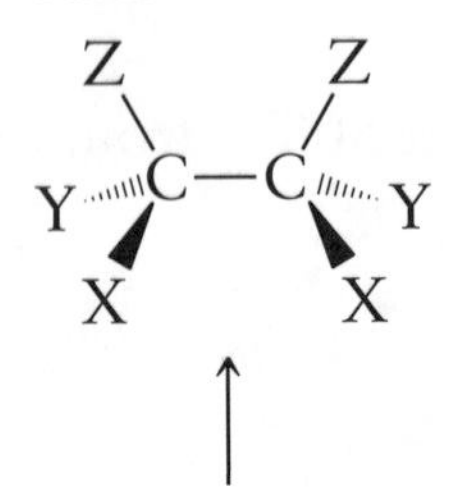

a mirror plane exists in the molecule

For comparison, the molecule

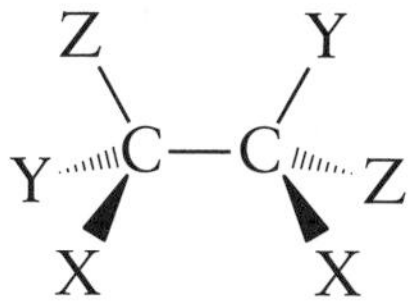

```
 Z      Y
  \    /
Y····C—C····Z
    ▼   ▼
   X     X
```

does have a nonsuperimposable mirror image and would exist as two members of an enantiomeric pair.

18.81 For a molecule such as 1,2-dichloro-4-diethylbenzene, $C_6H_3Cl_2(CH_2CH_3)$, 175.04 u, it is relatively easy to lose heavy atoms, such as chlorine, and groups of atoms, such as methyl and ethyl fragments. Molecules can also lose hydrogen atoms. In mass spectroscopy, P is used to represent the *parent ion,* which is the ion formed from the molecule without fragmentation. Fragments are then represented as $P - x$, where x is the particular fragment lost from the parent ion to give the observed mass. Because the mass spectrum will measure the masses of individual molecules, the mass of carbon used will be 12.00 u (by definition) because the large majority of the molecules will have all ^{12}C. The molar mass of H is 1.0078 u. Some representative peaks that may be present include:

Fragment Formula	Relation to Parent Ion	Mass (u)
$C_6H_3{}^{35}Cl_2(CH_2CH_3)$	P	174.00
$C_6H_3{}^{35}Cl^{37}Cl(CH_2CH_3)$	P	176.00
$C_6H_3{}^{37}Cl_2(CH_2CH_3)$	P	177.99
$C_6H_3{}^{35}Cl(CH_2CH_3)$	$P—Cl$	139.03
$C_6H_3{}^{37}Cl(CH_2CH_3)$	$P—Cl$	141.03
$C_6H_3{}^{35}Cl_2(CH_2)$	$P—CH_3$	158.98
$C_6H_3{}^{35}Cl^{37}Cl(CH_2)$	$P—CH_3$	160.97
$C_6H_3{}^{37}Cl_2(CH_2)$	$P—CH_3$	162.97
$C_6H_3{}^{35}Cl_2$	$P—CH_2CH_3$	144.96
$C_6H_3{}^{35}Cl^{37}Cl$	$P—CH_2CH_3$	146.96
$C_6H_3{}^{37}Cl_2$	$P—CH_2CH_3$	148.96
$C_6H_3{}^{35}Cl$	$P—CH_2CH_3—Cl$	109.99
$C_6H_3{}^{37}Cl$	$P—CH_2CH_3—Cl$	111.99

etc.

18.83 The presence of one bromine atom will produce in the ions that contain Br companion peaks that are separated by 2 u. Any fragment that contains Br will show this "doublet" in which the peaks are nearly but not exactly equal in intensity. Thus, seeing a mass spectrum of a compound that is known to have Br or that was involved in a reaction in which Br could have been added or substituted with such doublets, is almost a sure sign that Br is present in the compound. It is also fairly easy to detect Br atoms in the mass spectrum at 79 and 81 u, confirming their presence. If more than one Br atom is present, then a more complicated pattern is observed for the presence of the two isotopes. The possible combinations for a molecule of unknown formula with two Br atoms is $^{79}Br^{79}Br$, $^{79}Br^{81}Br$, $^{81}Br^{79}Br$, and $^{81}Br^{81}Br$. Thus, a set of three peaks (the two possibilities $^{79}Br^{81}Br$ and $^{81}Br^{79}Br$ have identical masses) will be generated that differ in mass by two units. The center peak, which is produced by the $^{79}Br^{81}Br$ and $^{81}Br^{79}Br$ combinations, will have twice the intensity of the outer two peaks, because statistically there are twice as many combinations that produce this mass. All modern mass spectrometers have spectral simulation programs that can readily calculate and print out the relative isotopic distribution pattern expected for any compound formulation, so that it is possible to easily match the expected pattern for a particular ion with the experimental result.

CHAPTER 19
ORGANIC CHEMISTRY II: FUNCTIONAL GROUPS

19.1 (a) RNH_2, R_2NH, R_3N (b) ROH (c) 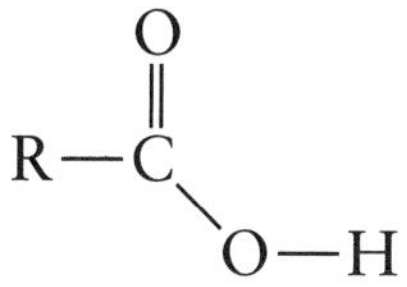 $R{-}C({=}O){-}O{-}H$ or RCOOH

(d) 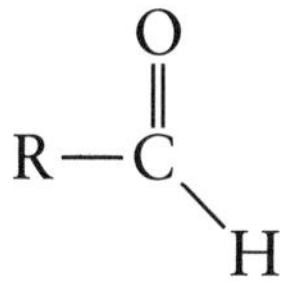 $R{-}C({=}O){-}H$ or RCHO

19.3 (a) ether; (b) ketone; (c) primary amine; (d) ester

19.5 (a) 2-iodo-2-butene; (b) 2,4-dichloro-4-methylhexane; (c) 1,1,1,-triiodoethane; (d) dichloromethane

19.7 (a) 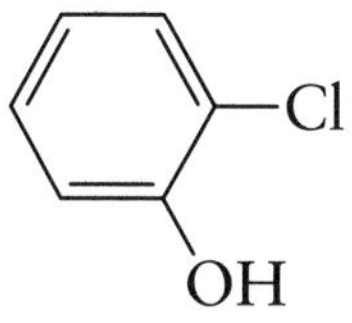 [benzene ring with Cl and OH on adjacent carbons], $C_6H_4Cl(OH)$, phenol; (b) $CH_3CH(CH_3)CH(OH)CH_2CH_3$, secondary alcohol; (c) $CH_3CH_2CH(CH_3)CH_2CH(CH_3)CH_2OH$, primary alcohol; (d) $CH_3C(CH_3)(OH)CH_2CH_3$, tertiary alcohol

19.9 (a) $CH_3CH_2OCH_3$; (b) $CH_3CH_2OCH_2CH_2CH_3$; (c) CH_3OCH_3

19.11 (a) butyl propyl ether; (b) methyl phenyl ether; (c) pentyl propyl ether

19.13 (a) aldehyde, ethanal; (b) ketone, propanone; (c) ketone, 3-pentanone

19.15 (a) 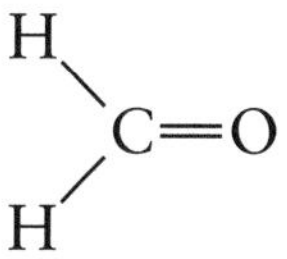 $(H)(H)C{=}O$ (b) $(CH_3)(CH_3)C{=}O$ (c) $(CH_3)(CH_3(CH_2)_3CH_2)C{=}O$

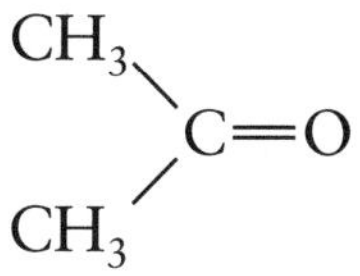

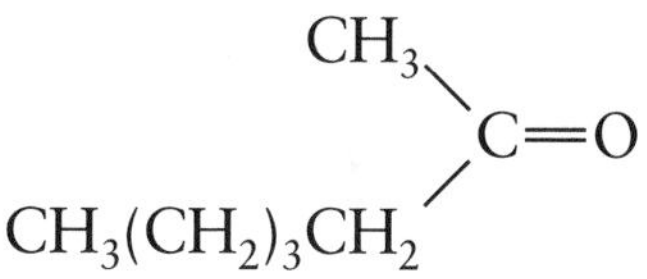

19.17 (a) ethanoic acid; (b) butanoic acid; (c) 2-aminoethanoic acid

19.19 (a) C_6H_5COOH (b) $H_3C{-}CH_2{-}CH(CH_3){-}CHCl{-}COOH$

(c) $H_3C{-}CH_2{-}CH_2{-}CH_2{-}CH_2{-}COOH$ (d) $H_2C{=}CH{-}COOH$

19.21 (a) methylamine; (b) diethylamine; (c) *o*-methylaniline or 2-methylaniline or *o*-methylphenylamine

19.23 (a) $C_6H_4(NH_2)(CH_3)$ (b) $(CH_3CH_2)_2N{-}CH_2CH_3$ (c) $[CH_3{-}N(CH_3)_2{-}CH_3]^+$

19.25 (a) The product is $C_6H_5(CH_3)CH(OH)$. (b) The mechanism must be S_N1, because an S_N2 mechanism would produce inversion of configuration; however, the product would still be optically active. The dissociation of the halide ion prior to addition of water allows the intermediate ion to form a racemic product.

19.27 Only (a) and (c) may function as nucleophiles, because they have lone pairs of electrons that will be attracted to a positively charged carbon center. CO_2 and SiH_4 have no lone pairs and cannot function as nucleophiles.

19.29 (a) ethanol; (b) 2-octanol; (c) 5-methyl-1-octanol. These reactions can be accomplished with an oxidizing agent such as acidified sodium dichromate, $Na_2Cr_2O_7$.

19.31 (a) 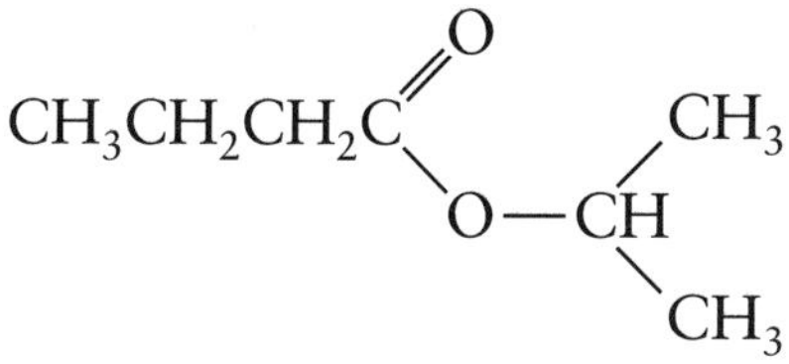 $CH_3CH_2CH_2C(=O)-O-CH(CH_3)_2$ (b) 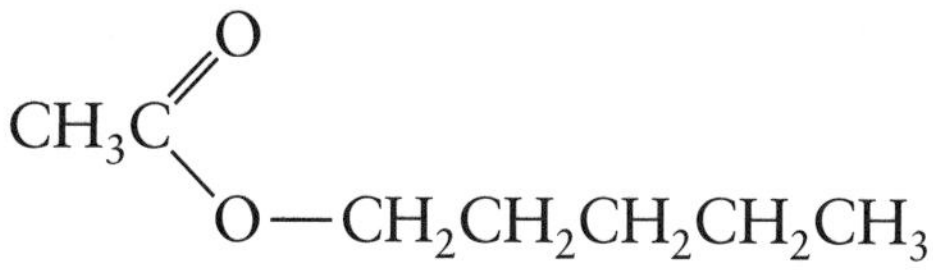 $CH_3C(=O)-O-CH_2CH_2CH_2CH_2CH_3$

(c) 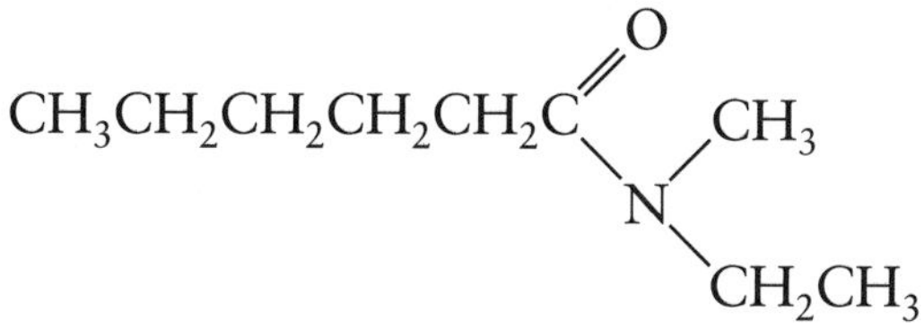 $CH_3CH_2CH_2CH_2CH_2C(=O)-N(CH_3)(CH_2CH_3)$ (d) 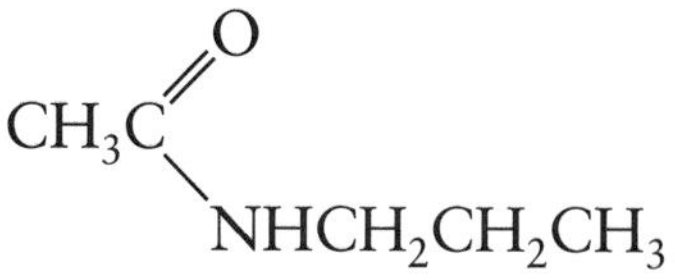 $CH_3C(=O)-NHCH_2CH_2CH_3$

19.33 The following procedures may be used:

(1) Use an acid-base indicator and look for a color change.

(2) $CH_3CH_2CHO \xrightarrow{\text{Tollens reagent}} CH_3CH_2COOH + Ag(s)$

(3) $CH_3COCH_3 \xrightarrow{\text{Tollens reagent}}$ no reaction

Procedure (1) distinguishes ethanoic acid from propanal and 2-propanone. (2) and (3) distinguish propanal from 2-propanone.

19.35 (a) 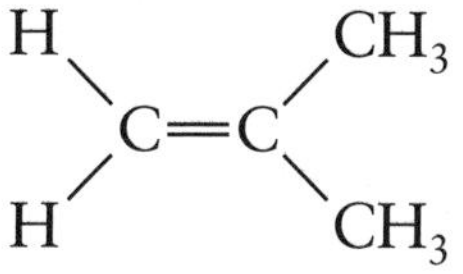 $(H)_2C{=}C(CH_3)_2$ $\quad -CH_2-C(CH_3)_2-CH_2-C(CH_3)_2-CH_2-C(CH_3)_2-$

(b) acrylonitrile: $CH_2{=}CH{-}CN$ $\quad -CH(CN)-CH_2-CH(CN)-CH_2-CH(CN)-CH_2-$

(c) 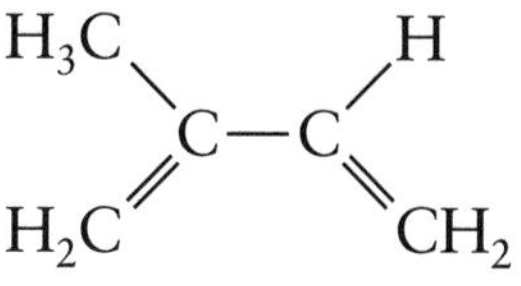 $H_2C{=}C(H_3C)-C(H){=}CH_2$

Isoprene

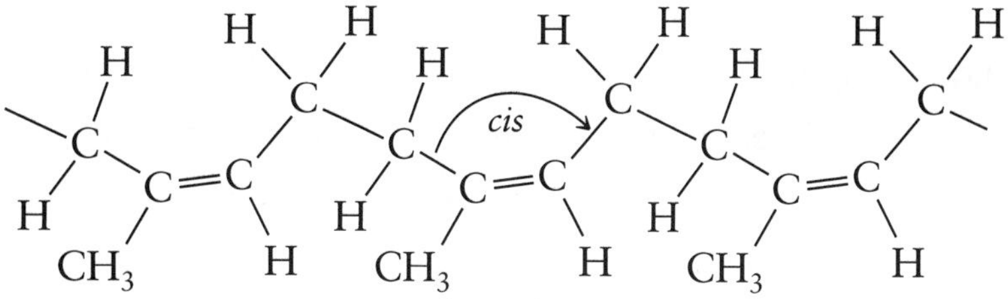

$-CH_2-C(CH_3){=}CH-CH_2-CH_2-C(CH_3){=}CH-CH_2-CH_2-C(CH_3){=}CH-CH_2-$ (cis)

cis version

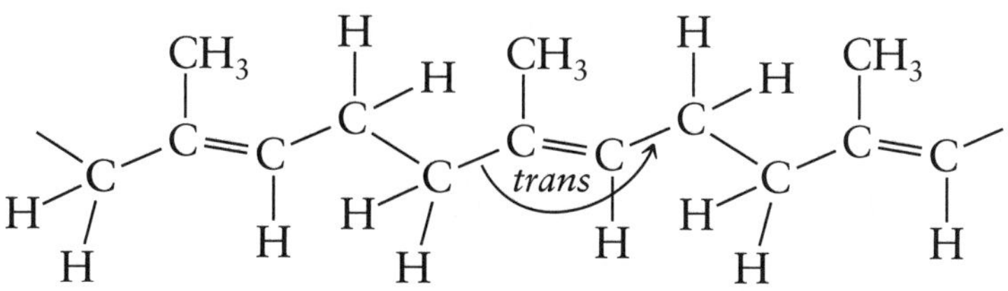

$-CH_2-C(CH_3){=}CH-CH_2-CH_2-C(CH_3){=}CH-CH_2-CH_2-C(CH_3){=}CH-$ (trans)

trans version

19.37 (a) 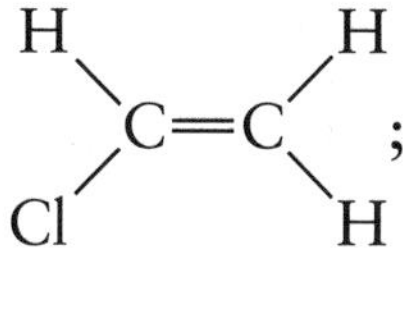 $(H)(Cl)C{=}C(H)_2$; (b) 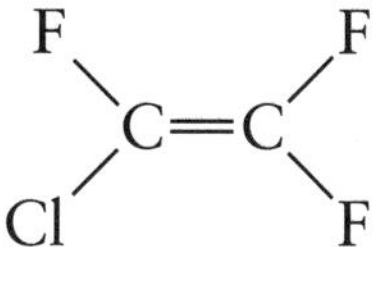 $(F)(Cl)C{=}C(F)_2$

19.39 (a) $-C(=O)-C(=O)-N(H)-(CH_2)_4-N(H)-C(=O)-C(=O)-N(H)-(CH_2)_4-N(H)-$

(b) $-C(=O)-CH(CH_3)-N(H)-C(=O)-CH(CH_3)-N(H)-$

19.41 An isotactic polymer is a polymer in which the substituents are all on the same side of the chain.

A syndiotactic polymer is a polymer in which the substituent groups alternate, from one side of the chain to the other.

An atactic polymer is a polymer in which the groups are randomly attached, one side or the other, along the chain.

19.43 block copolymer

19.45 Larger average molar mass corresponds to longer average chain length. Longer chain length allows for greater intertwining of the chains, making them more difficult to pull apart. This twining results in (a) higher softening points, (b) greater viscosity, and (c) greater mechanical strength.

19.47 Highly linear, unbranched chains allow for maximum interaction between chains. The greater the intermolecular contact between chains, the stronger the forces between them, and the greater the strength of the material.

19.49 The Lewis structures of the molecules are

(a) $H-C(H)(H)-C(H)(H)-C(H)(H)-C(=O)-O-H$

(b) $H-C(H)(H)-C(H)(H)-C(H)(H)-C(=O)-O-C(H)(H)-H$

(c) $H-C(H)(H)-C(H)(H)-C(H)(H)-C(H)(H)-O-O-H$

(d) $H-C(H)(H)-C(H)(H)-C(H)(H)-C(=O)-C(H)(H)-H$

Of these, only (c) is peroxide. (a) is a carboxylic acid, (b) is an ester, and (d) is a ketone.

19.51 (a) 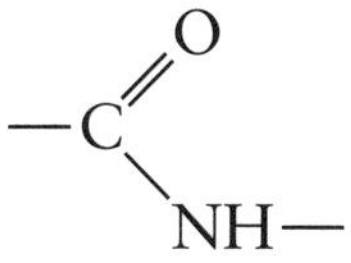 $-C(=O)-NH-$ (b) amide (c) condensation

19.53 Side groups that contain hydroxyl, carbonyl, amino, and sulfide groups are all potentially capable of participating in hydrogen bonding that could contribute to the tertiary structure of the protein. Thus, serine, threonine, tyrosine, aspartic acid, glutamic acid, lysine, arginine, histidine, asparagine, and glutamine satisfy the criteria. Proline and tryptophan generally do not contribute through hydrogen bonding, because they are typically found in hydrophobic regions of proteins.

19.55 $H-C(NH_2)(CH_2-C_6H_4-OH)-C(=O)-NH-CH_2-COOH$

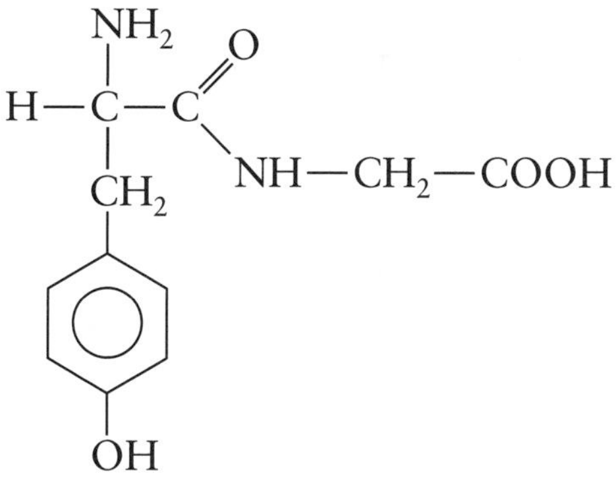

19.57 The functional groups are alcohols and aldehydes. The chiral carbon atoms are marked with asterisks (*).

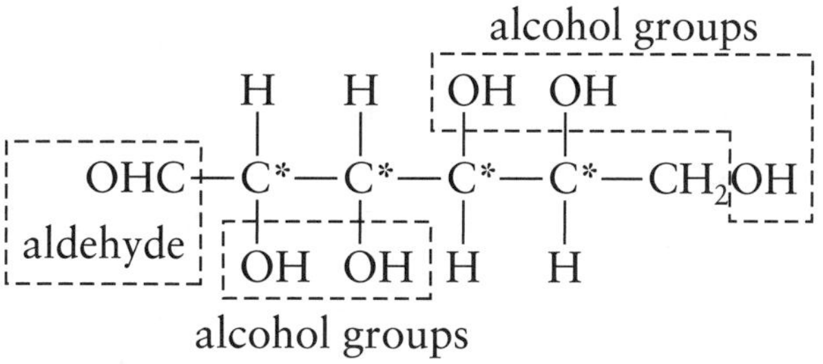

$OHC-C^*H(OH)-C^*H(OH)-C^*H(OH)-C^*H(OH)-CH_2OH$

(aldehyde: OHC; alcohol groups: the OH groups on the C* atoms and the CH_2OH group)

19.59 (a)
```
C A T G A G T T A
| | | | | | | | |
G T A C T C A A T
```
(b)
```
T G A A T T G C A
| | | | | | | | |
A C T T A A C G T
```

19.61 (a) $C_5H_5N_5O$ (b) $C_6H_{12}O_6$ (c) $C_3H_7NO_2$

19.63 (a) alcohol ($-OH$), ether ($-OCH_3$), aldehyde ($-CHO$)

(b) ketone ($>C=O$), alkene ($>C=C(CH_3)<$)

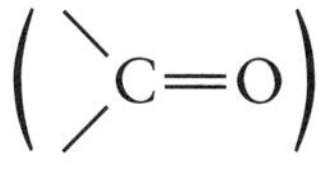

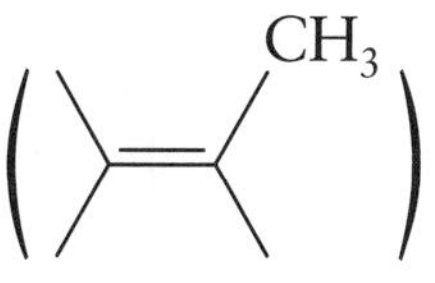

(c) tertiary amine ($>N-CH_3$), amide ($>N-C(=O)-$)

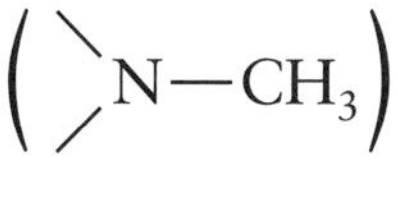

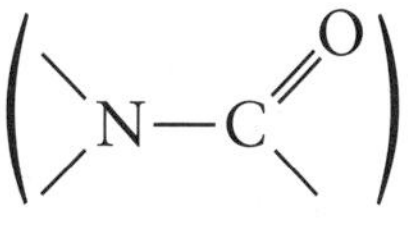

19.65 (a) carboxylic acid, ester

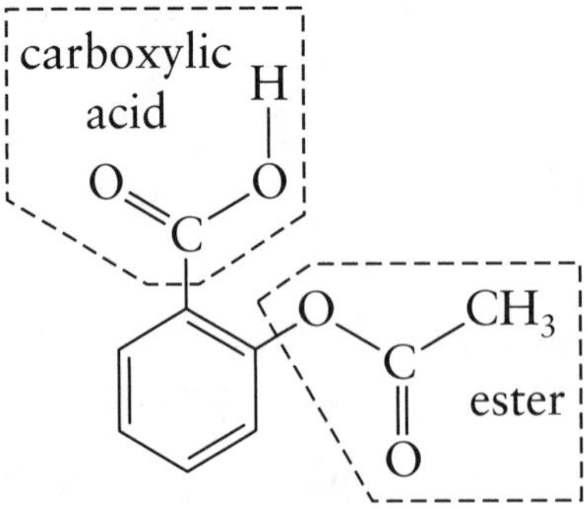

(b) ether, ketone, phenol, alkene

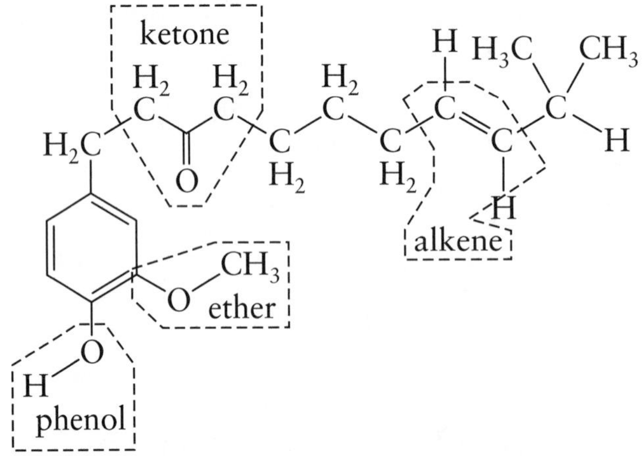

(c) aromatic amine, tertiary amine

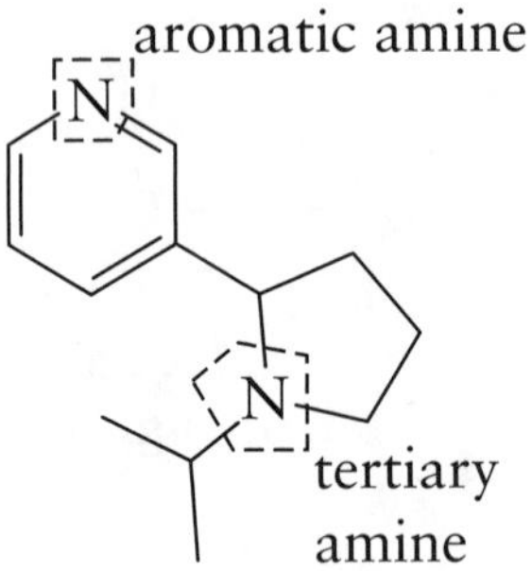

(d) ketone, alcohol, alkene

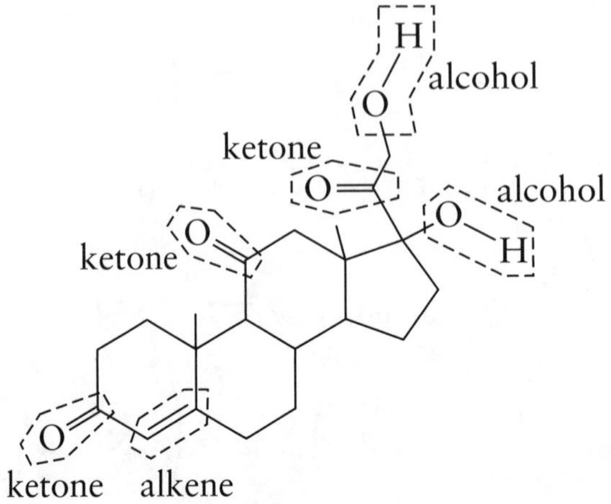

19.67 (a) CH_3, O (b) CH_3, OH, CH_3, O

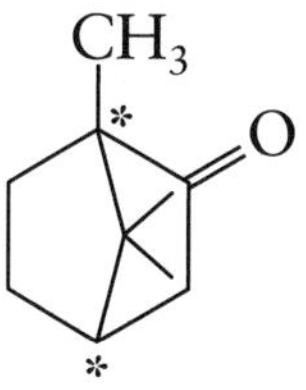

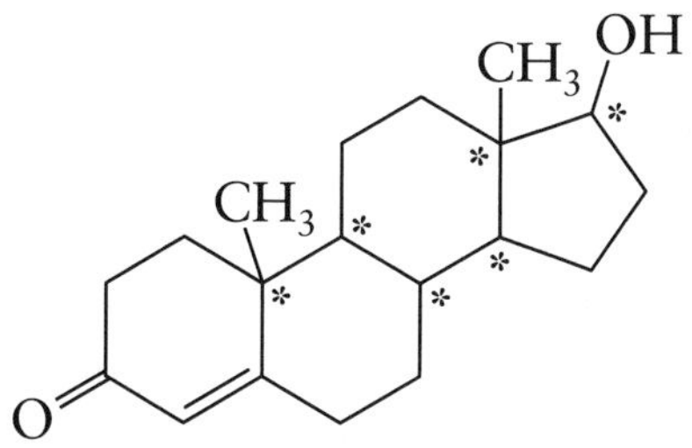

19.69 (a) H_2, C, CH, CH_2, O—CH, H_3C, N, CH_2, CH—CH, C, H_2, O=C, O—CH_3

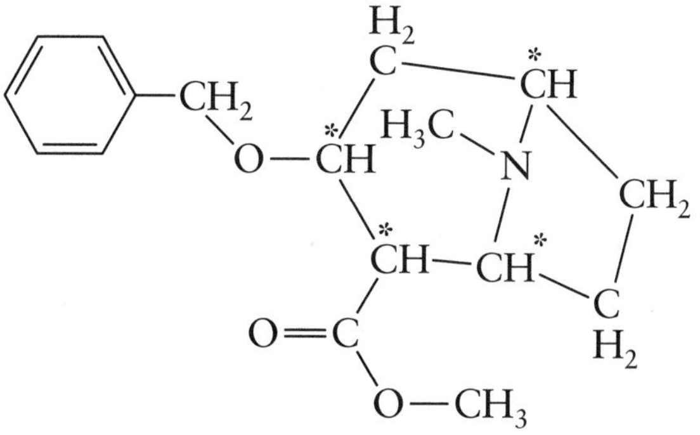

(b) O, O, H_2, O, C, CH_2, H_2C, CH, C, O, H_2, CH, O, O, CH_3

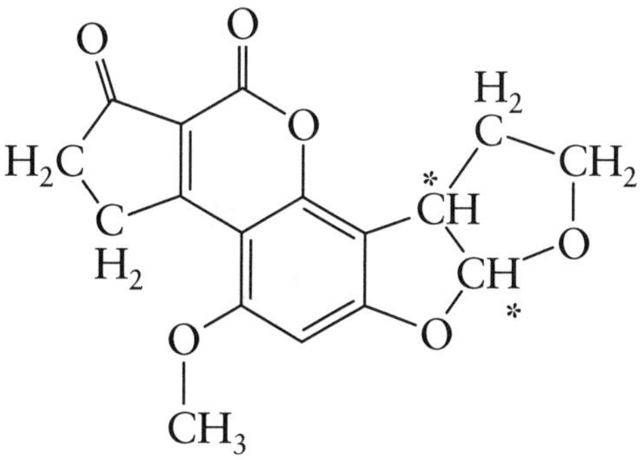

19.71 (a)

$$H{-}\overset{H}{\underset{H}{C}}{-}\overset{H}{\underset{H}{C}}{-}O{-}\overset{H}{\underset{H}{C}}{-}\overset{H}{\underset{H}{C}}{-}H \qquad H{-}\overset{H}{\underset{H}{C}}{-}\overset{H}{\underset{H}{C}}{-}\overset{H}{\underset{H}{C}}{-}\overset{H}{\underset{H}{C}}{-}O{-}H$$

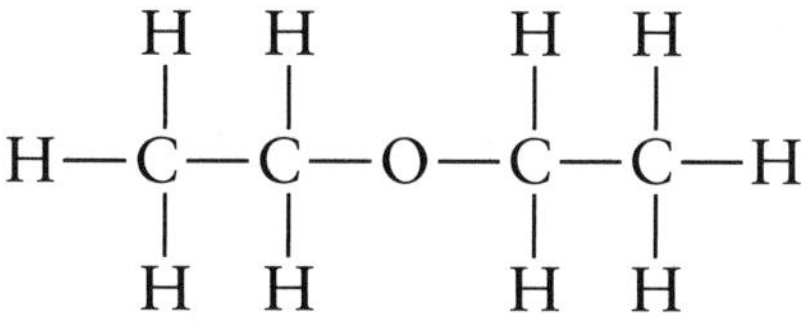

Diethyl ether 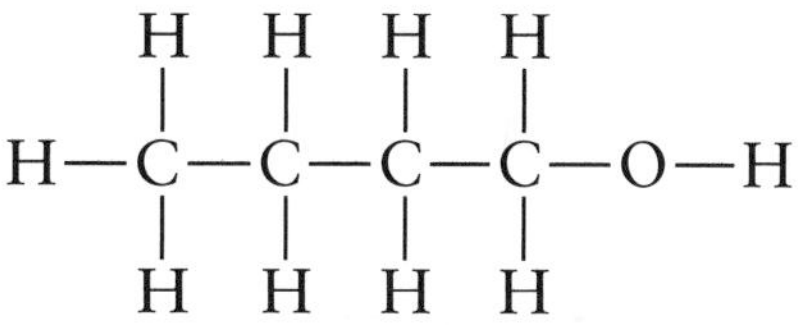 1-Butanol

(b) The principal forces between both of these compounds and water resulting in their solubility are London forces. Both molecules are likely to have very similar London forces with water because both contain the same atoms in a similar structural arrangement. However, 1-butanol can also undergo hydrogen bonding with itself, so the molecules are held together strongly in the liquid state, thereby resulting in a relatively high (117°C) boiling point.

---

* An asterisk (*) denotes a chiral carbon atom.

19.73 (a)

$$\begin{array}{l} H_2C{-}O{-}\overset{\displaystyle O}{\overset{\|}{C}}{-}(CH_2)_{16}CH_3 \\ \;| \\ HC{-}O{-}\overset{\displaystyle O}{\overset{\|}{C}}{-}(CH_2)_{16}CH_3 \\ \;| \\ H_2C{-}O{-}\overset{\displaystyle O}{\overset{\|}{C}}{-}(CH_2)_{16}CH_3 \end{array}$$

(b)

$$HO{-}C_6H_4{-}CH_2OH \xrightarrow[\text{organic solvent}]{Na_2Cr_2O_7(aq),\ H^+} HO{-}C_6H_4{-}CHO$$

19.75 (a) addition; (b) condensation; (c) addition; (d) addition; (e) condensation

19.77 (a) oxidation; (b) neither; (c) reduction; (d) oxidation; (e) oxidation

19.79 (a) $HOCH_2CH_2OH + 2\ CH_3(CH_2)_{16}COOH \longrightarrow$

$$CH_3(CH_2)_{16}\overset{\displaystyle O}{\overset{\|}{C}}{-}OCH_2CH_2{-}O{-}\overset{\displaystyle O}{\overset{\|}{C}}(CH_2)_{16}CH_3 + 2\ H_2O$$

(b) $2\ CH_3CH_2OH + HOOCCOOH \longrightarrow$

$$CH_3CH_2{-}O{-}\overset{\displaystyle O}{\overset{\|}{C}}{-}\overset{\displaystyle O}{\overset{\|}{C}}{-}O{-}CH_2CH_3 + 2\ H_2O$$

(c) $CH_3CH_2CH_2CH_2OH + CH_3CH_2COOH \xrightarrow{\Delta}$

$$CH_3CH_2\overset{\displaystyle O}{\overset{\|}{C}}OCH_2CH_2CH_2CH_3 + H_2O$$

19.81 Polyalkenes $<$ polyesters $<$ polyamides, due to the increasing strength of intermolecular forces between the chains. The three types of polymer have about the same London forces if their chains are about the same length. However, polyesters also have dipole forces contributing to the strength of intermolecular forces, and polyamides form very strong hydrogen bonds between their chains.

19.83 (a)

$$n\ (C_6H_5)HC{=}CH_2 \longrightarrow \left({-}\overset{\displaystyle H}{\overset{|}{\underset{\displaystyle C_6H_5}{\underset{|}{C}}}}{-}\overset{\displaystyle H}{\overset{|}{\underset{\displaystyle H}{\underset{|}{C}}}}{-}\right)_n$$

(b) n

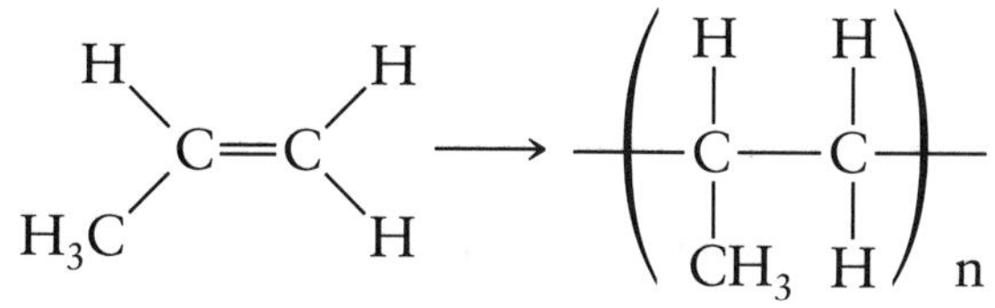

$$n\,\begin{matrix}H & & & & H\\ & \diagdown & & \diagup & \\ & & C{=}C & & \\ & \diagup & & \diagdown & \\ H_3C & & & & H\end{matrix} \longrightarrow \left(\begin{matrix} H & H \\ | & | \\ -C & -C- \\ | & | \\ CH_3 & H \end{matrix}\right)_n$$

(c) The backbones of all three polymers are the same. They differ simply in what is attached to the long carbon chain. Both styrene and propylene produce polymers that have organic groups attached to the side of the polymer, in the first case a phenyl ring and in the other, a methyl group. These groups are known as *side* groups or *side chains*. For ethylene, there is only one way in which the pure monomer may be polymerized, but for propylene and styrene, the asymmetry in the monomer means that the final polymer will be more complex. This is illustrated in the more extended structures shown below, for a general side group R. Notice that the polymer may be very regular or somewhat irregular in the way the R groups become attached, with respect to each other. The presence of regular or irregular structures affects the properties of the polymers as does the nature of the side chains—not only due to their size but also due to their shape, polarity, etc.

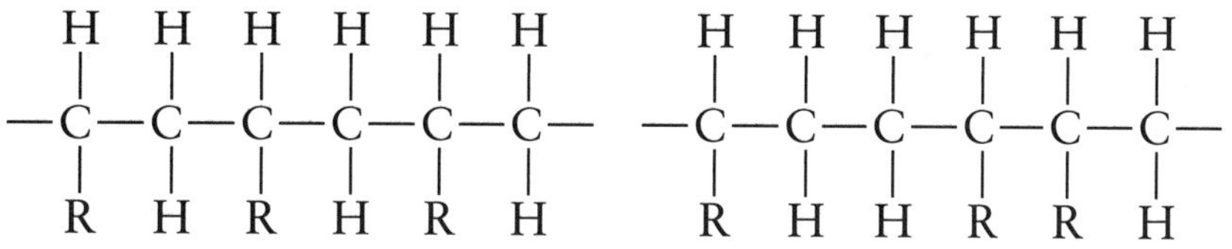

$$\begin{matrix} H & H & H & H & H & H \\ | & | & | & | & | & | \\ -C- & C- & C- & C- & C- & C- \\ | & | & | & | & | & | \\ R & H & R & H & R & H \end{matrix} \qquad \begin{matrix} H & H & H & H & H & H \\ | & | & | & | & | & | \\ -C- & C- & C- & C- & C- & C- \\ | & | & | & | & | & | \\ R & H & H & R & R & H \end{matrix}$$

19.85 (a) Primary structure is the sequence of amino acids along a protein chain. Secondary structure is the conformation of the protein, or the manner in which the chain is coiled or layered, as a result of interactions between amide and carboxy groups. Tertiary structure is the shape into which sections of the proteins twist and intertwine, as a result of interactions between side groups of the amino acids in the protein. If the protein consists of several polypeptide units, then the manner in which the units stick together is the quaternary structure.

(b) The primary structure is held together by covalent bonds. Intermolecular forces provide the major stabilizing force of the secondary structure. The tertiary structure is maintained by a combination of London forces, hydrogen bonding, and sometimes ion-ion interactions. The same forces are responsible for the quaternary structure.

19.87 The basic difference between polyester fabrics and fabrics such as wool, silk, and nylon, is that the latter group are all polyamides. The presence of the amide functional group, $-\underset{\underset{O}{\|}}{C}-\underset{\underset{H}{|}}{N}-$, allows for extensive cross-linking between polymer chains by hydrogen bonding with the amine group, $-N-H$, on one chain and the carbonyl group, $>C=O$, on another. Intermolecular attractions are thus greater between polyamide molecules than between polyester molecules. The less polar polyester molecules can more readily accept radon atoms among their chains.

19.89 HO— —CH_2

CH_2

HO NH_2

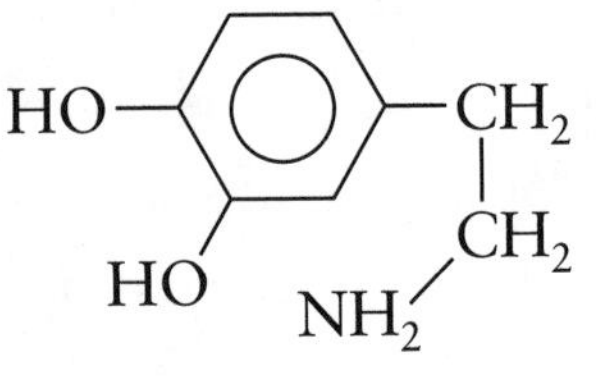

19.91 Condensation polymerization involves the loss of a small molecule, often water or HCl, when monomers are added together. 1,2-benzenedicarboxylic acid reacts with ethylene glycol to yield

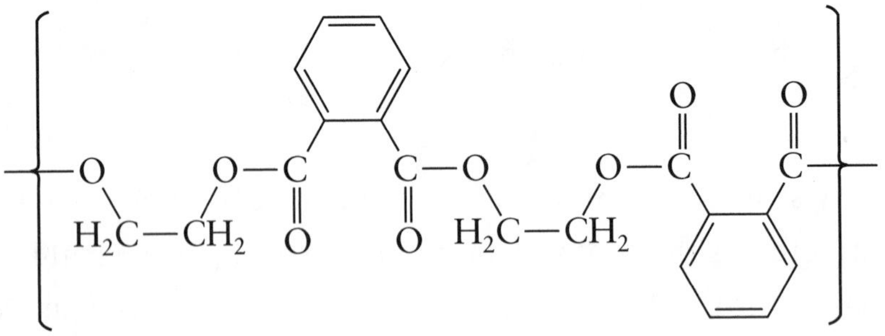

O O

O O—C C—O O—C C

H_2C-CH_2 O O H_2C-CH_2

Dacron:

O O

O O—C C—O O—C C

H_2C-CH_2 O O H_2C-CH_2

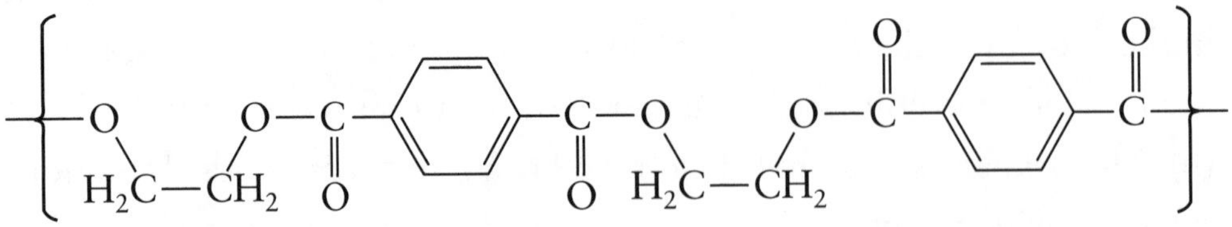

Dacron is more linear than the polymer obtained from benzene-1,2-dicarboxylic acid and ethylene glycol, so Dacron can be more readily spun into yarn.

19.93 Two peaks are observed with relative overall intensities 3 : 1. The larger peak is due to the three methyl protons and is split into two lines with equal intensities. The smaller peak is due to the proton on the carbonyl carbon atom and is split into four lines with relative intensities 1 : 3 : 3 : 1.

19.95 The Lewis structure of the molecule is

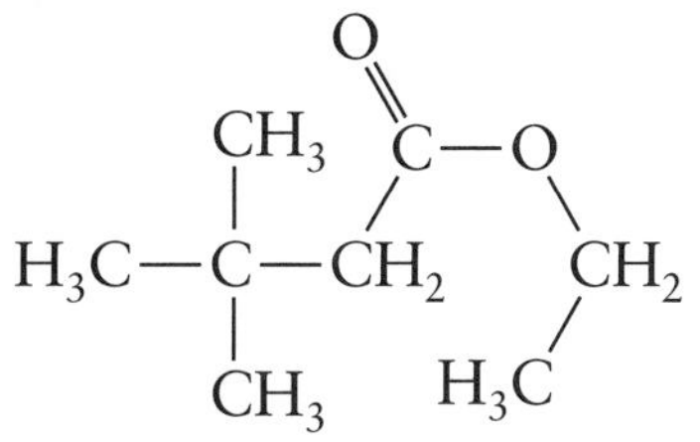

$$\begin{array}{ccccc} & & & \overset{O}{\underset{}{\|}} & \\ & CH_3 & & C-O & \\ & | & / & & \backslash \\ H_3C- & C & -CH_2 & & CH_2 \\ & | & & & / \\ & CH_3 & & H_3C & \end{array}$$

The peaks in the spectrum can be assigned, based upon the intensities and the coupling to other peaks. The hydrogens of the CH_3 unit of the ethyl group will have an intensity of 3 and will be split into a triplet by the two protons on the CH_2 unit. This peak is found at ca. 1.2 ppm. The CH_2 unit will have an intensity of 2 and will be split into a quartet by the three protons on the methyl group. This corresponds to the peak at 4.1 ppm. The CH_2 group that is part of the butyl function will have an intensity of two but will appear as a singlet because there are no protons on adjacent carbon atoms. This is the signal found at 2.1 ppm. The remaining CH_3 groups are equivalent and also will not show coupling. They can be attributed to the signal at 1.0 ppm. Notice that the peak that is most downfield is the one for the CH_2 group attached directly to the electronegative oxygen atom, and that the second most downfield peak is the one attached to the carbonyl group.

19.97 If one considers the reaction, the products should be those arising from substitution of hydrogen atoms on the propane by chlorine atoms. We would expect then to form a chloropropane or, perhaps a dichloropropane. Remember that in the halogenation of alkanes, substitution becomes more difficult as more halogen atoms are introduced. If we then consider the NMR spectrum, we see that there is one large peak that sees a single proton, because it is split into a doublet. There is also a weaker feature, corresponding most likely to one proton, which is split into a septet. This indicates that the proton sees six equivalent protons. The structure that is consistent with this spectrum is 2-chloropropane.

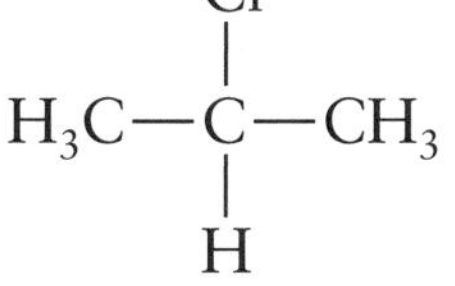

$$\begin{array}{c} Cl \\ | \\ H_3C-C-CH_3 \\ | \\ H \end{array}$$

19.99 (a) ^{13}C

(b) 1.11%

(c) No. The reason is that the probability of finding two ^{13}C nuclei next to each other is very low. A ^{12}C nucleus next to a ^{13}C nucleus will not interact with the ^{13}C nucleus because the ^{12}C nucleus has no spin. Because the natural abundance of ^{13}C is 1.11%, the probability of finding two ^{13}C nuclei next to each other in an organic molecule will be 0.0111×0.0111 or 1.23×10^{-4}. Although such coupling is possible, it is generally not observed because the signal is so much weaker than the signal due to the molecules with a single ^{13}C nucleus.

(d) Maybe. Because most of the carbon to which the protons are attached is ^{12}C, the bulk of the signal will not be split. The protons that are attached to the ^{13}C atoms will be split, but this amounts to only 1.11% of the sample, so the peaks are very small. Peaks that result from coupling to a small percentage of a magnetically-active isotope are referred to as satellites and may be observed if one has a very good spectrum.

(e) Yes. Although the splitting of protons by ^{13}C may not be observed because the amount of ^{13}C present is low, the opposite situation is not true. If a ^{13}C is attached to H atoms, the large majority of those H atoms will have a spin and so the ^{13}C will show fine structure due to splitting by the H atoms.